Association of Applied Biologists

Pests Pathogens and Vegetation

The role of weeds and wild plants
in the ecology of crop pests and diseases

Edited by J M Thresh

Associate Editors
J E Crosse
F H Jacob
F G W Jones
A F Posnette
M Solomon
M S Wolfe

The outcome of a meeting arranged at the University of York
15–17 April 1980 in collaboration with the British Ecological Society
and the Federation of British Plant Pathologists

Pitman Advanced Publishing Program
BOSTON · LONDON · MELBOURNE

PITMAN BOOKS LIMITED
39 Parker Street, London WC2B 5PB

PITMAN PUBLISHING INC
1020 Plain Street, Marshfield, Massachusetts

Associated Companies
Pitman Publishing Pty Ltd, Melbourne
Pitman Publishing New Zealand Ltd, Wellington
Copp Clark Pitman, Toronto

Library of Congress Cataloging in Publication Data

Main entry under title:

Pests, pathogens and vegetation.
Includes index.
1. Agricultural pests — Congresses. 2. Plant diseases — Congresses. 3. Weeds — Congresses. 4. Agricultural ecology — Congresses. I. Thresh, J M. II. Association of Applied Biologists. III. British Ecological Society. IV. Federation of British Plant Pathologists.
SB599.2.P47 632′.58 80-27419
ISBN 0-273-08498-4

Text set in 10/12 pt Linotron 202 Times, printed and bound in Great Britain at The Pitman Press, Bath

Contents

Introduction

> . . . and when it was morning the east wind brought the locusts. And the locusts went up over all the land of Egypt, and rested in all the coasts of Egypt: very grievous were they; before them were no such locusts as they, neither after them shall be such. For they covered the face of the whole earth, so that the land was darkened; and they did eat every herb of the land, and all the fruit of the trees which the hail had left; and there remained not any green thing in the trees, or in the herbs of the field, through all the land of Egypt. Exodus 10: 13–15

This passage from the Old Testament provides a graphic description of the sudden appearance of a swarm of locusts in ancient Egypt and of the devastation caused. Similar swarms are featured on the front cover of this volume and described by R. C. Rainey on page 327. They provide a striking example of the way in which serious problems can arise in agriculture when pests invade crops from areas of natural vegetation. Many disease outbreaks originate in a similar way and this phenomenon provided the initial stimulus for the long-standing interest of biologists in wild plants whilst they searched for the sources from which pests and pathogens spread to crops.

Wild plants have also been studied by agricultural scientists for other reasons. For example, wild plants or their progenitors are the ancestors of weeds and crop species and are of continuing interest to breeders as sources of, what J. A. Browning has referred to as, genetic 'spare parts' in attempts to produce improved varieties. More recently attention has turned to wild plants as a source of beneficial organisms in the search for biological control agents and in evaluating the complex effects of windbreaks, hedgerows, woodland and other patches of 'natural' vegetation in otherwise cultivated areas. Other workers have considered the role of wild plants in contributing to the genetic diversity of pests and pathogens, or in relation to the emergence of new races or biotypes that can overcome host resistance or withstand the application of pesticides.

Even more recently, and somewhat belatedly, plant pathologists have recognized that natural plant communities generally avoid the damaging epidemics that continue to occur and cause serious losses in agriculture. This has led to detailed comparisons between natural and agro-ecosystems and attempts are now being made to simulate the diversity found in nature and so utilize resistance mechanisms not previously exploited. Considerable progress has been made but there has been only limited input from the 'pure' ecologists who are not directly concerned with agriculture and who have contributed so greatly to studies on the abundance, distribution and regulation of animal populations.

Papers on all these topics were included in the symposium on 'The Role of Weeds and Wild Plants in the Ecology of Crop Pests and Diseases', held at the University of York, 15–17 April, 1980. The meeting was arranged jointly by the Association of Applied Biologists, the British Ecological Society and the Federation of British Plant Pathologists. It was planned along multidisciplinary lines with the aim of bringing together plant pathologists, zoologists, ecologists and weed biologists in a way that is seldom possible in these days of increasing specialization.

A multidisciplinary approach has also been adopted in preparing this proceedings volume, which contains expanded versions of thirty-six of the papers presented at York and three others by overseas contributors who were unable to attend the meeting. The three introductory papers are intended to provide a general background to those that follow. The opening one by J. R. Harlan will come as a revelation to those unfamiliar with the latest findings on the beginnings of agriculture and the evolution of crop plants. He surveys the wealth of varied evidence showing that agriculture has emerged in various forms at many different times and places, and that it has sometimes been adopted reluctantly and then abandoned.

In the following papers some of the most recent changes in agriculture are discussed by A. H. Bunting in a global context and by A. B. Trask for England and Wales. Both authors emphasize the trend towards increasingly intensive use of the best available agricultural land, whilst inferior areas are abandoned or put to other uses. They also stress the powerful economic forces that now influence agricultural policy and cropping practices at a time when there is an urgent need for a four-fold increase in food production to cope with the three-fold increase in human population projected over the next hundred years. These are challenging figures when considered in relation to the views of extreme conservationists and to FAO statistics (p. 26), showing that only 12% of the total land area of the world is devoted to arable or permanent crops, with a further 56% as permanent pasture or woodland and no less than 32% classed as 'other' land, much of it wilderness unsuitable for any form of agriculture or forestry. Such overall figures conceal great differences between regions: the proportion of agricultural land is large in Europe, India and the Far East as a whole and relatively small in Latin America, USSR, the Near East, China and Oceania. It is also small in Africa, where cereal production per head of the population has actually declined in recent years (p. 33).

The introduction of agriculture to new areas and the distribution of cultivated land in a matrix of natural or semi-natural vegetation give abundant opportunities for the movement of pests and pathogens between wild stands and crops. This is considered in the four central sections of the volume that deal with groups of pests or pathogens or with specific 'case histories'. Because of the wealth of potential material available it was necessary to impose some constraints on authors and they were asked to omit any detailed consideration of crop plants growing as weeds. Nevertheless, many different phenomena are described and the approach is not uniform as the papers range from comprehensive reviews to relatively short contributions. There are striking parallels between many of the contributions with frequent reference to 'boundary' or invasion effects due to local movement into crops from adjoining hedgerows or other areas of natural vegetation. However, there are great differences in the magnitude of the problems and in the distances involved, ranging from the hundreds of kilometres covered by birds and wind-borne spores or some insects to the short distances traversed by soil-inhabiting fungi and nematodes.

A recurring theme is the complex role played by hedgerows that are still such an obvious feature of much of the agricultural landscape of the British Isles. They provide windbreaks and shelter for crops and livestock and can be a major source of beneficial insects, mites and birds, yet they also harbour many noxious organisms including vertebrate and invertebrate pests, fungi, bacteria and viruses. Damaging outbreaks often occur first and spread most rapidly alongside hedgerows, but there are persuasive arguments for keeping hedges, quite apart from aesthetic considerations. This explains the generally neutral stance taken by the contributors to this volume in relation to the continuing controversy over hedgerow removal.

There are no such arguments over the need to control weeds to obtain the full benefits of advances in crop technology, and the sixth section of the volume deals with the impact of weeds and weed control on crop pests and diseases. In the opening paper J. D. Fryer summarizes the revolutionary changes in weed control and cropping practices that have occurred recently in many parts of the world following the introduction of selective herbicides. These are so effective and convenient that they have been adopted rapidly with little regard to any possible adverse effects on the crop environment or on the incidence of pests and diseases. Some of the complexities involved are considered by other contributors to this section and justify the suggestion, made by M. J. Way *et al.*, as long ago as 1956, that insect communities in and around agro-ecosystems have been affected much more by herbicides than by pesticides. Attention is now turning to a more precise evaluation of the role of weeds as sources of beneficial insects, with the possibility of achieving some control of pests by manipulating weeds as discussed by M. Altieri from experience in Colombia and Florida.

The major review of mammalian pests by L. M. Gosling is not supported by papers on particular pest species. Moreover, there are no papers dealing with the size and disposition of wild life refuges in relation to human settlements and agricultural areas. This topic was covered at York by S. K. Eltringham, who discusses the conflicts that can arise in Africa between conservationists and agriculturalists in his recent book *The Ecology and Conservation of Large African Mammals* (Macmillan, 1979).

Despite these omissions this volume contains a wealth of material on a topic not previously considered in such a comprehensive way. It should be a valuable source of reference for those working in several disciplines and for this reason all references are quoted in full.

The cover photograph of a swarm of Desert Locusts was kindly provided by C. Ashall of the Centre for Overseas Pest Research. Acknowledgments are also due to Academic Press, American Naturalist, Blackwell Scientific Publications, Cambridge University Press, *Journal of Zoology* and *Journal of Applied Zoology* for permission to reproduce or redraw figures.

It is a great pleasure to acknowledge the assistance given by M. Gratwick, J. Lee, S. McNeil, D. J. Royle and the Societies they represented in arranging the York meeting. I am also grateful to all authors, to E. J. Skene for editorial assistance and to the assistant editors for their hard work and collaboration in preparing the typescripts for submission to the publishers within eight months of the meeting.

J. M. Thresh
East Malling Research Station

Cross-references

There are numerous cross-references between the different papers in this volume. These are referred to and listed in the standard form except that the year of publication (*1981*) appears in italic print.

Section 1 **Introductory papers**

Ecological settings for the emergence of agriculture

Jack R Harlan
Crop Evolution Laboratory, Agronomy Department, University of Illinois, Urbana, IL 61801

Introduction

Studies on the origins and evolution of food production have increased greatly in the last two decades and particularly in recent years. Research in archaeology, palaeobotany, ethnobotany, geography, linguistics, botany, genetics, ecology and agronomy have all contributed to the mass of information. Crucial developments were the publications edited by Lee and DeVore (1968) and Ucko and Dimbleby (1969). These volumes challenged many traditional views on the economies of hunter-gatherers and the evolution of food production. Much of the available evidence had not previously been assimilated and researchers began to ask new questions or restate old questions differently. Field work became increasingly oriented towards the economy and ecology of food procurement systems.

The numerous reports from archaeological sites and anthropological field work have been assessed in a global or regional context (Murray, 1970; Tringham, 1971; Renfrew, 1973; Koslowski, 1973; Jennings, 1974a; Bender, 1975; Harlan *et al.*, 1976a; Cohen, 1977; Allen *et al.*, 1977; Reed 1977). Consequently there is now sufficient information available to consider the ecological settings of various kinds of agriculture in a comparative way.

A few scholars still view agriculture as a diffusible invention or discovery (Carter, 1977), yet the mass of information now available suggests otherwise (Bender, 1975; Cohen, 1977; Reed, 1977). Agriculture is a food procurement system that evolved slowly out of broad-spectrum 'mesolithic' hunting-gathering. It is not readily diffused between cultures and is usually adopted slowly and reluctantly. The concept of farming, involving planting and reaping, is and was well-understood by non-farming people and is not revolutionary. These views have been discussed elsewhere and this paper concentrates on the evolution of food production in an ecological context.

Diverse ecosystems are examined using very broad classes, which suffice to indicate the various pathways from mesolithic economies to agriculture. 'Mesolithic' is treated in the sense of a broad-spectrum hunting-gathering economy where plants and animals are important in the diet, together with any available aquatic resources. Although there are differences between the European mesolithic and the American archaic, they will be considered as roughly equivalent. The recent life styles of the Australian Aborigines, the Bushmen of southern Africa and most hunter-gatherers of North and South America are also similar. Whether agriculture emerged from these economies depended on the ecology and plant resources available.

The fundamental character of a subsistence agriculture is determined by the practices of hunter-gatherers long before agriculture itself emerges (Harlan, in press). Certain common features characterize hunter-gatherers throughout the world and these determine agricultural evolution. Hunter-gatherers seem to regard plants similarly in Africa, Australia, Asia, Europe or the Americas. They are attracted to similar species and develop similar methods of modifying vegetation, detoxifying poisonous plants and processing edible plant foods. They are likely to find similar sources of drugs, arrow poisons or fish poisons and use them analogously. In this paper common themes are identified according to a broad ecological classification of climate and vegetation.

Deserts and semi-deserts

Some 34% of the surface of the earth has a desert climate. Together with the desert fringes and semi-deserts these areas occupy nearly half the earth's land surface (Fig. 1).

The extensive deserts and semi-deserts of Asia and Africa have been exploited to some degree by nomadic herding tribes and these pasturalists continue to harvest food from wild plants. Various broad-spectrum hunting-gathering cultures have successfully colonized desert lands the world over and some still survive. The most studied have been the Bushmen of southern Africa, Aborigines of Australia, the Great Basin and Sonoran tribes of North America and the tribes of the Atacama, and Patagonia of South America.

Repetitive themes are conspicuous. Several tribes of the US desert southwest and Mexican northwest depended very heavily on mesquite (or algarroba) seeds and pods (*Prosopis juliliflora* and screwbean *P. pubescens*) (Barrows, 1971). In the dry Patagonia of South America the Huarpe and adjacent tribes used other species of *Prosopis* (Cooper, 1946). In parts of Australia a similar dependence developed on species of the related *Acacia*. Both *Acacia* and *Prosopis* seeds were harvested in the deserts of Africa and South Asia. Wherever they occurred, the seeds and pods of these thorny leguminous trees were gathered, ground, cooked and eaten. In South America, algarroba was also fermented for beer.

Grass seeds have also been utilized by desert gatherers, and often different species of the same genus were harvested on different continents. Five wild species of *Panicum* were harvested in North America (Yanovsky, 1936), and one was domesticated in Mexico (Gentry, 1942). Seven species were used in Africa (Jardin, 1967), four in Australia and two in Asia. At least five species of *Sporobolus* were harvested in North America, three in Africa and three in Australia. Several species of *Eragrostis* were gathered in North America, Australia and Africa. Six species are listed for Africa and one was domesticated (*E. tef*). *Eleusine* and *Dactyloctenium* were harvested in Australia, India and Africa with one species (*E. coracana*) domesticated. Many other grasses from less arid environments were used similarly. Hunter-gatherers invariably tend to exploit the same kinds of food resources (Harlan, 1975).

Figure 1 Sketch map of the world showing the main areas of desert (solid black) and temperate grassland (hatched).

Genera of other families of arid and semi-arid floras were also utilized. Different species of *Amaranthus, Chenopodium, Lepidium, Polygonum, Portulaca, Sesbania* and *Vitex* were harvested in Africa, North America, Asia, and Australia, and some of these in Europe and Oceania as well. Fruits of *Zizyphus* were collected in Africa, Asia and Australia. Harlan (1975) lists genera in which species were harvested in the wild in Australia and other congeneric species elsewhere.

Trends towards agriculture in the semi-desert Great Basin of North America have been observed by anthropologists. Seven of the nineteen tribes studied by Steward (1941) planted seeds of wild species that were harvested from natural stands and conserved. A patch of vegetation was burned in the fall and the seeds broadcast on the burned area the following spring. The most commonly seeded were species of *Chenopodium, Oryzopsis, Mentzelia* and *Sophora*, but none was ever domesticated. The Paiute of Owens Valley, California, irrigated but did not cultivate. The water was spread by canals over blocks up to 13 km^2 so as to increase production of useful wild plants including *Nicotiana attenuata* and species of *Salvia, Chenopodium, Helianthus, Oryzopsis* and *Eleocharis*. None of these plants was selected or domesticated, although *Nicotiana* was selected by some of the west coast hunter-gatherers (Steward, 1934; Lawton *et al.*, 1976).

Trends towards agriculture in Australia involved altering vegetation and replacing the heads of tubers in the holes made on harvesting so as to assure a subsequent crop. Fields of some wild tubers were dug up so extensively as to amount to tillage. None of these practices led to agriculture in the usual sense, but the flora was managed to increase the yield of the desired species (Harlan, 1975).

Desert environments are little suited for agriculture without irrigation, yet desert floras have contributed some useful cultigens including the date palm (*Phoenix dactylifera*), the doub palms (*Borassus* spp.), some edible cacti (*Opuntia* spp.) and *Agave* spp. The tepary bean (*Phaseolus acutifolius*), devil's claw (*Proboscidea parviflora*) and *Panicum sonorum* were domesticated in or near the Sonoran Desert. Among the cereals, it seems likely that pearl millet (*Pennisetum americanum*) was domesticated in the Sahara when it was less arid than now (Brunken *et al.*, 1977; Munson, 1976). The wild race of pearl millet still occurs in the Sahara. Barley has desert races (Harlan and Zohary, 1966) and, although wild barley extends into the oak woodlands of the Near East, it tolerates drought and heat better than the other wild cereals of the region. Recent findings of wild barley along the lower Nile dating to some 16 000 BC when that region was hyperarid suggest that barley could have originated from a desert flora (Wendorf *et al.*, 1979). Similarly, *Panicum miliaceum* might be derived from the desert margins or steppes of central Asia (Oestry-Stidd, unpublished).

Minor contributions from the desert flora include *Zizyphus spina-christi* (fruit), *Lophophora* spp. (peyote, drug), other cacti for their fruits and *Agave* spp. for pulque and fibre. Additional species including jojoba (*Simmondsia chinensis*) and guayule (*Parthenium argentatum*) are now attracting attention as potential crops.

Temperate grasslands

Temperate steppes and prairies cover vast regions (Fig. 1) but they have contributed very little to field agriculture until historic times. The grasslands originally abounded in game animals and hunting was the most convenient way of exploiting the resources. Primitive neolithic tools were generally inadequate to attack the sods produced by perennial grasses, and early farmers avoided such areas. Eventually, neolithic pastoral nomads began to exploit grasslands, but soil tillage was restricted to small-scale efforts. The soils of these formations are naturally rich and eventually yielded to animal traction and metal ploughs, but breaking-out the sod was so difficult that some of the best land was not cultivated until this century when tractors, steel ploughs and tile drainage became available.

The hunting peoples of the Great Plains of North America and the South American pampas survived into the ethnographic present. They were aware of farming techniques and some tribes drifted in and out of agricultural pursuits. Intrusive farming outposts occurred in the Great Plains along the Missouri River and some tribes ceased farming after the introduction of the horse made bison hunting more efficient and more pleasurable. Some tribes of the South American pampas practised small-scale agriculture. In Europe, agriculture was introduced by farming tribes from the Near East and local tribes often continued a mesolithic life-style long after agriculture became established. As the invading farmers moved along the Danube, they tended to stay on loess or forest soils and avoided the true steppe (Tringham, 1971; Koslowski, 1973). The Tripolye culture of the Ukraine became securely established in a grassland region, cultivating wheat and millet (Milisauskas, 1978).

The first steps towards farming the grasslands were taken around the margins and ecotones. On the drier margins, the swards were less dense and the lighter (sandier) soils could be managed with primitive implements. The banks of rivers traversing the grasslands could sometimes be farmed, as were the ecotones between grassland and woodland. The Yang-shao culture of China evolved a neolithic agriculture in a steppe-like environment, but on very tractable loess soils. *Setaria* and *Panicum* millets were the main crops of early Yang-shao (K. C. Chang, 1968).

The grasslands have contributed various forage plants (including *Medicago sativa* and various species of *Agropyron, Elymus* and *Bromus*) but few food plants. However, present-day wheats may have come indirectly from a steppe flora. Some time during the neolithic of the Near East, the genomes of tetraploid wheat combined with that of *Aegilops squarrosa*. This little weedy goatgrass is the only member of the genus with a continental distribution and the only one extending into the Central Asian steppes. It transformed a rather ordinary cereal into the most widely grown food crop on earth.

There are probably other contributions of steppe and prairie floras, but in many cases the true origins of such crops are uncertain. Buckwheat (*Fagopyrum*), common millet (*Panicum miliare*), Italian millet (*Setaria italica*) and *Cannabis sativa* are all possibilities.

Temperate forests and woodlands

The temperate forests of the southern hemisphere are much smaller than those in the north (Fig. 2) and have contributed little to agriculture. The main cultural systems are the mesolithic of Europe and China, the Jomon of Japan, the archaic of eastern North America and the cultures of the Pacific slope that survived into the ethnographic present. They include some remarkable similarities in human ecology and food procurement systems. All of these people hunted and trapped game, which was generally abundant. They also exploited the aquatic resources of fresh water streams and lakes or seas. Water fowl and other birds were used on a considerable scale and plants were harvested systematically. The vegetable diet varied greatly with emphasis on tree products. *Quercus, Castanea, Aesculus, Juglans, Corylus, Prunus, Morus, Diospyros, Malus* and *Pinus* were genera available throughout these regions. Other genera occurring in one or more geographic regions included *Ginkgo, Eriobotrya, Pyrus, Carya.* The food supply was highly diverse, stable and abundant.

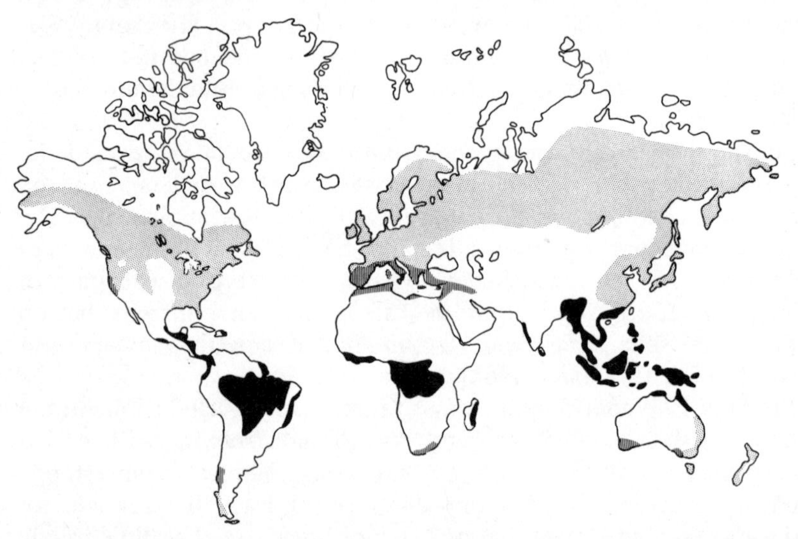

Figure 2 Sketch map of the world showing the main areas of tropical rain forest (solid black), temperate forest (stippled) and 'Mediterranean' vegetation (hatched).

Of the many tribes and cultures of this ecological formation, the Jomon of Japan is notable for a truly remarkable hunting–gathering efficiency. Although changing with time and developing regional differences, the culture lasted some 10 000 years. Fish and shellfish were very important dietary components in coastal or inland villages. The primary plant foods were acorns and nuts, although many other species were gathered. The Jomon lived in villages and produced the oldest known pottery, beginning a little before 10 000 BC (Kidder, 1968). The considerable population increase in middle Jomon seemed to coincide with the emergence of refined art styles and the development of a superior technology for detoxifying *Aesculus*.

The Japanese *Aesculus* is more poisonous than the North American species that can be detoxified by simple leaching. It contains tannins, saponins and alkaloids and requires an alkaline leach. The Jomon people learned how to burn specific hardwoods, to produce ashes that could be mixed with horsechestnut meal and so permit leaching until safe to eat. This technique made available an important food resource over much of Japan. The ancient methods of processing are still practised and there is a similar distribution of Jomon and modern usage of *Aesculus* (Watanabe, 1979). However, the detoxification of *Aesculus* is now done more for ceremonial purposes than for food.

The varied and stable food resources of Jomon Japan permitted the establishment of villages at about 7 km intervals over most of the main islands. This is a very high density for hunting-gathering peoples. Throughout the long history of the culture, the emphasis was on tree products. Shortly before the end of Jomon, however, there are palynological indications of considerable disturbance of the forests and the introduction of cultivated plants. Buckwheat pollen was identified (Tsukada, personal communication) and carbonized grains of barley. Rice was introduced just before the arrival of the Yayoi culture (*c.* 300 BC) that ended Jomon within a few centuries.

Across the Pacific, the west coast Indians developed similar subsistence patterns. Game, fish, shellfish and water fowl were important and the major plant food was usually acorns. Leaching facilities were developed for large-scale processing. Acorns were shelled, the meats pulverized and the meal leached, usually with hot water. Baskets were used in southern California and sand beds to the north. Ashes were sometimes used, especially in cold water leaching (Gifford, 1971). *Aesculus* nuts were sometimes processed, but on a lesser scale than in Japan. The diet included diverse plants and animals and consequently seems to have been stable and secure (Kroeber, 1971). The California Indians also developed high population densities. Some estimates indicate something like 300 000 people, and in some parts the density exceeded $4.4/km^{-2}$ or more than that of the agricultural Pueblo Indians of New Mexico and Arizona. There are many parallels between the west coast Indians and Jomon, although few Indian tribes used pottery, and it came very late.

The principles of plant cultivation were well known to west coast Indians and many of the tribes grew local species of tobacco. The most favoured species were *Nicotiana attenuata* and *N. bigelowii*. Special plots were prepared in the forest where logs were burned to provide ash fertilizer and the seeds were broadcast in the ash. Selection was made for high tar and nicotine and cultivated tobacco was clearly distinguished from wild forms (Harrington, 1932). In some tribes, tobacco was cultivated instead of harvested in the wild out of fear that the wild tobacco might have grown on a grave site and the ghost of the deceased might enter the body with the smoke. However, no food plants were cultivated, and the idea of planting and reaping tobacco did not change the food procurement system. Hunting and gathering continued to provide sustenance until the nineteenth century (Klimek, 1935).

The archaic of eastern North America may not have lasted as long as Jomon, but the Koster site in Illinois spans more than 6000 years. The earliest occupation there was about 7500 BC, but the traces are too few for much reconstruction. At 6500 BC, grinding and nutting stones are found together with mussel shell middens, nut shells and the bones of fish, deer and small mammals. *Carya pecan* was, by far, the most preferred nut at Koster, but acorn, walnut, hickory and hazel-nut were also used. The list of other carbonized seeds is lengthy, but the main ones were *Iva, Chenopodium, Amaranthus, Trifolium, Vitis* and *Datura* (presumably not for food) (Asch *et al.*, 1972; Struever and Holton, 1979).

Many other archaic sites have been excavated in eastern North America. The subsistence economy involved hunting deer and small mammals, catching fish and gathering molluscs, fowling, and using nuts on a large scale and in exploiting seeds of *Iva*, the chenopods, *Helianthus*, and some grasses. Seed size of some of the annuals increased with time and *Iva, Helianthus, Chenopodium* and *Phalaris caroliniana* may have been cultivated (Struever and Vickery, 1973). Cultigens from tropical America appeared, but did not seem to influence the life style. *Cucurbita* occurs as early as 2300 BC at a Missouri site (Chomko and Crawford, 1978). The bottle gourd appears by 1200 BC, maize by 200 BC and beans by 500 AD, but serious full-time farming came only with the Mississippian culture about 800–900 AD. The cultivation of plants seems to have been little more than a hobby for at least 2500 years. Like the Jomon, the archaic Indian did not need cultivated plants and agriculture was adopted very late.

The mesolithic of Europe can be reconstructed from many excavated sites and in several ways resembles the economies just described. Some cultures maintained a typically upper paleolithic hunting specialization; the people of Starr Car caught many roe deer and the Baltic Kunda sought elk. However, most of the cultures studied gradually evolved into broad-spectrum hunting-gathering economies exploiting mammals, birds, fish, shellfish and, presumably, plants (Cohen, 1977; Bender, 1975; Tringham, 1971).

The plant remains of the European mesolithic are inadequately documented. Hazel nuts are sometimes numerous, but little is known of the harvested plant products. One reason may be that the sites occupied were often on thin, acid soils where preservation can be poor. There were some attempts, perhaps independent, towards domestication of animals in late European mesolithic. Pig, elk, aurochs and deer were involved, but agriculture as a food procurement system was intrusive from the Near East. Farmers arrived in Greece by 6000 BC, colonized Yugoslavia, Bulgaria, parts of Rumania and Hungary during the period 5600–4500 BC, and reached Germany and Holland about 4500 BC. They tended to occupy the areas deliberately avoided by hunter-gatherers. These were the brown forest soils and alluvia in the Balkans and loess deposits in the Danube basin (Quitta, 1971). The two cultures long co-existed with some hunter-gatherers occasionally taking up farming and some farmers occasionally abandoning agriculture. The contributions of the European forest flora were correspondingly modest and late (Tringham, 1971; Koslowski, 1973).

Near Eastern woodlands

Floras that develop under winter rainfall and summer drought regimes have much in common around the world. These 'Mediterranean' formations (Fig. 2) look similar in the Americas and elsewhere, but there are profound differences in the importance of annuals. The woodlands of the Near East include a lush annual element, rich in large-seeded grasses and legumes. Findings show clearly that the emphasis of hunter-gatherers was on annual herbs. Aboreal products included *Prosopis*, pistachio, almond, grape, apricot and olive. Acorns were undoubtedly harvested in mesolithic times as they are now. More importantly, there are remains of wild barley, wild einkorn, wild emmer, vetch, flax and lentil in what appear to be pre-agricultural or proto-agricultural sites (Flannery, 1973; Harlan, 1975).

At Mureybat on the Euphrates there was evidence of large-scale use of wild einkorn and wild barley in the ninth millennium BC. Carbonized grains of wild cereals as well as cultivated races were found at Ali Kosh, Jarmo, Beidha, Çayönü, Hacilar and Tepe Guran (Renfrew, 1969). One of the most intriguing cultures of mesolithic character is the Natufian that ranged through the ninth and eighth millennia BC as represented by a number of sites from the Negev to Turkey. In some areas it persisted into the fifth millennium, long after agriculture had been well established elsewhere. Among the Natufian artifacts are many grinding stones and stone mortars, that were moveable or in bed rock. Sickle blades are abundant and sickle hafts occur. All the equipment was available for cereal harvest and utilization but plant remains have not been found and there is no evidence of actual cultivation. Presumably wild cereals were available in such abundance that it was unnecessary to cultivate them (Perrot, 1966, 1968). Even now, stands of wild cereals develop as dense as sown cultivated fields when protected from livestock.

The Near Eastern agricultural complex evolved from the mesolithic emphasis on annual grasses and legumes that coalesced into a viable food production system. It initially featured barley, emmer, einkorn, vetch, lentil, pea and flax. From the woodland flora came pistachio, almond, apricot, pear, pomegranate, grape, fig, poppy, safflower, mustardseed, rapeseed, fenugreek and others. Wild rye and wild oats are part of the flora, but they may have been domesticated much later from weed races that infested wheat and barley fields (Vavilov, 1926). The Near Eastern woodlands provided much of the germ-plasm for one of the earliest and most successful of all agricultures.

Tropical savannas and dry forest (long dry seasons)

This ecological zone (Fig. 3) is rich in harvestable food resources in Africa, fairly rich in Asia, and extremely variable in the Americas. Archaeological information on the prelude to agriculture is generally very scanty with only a few informative sequences.

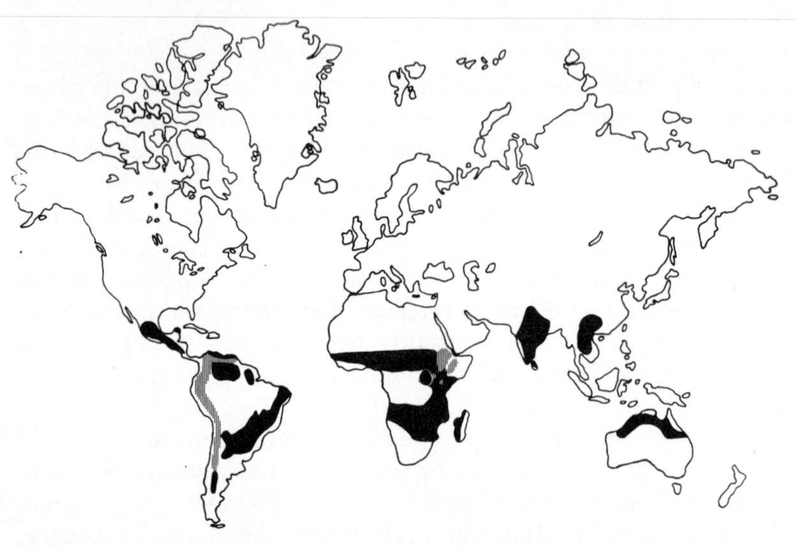

Figure 3 Sketch map of the world showing the main areas of tropical dry forest/savannah (solid black) and tropical highland (hatched).

Indigenous African agriculture is basically of savanna origin (Harlan *et al.*, 1976b). Such crops as sorghum (*Sorghum bicolor*), cowpea (*Vigna unguiculata*), Bambara groundnut (*Voandzeia subterranea*), fonio (*Digitaria exilis*), karité (*Butyrospermum*), African rice (*Oryza glaberrima*), lablab (*Dolichos lablab*), okra (*Abelmoschus*), *Hibiscus sabdariffa, Corchorus olitorius* and African yam (*Dioscorea rotundata*) were derived directly from the savanna flora. The progenitor of African rice (*O. barthii*) is an annual species adapted to water holes of the savanna that fill during the rains and later dry out. The crop moved into the forest zone after domestication. Yams are now the staple diet of some forest tribes, but the adaptation of annual tubers is for survival through long dry seasons and fires. Typical forest species of *Dioscorea* have perennial tubers. Even the oil palm (*Elaeis guineensis*) is a crop that requires full sunlight and, in the wild, was probably a plant of the forest–savanna ecotone (Harlan *et al.*, 1976a).

It is not known when plant domestication began and archaeological information of the relevant period is scarce. There was a time in African prehistory when men turned from big game hunting towards a greater exploitation of aquatic resources with camps along rivers and lake shores. Some lakes have since receded or dried up but there are harpoons and bones of fish, hippopotami, etc. Camps or village sites are widely scattered across the Sahara. They were occupied at times when the Sahara was less arid and there were numerous shallow playa lakes that enlarged during the rains and later contracted (Clark, 1970). Most of the Sahara sites dated to the sixth and fifth millennia BC have grinding stones and many have what appear to be sickle blades. The plant foods harvested and processed are as yet unknown, but barley has been proposed and the recent finds of Wendorf *et al.* (1979) are suggestive. However, barley is not a tropical crop and the evolution of native African domesticates has not yet been documented archaeologically. Sometimes, agriculture appears to have been introduced into a region without apparent change in the stone artifacts.

In Southeast Asia, a somewhat similar array of plants was domesticated from the savanna-dry forest flora. The origin of Asian rice (*Oryza sativa*) is more complex than that of African rice because in Asia the annual and perennial races hybridize freely, while in Africa the two are well-separated species. It is most likely, however, that the first attempts at growing rice would have used annual races adapted to rainy-season water holes as in Africa. Rice culture does not *begin* in deep-water flood plains (T-T. Chang, 1976). The Asian yams with annual tubers are adapted to the same conditions as African ones. A suite of leguminous and other plants evolved from this ecological zone in Asia, including: mung (*Vigna radiata*), urd (*V. mungo*), rice bean (*V. calcarata*), mat bean (*V. aconitifolia*), pigeonpea (*Cajanus cajan*), cucumber, Asian cotton, egg plant and sesame. Water hole plants other than rice include: lotus (*Nelumbium*), *Trapa, Eleacharis*, kodo millet (*Paspalum scrobiculatum*) and probably *Coix*.

Archaeological information as to where and when a south-east Asian plant complex came into use is still meagre, but such sites as Non Nok Tha and Ban Chiang in Thailand suggest that rice may have been cropped by 5000 BC or perhaps earlier. In general, Indian agriculture appears to be later than that in Thailand or the Near East and to have been greatly influenced by both (Vishnu-Mittre, 1977).

The ecological situation in Mesoamerica is extremely complex because of the mountainous terrain, and it is difficult to be certain of crop origins. Annuals adapted to a long dry season include some of the most important of American domesticates. Wild maize (*Zea mays* spp. *mexicana*) and wild beans (*Phaseolus vulgaris*) are adapted to open woodlands of steep mountain slopes at mid-elevations. The scarlet runner bean requires cooler and moister habitats. *Capsicum annuum, Cucurbita* spp., chia (*Salvia hispanica*), tomatle (*Physalis ixocarpa*), huaozontle (*Chenopodium nuttallii*) and upland cotton (*Gossypium hirsutum*) were domesticates from the local flora of this formation. The sweet potato (Nishiyama, 1971), avocado and papaya probably came from forest margins of the long dry season vegetation.

The situation in South America is distinct. The vast llanos and campos cerrados occupy immense reaches of Colombia, Venezuela and Brazil. The Chaco formation includes great areas of Brazil, Bolivia and Paraguay. They are awash during the rains and dry out in the dry season. Most of the soils are toxic with low pH and poisonous concentrations of metals. Few cultivated plants can tolerate these soils and the contribution of this flora to agriculture has been very limited. In more favourable sites, however, wild cassava, wild pineapple, wild peanuts and *Cucurbita* spp. occur. Wild beans and lima beans grow on the east slopes of the Andes. The forest margins have probably contributed guava, papaya, canna and other crops.

In Mesoamerica, there is archaeological information from Tamaulipas (MacNeish, 1958), Tehuacan (MacNeish, 1964; Byers, 1967), Oaxaca (Flannery *et al.*, 1967; Kirkby, 1973; Flannery, 1973) and the Valley of Mexico (Niederberger, 1979). In each case, the local evolution of an agriculture can be traced from broad-spectrum hunting-gathering economies. Plants of the local flora were being cultivated by the seventh millennium BC. Apparently a few plants were grown in summer camps where people assembled in relatively large groups for the rains. With the onset of the dry season, the groups dispersed into foraging bands. The nomadic life continued for several millenia after plants were first cultivated. Originally, the cultigens contributed very little to the diet and this at seasons when other food was abundant. It seem to have been more of a hobby than to avert hunger. Very slowly, over millennia, additional cultigens were grown and eventually contributed substantially to the diet. Obviously, growing plants solely for food was not a revolutionary technique.

The prelude to agriculture in South America is poorly documented, but there are informative sequences in Peru. The earliest settlers of South America were primarily big game hunters of grazing animals in open grasslands and the forests were occupied later. As in North America, large mammal species became extinct and by about 7000 BC there was a trend towards a wider exploitation base. A greater use of vegetable foods is indicated by the appearance of grinding tools (mortars, pestles, manos and metates). More attention was paid to fish and shellfish and the hunting-gathering economies drifed towards the mesolithic (Cohen, 1977).

Fully domesticated beans and lima beans appear at Guitarrero Cave, Peru *c.* 6000 BC (Kaplan *et al.*, 1973). Since the wild races occur today on the eastern slope of the Andes and this site is in an intermontane valley of the west slope, it seems likely that there were much earlier attempts at domestication. The evolution of agriculture on coastal Peru is well documented because of the abundance of occupational refuse that is superbly preserved. However, the region is too dry for agriculture without irrigation and the cultigens were all imported from elsewhere. Findings in coastal Peru merely reflect events that happened previously elsewhere. Lima beans and squash make their appearance *c.* 3000 BC. Further north in Ecuador maize agriculture was well established in the Valdivian culture *c.* 2500 BC (Lathrap, 1975).

The Chaco and the llanos were relatively poor in food resources and few traces can be found before pottery making began (Willey, 1971). Even the ceramic-producing cultures were mostly hunter-gatherers into historic times. Agriculture was clearly marginal and late in these regions and the flora contributed little or nothing.

Tropical rainforest (short dry seasons)

Rainforests (Fig. 2) are difficult environments for humans and these formations have been sparsely settled until recently, and some still support very low densities. The agricultural systems that evolved in Asia and the South Pacific resemble those in north-western South America and are rich in roots and fruits. The Asian wet forests contributed bananas and plantains, breadfruit, jack fruit, jambos, rambutam, litchi, lansium, mangosteen, citrus, tung oil, ginger, taro, *Alocasia, Cyrtospermum*, sago palm, pandanus, sugarcane and many others. The number of fruits is remarkable. From the American wet forests came cocoa (*Theobroma*), *Annona* spp., *Passiflora* spp., *Xanthosoma* spp., ramon (*Brosimum*), *Vanilla*, cashew, Brazil nut, guava, *Spondias* spp., *Casimiroa* spp., at least three species of *Solanum* with edible fruits and many other fruits and nuts.

The African rainforests were colonized by peoples of the savanna who adapted their crops as best they could to the new environment and relatively few forest plants were domesticated. The most important of these were not for food. African rainforests contributed arabica coffee, robusta coffee, cola, malaguette (*Afromormum*), akee apple (*Blighia*), *Sphenostylis* and other rather minor cultigens.

Slash and burn techniques were employed wherever rainforests were exploited. The archaeological information is extremely meagre relating to plant utilization. Findings in Thailand have indicated the many plants harvested by Hoabinhian people *c.* 6500 BC (Gorman, 1969, 1977; Yen 1977). In Papua-New Guinea there is evidence for the draining of swamps by an extensive canal system dating back to *c.* 7000 BC (Golson, 1977; White and Allen, 1980). The purpose of the drainage is uncertain but likely to be concerned with agriculture. The archaeological record in Amazonia indicates a rather late colonization, with the earliest known sites dated *c.* 2000 BC (Cohen, 1977).

Tropical highlands

The main contributions to agriculture in this category come from the Andes and the Ethiopian plateau with adjacent highlands of Kenya and Uganda (Fig. 3). The Andean derivatives include: quinoa (*Chenopodium quinoa*), cañihua (*C. pallidicaule*), chocho (*Lupinus mutabilis*), maca (*Lepidium meyenii*), oca (*Oxalis tuberosa*), yacón (*Polymnia sonchifolia*), potato (*Solanum tuberosum*), añu (*Tropaeolum tuberosum*) and ulluco (*Ullucus tuberosus*) (Léon, 1964). The only crop of global importance is the potato.

The East African highlands have contributed arabica coffee (*Coffea arabica*) and several relatively minor crops: finger millet (*Eleusine coracana*), tef (*Eragrostis tef*), noog (*Guizotia abyssinica*), ensete (*Ensete ventricosa*) and chat (*Catha edulis*).

These agricultures are probably relatively late, but they are not well dated. In the Ayacucho region of Peru, shifts in food procurement systems were detected with a decline in exploitation of large mammals, an increase in small mammal usage, and more dependence on plant foods (MacNeish, 1971; MacNeish *et al.*, 1975). Scheduled foraging and alternate use of wet and dry season camps were apparent by 7000 BC. Domesticated quinoa and squash seeds were found in refuse of the Piki phase 5800–4550 BC. In Ethiopia, there is one unpublished find of domesticated finger millet from a rock shelter that could date to the third millennium BC (W. D. Phillipson, personal communication; Hilu *et al.*, 1979). Well-preserved inflorescences of finger millet occurred in a cave near Durban, Natal, dated to 2500 BC (Davies and Gordon-Gray, 1977).

Sea coasts

A few plants originated from coastal floras and the most important is coconut. Sago palm and pandanus can also be included among tropical species. Radish, beet and cabbage all have wild races belonging to coastal floras but these seem to be relatively late contributions to agriculture.

Discussion

Each of the major temperate and tropical vegetation zones has provided at least some cultigens to agriculture. Understandably, the marginal extremes including deserts and true rainforests have made relatively minor contributions, although tropical forests have yielded some important species. Temperate grasslands resisted major agricultural development until historic times. Agriculture also came late to temperate forests and was then usually intrusive. This may have been because natural food resources were abundant and there was no need to change existing methods of procuring food. The most productive formations in terms of agricultural development were the Near Eastern woodlands and the tropical savannas and dry forests. Both formations are rich in species including trees and grasses. The savannas are probably richest in food plants near forest ecotones where rainfall is more reliable than in the drier thornbush savannas.

The annual floras from which so many major crops are derived mainly evolved under the constraints of long dry seasons. This is true for both Near Eastern woodlands and tropical savannas and dry forests. According to Whyte (1972), the monsoon climates of South Asia are post-Pleistocene developments and the annual habits of South Asian grasses and legumes evolved late. Indeed, the development of the annual habit may be so late that it might have been accelerated by disturbance of the habitat by livestock. The annual habit may evolve under other constraints as well, but it is primarily an adaptation to short growing seasons. Long, cold winters and short summer seasons are suitable for annuals and ephemerals are adapted to short rainy periods in dry climates.

Wright (1968, 1977) considered that a woodland flora was not present in the Near East at the close of the Pleistocene. It arrived by migration from unknown refugia shortly before there is evidence of wild cereal harvests. The role played by the annual habit of both grasses and legumes in this flora is most conspicuous. Climates with long dry seasons appear to be necessary for the most productive ecosystems for plant domestication.

It is now apparent why some parts of the world have contributed more cultigens than others and why certain regions seem to be localized centres of origin (Harlan, 1971). The Near Eastern woodlands look like a centre of origin because the flora from which crop plants were extracted has a limited distribution. There is little in the way of centres in Africa because of the vast ecosystem from which indigenous domesticates were derived (Harlan *et al.*, 1976a). The early agriculture of North China was based on loess and gives the appearance of a centre, while Southeast Asia and the South Pacific include vast areas of both long and short dry season climates. The long dry season flora of Mesoamerica is restricted, whereas the corresponding flora of South America is extensive.

Clearly, agricultures evolved out of broad-spectrum hunting-gathering systems. These systems were stable because many plants and animals were exploited. Some of them lasted for many millenia with little apparent change. There was stability in diversity. If the season was bad for some species, it might have been good for others and the hunter-gatherers had many options (Harris, 1973). As agriculture developed, a few select species were extracted from the flora for major cultivation. The options began to decrease and the number of food plants used by mankind has been decreasing ever since. Today, the human species depends on the performance of a few crops. The situation is potentially unstable and could be disastrous (Harlan, 1976).

Similarly, the genetic defences employed against pests and pathogens have been simplified with the evolution of modern agriculture. The defences of wild populations are extremely complex and consequently tend to be stable (Browning, 1974, *1981*). Diseases are always present, but generally do little damage. Hosts and their pathogens have co-evolved over geologic time and become mutually adjusted. In domesticating food plants, a portion of the genetic diversity is extracted from the wild populations and developed. The domesticated populations are simpler and the defences are less complex and, presumably, less stable.

P. R. Jennings (1974b) considers that crops are most difficult to improve in their centre of origin, where he suggests that:

(a) yields are lowest
(b) yield-limiting factors are most numerous and complex
(c) resistance to technical change is greatest
(d) the response to innovations is least.

If these basic principles are true, and generally I think they are, then much more is involved than complex genetic defences against pests and pathogens. In centres of origin of a crop, the domesticated populations occur together with wild and weedy relatives, all of which have co-evolved over a long period of time with each other and with human cultures. Relatively stable equilibria have developed involving hosts and parasites, cultural practices and human habits. The landrace populations are adapted to cultivation (the seedbed, soil fertility, sowing, weeding and harvesting, seed storage, etc.). The stabilized equilibria are complex and very difficult to modify without upsetting the balance and risking disaster. The farmers are keenly aware of this and their resistance to change has an ecological basis.

Agro-ecosystems far from centres of origin are less complete and simpler. They can be more easily modified and increases in yield can be achieved with fewer technological changes. However, the simpler genetic defences against pathogens and pests render crops more vulnerable to epidemic attack, a situation that rarely occurs in an unmodified centre of origin. There are many instances of major problems arising from the new encounters occurring when crops or their pests and pathogens are introduced to new areas. This leads to instability and several examples are considered in some detail by other contributors to this volume (e.g., Posnette, *1981*).

References

Allen, J., Golson, J. & Jones R. (Eds) (1977). *Sunda and Sahul: Prehistoric studies in Southeast Asia, Melanesia and Australia.* London: Academic Press.

Asch, N., Ford, R. I. & Asch, D. L. (1972). Paleoethnobotany of the Koster site: the archaic horizons. *Illinois State Museum. Report of Investigation* 24. *Illinois Valley Archaeological Research Paper* 6. Springfield, Illinois: State Museum.

Barrows, D. P. (1971). Desert plant foods of the Coahuilla. In *The California Indians*, pp. 306–314. R. F. Heizer & M. A. Whipple. Berkley: University of California Press.

Bender, B. (1975). *Farming in Prehistory: from Hunter-gatherer to Food Producers.* New York: St. Martin's Press.

Browning, J. A. (1974). Relevance of knowledge about natural ecosystems to

development of pest management programs for agro-ecosystems. *Proceedings American Phytopathological Society* **1,** 191–199.
Browning, J. A. (*1981*). The agroecosystem–natural ecosystem dichotomy and its impact on phytopathological concepts. In *Pests, Pathogens and Vegetation*, pp. 159–172. J. M. Thresh. London: Pitman.
Brunken, J., de Wet, J. M. J. & Harlan, J. R. (1977). The morphology and domestication of pearl millet. *Economic Botany* **31,** 163–174.
Byers, D. A. (Ed.) (1967). *The Prehistory of the Tehuacan Valley. Vols. 1 and 2.* Austin: University of Texas Press.
Carter, G. F. (1977). A hypothesis suggesting a single origin of agriculture. In *The Origins of Agriculture*, pp. 89–133. C. A. Reed. The Hague: Mouton.
Chang, K. C. (1968). *The Archaeology of Ancient China.* 2nd Edn. New Haven: Yale University Press.
Chang, T-T. (1976). The origin, evolution, cultivation, dissemination and diversification of Asian and African rice. *Euphytica* **25,** 425–441.
Chomko, S. A. & Crawford, G. W. (1978). Plant husbandry in prehistoric eastern North America: New evidence for its development. *American Antiquity* **43,** 405–408.
Clark, J. D. (1970). *The Prehistory of Africa.* New York: Praeger.
Cohen, M. N. (1977). *The Food Crisis in Prehistory.* New Haven: Yale University Press.
Cooper, J. M. (1946). The Patagonian and Pampean hunters. In *Handbook of South American Indians. Vol. 1. The Marginal Tribes*, pp. 127–176 J. H. Steward. Washington D. C.: US Government Printing Office.
Davies, O. & Gordon-Gray, K. (1977). Tropical African cultigens from Shongweni excavations, Natal. *Journal Archaeological Science* **4,** 153–162.
Flannery, K. V. (1973). The origins of agriculture. *Annual Review Anthropology* **2,** 271–310.
Flannery, K. V., Kirkby, A. V. T., Kirkby, M. J. & Williams, A. W. (1967). Farming systems and political growth in ancient Oaxaca, *Science* **158,** 445–454.
Gentry, H. S. (1942). Rio Mayo plants: a study of the flora and vegetation of the valley of Rio Mayo, Sonora. *Carnegie Institute Publication* No. 527.
Gifford, E. W. (1971). Californian balanophagy. In *The Californian Indians*, pp. 301–305. R. F. Heizer & M. A. Whipple. Berkeley: University of California Press.
Golson, J. (1977). No room at the top: Agricultural intensification in the New Guinea highlands. In *Sunda and Sahul: Prehistoric studies in Southeast Asia, Melanesia and Australia*, pp. 601–638. Eds J. Allen, J. Golson & R. Jones. London: Academic Press.
Gorman, C. (1969). Hoabinhian: a pebble tool complex with early plant associations in Southeast Asia. *Science* **163,** 671–673.
Gorman, C. (1977). *A priori* models and Thai prehistory: A reconsideration of the beginnings of agriculture in Southeastern Asia. In *The Origins of Agriculture*, pp. 321–355. C. A. Reed. The Hague: Mouton.
Harlan, J. R. (1971). Agricultural origins: centers and non-centers. *Science* **174,** 468–474.
Harlan, J. R. (1975). *Crops and Man.* Madison: American Society of Agronomy.
Harlan, J. R. (1976). The plants and animals that nourish man. *Scientific American* **238,** 89–97.

Harlan, J. R. (in press) Patterns of plant domestication. In *Symposium on the origin of Agriculture and Technology: West or East Asia? Moesgaard, Denmark*, Nov. 21–25, 1978. Scandinavian Institute of Asian Studies Monograph Series.

Harlan, J. R. & Zohary, D. (1966). Distribution of wild wheats and barley. *Science* **153,** 1074–1080.

Harlan, J. R., de Wet, J. M. J. & Stemler, A. L. B. (Eds) (1976a). *The Origins of African Plant Domestication.* The Hague: Mouton.

Harlan, J. R., de Wet, J. M. J. & Stemler, A. L. B. (1976b). Plant domestication and indigenous African agriculture. In *The Origins of African Plant Domestication,* pp. 3–19. Eds J. R. Harlan, J. M. J. de Wet & A. L. B. Stemler. The Hague: Mouton.

Harrington, J. P. (1932) Tobacco Among the Karuk Indians of California. *Smithsonian Institute, Bureau of American Ethnology Bulletin* No. 94.

Harris, D. R. (1973). The prehistory of tropical agriculture: an ethno-ecological model. In *The Explanation of Culture Change: Models in Prehistory*, pp. 391–417. C. Renfrew, Pittsburgh: University of Pittsburgh Press.

Hilu, K. W., de Wet, J. M. J. & Harlan, J. R. (1979). Archaeobotanical studies of *Eleusine coracana* spp. *coracana* (finger millet). *American Journal Botany* **66,** 330–333.

Jardin, C. (1967). *List of Foods Used in Africa.* Rome: FAO.

Jennings, J. D. (1974a). *Prehistory of North America.* 2nd edn. New York: McGraw-Hill.

Jennings, P. R. (1974b). Rice breeding and world food production. *Science* **186,** 1085–1088.

Kaplan, L., Lynch, T. F. & Smith, C. E. (1973). Early cultivated beans (*Phaseolus vulgaris*) from an intermontane Peruvian Valley. *Science* **179,** 76–77.

Kidder, J. E. (1968). *Prehistoric Japanese Arts: Jomon Pottery.* Tokyo: Kodansha International.

Kirkby, A. (1973). The use of land and water resources in the past and present valley of Oaxaca, Mexico. *University of Michigan Museum of Anthropology, Memoir* 5.

Klimek, S. (1935). Culture element distributions: I. The structure of California Indian culture. *University of California Publications in American Archaeology and Ethnology* **37,** 1–70.

Koslowski, S. K. (Ed.) (1973). *The Mesolithic of Europe.* Warsaw: Warsaw University Press.

Kroeber, A. L. (1971). The food problem in California. In *The California Indians*, pp. 297–300. R. F. Heizer & M. A. Whipple. Berkeley: University of California Press.

Lathrap, D. W. (1975). *Ancient Ecuador: Culture, Clay and Creativity*. Chicago: Field Museum of Natural History.

Lawton, H. W., Wilke, P. J., DeDecker, M. & Mason, W. M. (1976). Agriculture among the Paiute of Owens Valley. *Journal of California Anthropology* **3,** 13–50.

Lee, R. B. & DeVore, I. (1968). *Man the Hunter.* Chicago: Aldine.

León, J. (1964). Plantas alimenticias andinas. Lima: *Instituto Interamericano de Ciencias agricolas zona Andina, Boletin Tecnico.* No. 6.

MacNeish, R. S. (1958). Preliminary archaeological investigations of the Sierra de Tamaulipas. *Transactions American Philosophical Society* (n.s.), **48,** 1–210.

MacNeish, R. S. (1964). Ancient mesoamerican civilization. *Science* **143,** 531–537.

MacNeish, R. S. (1971). Early man in the Andes. *Scientific American* **224,** 36–46.

MacNeish, R. S., Patterson, T. C. & Browman, D. L. (1975). *The Central Peruvian Prehistoric Interaction Sphere.* Andover: Peabody Foundation.

Milisauskas, S. (1978). *European Prehistory.* New York: Academic Press.

Munson, P. J. (1976). Archaeological data on the origins of cultivation in the southwestern Sahara and their implications for West Africa. In *The Origins of Plant Domestication*, pp. 187–209. Eds J. R. Harlan, J. M. J. de Wet & A. L. B. Stemler. The Hague: Mouton.

Murray, J. (1970). *The First European Agriculture.* Chicago: Aldine.

Niederberger, C. (1979). Early sedentary economy in the basin of Mexico. *Science* **203,** 131–142.

Nishiyama, I. (1971). Evolution and domestication of the sweet potato. *Botanical Magazine, Tokyo* **84,** 377–387.

Perrot, J. (1966). Le gisement Natufian de Mallaha (Eynan), Israël. *L'Anthropologie* **70,** 437–484.

Perrot, J. (1968). La préhistoire palestinienne, *Supplement au dictionnaire de la Bible* **8,** 286–446.

Posnette, A. F. (1981). The role of wild hosts in cocoa swollen shoot disease. In *Pests, Pathogens and Vegetation*, pp. 71–78. J. M. Thresh. London: Pitman.

Quitta, H. (1971). Der Balkan als Mittler zwischen Vorderem Orient und Europa. In *Evolution und Revolution im Alten Orient und in Europa*, pp. 38–63. F. Schlette. Berlin: Akademie-Verlag.

Reed, C. A. (1977). *The Origins of Agriculture.* The Hague: Mouton.

Renfrew, J. M. (1969). The archaeological evidence for the domestication of plants: Methods and problems. In *The Domestication and Exploitation of Plants and Animals*, pp. 149–179. Eds P. J. Ucko & G. W. Dimbleby. London: Duckworth.

Renfrew, J. (1973). *Paleoethnobotany: the Prehistoric Food Plants of the Near East and Europe.* New York: Columbia University Press.

Steward, J. H. (1934). Ethnography of the Owens Valley Paiute. *University of California Publications of American Archaeology and Ethnology* **33,** 233–340.

Steward, J. H. (1941). Culture element distributions: XIII. Nevada Shoshoni. *University of California Anthropological Records* **4,** 4(2), 209–359.

Struever, S. & Holton, F. A. (1979). *Koster: Americans in Search of their Prehistoric Past.* Garden City: Anchor Press.

Struever, S. & Vickery, K. D. (1973). The beginning of cultivation in the Mid-West riverine area of the United States. *American Anthropologist* **75,** 1197–1220.

Tringham, R. (1971). *Hunters, Fishers and Farmers of Eastern Europe 6000–3000* BC. London: Hutchinson.

Ucko, P. J. & Dimbleby, G. W. (Eds) (1969). *The Domestication and Exploitation of Plants and Animals.* London: Duckworth.

Vavilov, N. I. (1926). Studies on the origin of cultivated plants. *Bulletin Applied Botany and Plant Breeding* (Leningrad) **16,** 1–243.

Vishnu-Mittre, (1977). Changing economy in ancient India. In *The Origins of Agriculture*, pp. 569–588. C. A. Reed. The Hague: Mouton.

Watanabe, Makoto (1979). Kodai iseki shutsudo no tochi no mi. (Archaeological remains of horse chestnut). *Bulletin of the Palaeological Association of Japan*, 63–74.

Wendorf, F., Schild, R., El Hadidi, N., Close, A. E., Kobusiewicz, M., Wieckowska,

H., Issawi, B. & Haas, H. (1979). Use of barley in the Egyptian late paleolithic. *Science* **205,** 1341–1347.

White, J. P. & Allen, J. A. (1980). Melanesian prehistory: some recent advances. *Science* **207,** 728–734.

Whyte, R. O. (1972). The Gramineae, wild and cultivated of monsoonal and equatorial Asia. I. Southeast Asia. *Asian Perspectives* **15,** 127–151.

Willey, G. R. (1971). *An Introduction to American Archaeology. Vol. 2.* Englewood Cliffs: Prentice Hall.

Wright, H. E. (1968). Natural environment of early food production north of Mesopotamia. *Science* **161,** 334–339.

Wright, H. E. (1977). Environmental change and the origin of agriculture in the Old and New Worlds. In *The Origins of Agriculture*, pp. 281–318. C. A. Reed. The Hague: Mouton.

Yanovsky, E. (1936). Food plants of the North American Indians. *Washington: USDA Miscellaneous Publication* No. 237.

Yen, D. E. (1977). Hoabinhian horticulture? The evidence and the questions from northwest Thailand. In *Sunda and Sahul: Prehistoric studies in Southeast Asia, Melanesia and Australia*, pp. 567–599. Eds J. Allen, J. Golson and R. Jones. London: Academic Press.

Changing patterns of land use: global trends

A H Bunting
Plant Science Laboratories, University of Reading, Whiteknights, Reading, Berks RG6 2AS

During most of the two million year history of hominids and men, there have been few of them. Their effects on vegetation by food-gathering, fire and other disturbances were probably local and transient, though they may well have created new secondary niches for the progenitors of crops, pests, pathogens and weeds.

Agriculture appeared in Middle America and the Middle East, about 10 000 years ago (Harlan, *1981*), and most of the plant cover of the earth has since been modified by man. For example, the lowland humid forests of the central basin of Zaire are a patchwork of secondary communities, in which the emergent trees testify to the even more majestic forests of former times. Moreover, much of the humid lowland forest of southern and eastern Nigeria consists of mixed systems of annual crops and economic trees. The forests of Latin America and Southeast Asia also include substantial areas of secondary vegetation. In West Africa, Chevalier's classic vegetation zones are artefacts of agricultural man, while the leguminous *Brachystegia–Julbernardia–Isoberlinia* woodlands which cover so much of the seasonally-arid parts of the continent appear to be largely secondary. During the last 150 years the wild forests of the north-eastern United States have engulfed the farms and settlements carved from the wilderness by early pioneers. Similarly, woodland now blankets former arable areas of southern France. Some of the most recent of such changes in Britain are discussed by Trask (*1981*).

Parts of the boreal and montane vegetation of the world, and perhaps some desert ecosystems, may have continued largely unchanged since the last major retreat of the ice. For the rest, the 'balance of nature' which has inspired both landscape painters and poets in Europe for centuries, is an artefact of man and his works.

Recent changes in land use

As a result of this global conversion about 1460 million ha, equivalent to 11% of the total land area of the earth, was arable land or carried permanent crops in 1977, and about 200 million ha were irrigated. 23% was permanent grassland (wild or cultivated) and 31% was forest and woodland (natural, planted or reserved for forestry) (FAO, 1979). The remaining 34% was referred to 'other' land (an undifferentiated balancing term derived by difference). It includes areas which are unused but potentially productive, roads, lanes, built-on areas and ornamental gardens, barren land (including saline, mountainous and desert), and that not included in other categories.

Tables 1 and 2 summarize FAO data on these areas, in 1961–65 and 1977, in different regions and selected countries. The data come from country reports or external estimates that vary in accuracy, and are particularly uncertain for China in 1961–65. Despite additional problems over definitions the figures provide acceptable evidence of large-scale trends, though the reasons are not always known.

Changes in areas

Arable area

In the world as a whole, the arable area appears to have increased between 1961–65 and 1977 by about 71 million ha (5.4%), from 1303 to 1374 million ha. Important increases were recorded in Latin America (particularly Brazil), in the developing countries of Africa and the Far East (especially India), and in North America and Oceania (mainly Australia). Important decreases occurred in both Western and Eastern Europe.

Area of permanent crops

The areas and net changes in area of permanent crops (including rubber, oilpalm, coconut, bananas and plantains, tea, coffee, cocoa, fruits, vines and olives) are comparatively small, but they are both economically and environmentally important. The world total increased from 76 to 88 million ha (15%), mainly in the developing countries with market economies. Malaysia and the Philippines added 1 million and 0.9 million ha (39 and 45%) to their permanently cropped land, whereas in India the area decreased by 1 million ha (15%).

Area of irrigated land

While the combined total of arable and permanent cropping increased by 83 million ha (6%), major irrigation increased by 49 million ha (33%). (The data do not appear to include small-scale irrigation from shallow or deep wells.) About 35 million ha (71%) of the increase was in the developing countries, particularly in the Far East (including both India and China), and 6 million ha were added in the USSR. Much of the new irrigation was on existing arable land which had previously been rain-fed, and it permitted important increases in output and in intensity of cropping, which are considered later.

Area of permanent pasture

Despite difficulties in definition, some of the contrasts and changes in pasture land presented in Table 2 are real and important. In the developed market economies permanent pasture decreased by 20 million ha (2%), mostly in North America. Among developing countries with market economies, Brazil reported an increase of 34 million (26%), largely used to raise beef for export to North America, for use mainly for convenience foods and hamburgers (Myers, 1980). The reported decrease in the Asian centrally planned economies was mostly in Mongolia, where it is balanced by an increase in 'other' land, suggesting that the change may have been in definition rather than in land use.

Area of forest and woodland

Between 1961–65 and 1977, the world appears to have lost 55 million ha net (1.3% of existing area) of forest and woodland to agriculture or pasture, felling for fuel and timber, and urban and other non-agricultural development. The largest absolute losses in the developed countries were in Oceania (mainly Australia). In developing countries, large losses are reported for Africa and Latin America, especially Brazil and Mexico. In contrast, considerable increases are reported in Western Europe and India (planting for timber and fuel) and in China. Extensive forests and woods have been established or encouraged in arid or mountainous areas of north China, partly at least to control erosion, run-off and air movement, and to regulate stream flow.

The losses are mostly small in proportion to total areas, yet they are bound to cause concern because forest and woodland protect soil and terrain, particularly in broken or mountainous country and on watersheds. Moreover, they are often both biologically and aesthetically interesting.

For the world as a whole, the rate of *net* loss from 1961–65 to 1977 (say 14 years) was about 7.5 ha minute^{-1}. In Brazil alone, the *net* rate was 2.5 ha minute^{-1}. The *gross* figure for regional losses is 93 million ha, which is equivalent to 13 ha minute^{-1}. Values ranging from 11 to 45 ha minute^{-1} have been reported for the rate of conversion in the forests of the humid tropics (Lanly and Clement, 1979; Myers, 1979, 1980). This is because conversion includes changes other than clear felling, such as turnover in shifting cultivation.

However, Lanly and Clement (1979) conclude that in Latin America more than 95% of the 'operable' closed hardwood forest is still undisturbed, and that 140 million ha (22%) of the 627 million ha of the total closed forest is not operable because it is on difficult terrain or is permanently inundated. Similarly, in Africa and Asia around 171 million ha (35%) of the closed hardwood forest is considered inoperable. It may seem unlikely, therefore, that these interesting and important formations will disappear from the earth in the foreseeable future.

Patterns of change

The global increase in the area of arable and permanent cropping between

Table 1 Land areas in 1977 and changes in areas of arable land, permanent crops and irrigated land between 1961–65 and 1977 and ratios of changes in irrigated and arable land (△ = change, ● = <0.5)

Region	All land	Arable land				Permanent crops				Irrigated land				Ratios irrigated/arable		
	1977	*1961–65*	*1977*	△	△%	*1961–65*	*1977*	△	△%	*1961–65*	*1977*	△	△%	*1961–65*	*1977*	△
World	13074	1303	1374	71	5	76	88	12	15	149	198	49	33	11	14	3
Developed market regions	3158	365	376	11	3	16	16	1	5	27	33	5	20	8	9	1
North America	1835	221	229	9	4	2	2	●	−8	15	18	3	18	7	8	1
Western Europe	373	93	83	−9	−10	12	13	●	4	7	9	2	31	7	11	3
Oceania	789	34	45	11	33	●	●	●	−4	1	2	●	38	4	4	●
Other	161	18	18	●	3	1	2	1	47	4	4	●	3	25	25	●
Developing market regions	6440	548	609	61	11	53	63	10	18	70	95	25	35	13	16	3
Africa	2329	153	168	15	10	11	14	2	20	1	2	1	60	1	1	●
Latin America	2020	92	116	24	26	23	27	4	16	8	13	4	52	9	11	2
Brazil	846	22	32	10	44	8	8	1	7	1	1		83	2	3	1
Mexico	192	22	22	−1	−3	1	2	●	12	3	5	2	72	13	23	10
Near East	1192	72	76	3	5	4	5	2	42	14	18	3	23	20	23	3
Far East	809	230	250	19	8	14	17	2	16	46	62	16	35	20	25	5
India	297	157	165	8	5	5	4	−1	−15	26	35	10	38	16	21	5
Indonesia	181	12	15	3	23	2	2	●	3	4	5	1	20	33	33	−1
Malaysia	33	3	3	●	9	2	3	1	39	●	●	●	44	8	11	3
Centrally planned regions	3476	390	388	−1	●	8	9	1	13	52	70	19	36	13	18	5
Asian	1149	119	117	−2	−2	1	1	●	29	40	50	10	25	34	43	10
China	930	110	106	−4	−3	1	1	●	58	38	49	10	26	35	46	11
E. Europe/USSR	2327	271	271	1	●	6	7	1	11	11	20	8	74	4	7	3
E. Europe	100	45	44	−1	−3	2	2	●	10	2	4	2	111	4	9	5
USSR	2227	225	228	2	1	4	5	●	11	10	16	6	66	4	7	3

Table 2 Changes in areas of permanent pasture, forest and woodland and 'other' land between 1961–65 and 1977. (To be read with Table 1; △ = change, ● = <0.5); populations in 1977

Region	*Permanent pasture*				*Forest and woodland*				*Other land*				*Population (millions)*
	1961–65	*1977*	△	△%	*1961–65*	*1977*	△	△%	*1961–65*	*1977*	△	△%	*1977*
World	3054	3058	4	●	4132	4077	−55	−1	4508	4477	−32	−1	4104
Developed market regions	903	883	−20	−2	907	884	−22	−2	968	997	30	3	768
North America	282	264	−18	−6	616	616	−1	●	714	723	10	1	240
Western Europe	75	72	−3	−4	115	126	10	+9	79	80	1	1	367
Oceania	459	464	6	1	145	114	−31	−21	151	165	15	10	17
Other	88	83	−5	−5	30	30	●	−2	25	29	4	17	144
Developing market regions	1411	1444	34	2	2142	2072	−70	−3	2286	2251	−34	−2	2028
Africa	689	686	−4	−1	568	540	−28	−5	908	922	14	2	338
Latin America	495	532	37	8	1061	1023	−38	−4	348	321	−27	−8	338
Brazil	132	166	34	26	527	509	−18	−3	157	130	−27	−17	116
Mexico	74	74	●	●	81	71	−10	−12	14	24	+10	76	63
Near East	188	190	1	1	141	140	−1	−1	787	782	−6	−1	195
Far East	37	35	−1	−4	331	327	−4	−1	197	181	−16	−8	1153
India	14	13	−1	−9	58	66	7	13	63	49	−13	−21	645
Indonesia	13	12	−1	−4	124	122	−2	−1	30	30	−1	−2	143
Malaysia	●	●	●	12	24	22	−2	−10	3	5	1	37	13
Centrally planned regions	740	730	−10	−1	1084	1120	37	3	1255	1228	−27	−2	1308
Asia	353	342	−12	−3	136	171	36	26	540	518	−22	−4	939
China	208	211	3	1	85	122	37	43	527	491	−36	−7	866
E. Europe/USSR	387	389	2	1	948	949	1	●	715	710	−5	−1	369
E. Europe	15	15	●	2	28	29	1	4	9	9	●	−2	110
USSR	372	374	2	1	920	920	●	●	706	701	−5	−1	259

Data (in millions of hectares and %) for the world, major regions and selected countries (from FAO, 1979).

1961–65 and 1977 totalled 83 million ha (0.67% of the entire land area). It came from forest and woodland and from 'other' land, which together cover 65% of the world's land area. These changes were associated with important increases in the irrigated area.

The regions and countries exhibit diverse patterns of change. In the United States and Canada net increases in arable and 'other' land, equivalent to about 1% of land area, are reflected by a decrease in permanent pasture. The extra arable land was largely used to grow cereals and soybeans for animal feed and export. Despite massive felling, the area of forest and woodland seems to have been maintained by new planting. In Western Europe, both arable and permanent pasture have decreased while forest, woodland and 'other' land have increased. This appears to continue the historic trend mentioned previously (p. 23) and is associated with current annual imports of around 20 million tons of cereals. In Australia, forest and woodland have decreased by about 4% of the land area in favour of arable, permanent pasture and 'other' land. The extra agricultural land has been used to produce cereals and beef for export.

In the developing countries with market economies, forest and woodland decreased while the areas of arable and permanent crops increased (average net total change 1.6% of land area) to meet increasing needs for food, income and foreign exchange. In Africa, where the total change affected about 1.3% of land area, permanent pasture also decreased, while 'other' land increased. This may reflect changes in classification, abandonment of marginal range grazing land in arid regions, and the encroachment of agriculture into grazing land. It also reflects decreases in the length of bush and grass fallows in shifting cultivation systems. In Latin America forest and 'other' land were used to increase both arable land and permanent pasture. The total change represented about 3.2% of land area, very largely in Mexico and particularly Brazil, where soybeans are grown on 7 million of the 10 million additional hectares of arable land. These trends may continue, as market demands for tropical hardwood timber, soya and beef are still increasing.

In the Far East market economies (2.6% overall net change in land use) the decrease in forest was small, and much of the increases in arable area came from 'other' land, presumably with the help of increased irrigation (9.7 million ha). In India (5.1%) 'other' land and some permanent pasture changed to arable and to forest and woodland which reflects the urgent need, both urban and rural, for fuel and timber. In the Near East 'other' land (desert and seasonal nomadic grazing) changed to arable (3.5 million ha, similar to the increase in irrigation), permanent crops and permanent pasture, giving a total change of 0.5% of the total land area. Malaysia added 1.0 million ha to permanent crops (oil palm, rubber and some cocoa) and 1.3 million ha to 'other' land (total change about 7.5% of land area), mainly at the expense of forest.

In Eastern Europe and the USSR, where the irrigated area increased by 8.5

million ha, 'other' land was the sole net source of increase in all other categories, affecting in total only about 0.2% of land area. In China, where total net change affected 4.3% of total land area, the main feature was the transfer of 36 million ha of 'other' land in the north and north west to forest and woodland. A small decrease in arable land was balanced by increases in permanent crops and grassland. Irrigation increased by 10 million ha and this has permitted more intensive cropping.

During this short period of about 15 years mankind has altered the vegetation of the earth on an impressive scale. The impetus has been provided by the needs and demands for food, timber, biological raw materials, rural income, commercial profit, and foreign exchange, and progress has been aided by an increase of about one-third in irrigated area. Proportionately the vegetation changes are still mostly small, but they have been of considerable significance for development in particular countries, especially Brazil, Mexico, China, India and Malaysia.

Changes in the use of arable land

The largest single component of recent changes in land use has been the world-wide increase in the area of arable land. It is important to consider how this has been used.

Table 3 lists the total harvested areas of the main crops and groups of crops in 1969–71 and 1977. Discontinuities in the time series, particularly for China, complicate comparisons over the longer time period covered by Tables 1 and 2. Nevertheless in those comparisons which seem reasonable, the trends appear to be similar in the two periods.

Apparent intensity of cropping

The global total of 1024 million ha for the harvested areas of the main crops in 1977 (Table 3) differs from the reported 1374 million ha of arable land detailed in Table 1. The ratio of the total harvested area to the total arable area (Table 4) is a crude index of the intensity of cropping. It varies widely between regions and countries, from 35% in Oceania to 174% in China. The reasons for the differences are diverse. Crop failures, temporary grass, forage and silage crops, annual medics and clovers (grown in rotation as nitrogen sources), cover crops and fallows are not included in the total harvested area although they are likely to be included in the arable area. These are probably the main reasons why none of the ratios exceeds 76% in the developed market economies, Eastern Europe, the USSR, Latin America, Africa and the Near East. Conversely, multiple cropping, aided by irrigation, is probably the main reason why the ratios for the Far East and China are larger, ranging from 90% to 174%. The small value for Malaysia, and the large one for Brazil, cannot readily be explained.

The ratios for 1961–65 (Table 4) are similar, but are usually smaller. The general use of arable land may have become more intensive in many countries during the past 20 years, particularly where irrigation has increased, but the statistics for many developing countries are probably not sufficiently reliable to sustain this interpretation fully.

Significance of different crops

Cereals were grown on nearly three-quarters of the total harvested area reported in the world in both 1969–71 and 1977 (Table 3). With the exception of North America, the proportions were larger (84–95%) in the developed economies (including the USSR), where most protein comes from animals fed in part on grain, than in the developing countries (70% or less). This is mainly because of the larger contribution of pulses, oilseeds and cotton in Africa and the Far East, of these crops plus soybean in China, and of soybean in Brazil. North America maintains both a large proportion of cereals (72–75%) and large areas of soybeans (16–20% of the harvested area). In the richer countries soybeans feed animals.

Table 3 Areas of principal arable crops harvested annually 1969–71 and 1977 (△ = change, ●<0.5)

	Cereals				*Roots and tubers*				*Pulses*			
Region	*1969–71*	*1977*	△	△%	*1969–71*	*1977*	△	△%	*1969–71*	*1977*	△	△%
World	702	753	51	7	48	47	●	●	77	81	4	5
Developed market regions	147	161	14	10	4	4	−1	−13	4	3	−1	−24
North America	77	91	13	17	1	1	●	−5	1	1	●	−1
Western Europe	47	45	−2	−2	3	3	●	−14	3	2	−1	−32
Oceania	12	15	3	25	●	●	●	−22	●	●	●	163
Other	11	11	●	2	●	●	●	−23	●	●	●	−45
Developing market regions	291	306	15	5	19	21	2	10	44	48	4	8
Africa	54	56	3	5	10	11	1	11	10	11	●	3
Latin America	47	51	4	8	4	4	●	3	7	8	1	11
Brazil	17	21	4	26	2	3	●	3	4	5	1	22
Mexico	10	10	●	4	●	●	●	9	2	2	●	−16
Near East	35	37	2	5	●	●	●	38	1	2	●	19
Far East	156	163	7	5	4	5	1	13	25	27	2	9
India	100	104	3	3	1	1	●	15	22	24	2	8
Indonesia	11	11	●	1	2	2	●	−14	1	1	●	25
Malaysia	1	1	●	5	●	●	●	4	●	●	●	●
Centrally planned regions	264	286	21	8	25	23	−2	−6	29	30	1	4
Asia	124	135	11	9	12	12	●	−1	21	23	2	8
China	115	125	10	9	12	11	●	−4	21	23	2	9
E. Europe and USSR	140	150	10	7	12	11	−1	−12	7	6	−1	−10
E. Europe	25	25	●	−1	4	4	●	−12	2	1	−1	−37
USSR	115	125	10	9	8	7	−1	−12	5	5	●	1

Data (in millions of hectares and %) for the world, major regions and selected countries (from FAO, 1979).

Changes in the harvested areas of crops other than cereals, pulses and soybeans were minor, though there was a remarkably consistent decrease in the harvested area of root and tuber crops in all developed regions (including Eastern Europe and the USSR) and China.

The pattern of change between 1969–71 and 1977 may be examined crudely by comparing the ratios of the harvested area of cereals to the total harvested area of all crops at each period. World-wide, there was no change, but the ratio decreased somewhat in North America, because the area of soybean increased, and it fell markedly in Brazil because of the substantial shift to soybean in an increasing total harvested area. The ratio increased consistently in all the centrally-planned economies.

Prospects for the future

Population and demand for agricultural products

The currently accepted perspective models for population growth (e.g. World Bank, 1978) assume that birth rates will tend to decline (as they are doing in

Soybean				*Other oil seeds and cotton*				*Sugar cane and beet*				*Total*			
1969–71	*1977*	Δ	Δ%	*1969–71*	*1977*	Δ	Δ%	*1969–71*	*1977*	Δ	Δ%	*1969–71*	*1977*	Δ	Δ%
35	49	14	40	69	71	2	2	19	22	4	20	950	1024	74	8
17	24	6	37	9	11	2	26	3	4	1	19	185	207	22	12
17	24	6	37	7	8	2	23	1	1	●	−3	104	125	21	20
●	●	●	516	1	2	1	41	2	2	1	27	56	53	−2	−4
●	●	●	600	●	●	●	47	●	●	●	33	12	16	3	26
●	●	●	−11	1	1	●	7	●	●	●	17	13	13	●	●
3	10	7	258	42	41	−2	−4	10	12	2	22	410	438	29	7
●	●	●	26	10	8	−1	−15		1	●	25	84	87	3	3
2	8	7	433	7	6	−1	−8	5	6	1	18	72	84	12	16
1	7	6	438	5	5	●	−8	2	2	1	33	31	42	11	36
●	●	●	134	1	●	●	−19	1	●	●	−6	13	13	●	1
●	●	●	313	3	4	●	9	●	1	●	44	40	43	3	7
1	1	●	25	22	22	●	1	4	5	1	25	213	224	12	5
●	●	●	3900	18	18	●	●	3	3	●	9	144	150	6	4
1	1	●	3	●	1	●	35	●	●	●	63	15	15	●	2
●	●	●	●	●	●	●	100	●	●	●	650	1	1	●	8
15	16	●	3	18	19	1	5	5	6	1	18	356	379	23	7
14	15	●	3	9	10	1	6	1	1	●	22	182	196	14	8
14	14	●	3	9	9	1	6	1	1	●	15	171	184	13	8
1	1	●	9	9	9	●	4	4	5	1	17	174	183	9	5
●	●	●	163	1	2	●	11	1	1	●	32	34	34	−1	−2
1	1	●	−9	7	8	●	3	3	4	●	12	140	150	10	7

the richer countries, and have begun to do in many of the poorer ones). They also assume that average expectations of life will increase to around 70–75 years world-wide, and that as a result of these trends the human population of the world may increase three-fold and reach a maximum in about 100 years. This suggests that it will be necessary to multiply the output of food, feed, fuel and timber four-fold over this period. Needs will be largest where birth rates have not yet begun to decline, where average expectations of life are now relatively short, and where environmental resources are already used relatively intensively in agriculture. Not surprisingly, both India and China recognize the need to achieve replacement or smaller birth rates as soon as possible.

Increasing the output of crops

Table 5 indicates some of the principal features of the very important changes

Table 4 *Left:* Total harvested areas as percentages of the total arable areas in 1961–65 and 1977. *Right:* Total harvested areas of cereals as percentages of total areas harvested in 1969–71 and 1977

	Harvested/arable			*Cereals/harvested*		
Region	*1961–65*	*1977*	*Change*	*1969–71*	*1977*	*Change*
World	69	74	5	74	74	●
Developed market economies	50	55	5	80	78	−2
North America	46	54	8	75	72	−2
Western Europe	62	64	2	84	84	●
Oceania	29	35	6	96	95	−1
Other	74	71	−3	85	86	2
Developing market economies	68	72	4	71	70	−1
Africa	48	52	4	63	64	1
Latin America	66	72	6	65	60	−5
Brazil	107	131	24	54	50	−9
Mexico	57	60	−3	74	77	2
Near East	52	57	5	86	85	−1
Far East	90	90	●	73	73	●
India	85	91	6	70	69	●
Indonesia	111	98	−13	74	74	●
Malaysia	20	38	18	93	90	−3
Centrally planned economies	89	98	9	74	75	1
Asia	135	167	32	68	69	1
China	139	174	35	67	68	1
E. Europe and USSR	68	67	−1	81	82	2
E. Europe	80	76	−4	74	75	1
USSR	66	66	●	82	84	1

Data for the World, major regions and selected countries (from FAO, 1979; ● = change <0.5%).

in the world output of cereals since 1948. The total output (including rice as paddy, not grain) increased between around 1950 and 1978 from under 700 to nearly 1600 million tons, because yield per ha per harvest increased 86% on a harvested area only 23% larger. Per head of population, world-wide output increased by more than a third.

Over the much shorter time span from 1969–71 to 1978, total output increased by 29% in North America and in India and China; and in North America and India these increases were due more to increases in yield per

Table 5 World and regional changes in cereal production, human population and output per head (FAO, 1972, 1979)

	Cereals				
Region	*Output (m tons)*	*Harvested area (m ha)*	*Average yield* (tons ha)*	*Output per head (tons)*	*Human Populations (millions)*
World					
1948–52	690	610	1.13	0.27	2513
1978	1596	757	2.11	0.37	4258
% increase (± 28 yr)	131	24	87	37	69
North America					
1969–71	244	77	3.16	1.08	226†
1978	315	85	3.68	1.30	242
% increase (± 8 yr)	29	10	17	20	7
China					
1969–71	210	115	1.83	0.25	826†
1978	270	131	2.07	0.29	933
% increase (± 8 yr)	29	14	13	14	13
India					
1969–71	111	100	1.11	0.20	551†
1978	143	104	1.38	0.22	648
% increase (± 8 yr)	29	4	24	10	18
Africa‡					
1969–71	43	54	0.80	0.15	283†
1978	47	57	0.83	0.13	355
% increase (± 8 yr)	9	7	4	−10	25

* Yield per harvest.
† 1970 data.
‡ Developing countries in Africa south of the Sahara.

harvest (from a very creditable initial level in North America) than in harvested area. Output per head of population increased by 20% in North America, 10% in India and 14% in China. The contrasting data for Africa exhibit something close to stagnation and indeed a decrease of 10% in output per head. Africa apart, these are heartening achievements, and they can be matched in a sufficient number of the smaller developing countries to provide distinct encouragement for the future.

Increasing output per unit of arable area

During the past 30 years increases in yield per harvest have evidently been more important in increasing output than increases in harvested area. This trend is likely to continue. In many developing countries the yields attained on the best farms and on experiment stations (the so-called practical potentials) are at least three or four times larger, even at this relatively early stage of scientific effort, than the average yields attained by producers. Research is likely to suggest ways to increase further the practical potential of individual crops and to increase the number of harvests per year, so as to maximize the number of days per year during which yield is accumulated. (Whether or not the appropriate technical means will in fact be realized sufficiently rapidly on a sufficiently large scale is a different and complex set of questions considered elsewhere (Bunting, 1979).)

Wood

In many developed and developing countries, it is as important to produce more wood, for fuel and timber, as it is to produce more food and other farm products. Some of this will be obtained by improved management of existing forest and woodland, but substantial areas of land will be needed for short-term plantations, particularly in the poorer developing countries without oil or gas.

Increasing arable area

Additional arable land is also likely to be needed, at least temporarily, particularly where rural populations are growing and the practical potential is difficult to attain sufficiently rapidly on an adequate scale. Estimates of the total area of land in the world which is suitable for arable or permanent cropping range from 1800 to 3200 million ha. The last figure was suggested by the US President's Science Advisory Committee (1967) and compares with a current area of around 1500 million ha.

Much of the potential additional land is in the developing countries of Latin America and Africa where less than 6–8% of the total land area is under arable or permanent crops. Much of it is in humid forest regions (UNESCO,

1978), where inappropriate methods of extraction, clearing and subsequent management of the terrain and soil have all too often led to substantial and even permanent damage in the past. The closed forest ecosystem is more or less stable because the vegetation cycles nutrients and water (Nye and Greenland, 1960), protects the surface of the soil from heat and from the impact of rain, and stabilizes run-off and stream flow. If the vegetation is suddenly removed most of the stock of plant nutrients in the system is removed with it, and the more mobile fractions of the nutrients remaining in the soil are likely to be leached. The break in transpiration may lead to waterlogging and flooding; heat leads to adverse physical and chemical changes at the surface; and the impact of raindrops and run-off can lead to major erosion of the soil and permanent damage to terrain.

However, research at the International Institute of Tropical Agriculture, Ibadan (Greenland, 1975), and elsewhere, has indicated safe and surprisingly simple ways of replacing forest by economic vegetation, and of managing soils and terrain for arable cropping in the humid tropics.

New land will be difficult to find in China (though less than 12% of land area is cropped in that country) and in the Far Eastern developing market economies, where 31% of land area (56% in India) is already arable. These regions will continue to depend for increased output on more intensive management of land and water. For example, improved water management can permit cropping during the rains on large areas of black soils in Central India which are, at present, too wet for cropping in that season. In China many farmers appear to be approaching an average of four crops a year.

Constraints on increasing arable area

Powerful constraints tend to oppose increases in arable area. Firstly, if the new land is not simply an extension of an existing agricultural area, both people and facilities (physical, social and administrative) will be necessary to operate and support the new area. Such ventures are usually difficult, costly and hazardous. The agricultural development of uninhabited and often waterless wilderness, or of desert, can be justified only when no other alternative is open, or at least likely to give a larger return on resources. Unlike the conquest of Everest, it is not to be undertaken simply because the wilderness or the desert is there.

Secondly, in all farming, mechanical energy for field operations is a limiting resource, whether it comes from heat engines or from the muscles of humans or animals. The mechanical energy needed to prepare, sow and weed land for a particular crop is similar whether yields are large or small. Irrigation water is also a limited and often costly resource. The amount of water evaporated by a crop depends almost entirely on its area and duration, and very little on productivity, so that similar amounts of water are required whether yield is small or large. Hence scarce resources of both mechanical energy and water are more efficiently used on large-yielding crops than on small-yielding ones. De Wit (1979) has also pointed out that nitrogen fertilizer is more effectively used by the larger crops produced in more favourable soil conditions. These considerations will be particularly telling where energy, water or fertilizer has to be purchased from outside the system or region. They also tend to favour multiple rather than single cropping.

Such considerations will tend to limit the extension of arable area and to put a premium on increases in output per unit of arable area. This has been the trend of the past 20 years, and indeed, in much of Europe, of the past 200 years.

As development proceeds, less productive land is likely to be abandoned as output increases from more productive land, more effectively used. This trend will be reinforced as cheaper supplies increase from other sources, economies become more diverse, younger people leave the land and the rural population ages and dwindles. The land so released may then become available for animal production, forestry, conservation and other purposes. Indeed part of it may eventually move out of management altogether and become wild, as in much of Europe and North America during the past century.

References

Bunting, A. H. (1979). Science and technology for human needs, rural development, and the relief of poverty. (Paper presented at an OECD Workshop on Scientific and Technical Cooperation with Developing Countries, Paris, April 1978). New York: *International Agricultural Development Service, Occasional Paper.*

de Wit, C. T. (1979). Efficient use of labour, land and energy in agriculture. *Agricultural Systems* **4,** 279–287.

FAO. (1972). *1971 Production Yearbook.* Rome: FAO.

FAO. (1979). *1978 Production Yearbook.* Rome: FAO.

FAO. (1980). *1979 Production Yearbook.* Rome: FAO.

Greenland, D. J. (1975). Bringing the green revolution to the shifting cultivator. *Science* **190,** 841–844.

Harlan, J. R. (*1981*). Ecological settings for the emergence of agriculture. In *Pests, Pathogens and Vegetation*, pp. 3–22. J. M. Thresh. London: Pitman.

Lanly, J. P. & Clement, J. (1979). Present and future forest and plantation areas in the tropics (FO: Misc/71/1, January 1979). Rome: FAO.

Myers, N. (1979). *The Sinking Ark.* Oxford and New York: Pergamon.

Myers, N. (1980). *Conversion of Tropical Moist Forests.* Washington D.C.: National Academy of Sciences.

Nye, P. H. & Greenland, D. J. (1960). The soil under shifting cultivation. Harpenden: *Commonwealth Bureau of Soils, Technical Communication* No. 51.

President's Science Advisory Committee (1967). The world food problem: report of the panel on the world food supply. Washington D.C.: The White House.

Trask, A. B. (*1981*). Changing patterns of land use in England and Wales. In *Pests, Pathogens and Vegetation*, pp. 39–49. J. M. Thresh. London: Pitman.
UNESCO (1978). Tropical forest ecosystems, a state of knowledge report prepared by UNESCO/UNEP/FAO. *Natural Resources Research XIV*. Paris: UNESCO.
World Bank (1978). *Development Review, 1978*. Oxford: Oxford University Press.

Changing patterns of land use in England and Wales

A B Trask
Agricultural Development and Advisory Service, Ministry of Agriculture, Fisheries and Food, Great Westminster House, Horseferry Road, London SW1

Agriculture contributes 2½% of the United Kingdom's gross domestic product but uses about 80% of the land area. In 1974 the value of agricultural and horticultural output was greater than that of the iron and steel, aerospace or gas industries, which places it high in the scale of importance in economic terms and emphasizes its dominant position as the major land user. The importance of agriculture in import saving has been a constant factor in the economic policies of successive governments and remains a key element of current policy. Farming will probably retain its position in the economy of the country if only because it supplies just over half the total food requirement. It is a buoyant industry which, unlike most others, is composed of numerous relatively small enterprises each of which must meet the chill winds of economic fortune and competition from within its own limited resources and potential.

The fortunes of farming have varied with those of the country as a whole in the post-war period, but its story has been one of prevailing success. It has a record of rising productivity achieved against continually inflating costs which is unrivalled by other industries.

The continued expansion of home production remains an important element of national economic policy and the drive will continue for total self-sufficiency in those commodities in which it has not already been achieved. As a country that relies on imports for so many essential needs, securing the future food supply must remain one of the main imperatives. It is safe to assume that the pressures upon agriculture which have caused major changes in the countryside in the post-war period will not abate. Before considering the consequences of these pressures in the years ahead it is necessary to look back into earlier years of this century to place in perspective the development of those features of the countryside whose alteration or disappearance has caused such widespread controversy.

Historic changes in land use

A frequent observation about our rural landscape is that it originated in the eighteenth century parliamentary enclosures. Some of it did, but enclosure of land in many parts of the country proceeded randomly from earliest times and in these areas there were often no Parliamentary Enclosure Acts. The practical significance of this difference in origins is that these earlier enclosures tended to produce fields of random and irregular shape according to historical patterns of ownership, occupation and topography. By contrast the parliamentary enclosures were nearly always of more regular and geometrical layout.

The conditions of agricultural prosperity which had triggered the parliamentary enclosures of the eighteenth and nineteenth centuries changed to deep economic depression for farming in the last quarter of the nineteenth century. There was some revival of fortunes in farming in the early part of the twentieth century and through the First World War, but agriculture returned to a state of severe depression by the early 1930s. The latter period was characterized by a shift of the arable areas into grass and livestock. Hall (1941) stated that in 1931 an estimated 85.2% of the total agricultural area of England and Wales was being utilized for livestock production. The total area of land in crops and grass excluding rough grazings in England and Wales was 10.3 million ha, having fallen from a peak of 11.2 million ha in 1890.

During this period uncontrolled sprawl of urban development into the countryside was taking 25 000 ha of land per year out of farming and it continued at this rate until 1939 and the outbreak of the Second World War (Best, 1977). The 1930s saw a decline in the area in arable cultivation almost balanced by a concurrent small rise in permanent grass and a larger increase in rough grazings. Some increase in livestock numbers occurred but the general downturn in agricultural activity in the inter-war depression left the nineteenth century countryside largely unaltered.

Between 1939 and 1945 the need to produce as much food as possible at home led to a reversal of the stagnation of the inter-war years. However, the framework of the countryside was little changed apart from ploughing of downland and heaths and the clearance by gangs of prisoners of war of scrub which had been allowed to invade much farmland during the depression. Field drainage played a prominent part in the war-time food production campaign but nonetheless such familiar features of the countryside as farm ponds were left largely undisturbed. Much of the land needed for war-time installations such as airfields was returned to farming use after the war.

Agricultural land classification

The ending of the Second World War in 1945 began a period of renewed heavy and increasing pressure on the country's limited stock of land. Persistent food shortages meant continuing need for maximum home food production and rising demands for development land for reconstructing the economy subjected farmland to pressure from all quarters. The post-war enactment of the Town and Country Planning Act 1947 extended control of development to the whole country and procedures were introduced requiring planning authorities to consult the Ministry of Agriculture, Fisheries and Food (MAFF) on proposals to take agricultural land for development. At that time the only available country-wide evaluation of land quality was Dudley Stamp's Land Utilization Survey carried out before and during the war. It was a valuable planning aid to MAFF in advising planning authorities on development proposals, but by the mid-1960s it had become thoroughly outdated and MAFF developed and carried out a new survey of land quality throughout the country, known as the Agricultural Land Classification (ALC) (Anon., 1966).

The survey was completed and maps published for England and Wales by 1974. It classifies land into five grades in a descending scale which reflects the varying degrees of physical limitations of land to agricultural use. Factors such as climate, relief and soil, which contribute to its physical make-up, determine the grades, and impermanent features such as fixed equipment, farm structure, accessibility and management are ignored. Grade 1 includes land with very minor or no physical limitations to agricultural use, in other words land which will produce consistently high yields and is highly versatile in cropping. At the other extreme the poorest land is Grade 5, which has very severe limitations due to adverse soil, relief or climate or a combination of them.

The survey revealed that land in agricultural use in England and Wales was distributed amongst the grades as follows:

Grade 1	2.8%
Grade 2	14.6%
Grade 3	48.9%
Grade 4	19.8%
Grade 5	13.9%

The figures show a relatively small proportion of land to be highly productive and versatile (Grades 1 and 2). Nearly half is Grade 3 land still valuable for a wide range of uses and the remaining third (Grades 4 and 5) suffers from severe or very severe limitations for agricultural use.

The ALC provides a basis for consistent advice to planning authorities on the quality of agricultural land proposed for development. It also supports the policy followed by all post-war governments to steer necessary development from higher quality land towards land of poorer quality or non-agricultural land wherever this is possible.

Recent losses of agricultural land

The rate at which land is lost from farming has proved very difficult to monitor for several reasons, but Best (1977) showed that the surge in demands on farmland since 1945 never reached the scale experienced in the 1930s. The only continuous set of statistics for transfers of farmland to urban use comes from the annual June returns made by farmers since 1922. The rate of loss increased in the 1920s and reached a peak of about 25 000 ha in the 1930s, when agriculture had touched its nadir. The post-war annual loss has been confined largely within the range of 13 000 to 20 000 ha in England and Wales.

The Strutt Committee (Anon., 1978) recorded that between 1970 and 1974 the annual loss of farmland to urban development was about 15 700 ha. However, the most recent five-year rolling average figure available for 1974–1978 shows a fall in the average loss of land for urban development to 11 600 ha per year. The Strutt Committee recorded a *total* average annual loss of land from farming between 1970 and 1974 of 30 600 ha. This figure includes a loss of 3900 ha to forestry and woodland and no less than 11 000 ha were attributed to 'other adjustments'. These figures are derived from farmers' own returns, which are frequently incomplete or inaccurate. Moreover, there is no guarantee that land shown in a return as lost from a farm to a non-agricultural use is actually used for the purpose indicated. The figures are consequently imprecise and best regarded as indicators of trends rather than accurate records of actual developments.

The post-war acceleration of losses of farmland to urban growth occurred in some unexpected patterns. Best (1977) commented:

> The early post-war years saw both urban growth and population increase going ahead more strongly at the core of the London region than in other parts of the country. By the 1960s, however, the configuration of urban expansion had altered a great deal and, confusingly, the newly emerging pattern now completely contradicted the main features of population change. Instead of continuing to maintain a fairly dominant position, London and the South East faded away in their rates of urban growth; whereas, in contrast the areas associated with the northern and midland conurbations rose to a marked pre-eminence. Indeed in the decade up to the early 1970s, northern counties such as Lancashire and Durham experienced actual decreases in total population while having among the highest conversion rates of farmland to urban use.

Best concluded that:

> The most rapid spread of urban areas at the expense of farmland is concentrated in a broad belt of country, stretching diagonally across England from Lancashire in the North West through the Midlands to London and the South East. Conversion rates in some sectors of this axis may reach up to 0.3 per cent per annum of the total regional area compared with the national average of about 0.1 per cent. Away from this urban

spine, however, large parts of the country have a very slow turnover of farmland to urban use, at a rate of around 0.05 per cent per annum or below.

Government, however, remains concerned at the rate of loss of farmland to development and regards even the present drain from our limited stock of good land as unacceptable.

Recent changes in woodland

The widespread clear felling of woodland carried out during the First World War to help withstand the submarine blockade led to the Forestry Act of 1919 under which the Forestry Commission was set up. The Commission was charged with the development of afforestation and the production and supply of timber in Great Britain. The Commission went ahead with acquiring land and planting and in England and Wales the total area of woodland (private and state) increased from 581 000 ha in 1924 to 920 000 ha in 1939.

During the inter-war years replanting by private owners lagged behind the rates of planting achieved by the Forestry Commission and the Second World War brought another wave of clear felling. After the war planting was resumed and the greater part of the new areas of forest was in Scotland. By 1979 the total area of woodland in England and Wales had reached 1 130 500 ha.

Table 1 A breakdown of land use in England and Wales: 1971

	million ha	*percentage*
Crops, grass and rough grazing	10.9	72.5
Total woodland	1.0	7.1
Inland water	0.1	0.6
Urban uses (estimate)	1.8	11.7
Other uses	1.2	8.0
Total surface area	15.0	100.0

Table 1 indicates the place occupied by woodland in the broad distribution of land uses in 1971, the latest year for which figures in all categories are available. Although the area of woodland continues to increase, both the Forestry Commission (FC; Anon., 1977) and the Centre for Agricultural Strategy (CAS; Anon., 1980a) have advocated major expansion of the forest area in Britain. Both organizations foresee difficulty in securing supplies of timber from overseas in the first half of the twenty-first century. The CAS forecasts that world demand for wood will exceed supply by the year 2000 and proposes the afforestation of a further 2 million ha of Great Britain over the next 50 years. The FC propose alternative levels of new planting — a further 1 million ha or a further 1.8 million ha.

Bowman (1980) stated:

> The land which is needed to attain a target of 60 000 ha of total planting a year has to be transferred from agriculture, from activities involving deer and game, and will involve areas which are used for water catchment, recreation and conservation. Some measure of the impact of a planting rate of 60 000 ha per year over 50 years can be gauged from the fact that the additional land required represents about 30 per cent of the total national area of rough grazing.

Both bodies recognize that much of the planting must inevitably be in Scotland, with some in the uplands of England and Wales.

Government has yet to respond to these reports.

Recent changes in agriculture

By 1938 the total arable area had declined to its lowest point of 3.6 million ha and the value of livestock output was estimated to be nearly two-thirds of total output (Hall, 1941). With the onset of war in 1939 and the subsequent continuing need for increased post-war food production the picture of total production and pattern of farming changed radically (*see* Table 2). The figures in Table 2 (Anon., 1935/1975) indicate sweeping changes in the composition and intensity of the farming industry though the proportions of

Table 2 Agricultural changes in England and Wales between 1935 and 1975

	1935	*1975*
Land (million ha)		
Arable land	3.8	5.5
Grassland	6.3	4.1
Rough grazing	1.7	1.2
Livestock (millions)		
Total cattle and calves	6.5	10.5
Total sheep and lambs	16.5	19.8
Total pigs	3.8	6.3
Total poultry	62.1	111.5
Labour (thousands)		
Total workers*	673	318

* Including casual and seasonal, but excluding salaried managers, farmers, partners and directors.

total output contributed by crops and livestock have remained unaltered since before the war. The increase in the arable area is almost wholly accounted for by the doubling of the cereal acreage and this expansion of the area and intensity of arable production has inevitably changed the farming landscape.

Changes in farming landscapes

Hedgerows

The post-war period has been one of regularly fluctuating economic conditions and farmers have mostly been obliged to meet continually rising costs out of increased efficiency. This has been achieved by enormous reductions in the number of people employed in agriculture and the progressive adoption of technical innovations. The most widely recognized manifestation of this has been the enlargement of fields in arable areas for more economic cultivation by removal of hedgerows and copses, etc. This has not occurred uniformly over the country, and some livestock rearing areas have largely escaped such changes. These are often represented as unprecedented but history suggests otherwise.

Hoskins (1955) comments in relation to the enclosure movement that 'the first effect of enclosure was to reduce the number of trees in thinly wooded country, for the new fences — hundreds of miles of them — required vast quantities of oak, elm and ash saplings for posts and rails'.

Enclosure was generally by private Act of parliament which appointed a special commissioner to superintend the provisions of an enclosure award for the designated parish. Although the awards generally required hedges to be planted within 12 months there was often considerable delay. This could be up to 50 years (Parker, 1975) and many hedges which should have been established were never planted. In some arable areas hedges were unpopular from the time they were established under enclosure awards; their removal commenced soon afterwards and has continued to the present day.

The post-war removal of hedges really began in the 1950s. There have been few systematic studies of its progress and so no record exists of the national pattern. However, Pollard *et al.* (1974) reviewed the information available then and studied aerial photographs of 6 million ha in the arable area of eastern England. This study revealed the following rates of hedgerow removal per year:

1946–54	1290 km
1954–62	3860 km
1962–66	5630 km
1966–70	3220 km

They commented in 1974 that 'the recent decline in hedge loss which has been recorded nationally may be due in part to the re-growth of hedges cut to the ground as immediately after cutting they will not be visible in aerial photographs'.

The spate of hedgerow removal in the 1950s and 1960s mainly affected the arable areas of the Midlands and eastern England. It can be expected that hedges which hamper the use of present day cultivation equipment will continue to be removed but MAFF surveyors consider that it has now declined to a low rate. In arable areas many old hedgerows which impeded cultivation by modern equipment have already been removed and the rising cost of clearance, for which no grant aid is available, makes it a less attractive means of increasing productivity than hitherto.

Field size and shape

A detailed study of field size and shape on large arable farms revealed that field shape does not materially affect the workable area within the boundary and that the primary factor affecting the efficiency of machinery within the field is not field area but average row length (Edwards, 1965). The return from lengthening rows beyond 460 m is fractional. The ideal shape of a piece of land for overall efficient working is a square, allowing efficient work in both directions. There is little advantage in greatly exceeding 20 ha, giving row lengths of approximately 460 m each way in a square field. No further work on field size has been done but the Edwards' conclusions remain valid today. However, Worthington (1979) considered 40 farms totalling 15 038 ha in a broad sweep of the country from Dorset to Lincolnshire extending as far west as Buckinghamshire and east into Norfolk. Amongst the landscape elements included in the study was field size and the 40 farms fell into five tenure groups (Table 3). The study covered a relatively small total area but the farms lay in mainly arable areas and the average field sizes are not so large as to suggest indiscriminate hedge removal from institutional or other land.

Table 3 Data on farm and field size from a survey in eastern and southern England (Worthington, 1979)

	Tenure group: (A)	(B)	(C)	(D)	(E)
Total area (ha)	5600	4401	3002	1243	790
Number of farms	10	10	10	5	5
Average farm size (ha)	560	440	300	248	158
Range of farm size (ha)	152–901	193–849	157–486	116–404	122–214
Average field size (ha)	13.7	11.3	10.1	8.5	7.3
Range of average field size (ha)	8–30	8–23	4–23	6–26	5–42

The tenure groups were: (A) new institution (direct farmed); (B) new institution (tenanted); (C) traditional institution (tenanted); (D) owner occupied; (E) private owner (tenanted).

Wild life and agriculture

In their study of hedges in relation to crop pests and beneficial insects, Pollard *et al.* (1974) concluded that there was no clear argument for or against hedges. On occasions they have had harmful influences, for instance before myxomatosis by harbouring rabbits and more recently in connection with the fireblight bacterium of apple and pear (Billing, *1981*). Arguments in their favour seem to be over-optimistic and as for the vast majority of pests, their influence is probably either negligible or nil. The case for preserving hedges must rest on other factors such as their value for landscape and wild life (*see also* van Emden, *1981*).

ADAS has been conducting trials of integrated and supervised pest control in dessert apple orchards with the aim of developing reduced spray programmes. This is consistent with the latest Report of the Royal Commission on Environmental Pollution (Anon., 1979), which advocated reducing pesticide usage to the minimum required for efficient food production. Moreover, ADAS proposes to carry out longterm trials on cereals comparing prophylactic pesticide programmes and supervised programmes of minimal pesticide application. The trials would investigate the effects of the contrasting regimes on yields, on the environment and the technical and management problems of a minimal spray programme under commercial conditions. The work will be carried out on large areas which would need to contain adequate wild life refuges in the form of hedgerows and copses. The trials should, inter alia provide firmer data than now exists for assessing the significance of such countryside features in the ecology of cropland.

The start of work aimed at illuminating the relationship between wild life and farming practice argues for the retention of wild life refuges in the farmed countryside, at least until that relationship is better understood. It supports the appeal to farmers by Moore (1979) not to destroy unnecessarily any existing habitat, whether wood, hedge, marsh, pond or rough grazing. This is emphatically the approach of MAFF in discharging its obligation under Section 11 of the Countryside Act 1968 to have regard to the desirability of conserving the natural beauty and amenity of the countryside when considering proposals by farmers such as land reclamation, etc.

Farm reservoirs

Much has been heard in recent years of loss of ponds and wetland from the countryside, but an offsetting trend has been the construction on farms of reservoirs for irrigation. A recent report (Anon., 1980b) estimates that 90 Mm^3 of water is used for irrigation in a dry year, three quarters of which comes from surface sources. Moreover, a MAFF survey in 1975 showed that 862 holdings in England and Wales had reservoirs. In 1977 and 1978 a total of 355 new farm reservoirs with a median capacity of 14 000 m^3 or about 3000 m^2 surface area (approximately $\frac{1}{3}$ ha) were constructed with the aid of MAFF grants.

The 1980 Report concluded:

> All the evidence points to a substantial increase in the demand for water by agriculture over the next 20 years, notably as a result of a greatly extended use of irrigation. We estimate that the period will see a doubling in the demand for water by the industry as a whole, and a quadrupling in the demand for irrigation. Whilst the increases are likely to be spread fairly evenly over all regions, there will perhaps be a rather greater increase proportionately in the East and South East of England.

The Report went on:

> As far as irrigation is concerned, the estimates should perhaps be regarded as the upper limit of the range of likely outcomes in reflecting the growth which we think is economically justifiable, if not necessarily attainable.

Certainly farm reservoirs are likely to become more frequently occurring features of the countryside and will increase the amount of water habitat for wildlife in some areas.

Future trends

Dwindling energy reserves are an important factor which will affect patterns of use of agricultural land in the remaining years of the twentieth century and beyond. Agriculture and horticulture together currently use about 4% of the energy available to consumers in the UK, both directly as fuel and indirectly as fertilizers, machinery, feedingstuffs and chemicals. Declining energy supplies, particularly of oil and gas, will impose progressive pressure on agriculture to make more effective use of residual supplies.

De Wit (1979) showed that added energy in the form of applied nitrogen is used more efficiently at higher yields. He concluded:

> However, although scarce, it seems likely that energy is available in at least reasonable quantities for a long time to come. In that case, agriculture may contribute to a sensible use of energy by developing in a direction where as high yields per hectare as possible are obtained from as small an area as possible by a reasonable number of highly skilled farmers.

His conclusion emphasizes the need to conserve for agriculture the higher quality land which responds best to nutrients and so makes the most economical use of available energy. Indeed, if energy supply to agriculture were to decline the value to the nation of the higher quality land would increase relative to poorer land because of its better potential for food production under simpler systems of farming which might then become necessary. How such change to more relaxed farming methods would be reflected in new ecological balances in the countryside is a fascinating field for speculation.

References

Anon. (1935/1975). *Agricultural Statistics, England and Wales*. London: Ministry of Agriculture, Fisheries and Food, HMSO.

Anon. (1966). *Agricultural Land Classification*. Technical Report No. 11. London: Ministry of Agriculture, Fisheries and Food.

Anon. (1977). *The Wood Production Outlook — a Review*. London: Forestry Commission, HMSO.

Anon. (1978). *Agriculture and the Countryside*. London: Advisory Council for Agriculture and Horticulture in England and Wales, HMSO.

Anon. (1979). *Agriculture and Pollution*. London: Royal Commission on Environmental Pollution, 7th Report. London: HMSO.

Anon. (1980a). *A Forestry Strategy for the UK*. CAS Report 6. Reading University: Centre for Agricultural Strategy.

Anon. (1980b). *Water for Agriculture: Future Needs*. London: Advisory Council for Agriculture and Horticulture in England and Wales, HMSO.

Best, R. H. (1977). Agricultural land loss — myth or reality? *The Planner, Journal of Royal Town Planning Institute*, **63,** 15–16.

Billing, E. (*1981*). Hawthorn as a source of the fireblight bacterium for pear, apple and ornamental hosts. In *Pests, Pathogens and Vegetation,* pp. 121–130. J. M. Thresh. London: Pitman.

Bowman, J. C. (1980). A Forestry Strategy for Great Britain. In *Proceedings of A National Forestry Policy Conference, 7th June 1979*, p. 16. London: Royal Society of Arts.

de Wit, C. T. (1979). Efficient Use of Labour, Land and Energy in Agriculture. *Agricultural Systems, Applied Science Publishers Ltd.*, **4,** 279–287.

Edwards, A. J. (1965). *Consideration of Field Size and Shape on the Large Arable Farm*, Technical Note for MAFF, East Midland Region, Nottingham (quoted in Pollard, E. *et al*. (1974) *ibid*.).

Emden, H. F. van (*1981*). Wild plants in the ecology of insect pests. In *Pests, Pathogens and Vegetation*, pp. 251–261. J. M. Thresh. London: Pitman.

Hall, A. D. (1941). *Reconstruction and the Land*. London: Macmillan.

Hoskins, W. G. (1955). *The Making of the English Landscape*. London: Hodder and Stoughton.

Moore, N. W. (1979). Nature Conservation Today. Paper read at *Conference on Conservation and Agriculture: A Balanced Approach*. Stoneleigh: National Agricultural Centre.

Parker, R. (1975). *The Common Stream*. London: Collins.

Pollard, E., Hooper, M. D. & Moore, N. W. (1974). *Hedges*. London: Collins.

Worthington, T. R. (1979). *The Landscapes of Institutional Landowners*. Cheltenham: A study for the Countryside Commission.

Section 2 **Plant pathogenic viruses, bacteria and fungi**

The role of weeds and wild plants in the epidemiology of plant virus diseases

J M Thresh
East Malling Research Station, Maidstone, Kent ME19 6BJ

Introduction

When studying the epidemiology and control of plant virus diseases it is important to identify the initial foci of infection from which spread occurs into or within crops. The role of weeds and wild plants was established in some of the earliest work on plant viruses in the United States (Boncquet and Stahl, 1917; Doolittle and Gilbert, 1919). Such studies have continued in many countries and the extensive literature has been assessed recently (Duffus, 1971; Bos, 1981).

No attempt has been made here at such comprehensive coverage and this paper concentrates on general principles, using selected examples to illustrate the various phenomena encountered.* Attention is drawn also to the advantages and general applicability of a multi-disciplinary, 'ecological' approach to epidemiology of the type adopted in some studies on the complex inter-relationships between viruses, vectors and their host plants.

The role of weeds and wild plants

In considering the overall importance of weeds and wild plants it is convenient to distinguish three quite different categories (Fig. 1):

(1) Wild species of little or no current epidemiological importance, although they were the original hosts of viruses now prevalent in crops.
(2) Weeds or wild plants playing a continuing and important role as alternative or perennating hosts of viruses infecting crops.
(3) Weeds or wild plants of crucial importance as the sole hosts of viruses that spread to but not between crop plants.

* Throughout this paper attention is restricted to diseases known or assumed to be caused by viruses. Virus-like diseases now associated with vector-borne mycoplasmas or spiroplasmas are not considered. Several of these pathogens have a wide host range including weeds or wild plants that play an important role in the epidemiology of diseases such as corn stunt and strawberry green petal; *see* Maramorosch and Harris (1979) for further details.

There is very inadequate evidence relating to the first category (Fig. 1 *left*), which is necessarily speculative and largely based on inference. This is because of the scant information on viruses of natural vegetation and on the early history of those now affecting crops. Nevertheless, there is little doubt that viruses can resemble other pathogens and pests in becoming established within crops after spread from wild species. This is apparent from various striking examples of virus diseases encountered when crops were first grown in new agricultural areas (Bennett, 1952; Thresh, 1980a). The occurrence of cocoa swollen shoot disease soon after the crop was introduced to the forest regions of West Africa is described by Posnette (*1981*), who provides persuasive evidence that the causal virus spread from indigenous tree species of the rain forest and adjoining savannah.

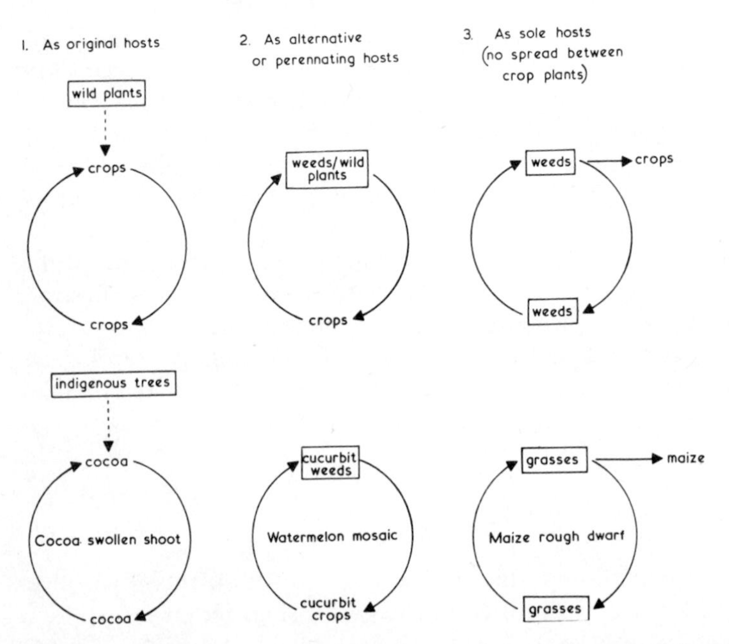

Figure 1 Diagrammatic representation of the role of weeds and wild plants in the epidemiology of plant virus diseases (top) with specific examples of each category (bottom) that are discussed in detail elsewhere in this volume (Adlerz, *1981*; Conti, *1981*; Posnette, *1981*).

In many other instances the initial transfer from wild to cultivated species seems to have occurred at an even earlier stage of agricultural development, long before virologists were available to monitor the situation (Thresh, 1980a). There are certainly many viruses that have long been present within crops with no evidence that they are in any way dependent on continuing spread from other species. For example, studies in various countries show that weeds and wild plants are not important sources of leafroll or virus Y in potato, except perhaps in the South American centre of origin (Jones, *1981*).

There is abundant evidence of the current role of weeds and wild plants as alternative hosts of viruses affecting crops (Fig. 1 *centre*). Biennial or perennial species are particularly important in facilitating the survival of viruses attacking annual plantings in areas with growing seasons restricted by prolonged drought or cold. Other viruses are transmitted frequently to the seed of common weed hosts that provides an effective means of dissemination between plantings and of survival for long periods in soil. This enables such viruses to spread far into new areas and to persist at sites already affected.

The third category (Fig. 1 *right*) is of special interest because of the crucial role of weeds as the sole hosts from which vectors are able to transmit viruses to crop plants. The few examples include maize rough dwarf and lettuce necrotic yellows viruses that are considered in detail elsewhere in this volume (Conti, *1981*; Martin and Randles, *1981*).

Spread into and within crops

In discussing other examples of non-cultivated species as virus sources a distinction is made according to their distribution in relation to crops (Fig. 2).

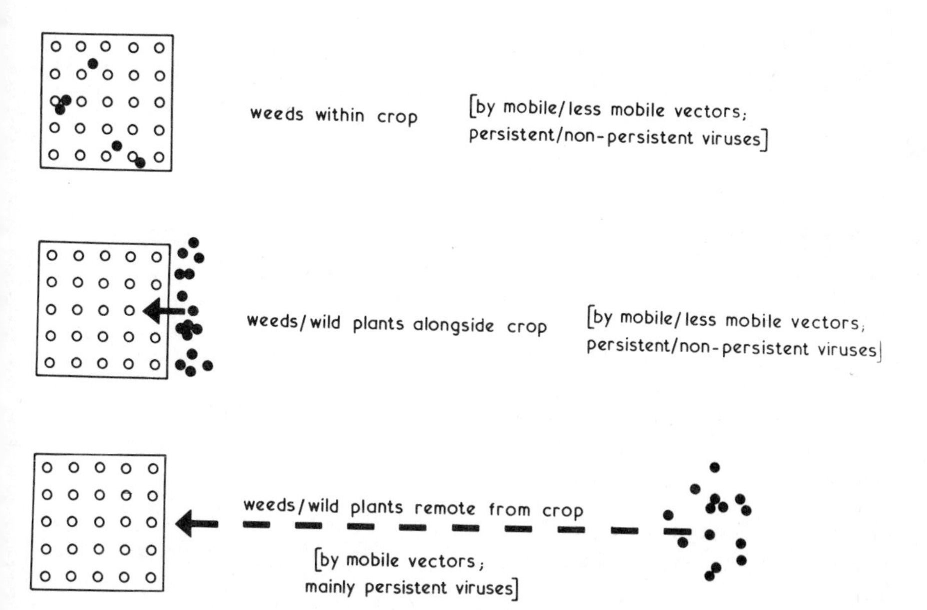

Figure 2 Diagrammatic representation of virus spread from weeds or wild plants within, alongside or remote from crop stands.

Infected weeds within plantings are particularly important and obviously exert the greatest 'infection pressure'. They act as foci for viruses that are transmitted persistently or non-persistently by vectors that do not have to be very mobile to perform effectively. By contrast, viruses that persist in highly mobile vectors are the only ones likely to be carried into crops from outside sources, unless these are nearby.

1 Spread from weeds within crops

Many viruses have weed hosts (Heathcote, 1970) and their importance in providing initial foci of infection within crops was apparent from early work

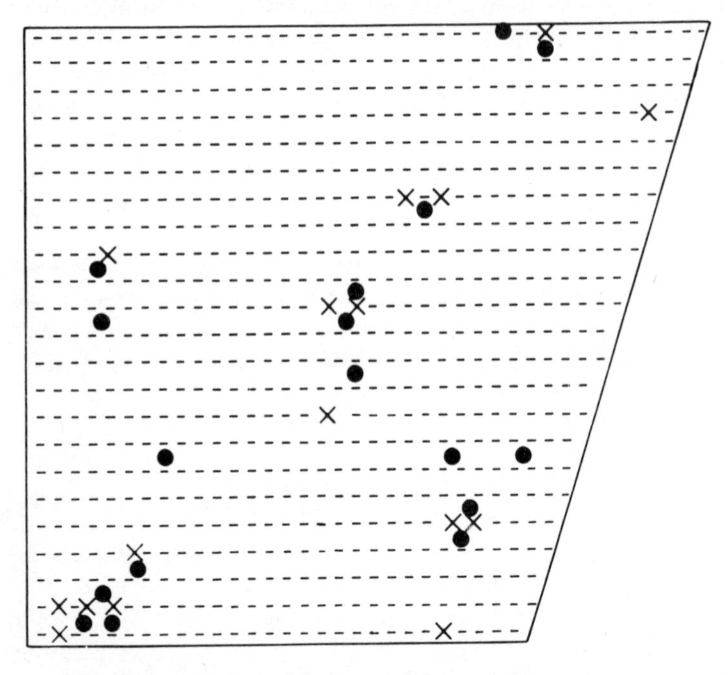

Figure 3 The initial spread of cucumber mosaic virus from mayweed (*Asclepias syriaca*) regenerating within a crop in Wisconsin (adapted from Doolittle and Walker, 1925).

on a virus of cucurbits in Wisconsin (Doolittle and Walker, 1925). In many spring plantings disease first appeared in scattered patches around infected plants of a perennial mayweed (*Asclepias syriaca*) that had overwintered and then regenerated (Fig. 3).

Similar instances have been reported in subsequent work on many other viruses with insect or nematode vectors. For example, cucumber mosaic virus is seed-borne in chickweed (*Stellaria media*) and other common weeds. These provide abundant overwintering sources from which aphids spread virus within plantings of lettuce (Tomlinson *et al.*, 1970) and tobacco (Häni, 1971). The abundance and longevity of infected weed seeds in arable soils suggest that the 'seed bank' could be an important factor contributing to the diversity and conservation of virus strains but this has not been investigated.

It is to be expected that the type and effectiveness of weed control measures will greatly influence the pattern, amount and sequence of virus spread. Such effects have received little attention, although they seem considerable and the underlying factors are undoubtedly complex. They range far beyond the consequences of simply removing hosts of viruses or their vectors. For example, there can be major changes in the weed flora (Fryer, *1981*) and the potential effects are illustrated by the increased prevalence of viruses in Californian lettuce crops associated with the increase in composite weeds following the widespread use of selective herbicides (Duffus, 1971). These materials can influence the physiology of crop stands and thus the overall environment for pests and pathogens (Atkinson and White, *1981*; Kavanagh, 1974).

In an early observation on the effects of weed control, the spread of sugar cane mosaic virus was found to be enhanced by repeated hoeing that controlled weed grasses but increased the movement of aphid vectors (Wolcott, 1928). A similar response occurred when pineapple plantings were cultivated to destroy weeds, including *Emilia sonchifolia*, the host from which thrips transmit yellow spot virus to pineapple (Carter, 1939). Weed control measures also influence the spread of groundnut rosette virus by aphids and African farmers have long known that losses are decreased if weeding is delayed (Hayes, 1932). This is explained by later studies on the aphids of groundnut and other crops, which showed that plants growing at wide spacing or in otherwise bare soil are more heavily colonized by incoming alates than plants at close spacing or amongst weeds (A'Brook, 1964; Smith, 1976; Way and Cammell, *1981*).

These findings suggest that the obvious advantages of controlling weeds with the effective pre-emergence herbicides now available are partially offset by the increased vulnerability of weed-free crops to aphid vectors and other insect pests. This can lead to increased vector populations and activity, even though weed hosts of viruses are largely eliminated. Another possibility is that controlling weeds could facilitate the spread of non-persistent viruses by decreasing the loss of inoculum that occurs whenever infective aphids probe the leaves of weeds that are immune to infection. However, direct or indirect effects of herbicides on the spread of arthropod-borne viruses have seldom been assessed. One of the few detailed studies showed the importance of the grass weeds of maize as hosts of maize rough dwarf virus and its planthopper vector. Virus and vector became increasingly difficult to control as the season progressed, and weed grasses appeared as the effectiveness of pre-emergence herbicides declined (Conti, 1976a, 1976b, *1981*).

In work on nematode-borne viruses the increased incidence of tobacco rattle virus in potato plantings kept free of weeds for two successive growing seasons was attributed to the lack of alternative hosts on which infective vectors could feed (Cooper and Harrison, 1973). Elsewhere, the use of pre-emergence herbicides delayed re-acquisition of other viruses from weeds by nematodes that had become non-infective after fallowing (Taylor and Murant, 1968; Murant, *1981*). There are also instances of herbicides directly influencing the susceptibility or response of crop plants to virus infection (*see* Heathcote, 1970; Kavanagh, 1974). These results emphasize the complex effects of weeds and weed control practices, which merit far more attention than they have yet received.

2 Spread from uncultivated plants around crops

It has long been apparent that many viruses are carried into crops by vectors originating or acquiring virus from uncultivated plants growing around the perimeter. For example, Wellman (1935) described how an aphid-borne virus of celery in Florida first appeared along the margins of plantings near clumps of *Commelina nudiflora* or other weed hosts of the virus (Fig. 4).

Other gradients of spread from nearby sources have been reported for many other viruses transmitted by aphids, whiteflies, thrips, leafhoppers, planthoppers, plant bugs or nematodes (Thresh, 1976). The precise form of the gradients depends on the type and mobility of the vectors and on the mode of transmission. Edge effects are most marked with viruses that do not persist in their aphid vectors or when soil-inhabiting nematodes spread viruses into crops as they move slowly from adjacent hedgerows (Murant, *1981*; Taylor and Thomas, 1968). With insect-borne viruses the number and disposition of the sources are important in relation to wind direction (Adlerz, 1974, *1981*; Quiot *et al.*, 1979). Spread is greater downwind than upwind, and increases when temperature and other seasonal conditions facilitate dispersal (Simons, 1957).

Various sources of viruses and/or their vectors have been located around crops, including trees, shrubs and herbaceous plants of related or unrelated species in weed stands or natural vegetation (Duffus, 1971; Bos, 1980). These can be the sole foci (Fig. 1 *right*) or they can be the primary sources for subsequent spread within plantings (Fig. 1 *centre*). However, a preponderance of infected plants around the perimeter does not necessarily mean that sources are nearby. This is because incoming vectors from afar often accumulate on the peripheral plants, especially to leeward of hedges, windbreaks or other physical barriers causing air turbulence and insect deposition (Lewis, 1969; Fig. 5).

A recurring feature in epidemiology is that peaks of spread coincide with the appearance of active winged migrants produced by crowded populations developing as host plants mature and senesce (Thresh, 1974). Much of the spread from uncultivated plants occurs in these circumstances and there are many reports of mass migrations from deteriorating habitats. For example, barley yellow dwarf virus first attracted attention when severe epidemics

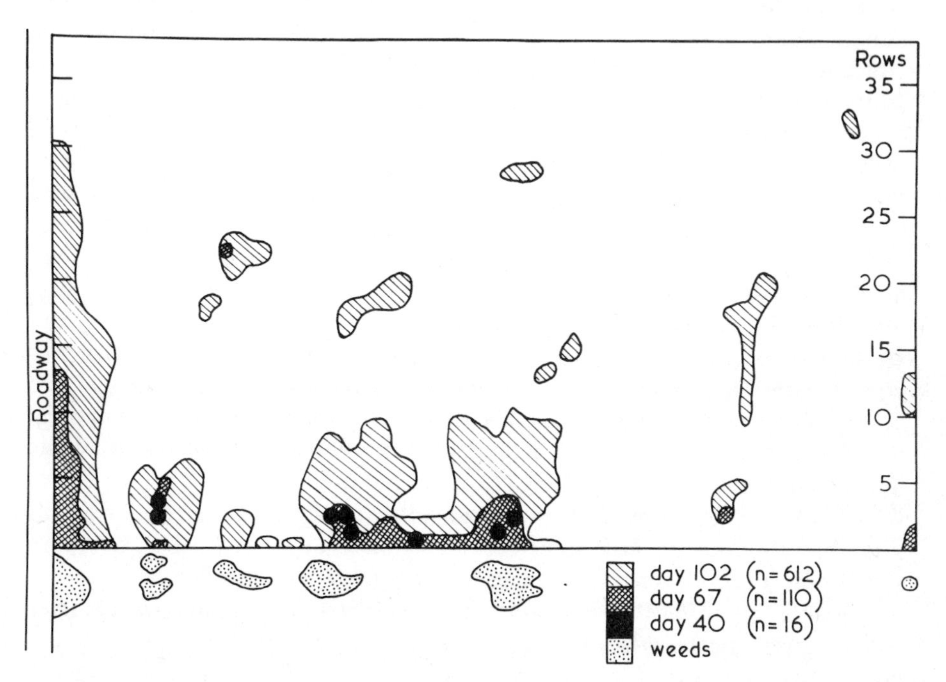

Figure 4 The spread of southern celery mosaic virus from clumps of *Commelina nudiflora* growing along one margin of a crop in Florida (adapted from Wellman, 1935). Shading indicates infection at different times after planting.

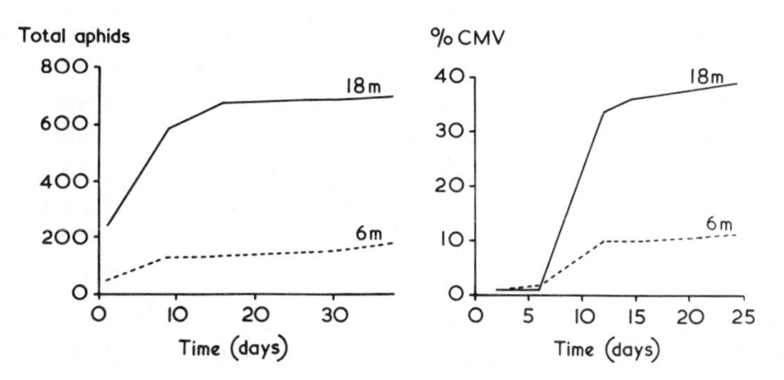

Figure 5 Catches of alate aphids (left) and spread of cucumber mosaic virus in tomato (right) at distances 6 m and 18 m from windbreaks 6 m high in south east France, (Quiot *et al.*, 1979).

occurred in California as aphids moved from wilting grasses to young cereal crops planted nearby (Oswald and Houston, 1953).

Such findings justify the attention given to physiological factors influencing the production of specialized migrants as in recent work on *Hyperomyzus lactucae*, which is the main aphid vector of lettuce necrotic yellows virus in Australia (Martin and Randles, *1981*). The virus does not spread between lettuce plants and it is carried into crops by aphids that breed and become infective whilst feeding on the common sowthistle (*Sonchus oleraceus*).

Attempts have been made to control lettuce necrotic yellows and many other viruses by eradicating or spraying the weeds around the perimeter of plantings (e.g. Stubbs *et al.*, 1963). These measures are justified for viruses that are seldom carried far and where there are major benefits from delaying the onset of disease. Early efforts were made to control insect-borne viruses of cucurbits in Wisconsin and of tobacco in Indonesia by removing weed hosts within 45–70 m (Doolittle and Walker, 1926; Thung, 1934 in Bos, 1981). Similarly with celery mosaic in Florida, where most of the spread was from weeds within 23 m of the crop boundary (Wellman, 1937) and control was greatly improved when 2,4-D became available (Townsend, 1947). The Florida situation was discussed by Vanderplank (1948) who suggested making plantings fewer, larger and more compact to decrease the proportion of plants in the vulnerable peripheral areas.

The margins of crops or adjacent uncultivated areas have received considerable attention in attempts to control many other virus diseases. It is sometimes more effective to control vectors on weeds around crops rather than within them. In other instances the outside rows are planted with barrier crops or particularly resistant varieties, or they are ploughed-in after the main influx of vectors (Coons *et al.*, 1958; Thresh, 1976). They may also be given additional sprays or special protection. For example, it is a requirement of several official fruit and hop certification schemes in England and Wales that nurseries must be planted at least 10 m away from hedgerows harbouring populations of virus-vector nematodes (Cotton, 1979).

Many epidemiological studies have emphasized the importance of weed hosts of viruses and/or their vectors growing along railways, roads, footpaths or contour strips. The banks of drainage, irrigation or flood control channels are particularly important in permitting dense stands of weeds to develop around and sometimes within crops, as in the vegetable-growing areas of Florida (Adlerz, *1981*). Such sites provide a protected environment where weeds and the vectors they support can thrive, even when conditions are too dry or too cold for survival elsewhere. For example, some of the irrigation canals in Washington State are fed from underground springs that maintain the water temperature 22–32°C above air temperature during the cold winter months (Wallis and Turner, 1969). This facilitates the survival of weed hosts of beet western yellows virus and its main aphid vector (*Myzus persicae*). Major efforts are made to burn-off the weeds growing along the canal banks as the aphids overwintering anholocyclically in these areas are far more important in spreading virus than the much greater populations originating from eggs laid on peach trees in nearby orchards.

Work on *M. persicae*, *H. lactucae*, and other aphids in less extreme climates provides further evidence of the crucial importance of populations overwintering on herbaceous weeds (Eastop, *1981*). These lead to an earlier and more prolonged migration in the spring than occurs from woody hosts and the migrants are more likely to be infective. Consequently the spread of aphid-borne viruses of sugar beet tends to be greater after mild winters than after severe ones, when there is greater mortality of aphids and weed hosts (Watson, 1967). This relationship is used in pre-planting forecasts of the incidence of beet yellowing viruses from the number of freezing days during the winter months. There is likely to be a similar correlation between the spread of certain insect-borne viruses in the arid tropics and the severity of the dry season. The possibility of establishing predictive relationships has been considered in relation to the survival of crop debris and 'volunteers' (e.g. Storey and Bottomley, 1928) but not for wild hosts of viruses or their vectors except in the work on maize streak discussed later (p. 117).

3 Spread from remote sources

Sugar beet curly top is the best-known and most-studied of the viruses regularly carried far into crops by vectors moving from weed sources that are sometimes remote from agricultural areas. It was first encountered in early attempts to establish beet production in the western United States, where devastating losses led to abandoned fields and factory closures (Carter, 1930). A characteristic feature of curly top is that it occurs suddenly and extensively over large areas, soon after a major influx of the beet leafhopper (*Circulifer tenellus*). This was soon identified as the only vector and detailed investigations began on the origins of the serious infestations that occur, even in areas where the leafhopper cannot overwinter (Severin, 1919).

The overwintering hosts of virus and vector were found to be diverse weeds, including species of the Chenopodiaceae believed to have been introduced to North America, in some instances with animals accompanying some of the original immigrants from the Mediterranean (Hallock and Douglass, 1956). Dense stands of weeds occur over very large areas during the winter months, wherever moisture and temperature conditions permit (Fig. 6). Enormous populations of leafhoppers develop and these migrate to beet and other crops in spring when the weed hosts mature, senesce and in some areas die-out with the onset of the summer drought (Fig. 7). The colonizers breed on beet and some other crops during the summer, thriving particularly well on plants infected with curly top (Carter, 1930). However, there is no multiplication on tomato, melon, bean or some other crops, that are infected solely by spread from outside sources. This can lead to almost total infection, as recorded in Idaho bean fields (Annand *et al.*, 1932) and in the particularly severe Californian epidemics in tomatoes during 1950, 1956, 1966 and 1977.

At some sites and in some seasons beet leafhoppers persist throughout the year on crops or weeds within the agricultural districts. However, the crucial importance of the outlying breeding grounds has long been apparent and explains the considerable attention given to the colonization of these areas by weeds. Over whole regions they have supplanted the indigenous vegetation of shrubs and perennial grasses that are not hosts of curly top or its vector.

Figure 6 Monitoring populations of the beet leafhopper (*Circulifer tenellus*) on weed stands in one of the extensive desert breeding grounds of California (courtesy of the California Department of Agriculture).

Ecological investigations in several States have revealed a complex situation, largely determined by cropping practices and land-use. Much of the original vegetation was destroyed by burning, clearing and ploughing. Many of these areas have been abandoned or farmed intermittently according to prevailing economic conditions. Others are seriously trampled and overgrazed by livestock. Transient stands of weeds develop during the winter or summer months and the land becomes vulnerable to dust storms and gulley erosion, especially where there is extensive damage by rodents (Piemeisel *et al.*, 1951).

Vast, continually changing areas of weeds remain and it is difficult to protect crops adequately by using insecticides. Additional control measures are necessary including the manipulation of planting dates and the use of tolerant or resistant varieties of beet and other vulnerable crops. Attempts have also been made to combat the leafhoppers at source using strikingly different approaches (Duffus, 1980). The long-term 'ecological' strategy involves improved systems of cropping and range management to avoid weed infestations of arable land or to permit the growth of useful grasses.

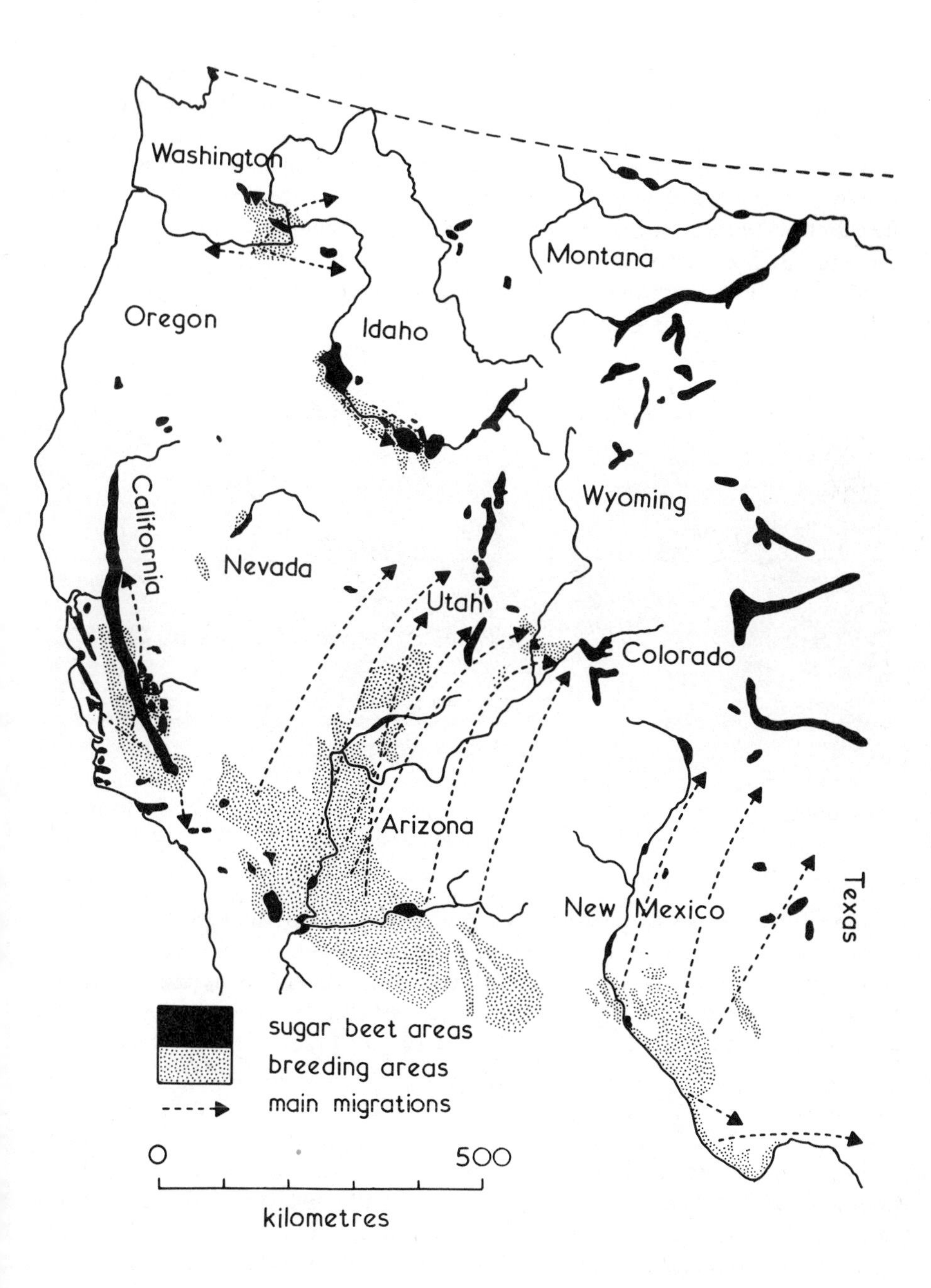

Figure 7 The long distance spread of beet curly top virus by leafhoppers flying from breeding grounds in western USA (adapted from Douglass and Cook, 1954).

Reseeding with perennial grasses is advocated for good soils, whilst others are allowed to regenerate naturally by protecting from fire, rodents and overgrazing (Piemeisel, 1954).

There is little published information on the extent to which these methods have been adopted, except in parts of Idaho where 100 000 ha had been sown with grasses by 1972, covering an estimated 77% of the total leafhopper breeding area (Knipling, 1979). Such methods do not appear to have been widely adopted elsewhere and there is little collaboration between the livestock producers of the rangelands and those farming the irrigated valleys. Far greater use has been made of pesticides in the breeding grounds using funds contributed by the sugar companies, State Departments of Agriculture or beet and vegetable growers.

Figure 8 An aircraft being supplied with fuel and insecticide for aerial spraying operations against the beet leafhopper in the desert breeding grounds of California (courtesy of the California Department of Agriculture).

State-financed operations began in California in 1943 and were intensified after the serious epidemics of 1950. Between 1950 and 1956, over 70 000 ha were sprayed annually with DDT in diesel oil, which was applied by range vehicles or from the air, mainly during October (Fig. 8). The weed areas to be sprayed were selected during summer surveys of weed populations over a region of 1280 km^2. Moreover, the control teams annually cut 1525–17 454 ha of weeds and treated those along 2326–6315 km of ditches and road verges (Armitage, 1957).

Similar campaigns have been mounted in Idaho and elsewhere. These continue, although it is impossible to assess the success of the operations because there are no comparable untreated areas (Duffus, 1980). However, overall leafhopper populations on beet in Idaho declined following the use of insecticides in the desert breeding areas, with a further decline after the reseeding programme had begun (Knipling, 1979). DDT has now been withdrawn and malathion is an inadequate substitute because of its limited persistence. The search continues for more effective insecticides and for biological control agents of the beet leafhopper, which is an introduced pest with few natural enemies in the United States.

Many other insect vectors exploit ephemeral habitats and resemble the beet leafhopper in their migratory habits. This suggests that other viruses might also be carried far from deteriorating weeds or wild hosts. The possibility has received little attention except in Central Africa, where several *Cicadulina* leafhoppers of the veld grasslands are vectors of maize streak virus (Rose, 1978; Conti, *1981*). During the summer rainy season leafhoppers are widely distributed at low density throughout the extensive areas of grassland and the predominant forms are long-bodied ones that seldom disperse far. Active, short-bodied forms that can fly for relatively long periods become numerous with the onset of the dry season. Extensive flight then occurs with shallow dispersal gradients over considerable distances to grasses in moist low-lying areas and to irrigated crops. Rain-fed crops are seldom affected by streak except immediately around the margins of plantings near major sources of infection. By contrast, there is much spread to irrigated crops sown during the main flight period, even if the plantings are isolated or only a small proportion of the migrants has acquired virus from grasses. Losses are particularly great after late rains in the wet season have extended the breeding period on grasses and led to prolonged flight activity. Hence the correlation between total April–June rainfall and subsequent suction trap catches of leafhoppers, which is useful in predicting years favourable for maize streak epidemics (Rose, 1972).

These findings could be relevant to the situation in many other crops of regions with markedly seasonal climates. Moreover, they provide a further striking example of the way in which virus spread is closely related to the ecology and behaviour of the vectors and their host plants. There are likely to be other instances awaiting discovery, or occurring when new crops are introduced along the dispersal routes of indigenous insect pests already established in the native flora.

The ecological approach to virus disease epidemiology

Several authors at different times and for diverse reasons have stressed the advantages of an 'ecological' multi-disciplinary approach to epidemiology. The involvement of virologists with vectors, weeds and wild plants has provided a particularly important stimulus in developing these attitudes. This was implicit in some of the earliest publications on insect-borne viruses in the United States (Boncquet and Stahl, 1917; Doolittle and Walker, 1925) and was stated quite explicitly by Carter (1930), who attempted to forecast the incidence of sugar beet curly top in Idaho from pre-planting observations on the overwintering of the leafhopper vector and its weed hosts. In subsequent work on pineapple yellow spot in Hawaii, Carter (1939) wrote of virus, vector and plant hosts as 'the inseparable ecological trinity' and showed that infected weeds lived longer than healthy ones and were more favourable as hosts of the thrips vector. From these experiences and later visits to West Africa he went on to advocate an equally comprehensive approach to all aspects of plant virus disease epidemiology (Carter, 1961, 1973).

This has been the policy adopted in crucial phases of the continuing work on swollen shoot and other diseases of cocoa in West Africa (Johnson, 1962; Tinsley, 1964). For example, the role of wild hosts was studied at an early stage (Posnette *et al.*, 1950) and over 120 insect species were found to be important in the entomology of swollen shoot (Strickland, 1951). There is a complex inter-relationship between populations of the mealybug vectors and the predominant ants. Some ant species tend mealybugs, whereas others are hostile and the antagonistic groups occupy separate territories creating continually changing 'mosaics' that are related to the vegetation present (Majer, 1972; Leston, 1973).

There have been similarly wide-ranging studies on some other viruses, including several of those with aphid vectors affecting sugar beet, potato and lettuce. The continuing work on viruses with soil-inhabiting vectors has also involved close collaboration between virologists, mycologists and nematologists (Harrison, 1977). However, considering the entire field of epidemiology it is clear that the necessary multi-disciplinary teams have seldom been established or maintained for long. Virological studies have rarely been closely coordinated with those of weed biologists, plant breeders, agronomists, agricultural meteorologists or zoologists working on vectors. Other difficulties are due to the current preoccupation of virologists with the detailed properties of viruses at the expense of field studies on the diseases they cause. In many countries plant virus disease epidemiology is being neglected or left largely to zoologists.

Such a restricted approach was criticized by Kennedy (1951), who stressed that direct or indirect effects of viruses on the behaviour and ecology of their vectors are likely to occur more frequently than suggested by the available information. Despite this early warning, the lack of any radical change in attitudes continues to impede progress and accounts for some of the difficulties encountered in attempts to control virus diseases. The situation is unlikely to improve unless there is a greater awareness of the advantages of multi-disciplinary studies (Thresh, 1980b). The need for such studies and the potential benefits of an ecological approach are apparent from much of the work on weeds and wild plants considered in the foregoing sections and elsewhere in this volume.

References

A'Brook, J. (1964). The effect of planting date and spacing on the incidence of groundnut rosette disease and of the vector, *Aphis craccivora* Koch., at Mokwa, Northern Nigeria. *Annals of Applied Biology* **54,** 199–208.

Adlerz, W. C. (1974). Wind effects on spread of watermelon mosaic virus 1 from local virus sources to watermelon. *Journal of Economic Entomology* **67,** 361–364.

Adlerz, W. C. (*1981*). Weed hosts of aphid-borne viruses of vegetable crops in Florida. In *Pests, Pathogens and Vegetation*, pp. 467–478. J. M. Thresh. London: Pitman.

Annand, P. N., Chamberlin, J. C., Henderson, C. F. & Waters, H. A. (1932). Movements of the beet leafhopper in 1930 in Southern Idaho. *United States Department of Agriculture, Washington D.C., Circular* No. 244.

Armitage, H. M. (1957). Report on sugarbeet leafhopper — curly-top virus control in California. *California Department of Agriculture, Bulletin* No. 46.

Atkinson, D. & White, G. C. (*1981*). The effects of weeds and weed control on temperate fruit orchards and their environment. In *Pests, Pathogens and Vegetation*, pp. 415–428. J. M. Thresh. London: Pitman.

Bennett, C. W. (1952). Origin and distribution of new or little known virus diseases. *Plant Disease Reporter*, Supplement 211, 43–46.

Boncquet, P. A. & Stahl, C. F. (1917). Wild vegetation as a source of curly-top infection of sugar beets. *Journal of Economic Entomology* **10,** 392–397.

Bos, L. (1981). Wild plants in the ecology of virus diseases. In *Plant Diseases and Vectors: Ecology and Epidemiology*, pp. 1–28. K. Maramorosch & K. F. Harris. New York, San Francisco and London: Academic Press.

Carter, W. (1930). Ecological studies of the beet leafhopper. *United States Department of Agriculture, Technical Bulletin* No. 206.

Carter, W. (1939). Populations of *Thrips tabaci*, with special reference to virus transmission. *Journal of Animal Ecology* **8,** 261–276.

Carter, W. (1961). Ecological aspects of plant virus transmission. *Annual Review of Entomology* **6,** 347–370.

Carter, W. (1973). *Insects in Relation to Plant Disease.* 2nd Edn. London and New York: Wiley-Interscience.

Conti, M. (1976a). Epidemiology of maize rough dwarf virus. II. Role of the different generations of vector in virus transmission. *Agriculturae Conspectus Scientificus* **39,** 149–156.

Conti, M. (1976b). Epidemiology of maize rough dwarf virus. III. Field symptoms, incidence and control. *Maydica* **21,** 165–175.

Conti, M. (*1981*). Wild plants in the ecology of hopper-borne viruses of grasses and cereals. In *Pests, Pathogens and Vegetation*, pp. 109–119. J. M. Thresh. London: Pitman.

Coons, G. H., Stewart, D., Bockstahler, H. W. & Schneider, C. L. (1958). Incidence of savoy in relation to the variety of sugar beets and to the proximity of wintering habitat of the vector, *Piesma cinerea. Plant Disease Reporter* **42,** 502–511.

Cooper, J. I. & Harrison, B. D. (1973). The role of weed hosts and the distribution and activity of nematodes in the ecology of tobacco rattle virus. *Annals of Applied Biology* **73,** 53–66.

Cotten, J. (1979). The effectiveness of soil sampling for virus-vector nematodes in MAFF certification schemes for fruit and hops. *Plant Pathology* **28,** 40–44.

Doolittle, S. P. & Gilbert, W. W. (1919). Seed transmission of cucurbit mosaic by the wild cucumber. *Phytopathology* **9,** 326–327.

Doolittle, S. P. & Walker, M. N. (1925). Further studies on the overwintering and dissemination of cucurbit mosaic. *Journal of Agricultural Research* **31,** 1–58.

Doolittle, S. P. & Walker, M. N. (1926). Control of cucumber mosaic by eradication of wild host plants. *United States Department of Agriculture, Bulletin* No. 1461.

Douglass, J. R. & Cook, W. C. (1954). The beet leafhopper. *United States Department of Agriculture, Washington D.C., Circular* No. 942.

Duffus, J. E. (1971). Role of weeds in the incidence of virus diseases. *Annual Review of Phytopathology* **9,** 319–340.

Duffus, J. E. (1980). Curly top virus control. In *Proceedings of the XIth International Congress on Plant Protection*. Minneapolis: Burgess Publishing Company (in press).

Eastop, V. F. (*1981*). Wild hosts of aphid pests. In *Pests, Pathogens and Vegetation*, pp. 285–298. J. M. Thresh. London: Pitman.

Fryer, J. D. (*1981*). Weed control practices and changing weed problems. In *Pests, Pathogens and Vegetation*, pp. 403–414. J. M. Thresh. London: Pitman.

Hallock, H. C. & Douglass, J. R. (1956). Studies of four summer hosts of the beet leafhopper. *Journal of Economic Entomology* **49,** 388–391.

Häni, A. (1971). Zur epidemiologie des gurkenmosaikvirus in Tessin. *Phytopathologische Zeitschrift* **72,** 115–144.

Harrison, B. D. (1977). Ecology and control of viruses with soil-inhabiting vectors. *Annual Review of Phytopathology* **15,** 331–360.

Hayes, T. R. (1932). Groundnut rosette disease in the Gambia. *Tropical Agriculture* (Trinidad) **9,** 211–217.

Heathcote, G. D. (1970). Weeds, herbicides and plant virus diseases. In *Proceedings 10th British Weed Control Conference*, pp. 934–941.

Johnson, C. G. (1962). The ecological approach to cocoa disease and health. In *Agriculture and Land Use in Ghana*, pp. 348–352. J. B. Wills. London, Accra and New York: Oxford University Press.

Jones, R. A. C. (*1981*). The ecology of viruses infecting wild and cultivated potatoes in the Andean region of South America. In *Pests, Pathogens and Vegetation*, pp. 89–107. J. M. Thresh. London: Pitman.

Kavanagh, T. (1974). The influence of herbicides on plant disease. II. Vegetables, root crops and potatoes. *Scientific Proceedings of the Royal Dublin Society, Series B*, **3,** 251–265.

Kennedy, J. S. (1951). A biological approach to plant viruses. *Nature* (London), **168,** 890–894.

Knipling, E. F. (1979). The basic principles of insect population suppression and management. *United States Department of Agriculture, Washington D.C., Handbook* No. 512.

Leston, D. (1973). The ant mosaic — tropical tree crops and the limiting of pests and diseases. *PANS* **19,** 311–341.

Lewis, T. (1969). Factors affecting primary patterns of infestation. *Annals of Applied Biology* **63,** 315–317.

Majer, J. D. (1972). The ant mosaic in Ghana cocoa farms. *Bulletin of Entomological Research* **62,** 151–160.

Maramorosch, K. & Harris, K. F. (1979). *Leafhopper Vectors and Plant Disease Agents*. New York and London: Academic Press.

Martin, D. K. & Randles, J. W. (*1981*). Interrelationships between wild host plant and aphid vector in the epidemiology of lettuce necrotic yellows. In *Pests, Pathogens and Vegetation*, pp. 479–486. J. M. Thresh. London: Pitman.

Murant, A. T. (*1981*). The role of wild plants in the ecology of nematode-borne viruses. In *Pests, Pathogens and Vegetation*, pp. 237–248. J. M. Thresh. London: Pitman.

Oswald, J. W. & Houston, B. R. (1953). Host range and epiphytology of the cereal yellow dwarf disease. *Phytopathology* **43,** 309–313.

Piemeisel, R. L. (1954). Replacement control; changes in vegetation in relation to control of pests and diseases. *Botanical Review* **20,** 1–32.

Piemeisel, R. L., Lawson, F. R. & Carsner, E. (1951). Weeds, insects, plants, diseases and dust storms. *Scientific Monthly* **73,** 124–128.

Posnette, A. F. (*1981*). The role of wild hosts in cocoa swollen shoot disease. In *Pests, Pathogens and Vegetation*, pp. 71–78. J. M. Thresh. London: Pitman.

Posnette, A. F., Robertson, N. F. & Todd, J. McA. (1950). Virus diseases of cacao in West Africa. V. Alternative host plants. *Annals of Applied Biology* **37,** 229–240.

Quiot, J. B., Verbrugghe, M., Labonne, G., Leclant, F. & Marrou, J. (1979). Ecologie et épidémiologie du virus de la mosaïque du concombre dans le sud-est de la France. IV. Influence des brise-vent sur la répartition des contaminations virales dans une culture protégée. *Annales de Phytopathologie* **11,** 307–324.

Rose, D. J. W. (1972). Times and sizes of disposal flights by *Cicadulina* species (Homoptera: Cicadellidae), vectors of maize streak disease. *Journal of Animal Ecology* **41,** 495–506.

Rose, D. J. W. (1978). Epidemiology of maize streak disease. *Annual Review of Entomology* **23,** 259–282.

Severin, H. H. P. (1919). Investigations of the beet leafhopper (*Eutettix tenella* Baker) in California. *Journal of Economic Entomology* **12,** 312–326.

Simons, J. N. (1957). Effects of insecticides and physical barriers on field spread of pepper veinbanding mosaic virus. *Phytopathology* **47,** 139–145.

Smith, J. G. (1976). Influence of crop background on aphids and other phytophagous insects on Brussels sprouts. *Annals of Applied Biology* **83,** 1–13.

Storey, H. H. & Bottomley, A. M. (1928). The rosette disease of peanuts *Arachis hypogaea* L. *Annals of Applied Biology* **15,** 26–45.

Strickland, A. H. (1951). The entomology of swollen shoot of cacao. I. The insect species involved, with notes on their biology. *Bulletin of Entomological Research* **41,** 725–748.

Stubbs, L. L., Guy, A. D. & Stubbs, K. J. (1963). Control of lettuce necrotic yellows virus disease by the destruction of common sowthistle (*Sonchus oleraceus*). *Australian Journal of Experimental Agriculture and Animal Husbandry* **3,** 215–218.

Taylor, C. E. & Murant, A. F. (1968). Chemical control of raspberry ringspot and tomato black ring viruses in strawberry. *Plant Pathology* **17,** 171–178.

Taylor, C. E. & Thomas, P. R. (1968). The association of *Xiphinema diversicaudatum* (Micoletsky) with strawberry latent ringspot and arabis mosaic viruses in a raspberry plantation. *Annals of Applied Biology* **62,** 147–157.

Thresh, J. M. (1974). Vector relationships and the development of epidemics: the epidemiology of plant viruses. *Phytopathology* **64,** 1050–1056.

Thresh, J. M. (1976). Gradients of plant virus diseases. *Annals of Applied Biology* **82,** 381–406.

Thresh, J. M. (1980a). The origins and epidemiology of some important plant virus diseases. *Applied Biology* **5,** 1–65.

Thresh, J. M. (1980b). An ecological approach to the epidemiology of plant virus diseases. In *Comparative Epidemiology*, pp. 57–70. J. Kranz and J. Palti. Wageningen: Pudoc.

Thung, T. H. (1934). Bestrijding der krul- en kroepoekziekten van tabak. *Mededeeling,* No. 78, *Proefstation voor Vorstenlandsche Tabak, Klaten, Java* (cited in Bos, 1981).

Tinsley, T. W. (1964). The ecological approach to pest and disease problems of cacao in West Africa. *Tropical Science* **6,** 38–46.

Tomlinson, J. A., Carter, A. L., Dale, W. T. & Simpson, C. J. (1970). Weed plants as sources of cucumber mosaic virus. *Annals of Applied Biology* **66,** 11–16.

Townsend, G. R. (1947). Celery mosaic in the Everglades. *Plant Disease Reporter* **31,** 118–119.

Vanderplank, J. E. (1948). The relation between the size of fields and the spread of plant disease into them. I. Crowd diseases. *Empire Journal of Experimental Agriculture* **16,** 134–142.

Wallis, R. L. & Turner, J. E. (1969). Burning weeds in drainage ditches to suppress populations of green peach aphids and incidence of beet western yellows disease in sugar beets. *Journal of Economic Entomology* **62,** 307–309.

Watson, M. A. (1967). Epidemiology of aphid-transmitted plant-virus diseases. *Outlook on Agriculture* **5,** 155–166.

Way, M. J. & Cammell, M. E. (*1981*). Effects of weeds and weed control on invertebrate pest ecology. In *Pests, Pathogens and Vegetation*, pp. 443–458. J. M. Thresh. London: Pitman.

Wellman, F. L. (1935). Dissemination of southern celery mosaic virus on vegetable crops in Florida. *Phytopathology* **25,** 289–308.

Wellman, F. L. (1937). Control of southern celery mosaic in Florida by removing weeds that serve as sources of mosaic infection. *United States Department of Agriculture, Washington, D.C., Technical Bulletin* No. 548.

Wolcott G. N. (1928). Increase of insect transmitted plant disease and insect damage through weed destruction in tropical agriculture. *Ecology* **9,** 461–466.

The role of wild hosts in cocoa swollen shoot disease

A F Posnette
East Malling Research Station, Maidstone, Kent ME19 6BJ

Cocoa, *Theobroma cacao*, was introduced into mainland West Africa from South America via the two islands San Thomé (Portuguese colony) and Fernando Po (Spanish) during the nineteenth century. Early in the twentieth century further introductions were made directly to the British colonies from Trinidad and Venezuela via Kew Gardens. Although Wardian cases were sometimes used, the plants almost certainly left the country of origin as seed. As none of the known cocoa swollen shoot viruses are seed-borne they are unlikely to be exotic in West Africa.

Cocoa is generally regarded as an under-storey forest tree and traditionally it has been grown under the shade of taller trees. These were either specially planted, as in the plantations of South America and the Caribbean, or survivors of the indigenous forest complex, as in the West African farms. In Ghana and Nigeria cocoa planting expanded rapidly in the early years of the twentieth century, following the growth of the market for chocolate in Europe and North America and the decline of production in South America because of fungus diseases. The first cocoa farms were planted near ports but the gradual enlargement of the area under cocoa depended on the extension of railways and road systems. By 1920 contiguous cocoa farms occupied much of the forested area of the eastern part of Ghana. In 1943 farmers could remember a patch of cocoa trees dying some 20 years earlier in the manner of swollen shoot disease. The first authentic record of swellings was in 1936, by which date the disease was widespread in the Eastern Province.

The known strains of the cocoa swollen shoot virus (CSSV) differ greatly in pathogenicity. Only four have been studied serologically (Kenton and Legg, 1971) and the results, together with differences in various mealybug vector species (pseudococcids) and host plants, provide some justification for treating the Ghana viruses as three groups. Less is known about the Nigerian and other West African isolates. They differ from the Ghana isolates in symptomatology but form a similar spectrum.

(1) Western Ghana strains of CSSV

The extreme south-western part of Ghana has the highest rainfall in the country and the greatest area of remaining rain forest. Cocoa farms are scattered among areas of forest, including some reserves in which a few cocoa farms were made either clandestinely or before the reserves were demarcated. The unique feature of this region and the adjoining south-eastern border of the Ivory Coast is the occurrence of *Cola chlamydantha* (Fig. 1). It is an under-storey forest tree that is often left when clearings are made for farms. The straight, unbranched stem of hard wood is used to make pestles for pounding cassava and other foods into flour.

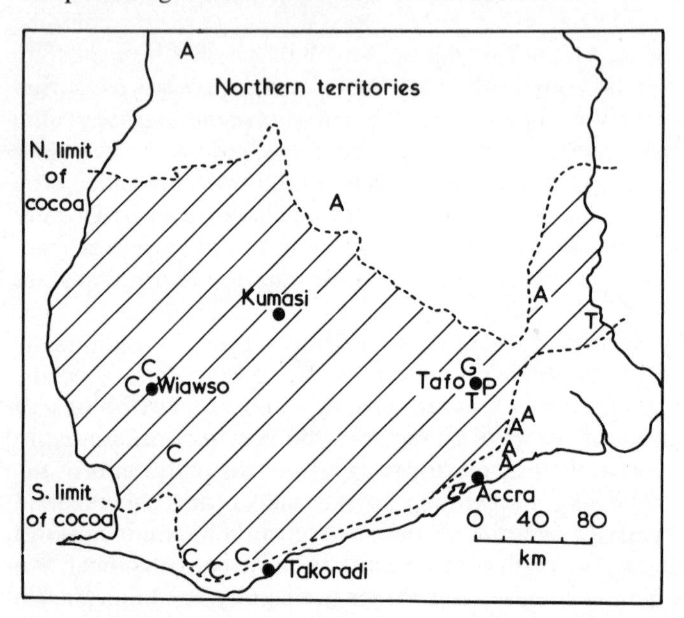

Figure 1 Sketch map of southern Ghana showing the main cocoa-growing areas and sites where indigenous trees have been found naturally infected with cocoa swollen shoot virus (adapted from Thresh, 1980). A = *Adansonia digitata* (baobab). C = *Cola chlamydantha.* G = *Cola gigantea.* P = *Ceiba pentandra* (silk cotton/kapok). T = *Sterculia tragacantha.*

Trees of *C. chlamydantha* occur singly or in colonies, the latter when growing beneath tall top-storey species (such as the kapok or silk cotton *Ceiba pentandra*) being used as roosts by fruit bats. The fruit of *C. chlamydantha* is borne directly on the trunk, as in cocoa; the husk splits open when ripe to expose the seeds, which have a fleshy aril on which fruit bats feed. The fruits are frequently infested with the virus vectors *Planococcus njalensis*, which enter when the husk opens and feed on the cotyledons of the seed (Posnette, 1960).

C. chlamydantha trees are frequently infected with CSSV, both in cocoa farms and in forests several miles away from cocoa (Dale and Attafuah, 1957; Posnette *et al.*, 1950; Todd, 1951; Tinsley, 1955). In a survey of virus infection in *C. chlamydantha* trees Todd found infected trees in about a tenth of the localities where cocoa was rare and not infected. Tinsley (1971) reported that 35% of cocoa farms had *C. chlamydantha* trees in or near them and swollen shoot was more prevalent in these than in farms without *C. chlamydantha*.

The association between infected *C. chlamydantha* and swollen shoot in cocoa seems clear. The question is whether the cocoa became infected from the cola or vice versa. The occurrence of infection in cola over a wider area than that in cocoa indicates that the former was the source, although the cola may itself have been infected from another species. If cocoa were the original host in western Ghana, one must suppose that the virus was brought there with planting material from Ashanti or the Eastern Province. Contrary circumstantial evidence is the difference in virus strains (Anon., 1949) and the absence of virus from the earliest cocoa planted in the Western Region. This was at the agricultural station at Wiawso where cocoa was established using seed from Aburi Botanic Gardens where virus infection has never been found.

(2) Eastern Ghana strains of CSSV

Swollen shoot disease is widespread in the Eastern Province and great variation occurs in virus pathogenicity as expressed in symptoms. Virulence may differ between isolates from neighbouring outbreaks and even the same tree (Posnette and Todd, 1955). The 'type' virus, usually termed New Juaben, is the most virulent and the most readily transmitted by the vector mealybugs, probably because it reaches a higher concentration in cocoa phloem than milder strains. The New Juaben strain has been used more than others in host range studies and it has been transmitted to all the 28 species known to be susceptible to any of the isolates of CSSV (Tinsley and Wharton, 1958). All susceptible species are in the Tiliales, all but one genus being in the Bombacaceae and Sterculiaceae. The exception was *Corchorus* spp. (Tiliceae), which proved to be even more sensitive than cocoa to eight isolates. It is a small shrub which these authors considered 'unlikely to be important as natural hosts'.

As swollen shoot disease is so prevalent in cocoa in the Eastern Region of Ghana, many trees of the host species must have become infected from cocoa, and which species was the original source cannot be determined. Very few trees of indigenous species have been shown to be infected, probably because of the difficulty of proving infection in the chronic stage. After initial infection, when some but not all species develop leaf symptoms, infected trees appear healthy and mealybugs transferred from them to cocoa do not transmit virus. After coppicing, however, new growth may show transient symptoms and the virus can then be transmitted (Dale and Attafuah, 1957; Posnette *et al.*, 1950). Regenerating stumps from fallen trees may be more frequent sources of virus than undamaged trees.

Perhaps the most likely candidate for the role of original host of this group of CSSV strains is *Sterculia tragacantha*. Seedlings of this species show no symptoms when infected and yet the virus is not difficult to transmit from them. The two known instances of natural infection were both associated with diseased cocoa from which they could have become infected. One was found on the Research Institute at Tafo in cocoa in an area where swollen shoot had recurred after several attempts at eradication (Legg and Agbodjan, 1969). The other was in an isolated outbreak near the Togo border (Owusu and Lovi, 1970). Trees of *S. tragacantha* are very common in the secondary forest and as survivors in cocoa farms.

Other candidates as sources of infection, either primary or secondary (i.e. after being infected with virus from cocoa), are *Ceiba pentandra* and *Cola gigantea*. The only example of the latter (reported as *C. cordifolia*) being naturally infected was a very young seedling with leaf symptoms growing beneath healthy cocoa. As it would be unusual for a young cocoa seedling to have been infected in these circumstances, it seems probable that the *C. gigantea* was infected from the parent tree by mealybugs feeding on the bean after the pod had opened, as occurs in *C. chlamydantha*.

The Savannah 'mottle leaf' group of CSSV

The type strain of this group was called mottle leaf to distinguish it from swollen shoot because it was characterized by conspicuous red mottle of the young foliage followed by interveinal chlorosis and the absence of stem swellings. The type location, Kpeve, on the Togo side of the Volta River, has intermediate forest/savannah vegetation and other occurrences of this group in cocoa (e.g., Alaparun in Nigeria) have been found in similar ecological conditions.

The experimental host range of virus strains in this group is more restricted than that of the others and they do not infect *C. chlamydantha*, *C. gigantea* or *S. tragacantha*. They commonly infect baobab trees *Adansonia digitata* (Fig. 2) growing in the savannah zone fringing the north and the south of the cocoa-growing forest area of Ghana (Attafuah and Tinsley, 1958). Mealybug transfers demonstrated infection in 13 of 96 trees tested. In cocoa the virus isolates varied in the symptoms they caused, but all were similar to those of the cocoa mottle leaf group. Legg and Bonney (1967) found that the baobab and cocoa isolates have the same host range.

Attafuah and Tinsley (1958) concluded that because 'infected *A. digitata* were found only in areas remote from the forest it is unlikely that this species was an initial source of virus infection for cacao'. However, there are few baobabs within the main cocoa-growing areas and similar logic would imply that the baobab trees were not infected by virus spread from cocoa. The occurrence of the comparatively rare mottle leaf type of CSSV in cocoa outside or at the fringe of the forest ecological zone suggests either a common source or occasional spread from baobab to cocoa. Spread of the virus between baobabs, widely scattered as they often are, must be affected by wind-borne mealybugs, as postulated for spread over long distances in cocoa.

Figure 2 Left. A baobab tree (*Adansonia digitata*) growing near the Volta river in Ghana with scrub including cocoa trees in the background.

Right. A large silk cotton tree (*Ceiba pentandra*) surviving from a former rain forest area of the Eastern Province of Ghana with cocoa growing in the background below other forest trees.

An herbaceous host?

Recently investigations in Togo have shown by electron microscopy that virus particles resembling CSSV occur in *Commelina lagosensis*. This is a common monocotyledonous plant invading cocoa farms when light intensity at soil level increases as the cocoa leaf canopy degenerates. If this virus proves to be CSSV, this herbaceous weed would be a potential reservoir of virus, or relay host, from which cocoa farms could be re-infected after diseased cocoa trees have been eradicated (Amefia *et al.*, 1981).

Discussion

The role of wild host plants in the epidemiology of cocoa swollen shoot disease has been difficult to investigate because of the failure to transmit the virus mechanically to or from them. An additional difficulty is that transmission by vectors is erratic. The less pathogenic strains seem to be in lower concentration in cocoa than the more virulent ones. In most wild host plants even the strains that are virulent in cocoa attain only a low concentration in the chronic phase, and probably the virus is often localized in the plant.

Because of these difficulties some susceptible species were apparently immune in early tests but were infected in later ones. Although some of the differences between strains in their host range may merely reflect differences in concentration/infectivity, there are some species that all investigators have found to be differential hosts. Grouping strains on the basis of host range is supported by differences in vector species and symptoms produced in cocoa (Bald and Tinsley, 1970).

The practical difficulties in proving the susceptibility/immunity of young plants are enhanced when testing large forest trees for infection. It can be no coincidence that the two wild hosts in which most infected trees have been found (*A. digitata* and *C. chlamydantha*) are those in which the availability of virus to vectors is highest and does not decline greatly with the duration of infection. This is consistent with the continued production of symptoms in many trees. Some apparently healthy *C. chlamydantha* trees were infected, but all the *A. digitata* that were proved to be infected showed symptoms. The ready transmissibility of virus from these two wild hosts should make them more potent sources of infection for cocoa growing nearby. Undoubtedly many outbreaks of cocoa swollen shoot disease have originated from *C. chlamydantha* in the densely forested western part of Ghana. Few outbreaks of cocoa mottle leaf have been found and this is probably because cocoa is rarely grown near baobab trees as their ecological requirements differ.

The eastern group of CSSV strains that occur in the absence of *C. chlamydantha* and *A. digitata* are more prevalent than the Western Province group. They were probably transferred to cocoa at an earlier date and the main affected area is much more densely planted with cocoa. Many farms are contiguous, providing favourable conditions for spread from cocoa to cocoa over hundreds of square miles. The virulence of the most prevalent strain also accelerates spread because the vectors migrate from dying trees to better feeding sites. Susceptible wild trees growing in or near cocoa farms must be expected to become infected from cocoa and subsequently to remain as sources of fresh outbreaks in new cocoa farms established on the same site after the destruction of the previous ones. It is not possible to determine whether a wild tree many years old was first infected from cocoa or from other wild trees before cocoa was planted near to it.

Despite the intensive efforts by several investigators, only a few kapok trees have been shown to be infected because of the extremely low rate of virus transmission by vectors. For example, 25 000 mealybugs were transferred from two trees known to be infected to over 1000 cocoa seedlings, none of which became infected (Anon., 1951). From such results the role of *C.*

pentandra as a source of virus for cocoa would seem to be insignificant were it not one of the commonest wild trees in cocoa farms in eastern Ghana.

S. tragacantha is also a prevalent species, though less numerous than *C. pentandra*, in cocoa farms and virus has been readily transmitted to cocoa from symptomless trees experimentally infected. Naturally infected *S. tragacantha* trees have been found and this species retains a relatively high virus content and yet develops no symptoms. It is a strong candidate for speculation as the original host of the eastern Ghana group of CSSV strains.

Although considerable efforts have been made to investigate the wild hosts of CSSV, evidence that they play an important role in frustrating efforts to control the disease in cocoa is still equivocal. Widespread destruction of the known host species is impracticable with the current practices of cocoa farming. Clear felling of the forest before planting is inadvisable on ecological grounds and would have to be followed by selective eradication of regenerating stumps of susceptible species and their seedlings.

When a swollen shoot outbreak of recent origin is found in cocoa remote from known infections, an intensive search for a local source among susceptible wild species would seem a logical concomitant of the eradication of infected cocoa trees. Unless the source is also removed a recurrence of the disease should be expected. Nevertheless isolated outbreaks have been eradicated without re-infection, suggesting that spread from wild hosts (other than *C. chlamydantha*) to cocoa is infrequent, for reasons that experiments with vectors have indicated.

References

Amefia, Y. K., Brunel, J., Delecolle, B. & Partiot, M. (1981). In *Proceedings 7th International Cocoa Research Conference, Douala, Cameroon* (in press).

Anon. (1949). The Western Province Viruses. *Report of the West African Cocoa Research Institute for 1947–48*, pp. 11–13.

Anon. (1951). The search for alternative hosts in the field. *Report of the West African Cocoa Research Institute for 1949–50*, pp. 30–31.

Attafuah, A. & Tinsley, T. W. (1958). Virus diseases of *Adansonia digitata* L. (Bombacaceae) and their relation to cacao in Ghana. *Annals of Applied Biology* **46,** 20.

Bald, J. G. & Tinsley, T. W. (1970). A quasi-genetic model for plant virus host ranges. IV. Cacao swollen shoot and mottle leaf viruses. *Virology* **40,** 369–378.

Dale, W. T. & Attafuah, A. (1957). Virus research: the host range of cacao viruses. *Report of the West African Cocoa Research Institute for 1955–56*, pp. 28–30.

Kenton, R. H. & Legg, J. T. (1971). Serological relationships of some viruses from cocoa (*Theobroma cacao* L.) in Ghana. *Annals of Applied Biology* **67,** 195–200.

Legg, J. T. & Agbodjan, F. X. (1969). Swollen shoot disease. *Report of the Cocoa Research Institute of Ghana for 1967–68*, pp. 23–25.

Legg, J. T. & Bonney, J. K. (1967). The host range and vector species of viruses from

Cola chlamydantha K. Schum., *Adansonia digitata* L. and *Theobroma cacao* L. *Annals of Applied Biology* **60,** 399–403.

Owusu, G. K. & Lovi, N. K. (1970). Study of swollen shoot outbreaks. *Report of the Cocoa Research Institute of Ghana for 1968–69*, p. 33.

Posnette, A. F. (1960). Some aspects of virus spread among plants by vectors. In *Proceedings of the 7th Commonwealth Entomology Conference*, pp. 162–165. London: Commonwealth Institute of Entomology.

Posnette, A. F. & Todd, J. M. (1955). Virus diseases of cacao in West Africa. IX. Strain variation and interference in Virus 1A. *Annals of Applied Biology* **43,** 433–453.

Posnette, A. F., Robertson, N. F. & Todd, J. M. (1950). Virus diseases of cacao in West Africa. V. Alternative host plants. *Annals of Applied Biology* **37,** 229–240.

Thresh, J. M. (1980). The origins and epidemiology of some important plant virus diseases. *Applied Biology* **5,** 1–65.

Tinsley, T. W. (1955). Virus infection in *Cola chlamydantha*. *Report of the West African Cocoa Research Institute for 1954–55*, p. 32.

Tinsley, T. W. (1971). The ecology of cocoa viruses. 1. The role of wild hosts in the incidence of swollen shoot virus in West Africa. *Journal of Applied Ecology* **8,** 491–495.

Tinsley, T. W. & Wharton, A. L. (1958). Studies on the host ranges of viruses from *Theobroma cacao* L. *Annals of Applied Biology* **46,** 1–6.

Todd, J. M. (1951). An indigenous source of swollen shoot disease of cacao. *Nature* (London) **167,** 952–953.

The possible role of amenity trees and shrubs as virus reservoirs in the United Kingdom

J I Cooper
Natural Environment Research Council, Institute of Virology,
5, South Parks Road, Oxford OX1 3UB

To assess the role of amenity trees and shrubs of wild and cultivated species as virus reservoirs, it is necessary to know the frequency of natural infection, the extent of experimental susceptibility, how rapidly the agents spread, presence and habits of vectors, etc. Unfortunately, in many instances all that is known is that infected woody plants occur near susceptible commercial or garden crops. Restricting consideration to the trees and shrubs winter-hardy in UK, knowledge about the infecting agents is still fragmentary, being based, in many instances, on accidental/incidental discoveries. Nevertheless, examination of the hitherto scattered data collated by Schmelzer (1977) and Cooper (1979) indicates the potential. Collectively these publications list most of the relevant literature and the number of additional citations listed here has been restricted by space limitations.

Agents associated with virus-like diseases

Woody perennials are so long-lived that they are vulnerable to infection for protracted periods. Many have been propagated vegetatively for decades and predictably there are numerous records of virus-like diseases that have not been adequately investigated. Thus, it is impossible presently to assess the reservoir role of the 58 species (Table 1) that are known hosts of agents that have only been transmitted experimentally by grafting. Nothing is known concerning their transmission or other properties. However, one agent that is normally detected by graft-transmission tests deserves special mention because it seems to spread naturally as well as by vegetative propagation. Little cherry disease was first recognized in fruiting plantations in British Columbia where its spread was originally attributed to leafhopper species transiently feeding but not breeding on sweet/sour cherries. The source of this and other economically damaging epidemics in Western North America was traced to symptomless flowering cherries (*Prunus serrulata*) originating in Japan. In the UK, where the disease in fruit crops was recognized about 1955, ornamental

Prunus spp. are commonly infected and unquestionably pose a threat (e.g. Cammack, 1966).

The electron microscope has been used frequently to investigate virus-like diseases. Unfortunately, in the absence of additional information it is impossible to assess the significance of such findings as the occurrence of rhabdovirus-like particles in foliar cells of *Euonymus japonicus* and *Laburnum anagyroides*, or the presence of tubular virus-like structures in sap from conifer needles.

Table 1 Natural infection of amenity tree/shrub hosts by viruses and other graft-transmissible agents

Viruses/other agents	*Tree/shrub hosts reported*		
	Number of species	*Number of genera*	*Number of families*
Cucumo, alfalfa mosaic and broad bean wilt viruses	42	35	22
Nepo, tobra and tobacco necrosis viruses	61	42	26
Ilar and cherry leaf roll viruses	31	18	10
Graft-transmitted agents	58	33	22

Mollicute-like organisms* have been detected by electron microscopy in tissues from some 200 plant species including, in warm-temperate or subtropical parts of the world, diverse trees including *Aralia spinosa* (Dale, 1979), *Carya* spp., *Cornus* spp., *Eucalyptus* spp., *Fraxinus* spp., *Juglans* spp., *Larix decidua, Morus alba, Myrica carolinensis, Paulownia* spp., *Pseudotsuga taxifolia, Quercus robur, Robinia pseudoacacia, Rosa* spp., *Salix* spp. and *Ulmus* spp. There is evidence that in the USA the agent of elm phloem necrosis disease is spread by a leafhopper (*Scaphoideus luteolus*). Similarly in Japan a leafhopper (*Hishimonus sellatus*) is the vector of the agent causing mulberry dwarf. Furthermore, there has been recent speculation (Thomas and Donselman, 1979) concerning the role of amenity/wild palms as sources of agents associated with coconut lethal yellowing disease in the Caribbean region, but there is neither direct proof of the pathogenicity of mollicute-like organisms nor knowledge about their spread between genera/species. In any event, the known distribution of mollicute-like organisms suggests that the risks of spread are greatest when and where temperatures are higher than normal in the UK, where roses are presently the only recorded amenity hosts.

* Mollicute-like organisms are a group of gram-negative pleomorphic prokaryotes distinct from bacteria yet containing both ribose and deoxyribose nucleic acids. Some (class Mollicutes *sensu* Freundt, 1974) are bounded by triple-layered unit membranes (*c.* 8 nm in thickness), whereas others somewhat resemble Rickettsiae and possess a cell wall *c.* 25 nm in thickness.

Apple chlorotic leafspot virus

Alternative means are available for detecting apple chlorotic leafspot virus (CLSV), but graft transmission tests are routinely used and have revealed agents resembling CLSV in numerous hosts other than *Malus*, e.g. *Amelanchier* sp., *Chaenomeles japonica*, *Crataegus* spp., *Cydonia oblonga*, *Mespilus germanica, Prunus* spp., *Sorbus tianshanica.* Although there is as yet no evidence that CLSV is seed-borne and no natural vector is known, some infection of previously healthy pome fruit trees has been observed (Posnette and Cropley, 1964). Because the virus is frequent (*c.* 25%) in hawthorn hedge plants, probably grown as seedling 'quicks', it is likely that amenity stock poses a threat. The magnitude of the threat is difficult to assess though probably small. CLSV seems to exist in nature as a range of types differing in their abilities to cross graft unions between species and in the symptoms that they cause in graft-infected woody indicator hosts. However, taking the experience of, for example, Bernhard and Dunez (1971) at face value and neglecting the possibility that unknown latent viruses in the experimental material influenced the result, the differences might be assumed to indicate an inefficient mechanism of virus spread which tended to facilitate 'in breeding' within discrete host/virus populations. Possibly natural root grafts assist virus spread in closely planted hedges/gardens. Such grafts have been recognized surprisingly often and Graham and Bormann (1966) recorded them in 150 species of woody plant; intraspecific fusion being common in both angiosperms and gymnosperms. A more detailed knowledge is needed of the properties of CLSV isolates from individual trees in hedges and from geographically scattered natural hosts of the same genus/species.

Other viruses with rod-shaped particles

Sap-transmissible viruses assignable to 13 of the 24 established groups infecting higher plants have been detected in amenity trees and shrubs (Matthews, 1979). On sparse evidence, I feel justified in dismissing as trivial, the role of woody perennials as reservoirs for some agents with virus-like particles morphologically resembling carla, tobamo and potex viruses. Carla-like viruses have been recognized frequently in daphne, privet, honeysuckle, mulberry, passionflower, poplar and elder. It is likely that vegetative propagation explains much of this infection. However, in elder a virus having serological affinities with carnation latent virus was transmitted from wild and, therefore, probably seed-derived *Sambucus nigra* (morphologically similar particles were noted in seedling *S. racemosa*). Generally, there are few or no serological affinities between different members of the carlavirus group, suggesting 'inbreeding' consequent upon restricted natural spread. It may be relevant to note that *Aphis sambuci* specifically alternates between *Dianthus* spp. and *Sambucus* spp. (Eastop, personal communication). There may be a similar common feature linking between honeysuckle and poplar as a carlavirus from *Lonicera* (supplied from the Netherlands by F. A. van der Meer) is related serologically to an isolate from *Populus x euramericana* cv. OP226, (Cooper, unpublished data).

Tobamo viruses occur in large concentrations in infected plants and like potex viruses (recognized in daphne and hydrangea) may in some instances spread when healthy plants contact infected ones. These viruses are readily detectable and have moderately wide experimental host ranges, yet they have not yet been detected in many amenity trees/shrubs and cannot be common. Neglecting the possibility that records reflect accidental glasshouse contamination, there is evidence that tobamo-like viruses naturally infect *Daphne* spp., *Fraxinus americana, Lithocarpus densiflorus, Malus platycarpa, Quercus agrifolia, Q. phellos, Rosa* spp. and *Wistaria sinensis.*

Amenity trees/shrubs may be reservoirs of some potyviruses. In south/central Europe, the economically damaging plum pox virus has been detected in *Lycium halimifolium, Prunus spinosa* and *P. glandulosa.* However, it is difficult to assess whether these hosts are important reservoirs or merely indicators of virus 'spilling over' from fruit crops or other sources. They may at times fall into both categories, as with leguminous trees/shrubs such as *Robinia pseudoacacia* and *Cladrastis lutea* which are known to harbour a potyvirus (bean yellow mosaic virus), able to infect related food crops. Insufficient information is available concerning the possible relationships, of poty-like viruses isolated from *Daphne* spp., *Maclura pomifera, Populus tremula, P. tremuloides* and *P. x euramericana* to justify speculation.

Pollen/seed-borne viruses

Cherry leaf roll virus (CLRV) naturally infects species of *Betula, Cornus, Juglans, Ligustrum, Ptelea, Prunus, Rheum, Sambucus* and *Ulmus.* It is ascribed to the nepovirus group but in its ecology it is most akin to ilarviruses. Spread between plants is probably largely limited to species within a genus. Properties of the viral coat protein (reflected to some extent in antigenic analysis) greatly influence the adsorption of particles to vector surfaces and therefore have epidemiological significance. With CLRV, each host genus normally contains a distinctive serotype and different species within a genus (e.g. *Sambucus canadensis*, *S. nigra* and *S. racemosa*) or geographically scattered individuals of a species yield isolates having few, if any, antigenic determinants not shared. In rare instances, slight antigenic variations between CLRV isolates from walnut, dogwood and elder have been recorded but overall the data contrast with those relating to typical members of the nepovirus group in which vector nematode species, rather than the natural host, select the serological characteristics of the infecting viruses. Notwithstanding reports of experimental CLRV transmission by nematodes, soil-inhabiting vectors have not been consistently associated with natural sites of infection. Compelling evidence suggests that pollen or some other extremely genus-specific vector is the most important method of natural spread between plants.

Ilarviruses (including prune dwarf) are most commonly recorded in rosaceous hosts grown for fruit. They also naturally infect flowering cherries (many cultivars seem wholly infected; Ramaswamy and Posnette, 1972; Colin and Verhoyen, 1975) and roses as well as *Aesculus* spp. (Sweet and Barbara, 1979), *Betula* spp., *Chaenomeles lagenaria* (Sweet and Campbell, 1976), *Corylus avellana, Ligustrum japonicum, Pyracantha rogersiana, Syringa vulgaris, Ulmus carpinifolia* and *U. glabra.* Some ilarviruses are seed and pollen-borne but no alternative means of natural transmission is known and their prevalence in fruiting and ornamental stock is mainly attributable to vegetative propagation of infected plants. Prunus necrotic ringspot, apple mosaic and rose mosaic form parts of a continuum of isolates with serological affinities. Surprisingly, apple mosaic serotypes naturally infect plum, rose, horse chestnut, hop and birch. It is unlikely that this serotype originated separately in these four very different plant families. Probably this ilarvirus, and perhaps others, can spread between genera. Conceivably virus-carrying pollen contaminates flowers and produces abortive germ tubes that facilitate infection and virus spread between disparate species/genera. Another possibility is that mechanical transmission occurs naturally when adjacent pollen-coated leaves rub each other. Whatever the mechanism, the potential reservoir is surprisingly large. In some localities ilarviruses naturally infect up to 50% of trees of wild North American cherries (*Prunus americana, P. pennsylvanica, P. serotina* and *P. virginiana*; Fulton, 1961). By contrast, in Europe, fewer than one in ten of the wild hosts (e.g. *P. spinosa, P. padus, P. serotina* and *P. virginiana*) are known to be similarly infected (e.g. Schimanski *et al.*, 1975; Sweet, 1980). Interestingly, prune dwarf serotypes were recognized about three times more frequently in *P. spinosa*, than prunus necrotic ringspot serotypes. Because prune dwarf was isolated somewhat the more commonly from flowering cherries (e.g. Ramaswamy and Posnette, 1972) amenity *Prunus* spp. may be more important as reservoirs of prune dwarf than of some other ilarviruses. Pending more detailed comparisons, it is probably unwise to assume that, for example, fruit crops were the reservoirs for apple mosaic, infecting hazel, birch and horse chestnut or that elms and lilacs are reservoirs that might hazard rosaceous food crops. For these, wild and ornamental rosaceous hosts are undoubtedly a threat. However, because elm mottle isolates from lilac and elm possess common antigens, it is likely that species in these two disparate genera are, at least occasionally, sources of infection for each other.

Soil-borne viruses

Food plants and others are probably most at risk from amenity trees/shrubs acting as reservoirs for viruses having soil-inhabiting vectors (e.g. tobra, tobacco necrosis and nepoviruses, excepting CLRV). These viruses have been recorded more frequently than others, and in a greater variety of amenity trees/shrubs (Table 1). This is partly due to the inherent stability of their particles and to other properties facilitating detection. Amenity trees/shrubs seem important reservoirs for another reason. Vectors that are slow-moving such as soil-inhabiting nematodes and fungi only spread viruses over short distances. However, in this context, it should be noted that tree roots explore the soil widely and poplars may have roots extending up to 70 m from the trunk. Nearness is also a requirement for the natural dissemination of aphid-borne viruses that are non-circulative and persist for only short periods in their vectors. However, the critical distance cannot be judged accurately because it varies with the ambient temperature, and other factors influencing virus concentration and vector activity (Adlerz, *1981*).

The spread of soil-borne viruses necessarily depends on the presence of appropriate vectors. There are local discontinuities, but surveys have revealed *Xiphinema* and *Longidorus* nematodes in most commercial nurseries, as well as urban parks and gardens (Sweet, 1975; Cooper, unpublished). Hedgerows generally, and in one instance a particular species (*P. spinosa*), have been implicated as foci of arabis mosaic virus in hop and strawberry crops or of strawberry latent ringspot virus in raspberries (Fig. 1. of Murant, *1981*). Arabis mosaic virus has been frequently isolated from amenity trees/shrubs, presumably reflecting vector preferences. However, infection with this virus is by no means common in wild trees. At one site in Avon, arabis mosaic virus was detected in foliage of only four of six adjacent ash trees that had grown for 50 years in soil that is now, and was probably throughout their life, infested with virus-carrying nematodes. However, such a slow rate of infection may not be typical. Thus van Hoof and Caron (1975) noted a small amount of new infection of roses within a year. Similarly, I detected much infection with arabis mosaic, tomato black ring, tobacco rattle and tobacco necrosis viruses a year after planting unrooted poplar cuttings. In contrast to the general experience with nepoviruses, tobacco rattle and tobacco necrosis viruses have been recorded from foliage of few amenity tree/shrub species. Because of the site of inoculation, infection with these viruses is often confined to roots where it escapes notice, yet is well placed to augment herbaceous weed sources of infection. Viruses that are seed-borne in amenity trees/shrubs such as arabis mosaic in ash, rose, privet, daphne, etc., contribute to the general burden (especially if used as rootstocks) but I suspect that the reservoir role of tree seedlings is trivial compared with herbaceous weed species.

Non-circulative aphid-borne viruses

Root-restricted infection with viruses having soil-inhabiting vectors might well escape detection. However, the comparative rarity of recorded infection of trees by viruses that are non-circulative in aphid vectors cannot be explained in the same way. Alfalfa mosaic, broad bean wilt and cucumo viruses naturally infect a wide variety of amenity trees/shrubs (Table 1) but, considering the ubiquity of polyphagous vectors and non-tree sources of the viruses, the incidence of infections is small (as also in bush fruit, *Ribes* and *Rubus* spp.). Indeed, my own tests on hundreds of specimens of ash, birch, walnut and poplar revealed other viruses (arabis mosaic, poplar mosaic, cherry leaf roll), but not cucumber mosaic or alfalfa mosaic. Discontinuous distribution, as with other viruses in trees, is an important factor that makes detection difficult. Components of tree/shrub foliar sap may preferentially inhibit detection of some viruses by mechanical means, but this seems an inadequate explanation. Indeed, Tremaine (1976) did not observe any natural spread of cucumber mosaic virus from infected peach seedlings to adjoining peach or sour cherry in ten years, suggesting that amenity trees/shrubs are unimportant reservoirs, at least for this virus.

Conclusion

Disregarding some of the pitfalls, I have generalized and speculated when attempting to assess the hazard which amenity trees and shrubs present with respect to the differing agents now known. There is a paucity of data and that which is available may not be representative. At least 13 viruses have been recognized in *Daphne* spp. and a similar situation is in prospect for *Rosa* spp., but there is little knowledge concerning the occurrence much less the prevalence of viruses that undoubtedly infect native trees such as oak, beech, lime, alder, willow, etc. Because viruses are sometimes responsible for the horticultural characteristics of a plant that must thereafter be propagated vegetatively, gardeners tend, unknowingly, to facilitate the accumulation of diverse virus-like agents near susceptible crops. There is a risk of direct spread but, in general, I suspect that ephemeral 'weed' hosts are more potent and they are certainly more numerous. Amenity trees/shrubs may act more subtly, providing reservoirs of genetic variability within the virus populations that they have accumulated over centuries of growth.

References

Adlerz, W. C. (*1981*). Weed hosts of aphid-borne viruses of vegetable crops in Florida. In *Pests, Pathogens and Vegetation*, pp. 467–478. J. M. Thresh. London: Pitman.

Bernhard, R. & Dunez, J. (1971). Le virus du chlorotic leaf spot. Contamination de différentes éspèces de *Prunus*. Populations de virus. *Annales de Phytopathologie, hors serie* **71–2,** 317–336.

Cammack, R. H. (1966). Little cherry virus in ornamental cherry. *Plant Pathology* **13,** 31–33.

Colin, J. & Verhoyen, M. (1975). Identification des virus chez les *Prunus* ornamentaux en Belgique. *Reviu de l'Agriculture* **28,** 935–951.

Cooper, J. I. (1979). *Virus Diseases of Trees and Shrubs.* Cambridge: Institute of Terrestrial Ecology.

Dale, J. L. (1979). Mycoplasma-like organism observed in *Aralia spinosa* trees. *Plant Disease Reporter* **63,** 472–474.

Freundt, E. A. (1974). The mycoplasmas. In *Bergey's Manual of Determinative Bacteriology*, 8th Edn. pp. 929–955. R. E. Buchanan and N. E. Gibbons. Baltimore: The Williams and Wilkins Co.

Fulton, R. W. (1961). Characteristics of a virus endemic in wild prunus. *Tidsskrift for Planteavl* **65,** 147–150.

Graham, B. F. & Bormann, F. H. (1966). Natural root grafts. *Botanical Review* **32,** 255–292.

Hoof, H. A. van & Caron, J. E. A. (1975). Strawberry latent ringspot virus infested fields of *Rosa rugosa* produced healthy cuttings but diseased rootstocks for standard roses. *Mededelingen van de Landbouwhoogeschool en opzoekingsstations van de Staatde Gent* **40,** 759–763.

Matthews, R. E. F. (1979). Classification and nomenclature of viruses. Third report of the International Committee on Taxonomy of Viruses. *Intervirology* **12,** (3–5), 129–296.

Murant, A. T. (*1981*). The role of wild plants in the ecology of nematode-borne viruses. In *Pests, Pathogens and Vegetation*, pp. 237–248. J. M. Thresh. London: Pitman.

Posnette, A. F. & Cropley, R. (1964). Natural spread of chlorotic leaf spot virus in apple and quince. *Report of East Malling Research Station for 1963*, pp. 115–116.

Ramaswamy, S. & Posnette, A. F. (1972). Yellow mottle disease of ornamental cherries caused by a strain of prune dwarf virus. *Journal of Horticultural Science* **47,** 107–112.

Schimanski, H.-H., Schmelzer, K., Kegler, H. & Albrecht, H.-J. (1975). Wildavachsende *Prunus*-Arten der Untergattungen *Prunophora* und *Padus* als naturliche Wirtspflanzan für Kirschenringflecken-viren. *Zentralblatt für Bakteriologie, Parasitenkunde, Infektionskrankheiten und Hygiene* (Abteilung II) **130,** 109–120.

Schmelzer, K. (1977). Zier-, Forst- und Wildgeholze. In *Pflanzliche Virologie*, 3rd Edn., **3,** pp. 276–405. M. Klinkowski, K. Schmelzer, & D. Spaar. Berlin: Akademie–Verlag.

Sweet, J. B. (1975). Soil-borne viruses occurring in nursery soils and infecting some ornamental species of Rosaceae. *Annals of Applied Biology* **79,** 49–54.

Sweet, J. B. (1980). Hedgerow hawthorn (*Crataegus* spp.) and blackthorn (*Prunus spinosa*) as hosts of fruit tree viruses in Britain. *Annals of Applied Biology* **94,** 83–90.

Sweet, J. B. & Barbara, D. J. (1979). A yellow mosaic disease of horse chestnut (*Aesculus* spp.) caused by apple mosaic virus. *Annals of Applied Biology* **92,** 335–341.

Sweet, J. B. & Campbell, A. I. (1976). Pome fruit virus infections of some woody ornamental and indigenous species of Rosaceae. *Journal of Horticultural Science* **51,** 91–97.

Thomas, D. L. & Donselman, H. M. (1979). Mycoplasma-like bodies and phloem degeneration associated with declining Pandanus in Florida. *Plant Disease Reporter* **63,** 911–916.

Tremaine, J. H. (1976). Cucumber mosaic virus infection in *Prunus.* In *Virus Diseases and Non-infectious Disorders of Stone Fruits in North America,* pp. 238–239. United States Department of Agriculture Handbook 437.

The ecology of viruses infecting wild and cultivated potatoes in the Andean region of South America

R A C Jones
MAFF Harpenden Laboratory, Hatching Green, Harpenden, Herts AL5 2BD

Centre of origin of the potato crop

Wild potatoes were brought into cultivation by the indigenous Indian peoples of the Andes at least 2000 years before the Spanish conquest. That the potato was a staple food in Peru and Bolivia in pre-Columbian times is amply depicted in the archaeological record. Excavations at sites from the Early-Tihuanaco period (around 250 BC) revealed remains of dehydrated potatoes (chuño) and representations of the crop in pottery occur in the Proto-Chimu period (around 500 AD) and thereafter in ceramics up to late Inca times (Salaman, 1949; Bushnell, 1963). Because the greatest numbers both of native cultivars (=land races) and of wild potato species occur in the Lake Titicaca region (Altiplano) of southern Peru and northern Bolivia, this part of the Andes is generally considered to be the centre of domestication of the crop.

About 200 wild tuber-bearing *Solanum* species are known and these occur in the Americas from south-western USA, southward through Mexico, Central and South America almost to Cape Horn (Hawkes, 1947; Correll, 1962). They occur in diverse habitats including arid deserts, sub-tropical forests and maritime regions but are most typical of high mountains. Underground tubers provide an efficient means of survival through unfavourable conditions, whereas long distance dispersal is mainly by true seed.

Numerous native cultivars are still grown in the Andean region, particularly in the more isolated mountain valleys. These cultivars are extremely diverse in characters such as leaf morphology, flower colour, tuber colour, size and shape. They belong to eight different species, four of which are diploid (*S. ajanhuiri*, *S. goniocalyx*, *S. phureja* and *S. stenotomum*), two triploid (*S. chaucha* and *S. juzepczukii*), one tetraploid (*S. tuberosum* sub. sp. *andigena*) and one pentaploid (*S. curtilobum*) (Hawkes, 1963). The *andigena* types are the most numerous and diverse. Outside South America only the rather uniform *S. tuberosum* sub. sp. *tuberosum* group, derived from relatively few introductions of *andigena*, is cultivated. Native cultivars arose in the Andes from natural intra- and inter-specific hybrids which were then selected by man. Salaman (1949) commented that 'it is to the great variety in the climatic and soil environment, resulting from the diversity of depth and direction of the mountain valleys that we ascribe the high degree of variation and differentiation which aided by selection both natural and controlled led to their establishment'.

Native cultivars are grown mainly by subsistence farmers and they survive perfectly well without the centralized 'seed' tuber multiplication programmes needed to prevent most modern cultivars from degenerating due to virus diseases. They are grown mostly in small fields, often in potato cultivar and species mixtures and sometimes mixed with other root or tuber crops, e.g., arracacha (*Arracacia xanthorrhiza*), ollucu (*Ullucus tuberosus*), oca (*Oxalis tuberosa*) and mashua (*Tropaeolum tuberosum*) (Kay, 1973). Crop rotations are usually short or absent but the plants are protected from competition with wild potato species. These conditions are intermediate ecologically between the wild situation and potato cultivation using uniform single cultivar stands and modern agricultural practices.

In the last 40 years, the traditional native cultivars have gradually been displaced in many areas of the Andean region by modern higher yielding cultivars either bred locally or, in the case of Chile, imported from Europe. From the standpoint of gene conservation, this trend is alarming (Hawkes, 1973). Modern cultivars are mostly grown using modern agricultural practices and simple 'seed' certification programmes.

The potato viruses found

Viruses that infect potato naturally form two ecological groups: (1) those that depend primarily on potato for survival and (2) those that are predominantly viruses of other plants although sometimes infecting potato. The first group presumably developed over geological rather than agricultural time in co-evolution with wild tuberous *Solanum* species. When native cultivars were selected from wild species the association with some of the viruses continued and traffic in 'seed' potato tubers resulted in wide dissemination. After the Spanish conquest, some viruses were introduced to Europe and elsewhere in tubers, but others still seem to be restricted to South America. Those viruses and virus-like diseases which infect cultivated potato in other regions but not in South America, are mostly ones with wide natural host ranges. They presumably spread to the crop from other species after potatoes were introduced.

The viruses and virus-like diseases known to affect potatoes in the Andes, the abbreviations used hereafter for them and their known distribution are listed in Table 1. All are almost certainly indigenous to the region except spindle tuber viroid which may have been introduced recently from North America. With few exceptions, they depend mainly or entirely on potato for survival. Viruses which affect potato elsewhere in South America but have not yet been found in potato in the Andes include tomato spotted wilt, tobacco streak, sugar beet curly top, alfalfa mosaic and potato mosaico deformante. Little is known about the last of these but the others are all viruses prevalent in other plants. Other viruses or pathogens causing virus-like diseases which infect potato outside the South American sub-continent but have not yet been found there in the crop include tobacco rattle, tobacco necrosis, cucumber mosaic, tomato black ring, potato stunt, potato yellow dwarf, potato witches' broom and potato haywire. Apart from tobacco rattle, all these are unimportant and of limited occurrence in potato. The first four all have very wide host ranges, little is known about potato stunt and the last three are occasionally spread to potato from other plants by leafhoppers.

Breeders have been very active over the last 30 years in screening collections of wild potato species from different parts of the Americas for sources of resistance to the more economically important viruses (particularly PLRV, PVX, PVY, PVA, PVS and PVM) (Cockerham, 1970; Howard, 1974; Clark, 1963; H. Ross, 1966; R. W. Ross and Rowe, 1969; Bagnall, 1972). However, little attention has been given to the viruses occurring naturally in them. The few studies that have been made were done for quarantine purposes before collections were incorporated into national gene banks or passed to breeders. Kahn *et al.* (1963) found the following in wild species introduced from Mexico: PLRV in *S. brachycarpum*, PVY in *S. pinnatisectum*, PVA in *S. bulbocastanum*, and PVX in *S. brachycarpum*, *S. bulbocastanum*, *S. sambucinum* and *S. stoloniferum*. Other quarantine studies do not state the species in which particular viruses were found, but those detected in wild species collections include PLRV, PVX, PVY, PVA, PVS and 'unidentified viruses' (Cockerham *et al.*, 1963; McKee, 1964; Kahn and Monroe, 1970). The only other published work on viruses of wild potatoes is that on WPMV which affects *S. chancayense* in the lomas vegetation of the Peruvian coastal desert (Jones and Fribourg, 1979). Wild potato species obviously contain viruses already known in cultivated potatoes and others peculiar to themselves, some of which probably resemble WPMV in being adapted to survive with their hosts in specialized habitats.

Compared with the scanty information on viruses in wild species, much is known about the occurrence of the common viruses of cultivated potatoes in the Andes. The incidence of PLRV, PVX, PVY, PVA, PVS and PVM has been studied in collections of native cultivars from all parts of the region (Silberschmidt, 1961; Khan *et al.*, 1963, 1967; McKee, 1964; Monasterios, 1965; Brücher, 1969). Moreover, McKee (1964) found no connection between the ploidy of native cultivars and incidence of PVX, PVY and PVS. More recent information on the incidence of PLRV, PVX, PVY and PVS in both native and modern cultivars was provided by work at the International Potato Centre, Peru (Fribourg, 1975; Rodriguez and Jones, 1978; Moreira, 1978; Moreira *et al.*, 1980; Santillan, 1979). Table 1 lists key references to

Table 1 Viruses and virus-like diseases affecting potato in the Andean region of South America: occurrence and modes of survival and spread

Virus (abbreviation)	Proven occurrence	Transmission by: vectors*	Transmission by: contact	Transmission by: tubers	Transmission by: true seed	Proven alternative hosts	Key references
(a) *Viruses of worldwide distribution*							
Potato virus A (PVA)	Bolivia, Chile, Peru	Aphids (NP)	–	Complete	–	—	54†
Potato virus Y (PVY)	All Andean countries	Aphids (NP)	–	Complete	–	Tomato, tobacco, pepper and various solanaceous weeds	37; Hille Ris Lambers (1972)
Potato virus S (PVS)	All Andean countries	Aphids (NP) (some strains)	+	Complete	–	Pepino (*S. muricatum*)	60; Bode and Weidemann (1971)
Potato virus M (PVM)	Bolivia, Chile, Peru	Aphids (NP)	±	Complete	–	—	87; Bode and Weidemann (1971)
Potato virus X (PVX)	All Andean countries	None known	+	Complete	–	Tomato, pepper	4; Wright (1974)
Potato aucuba mosaic (PAMV)	Bolivia‡	Aphids (NP)	±	Complete	–	—	98
Potato leaf roll (PLRV)	All Andean countries	Aphids (P)	–	Almost complete	–	Tomato, *Datura stamonium* and *Physalis* spp.	36; Hille Ris Lambers (1972)
Potato mop-top (PMTV)	Bolivia, Peru	Fungus	–	Partial	–	—	138; Salazar and Jones (1975)

(b) *Viruses restricted to the Andes*							
Andean potato latent (APLV)	Bolivia, Colombia, Ecuador, Peru	Beetle	+	Partial	+	—	Fribourg *et al.* (1977a); Koenig *et al.* (1979)
Andean potato mottle (APMV)	Bolivia, Peru	Beetle?	+	Complete	–	—	203; Fribourg *et al.* (1977b)
Potato virus T (PVT)	Bolivia, Peru	None known	–	Almost complete	+	—	187; Salazar and Harrison (1978a)
Potato yellow vein (PYVV)	Colombia, Ecuador	Whitefly?	–	Partial	?	—	Diaz Moreno (1966); Smith (1957)
Tobacco ringspot: Andean potato calico strain (TRSV-Ca)	Peru	Nematode	–	Partial	?	Oca	Fribourg (1977); Salazar and Harrison (1978b)
Arracacha virus B, oca strain (AVB-O)	Peru	None known	–	?	?	Oca	Jones and Kenten (1981); R. A. C. Jones (unpublished)
Wild potato mosaic (WPMV)	Peru	Aphid (NP)	–	Complete	?	—	Jones and Fribourg (1979)
(c) *Virus-like diseases*							
Potato spindle tuber viroid (PSTV)	Peru‡	None known	+	Complete	+	Tomato	66; Fernow *et al.* (1970)
Aster yellows mycoplasm (AYM)	Peru	Leafhoppers	–	Little	–	Various crop plants and weeds	O'Brien and Rich (1976)

* NP = non-persistent and P = persistent.
† Numbers refer to CMI/AAB *Descriptions of Plant Viruses*.
‡ Detected only in breeding lines (PSTV), preliminary identification only (PAMV).

work on the incidence of PMTV, APLV, APMV, PVT, PYVV and TRSV-Ca in native and modern cultivars. Fribourg (1980) reviewed published work on the distribution of viruses in cultivated potatoes in Latin America, including the Andean region. In summary: PLRV, PVX, PVY and PVS are common throughout the Andes; APLV is common in much of the region; APMV is common in Peru and probably elsewhere; PYVV is common in Ecuador and southern Colombia; PVA and PVM are less common but probably widely distributed; PMTV, TRSV-Ca and AYM are common only in certain localities; PVT is relatively uncommon; AVB-O and PAMV have each only been found once; and PSTV has so far only been confirmed in potato lines (which include ones introduced from North America) belonging to the International Potato Centre in Peru.

At the centre of origin of a crop great diversity might be expected not only in the total number of viruses affecting it, but also within each virus. This aspect has been little investigated, but there is already clear evidence of novel strains. Thus, the Andean strain of PVS, distinguished by its systemic invasion of *Chenopodium quinoa* (Hinostroza, 1973), occurs in all countries from Chile (Isle of Chiloe) northward to Colombia (Santillan, 1979; Santillan *et al.*, 1980). Similarly, a strain of PVX (X_{HB}) distinguished by its failure to produce local lesions in *Gomphrena globosa* and by its resistance-breaking properties is common in native cultivars in Bolivia (Moreira *et al.*, 1980). Moreover, the tobacco veinal necrosis strain of PVY, which is common in the Andes, is thought to have spread to Europe from this region only in the last 30–40 years (Brücher, 1969; Todd, 1961).

Methods of virus spread and survival

The strategies that individual potato viruses have for survival and spread are varied. The important features are spread by vectors and contact, passage through tubers and true seed, and infection of alternative natural hosts (as summarized in Table 1). For the common potato viruses of world-wide distribution, most of this information comes from work in Europe and North America. Little ecological work has been done in the Andes with viruses of either wild or cultivated potato. There is a general lack of glasshouse and laboratory facilities suitable for virological investigations, and of adequately trained personnel. Also, ecological studies necessarily have a much lower priority than work on control measures in those research programmes, such as that of the International Potato Centre, that are located in the region and are funded to solve immediate practical problems. A major difficulty encountered in doing ecological work is that communications are hampered by poor roads and the high mountain passes up to 5000 m (17 000 ft) above sea level are frequently affected by landslides.

Spread by contact and by vectors

In quarantine studies on new collections, the overall infection with common potato viruses was significantly greater in native cultivars than in wild species, especially with PVX which is solely contact-transmitted (McKee, 1964; Kahn and Monroe, 1970). The explanation offered was that close planting and human disturbance in cultivation facilitate virus spread either by insect vectors or contact, whereas the scattered distribution and lack of human involvement with wild potato plants provide fewer opportunities for spread, especially by contact. McKee (1964) also found that PVY was commoner in samples collected in Peru and Bolivia at 2130–2740 m above sea level than in ones collected at 3650–4260 m. Similarly, the incidence of PLRV in Peru was greater in fields at 1900–3000 m than at 3200–3800 m (Rodriguez and Jones, 1978). In Peru and Bolivia, potatoes are mostly grown at 2000–4250 m. Frost is the limiting factor above this zone and other crops more suited to sub-tropical conditions are grown below it. *Myzus persicae* is common below 3000 m during the rainy (= potato growing) season but becomes less frequent in the cooler higher altitudes. The prevalence of PVY and PLRV therefore coincide with that of *M. persicae*, their principal vector. Santillan (1979) showed that the Andean strain of PVS, which is common throughout the region, was transmitted more readily by *M. persicae* than the strain(s) found elsewhere, which are generally considered to be almost entirely contact-transmitted.

PMTV is the only soil-borne virus known to infect potato in both the Andes and other parts of the world. It is transmitted by the fungus *Spongospora subterranea* which occurs in localities with cool temperatures and relatively high rainfall during the growing season. Such conditions also seem to determine the distribution and prevalence of PMTV in Peru (Salazar and Jones, 1975).

Specific vectors have been identified for only two of the six viruses that infect cultivated potato in the Andes but not elsewhere (Table 1). APLV is transmitted by the flea beetle *Epitrix* sp. (Jones and Fribourg, 1977) and TRSV-Ca by the nematode *Xiphinema americanum* (C. E. Fribourg and P. Jatala, unpublished). *Epitrix* sp. did not transmit APMV (Fribourg *et al.*, 1977b), but its vector is probably another leaf-feeding beetle because it is one of the comoviruses, which are typically beetle-transmitted. Spread of PYVV has been associated with whiteflies (Smith, 1957; Diaz Moreno, 1966). All that is known with PVT and AVB-O is that *M. persicae* does not transmit them. APLV and APMV are both readily transmitted by contact and this is probably their normal means of spread within potato fields.

M. persicae is the vector of WPMV which infects *Solanum chancayense* in its specialized habitat in the lomas of the Peruvian coastal desert (Jones and Fribourg, 1979). The lomas are hilly areas where low cloud during the cool 'winter' season provides sufficient moisture to support the growth of short-lived plants adapted to this habitat, including *S. chancayense* and several other wild potato species. *S. chancayense* plants grow in small groups among rocks and WPMV is spread between groups by winged *M. persicae*, the virus surviving between 'winters' in dormant tubers underground.

Transmission through tubers

Native cultivars are usually somewhat localized in distribution, whereas 'seed' tubers of modern cultivars are often transported considerable distances. Infection of tubers is therefore important in the survival of viruses between growing seasons, and also in long distance dissemination. Viruses that only pass to a proportion of the progeny tubers produced by an infected plant are gradually eliminated unless reinfection occurs (Cooper *et al.*, 1976). Consequently, those viruses that depend primarily on potato for survival and spread are usually transmitted through all, or almost all, the tubers from infected plants (Table 1). PMTV is exceptional in that it can survive for many years at infested sites in dormant resting spores of its fungus vector (Jones and Harrison, 1972), so that passage through all tubers is not so important.

Passage of TRSV-Ca, APLV and PYVV through progeny tubers is only partial and AYM passes through relatively few (Table 1). AYM is primarily a pathogen of other plants but is sometimes spread from them to potato by its leafhopper vector and TRSV-Ca is one of the nepoviruses, a group characterized by wide natural host ranges. Whether APLV and PYVV are primarily viruses of potato is unknown, but both presumably need additional means of surviving between growing seasons other than in tubers. The extent to which AVB-O passes through tubers is unknown but potato may be a host of only secondary importance because systemic movement in potato plants is erratic (Jones and Kenten, 1981).

Transmission through true seed

Passage of viruses through true seed is unimportant in a vegetatively propagated crop like potato, except in breeding new cultivars. By contrast, it could be significant in wild species dependent on true seed for dispersal. None of the common potato viruses of world-wide distribution is seed-borne in *S. tuberosum*, but APLV and PSTV are, as is PVT in *S. demissum* (Table 1). There is no information on whether TRSV-Ca, PYVV, WPMV and AVB-O are seed-borne in potato.

Alternative hosts

In the Andean region, there is a great range of both tuber-forming and non-tuber-forming *Solanum* spp. and also of other solanaceous species, including ones in genera such as *Datura*, *Physalis*, *Lycopersicon*, *Nicandra* and *Nicotiana*. Several of these are susceptible to many viruses and are standard virological indicator hosts. They include *Datura stramonium*, *Nicandra physaloides*, *Lycopersicon chilense*, *Physalis peruviana*, *Nicotiana debneyii*, *N. glutinosa* and *N. rustica*. Moreover, tomato, pepper, tobacco and pepino (*S. muricatum*) are common solanaceous crop plants at lower altitudes while common crops at higher altitudes include the virus indicators *Chenopodium quinoa* and *Amaranthus edulis*. The scope for other plants to act as alternative hosts and reservoirs of infection for potato viruses is therefore enormous (Table 1) and a thorough search is needed in weeds and other crops in potato growing areas. Among non-solanaceous crops, oca, ollucu and mashua are of particular interest as they are often interplanted with potatoes (p. 90) and oca is already known as a natural host of TRSV-Ca and AVB-O (Jones and Kenten, 1981). The importance of alternative hosts as reservoirs of particular viruses obviously depends to some extent on the host ranges of their vectors. *M. persicae*, *S. subterranea*, *Epitrix* sp. and *X. americanum* have relatively wide host ranges suggesting that the viruses they transmit can be spread readily between different species.

Symptom expression

In general, symptoms of a particular virus are expressed most clearly in infected plants when climatic and other conditions favour optimum multiplication of that virus. Optimum multiplication in turn reflects adaptation to environment. Bright yellow leaf symptoms are very distinct when potatoes infected with PMTV, APMV or TRSV-Ca are grown at high altitude. By contrast, PMTV causes symptomless infection and APMV mainly a mottle when infected plants are grown during the cool season in the irrigated desert plantings of the Peruvian coast (Salazar and Jones, 1975; Fribourg *et al.*, 1979; Fribourg, 1980). Similarly, TRSV-Ca usually causes no symptoms on the coast, producing yellowing in only one cultivar (Fribourg, 1977). Conditions at high altitude, which are cooler, especially at night, thus favour the production of yellow symptoms. They also enhance expression of symptoms (mosaic, rugosity and vein netting) in APLV-infected plants (Jones and Fribourg, 1978). These four viruses seem well adapted to the cool conditions typical of Andean potato growing areas. The only one of them known to occur outside South America, PMTV, is restricted to cool regions elsewhere.

Role of resistance genes

Resistance to the common potato viruses of world-wide distribution has been sought in a wide spectrum of wild potato species and in cultivated types (Cockerham, 1970; Howard, 1974; Ross and Rowe, 1969; Bagnall, 1972;

Clark, 1963; H. Ross, 1966). This resistance presumably arose in wild species during evolution to protect them against the viruses commonly encountered. The resistance now found in cultivated potatoes must have been inherited from wild ancestors, since it could hardly have arisen in the short evolutionary time they have existed.

Vanderplank (1968) classified all disease resistance in plants as horizontal or vertical. This dichotomy is well illustrated by the resistance in wild and cultivated potato species to common potato viruses. Horizontal resistance to infection by vectors or contact occurs against PLRV, PVY, PVA, PVX, PVS and PVM and is polygenically inherited. It can always be overcome when inoculation 'pressure' is sufficient but is not strain specific (Wiersema, 1972; Davidson, 1973; Chuquillanqui and Jones, 1980; H. Ross, 1978 and personal communication). The two main kinds of vertical resistance are hypersensitivity and immunity (= extreme resistance), both of which are controlled by single genes. Genes specifying a mainly localized hypersensitivity operate against PVX, PVY, PVA and PVS, and against different PVX and PVY strains. For PLRV and PVM only a slower acting systemic hypersensitivity has been found (Cockerham, 1970; Baerecke, 1967; Wiersema, 1972; H. Ross, 1978 and personal communication). The mainly localized hypersensitivity confers full resistance in fields of cultivated potato, as infected plant parts rapidly die and there is no spread of infection. By contrast, the slower acting systemic hypersensitivity permits some spread of infection in cultivation as whole plants are affected and death is only gradual. It is undoubtedly more effective in the wild where plants occur in scattered groups and not in dense uniform stands. Genes for immunity to PVX and to PVY + PVA are known (Cockerham, 1970; Munoz *et al.*, 1975). These genes have withstood challenge by all strains tested in Europe and North America, but strain X_{HB} from Bolivia readily overcomes the PVX immunity present in *S. acaule* and in *S. tuberosum* (Moreira *et al.*, 1980).

Tolerance is a distinct defence mechanism inherited independently from resistance (Wiersema, 1972). A potato plant is considered tolerant to infection when symptoms are slight or absent and tuber yield is hardly affected. Tolerance obviously helps native cultivars to survive (*see* below), but is probably less significant in wild species that depend mainly on true seed for dispersal rather than on tubers. Thus, because only few potato viruses are seed-borne, most wild potato plants derived from seed are initially healthy. In contrast, all plants derived from the tubers of mother plants infected with the common potato viruses are themselves infected, so tolerance is advantageous if they are to compete successfully.

In wild plant populations in which hosts and pathogens have co-evolved over long periods there is likely to be a balanced dynamic equilibrium between them that ensures survival of both. Genes for vertical resistance occur dispersed at random in the population and are supplemented by ones for horizontal resistance present to some degree in all plants. These resistance genes and the scattered distribution of wild plants combine to restrict the pathogen, thus preventing devastating outbreaks of disease. Survival of the pathogen is ensured by its diversity and ability to overcome host vertical resistance. Although these general principles have been developed mainly from work with fungal pathogens (Robinson, 1976; Browning, 1974, *1981*; Dinoor, *1981*), they can also be applied to viruses. Thus the spatial separation between individuals or groups of plants typical of wild potatoes combined with genes for hypersensitivity and/or immunity dispersed at random in the population and ones for horizontal resistance to infection in all plants would prevent devastating virus disease outbreaks from developing. Survival of the virus would be ensured by the presence of a spectrum of strains sufficient to overcome all vertical resistance genes present, or by being able to mutate readily to counteract any new ones that are encountered.

The cultivation of modern potato cultivars in the Andes using advanced agricultural practices presents a very different picture in which the balance between host and virus has been disrupted. Modern cultivars are grown in dense uniform stands often in large fields and are given various agronomic treatments to ensure maximum productivity. These cultivars have been bred almost entirely for yield with little or no regard to virus resistance. Although some hypersensitivity genes persist amongst them, e.g., to PVX (Fribourg, 1975; Moreira, 1978), there has been a serious loss of resistance, particularly horizontal resistance, as with PLRV (Jones, 1978). The survival of these modern cultivars now depends entirely on man's control measures as is amply demonstrated by their rapid degeneration and virtual disappearance in Peru once they cease to be multiplied within the official 'seed' certification programme.

Subsistence cultivation of native cultivars in the more remote parts of the Andes is of particular interest as it represents a primitive agricultural system. A balance between host and viruses is retained and man's influence is mainly restricted to storing and planting 'seed' tubers, weeding and use of dung as fertilizer. The plants are grown in dense stands but these are rarely uniform as potato cultivar or species mixtures are usual. Also, the potatoes are often interplanted with other crops, the fields used are very small and crop rotations are short or absent. Moreover, although the common potato viruses (especially PVX, PVY and PVS) are almost universally present in fields of native cultivars, no precautions are taken to select healthy 'seed' tubers. Natural selection for virus resistant and tolerant types would obviously be stringent in these circumstances and is perhaps greater than in the wild because of the close plant spacings used. Genes for vertical resistance would therefore be expected to occur frequently, with strong horizontal resistance and tolerance in most plants. A wide spectrum of virus strains, however, would help to ensure survival of any particular virus.

Although relatively little information is available on the occurrence of vertical resistance genes in native cultivars of other species, at least eight such genes have been located in *S. andigena* (Table 2). Little is known of the incidence of most of these, but immunity to PVX occurs frequently (Cockerham, 1970). Tolerance to infection seems widespread among native cultivars but with horizontal resistance it is known only that the expected high levels of infection resistance to PLRV exist in *S. andigena* (Jones, 1978; Chuquillanqui and Jones, 1980) and in *S. phureja* (Clark, 1963). Evidence

Table 2 Resistance genes determining strong hypersensitivity and immunity in *Solanum tuberosum* sub. sp. *andigena*

Virus	*Gene*	*Effective against*	*Selected references*
	Strong hypersensitivity		
PVA	Na*	All strains	Cockerham (1970)
PVS	Ns	All strains	Baerecke (1967)
PVX	Nb	Strain groups 1 and 2	Cockerham (1955, 1970); Fribourg (1975)
	Nx	Strain groups 1 and 3	Cockerham (1955, 1970); Fribourg (1975)
	Rx(acl)	Strain groups 1–3†	Cockerham (1970)
PVY	Nc*	Strain C only	Cockerham (1970)
	Immunity		
PVX	Rx	All strains except X_{HB}	Cockerham (1970); Moreira *et al.* (1980)
PVY + PVA	Ry	All strains	Cockerham (1970); Munoz *et al.* (1975)

* Cockerham (1970) quotes Na and Nc as probably occurring in *andigena*; both occur in *tuberosum* which is derived from *andigena*.

† Rx(acl) in *andigena* has not yet been tested with either group 4 or X_{HB} strains.

that virus strains occur in native *S. andigena* that can counteract all the known vertical resistance genes present in it comes from work with PVX. Thus PVX strain groups 2, 3 and 4, which overcome Nx, Nb and Nx + Nb, respectively (Cockerham, 1955, 1970), have all been found (McKee, 1964; Cockerham and Davidson, 1963; Fribourg, 1975; Moreira *et al.*, 1980) and strain X_{HB} which overcomes Nx, Nb and Rx was common in Bolivian *S. andigena*, occurring in 7% of the clones in a collection (Moreira *et al.*, 1980).

Evidence of gene-for-gene relationships in potato is already available for *Phytophthora infestans* (Robinson, 1976) and for the potato cyst nematodes *Globodera pallida* and *G. rostochiensis* (F. G. W. Jones, 1979). There is little doubt that similar relationships also account for the ecological balance between vertical resistance genes and virus strains both in wild potato populations and in native cultivars. Table 3 shows how PVX strains would interact with host vertical resistance genes on the gene-for-gene model. For example, at opposite ends of the range, group 1 strains would possess dominant complementary genes which are incompatible with host vertical resistance genes Nx, Nb and Rx, whereas strain X_{HB} would have compatible recessive alleles of all three viral genes. Similarly, PVA and PVY strain C would possess dominant complementary genes which are incompatible with host genes Na and Nc respectively, whereas PVY ordinary strain would have compatible recessive alleles (Davidson and Butzonitch, 1978; Cockerham, 1970).

Table 3 Interaction of strains of potato virus X with host vertical resistance genes (A), and the presumed basis of resistance (B)

		Host genes			
(A)	*Virus strain group*	*None*	*Nx*	*Nb*	*Rx*
1	*+	−	−	−	
	2	+	+	−	−
	3	+	−	+	−
	4	+	+	+	−
	†5	+	−	−	+
	†6	+	+	−	+
	†7	+	−	+	+
	8(=X_{HB})	+	+	+	+

(B)	*Major resistance gene in host*	*Complementary viral gene*	*Effect*
	h	v	Compatible, susceptible
	h	V	Compatible, susceptible
	H	v	Compatible, susceptible
	H	V	**Incompatible, resistant

* +, susceptible; −, resistant
† hypothetical strains not yet found
** Products from alleles H and V interact limiting or preventing virus invasion.

That PVX strains can readily mutate to overcome host vertical resistance genes was shown by grafting potato scions infected with a group 2 strain onto plants of potato cultivars carrying either Nb or Nx + Nb. Typical top-necrotic responses developed. When progeny plants grown from tubers taken from the plants with Nb were tested, they contained not group 2 but PVX strain group 3. Similarly, a proportion of the progeny plants from plants carrying both genes contained strain group 4 (Rozendaal, 1966). Evidently the virus had mutated in both instances to overcome Nb, but ability to overcome Nx had been lost in the absence of this gene. Thus, on the gene-for-gene hypothesis, presence of Nb in the grafted plants had resulted in selection of a compatible allele in its complementary viral gene. Moreover, the compatible allele in the viral gene complementary to Nx had been retained in the presence of this gene, whereas reversion to the dominant incompatible viral allele had occurred in its absence. Grafting is a much more potent inoculation procedure than inoculation by contact or vectors. Nevertheless, loss of a compatible allele in the absence of selection for it exerted by the corresponding host gene may be particularly significant. Thus mutations in the virus, which result in production of compatible alleles, may be associated with deleterious changes in fitness which decrease competitive ability, so that in the absence of the complementary host gene such strains become superseded by those with dominant incompatible alleles. That resistance-breaking strains of certain other viruses cannot compete with normal strains in the absence of host vertical resistance genes has already been suggested (Pelham, 1972; Hanada and Harrison, 1977; Russell, 1978). A possible explanation is that because viral nucleic acid has space to code for only a few genes, each individual gene has several functions. Consequently, any mutation in a complementary viral gene to overcome a host gene would inevitably affect the efficiency of the other functions it controls.

The selection pressures on PVX strains among native cultivars in Bolivia clearly favour widespread occurrence of the resistance-breaking strain X_{HB} and this strain must have been disseminated in the past with potato introductions and germplasm collections. Many *S. tuberosum* cultivars carry Nb, some Nx and a few carry both, but Rx is absent (Cockerham, 1970). Failure of X_{HB} to become established outside the Andes could be due to its inability to compete with other strains in plant populations lacking Rx. Similarly, that group 4 strains seldom occur in *S. tuberosum* could be because they are unable to compete when few cultivars are of genotype NxNb. That group 3 is commoner than group 2 in *S. tuberosum* (Cockerham, 1955) reflects the more frequent presence of Nb (Cockerham, 1970).

Factors other than resistance and tolerance that are likely to be important in maintaining a natural balance between host and virus are: (1) virus strains of low virulence that cause little damage; (2) resistance to vectors; and (3) changes in distribution of vectors. There can be little doubt that (1) and (2) both play important roles but they are beyond the scope of this article. With regard to (3), however, the suggestion that *M. persicae* may have first encountered potato only about 400 years ago (Eastop, 1977) would, if correct, be particularly significant. Although several of the non-persistently transmitted viruses of potato are readily transmitted by different aphid species (Hille Ris Lambers, 1972), *M. persicae* is the most important of these, and for PLRV, which is persistently transmitted, only one other vector species (*Macrosiphum euphorbiae*) is known. Introduction of such an efficient vector to South America in such recent evolutionary time would inevitably have disrupted the natural equilibrium between potato and its aphid-transmitted viruses in favour of the latter. However, there is no clear evidence that this has occurred. Indeed, it seems far more likely that *M. persicae* reached South America much earlier, that it was present from the start of the co-evolution of aphid-transmitted viruses with wild potato species and that these viruses became adapted to it as their principal vector over a long period of evolutionary time.

Epilogue

At their centre of origin in the Andean region of South America, cultivated potato species are hosts of several viruses which co-evolved with wild potatoes over geological time and which depend on potato for their survival. At least some strains of some of these viruses have been introduced with the potato to other parts of the world. In these other regions, the potato encountered other viruses or pathogens causing virus-like diseases that spread to it from other plants. However, none of these seem well-adapted to the crop which remains a host of secondary or minor importance for them.

Further work is needed to determine the potential for spread of the known viruses and virus strains that have not yet become established outside the Andes so as to avoid problems of the type that arose when the tobacco veinal necrosis strain of PVY escaped from potato material introduced into Europe 30–40 years ago. Some of these viruses must have been introduced in the past and failed to survive but they may eventually become established. There is also a threat from other Andean potato viruses which may never have been introduced and this may include some as yet undiscovered ones, as there is little doubt that further potentially harmful types await discovery in the isolation of remote Andean valleys. Possible reasons for the failure of some viruses to become established elsewhere include: (1) lack of appropriate vectors; (2) lack of the solanaceous and other alternative weed and crop hosts abundant in the Andes; and (3) failure of some viruses to spread to all the tubers produced by infected plants. Temperate regions would seem at greatest risk from types transmitted by aphids or by contact. Clearly, the risks of introduction and escape are increasing due to the growth of tourism and commerce in Andean countries and to the increased exchange of plant material. Those distributing material from potato gene banks have a particularly heavy responsibility to ensure that harmful viruses and virus strains are not also unwittingly distributed. PYVV, APLV and APMV are examples of such viruses that have already been found in gene banks outside the Andes.

A thorough investigation of the ecology of the common potato viruses in wild species and native cultivars would undoubtedly provide clues on ways of improving potato production in advanced agricultural conditions; for example, on how to breed cultivars which combine high quality and yield with sufficient multiple virus resistance to obviate much of the expense and labour required to control viruses by 'seed' certification schemes. The breakdown of vertical resistance following the emergence of new strains emphasizes that breeders should beware of neglecting horizontal resistance. Native cultivars, especially *S. andigena*, are much easier to use in breeding than wild species. They should be further examined to identify additional sources of resistance and breeders should be encouraged to exploit them far more than they have in the past in breeding for virus resistance.

References

Baerecke, M. L. (1967). Uberempfindlichkeit gegen das S-Virus der Kartoffel in einem bolivianischen Andigena-Klon. *Züchter* **37,** 281–286.

Bagnall, R. H. (1972). Resistance to potato viruses M, S, X and the spindle tuber virus in tuber-bearing *Solanum* species. *American Potato Journal* **49,** 342–348.

Bode, O. & Weidenmann, H. L. (1971). Untersuchungen zur Blattlausübertragbarkeit von Kartoffel-M und S-Virus. *Potato Research* **14,** 119–129.

Browning, J. A. (1974). Relevance of knowledge about natural ecosystems to development of pest management programs for agro-ecosystems. *Proceedings of the American Phytopathological Society* **1,** 191–199.

Browning, J. A. (*1981*). The agroecosystem-natural ecosystem dichotomy and its impact on phytopathological concepts. In *Pests, Pathogens and Vegetation*, pp. 159–172. J. M. Thresh. London: Pitman.

Brücher, H. (1969). Observations on origin and expansion of Y^N-virus in South America. *Angewandte Botanik* **43,** 241–249.

Bushnell, G. H. S. (1963). *Peru*. 2nd Edn. London: Thames and Hudson.

Chuquillanqui, C. & Jones, R. A. C. (1980). A rapid technique for assessing the resistance of families of potato seedlings to potato leaf roll virus. *Potato Research* **23,** 121–128.

Clark, R. L. (1963). Leaf roll resistance in some tuberous solani under controlled aphid inoculations. *American Potato Journal* **40,** 115–120.

Cockerham, G. (1955). Strains of potato virus X. In *Proceedings of the Second Conference on Potato Virus Diseases, Lisse-Wageningen, 1954*, pp. 89–92.

Cockerham, G. (1970). Genetical studies on resistance to potato viruses X and Y. *Heredity* **25,** 309–348.

Cockerham, G. & Davidson, T. M. W. (1963). Note on an unusual strain of potato virus X. *Scottish Plant Breeding Station Record for 1963*, pp. 26–29.

Cockerham, G., Davidson, T. M. W. & Macarthur, A. W. (1963). Report on the virus content and the reactions to viruses X, S and Y of tuber-bearing Solanums collected by the Birmingham University Expedition to Mexico and Central America, 1958. *Scottish Plant Breeding Station Record for 1963*, pp. 30–34.

Cooper, J. I., Jones, R. A. C. & Harrison, B. D. (1976). Field and glasshouse experiments on the control of potato mop-top virus. *Annals of Applied Biology* **83,** 215–230.

Correll, D. S. (1962). *The Potato and its Wild Relatives*. Renner: Texas Research Foundation.

Davidson, T. M. W. (1973). Assessing resistance to leaf roll in potato seedlings. *Potato Research* **16,** 99–108.

Davidson, T. M. W. & Butzonitch, I. P. (1978). The grouping of some strains of virus Y in relation to resistant cultivars. *Abstracts of papers, 7th Triennial Conference of the European Association of Potato Research, Warsaw*, p. 161.

Diaz Moreno, J. (1966). Incidencia del virus del amarillamiento de venas en papa en el Ecuador y su transmissión a través de los tubérculos. *Turrialba* **16,** 15–24.
Dinoor, A. (*1981*). Epidemics caused by fungal pathogens in wild crop plants. In *Pests, Pathogens and Vegetation*, pp. 143–158. J. M. Thresh. London: Pitman.
Eastop, V. F. (1977). Worldwide importance of aphids as virus vectors. In *Aphids as Virus Vectors*, pp. 3–47. K. F. Harris & K. Maramorosch. London: Academic Press.
Fernow, K. H., Peterson, L. C. & Plaisted, R. L. (1970). Spindle tuber virus in seeds and pollen of infected potato plants. *American Potato Journal* **47,** 75–80.
Fribourg, C. E. (1975). Studies on potato virus X strains isolated from Peruvian potatoes. *Potato Research* **18,** 216–226.
Fribourg, C. E. (1977). Andean potato calico strain of tobacco ringspot virus. *Phytopathology* **67,** 174–178.
Fribourg, C. E. (1980). Historia y distribución de virus de papa en America Latina. *Fitopatologia* **15,** 13–24.
Fribourg, C. E., Jones, R. A. C. & Koenig, R. (1977a). Host plant reactions, physical properties and serology of three isolates of Andean potato latent virus from Peru. *Annals of Applied Biology* **86,** 373–380.
Fribourg, C. E., Jones, R. A. C. & Koenig, R. (1977b). Andean potato mottle, a new member of the cowpea mosaic virus group. *Phytopathology* **67,** 969–974.
Fribourg, C. E., Jones, R. A. C. & Koenig, R. (1979). Andean potato mottle virus. CMI/AAB *Descriptions of Plant Viruses* No. 203.
Hanada, K. & Harrison, B. D. (1977). Effects of virus genotype and temperature on seed transmission of nepoviruses. *Annals of Applied Biology* **91,** 101–106.
Hawkes, J. G. (1947). Some observations on South American potatoes. *Annals of Applied Biology* **34,** 622–631.
Hawkes, J. G. (1963). A revision of the tuber-bearing Solanums. *Scottish Plant Breeding Station Record for 1963*, pp. 76–181.
Hawkes, J. G. (1973). Potato genetic erosion survey—preliminary report. In *Planning Conference on Germplasm Exploration and Taxonomy of Potatoes*, pp. 36–48. Lima: International Potato Centre.
Hille Ris Lambers, D. (1972). Aphids: their life cycles and their role as virus vectors. In *Viruses of Potatoes and Seed-potato Production*, pp. 36–56. J. A. de Bokx. Wageningen: Centre for Agricultural Publishing and Documentation.
Hinostroza, A. M. (1973). Some properties of potato virus S isolated from Peruvian potatoes. *Potato Research* **16,** 244–250.
Howard, H. W. (1974). Position paper on the utilisation of genetic resources. In *Planning Conference on the Utilisation of Genetic Resources of the Potato*, pp. 89–132. Lima: International Potato Centre.
Jones, F. G. W. (1979). The problems of race-specificity in plant resistance breeding. In *Proceedings of the British Crop Protection Conference — Pests and Diseases, Brighton* **3,** pp. 741–752.
Jones, R. A. C. (1978). Progress in leaf roll resistance work at the International Potato Centre. In *Planning Conference on Developments in the Control of Potato Viruses,* pp. 15–26. Lima: International Potato Centre.
Jones, R. A. C. & Fribourg, C. E. (1977). Beetle, contact and potato true seed transmission of Andean potato latent virus. *Annals of Applied Biology* **86,** 123–128.
Jones, R. A. C. & Fribourg, C. E. (1978). Symptoms induced by Andean potato latent virus in wild and cultivated potatoes. *Potato Research* **21,** 121–127.

Jones, R. A. C. & Fribourg, C. E. (1979). Host plant reactions, some properties and serology of wild potato mosaic virus. *Phytopathology* **69,** 446–449.

Jones, R. A. C. & Harrison, B. D. (1972). Ecological studies on potato mop-top virus in Scotland. *Annals of Applied Biology* **71,** 47–57.

Jones, R. A. C. & Kenten, R. H. (1981). A strain of arracacha virus B infecting oca (*Oxalis tuberosa*; oxalidaceae) in the Peruvian Andes. *Phytopathologische Zeitschrift* **100,** 88–95.

Kahn, R. P., Hewitt, W. B. *et al.* (1963). Detection of viruses in foreign plant introductions under quarantine in the United States. *Plant Disease Reporter* **47,** 261–265.

Kahn, R. P. & Monroe, R. L. (1970). Virus infection in plant introductions collected as vegetative propagations: 1. wild vs. cultivated *Solanum* species. *FAO Plant Protection Bulletin* **18,** 97–101.

Kahn, R. P., Monroe, R. L. *et al.* (1967). Incidence of virus detection in vegetatively propagated plant introductions under quarantine in the United States, 1957–1967. *Plant Disease Reporter* **51,** 715–719.

Kay, D. E. (1973). *TPI Crop and Product Digest No. 2. Root Crops.* London: The Tropical Products Institute.

Koenig, R., Fribourg, C. E. & Jones, R. A. C. (1979). Symptomatological, serological and electrophoretic diversity of isolates of Andean potato latent virus from different regions of the Andes. *Phytopathology* **69,** 748–752.

McKee, R. K. (1964). Virus infection in South American potatoes. *European Potato Journal* **7,** 145–151.

Monasterios, T. (1966). Presence of viruses in Bolivian potatoes. *Turrialba* **16,** 257–260.

Moreira, A. (1978). Nuevo strain de virus X (X_{HB}) de Bolivia que rompe resistencia en papa y no causa lesiones locales en *Gomphrena globosa. MSc Thesis,* Universidad Nacional Agraria, Lima.

Moreira, A., Jones, R. A. C. & Fribourg, C. E. (1980). Properties of a resistance-breaking strain of potato virus X. *Annals of Applied Biology* **95,** 93–103.

Munoz, F. J., Plaisted, R. L. & Thurston, H. D. (1975). Resistance to potato virus Y in *Solanum tuberosum* spp. *andigena, American Potato Journal* **52,** 107–115.

O'Brien, M. J. & Rich, A. E. (1976). Potato diseases. *USDA/ARS Agricultural Handbook* No. 474.

Pelham, J. (1972). Strain-genotype interaction of tobacco mosaic virus in tomato. *Annals of Applied Biology* **71,** 219–228.

Robinson, R. A. (1976). *Plant Pathosystems.* Berlin: Springer-Verlag.

Rodriguez, A. & Jones, R. A. C. (1978). Enanismo amarillo disease of *Solanum andigena* potatoes is caused by potato leaf roll virus. *Phytopathology* **68,** 39–43.

Ross, H. (1966). The use of wild *Solanum* species in German potato breeding of the past and today. *American Potato Journal* **43,** 63–80.

Ross, H. (1978). Methods for breeding virus resistant potatoes. In *Planning Conference on Developments in the Control of Potato Viruses*, pp. 93–114. Lima: International Potato Centre.

Ross, R. W. & Rowe, P. R. (1969). Inventory of tuber-bearing *Solanum* species. *Bulletin of the College of Agriculture and Life Sciences, University of Wisconsin* No. 533.

Rozendaal, A. (1966). Potato grafting as a source of virus strains dangerous for the crop. In *Proceedings of the 3rd Triennial Conference of the European Association for Potato Research, Zurich*, pp. 231–233.

Russell, G. E. (1978). *Plant Breeding for Pest and Disease Resistance*. London: Butterworth.

Salaman, R. N. (1949). *The History and Social Influence of the Potato*. Cambridge: Cambridge University Press.

Salazar, L. F. & Harrison, B. D. (1978a). Host range, purification and properties of potato virus T. *Annals of Applied Biology* **89,** 223–235.

Salazar, L. F. & Harrison, B. D. (1978b). Host range and properties of potato black ringspot virus. *Annals of Applied Biology* **90,** 375–386.

Salazar, L. F. & Jones, R. A. C. (1975). Some studies on the distribution and incidence of potato mop-top virus in Peru. *American Potato Journal* **52,** 143–150.

Santillan, F. W. (1979). Estudio comparativo de once aislamientos de virus S de la region Andina. *MSc Thesis*, Universidad Nacional Agraria, Lima.

Santillan, F. W., Fribourg, C. E. & Jones, R. A. C. (1980). Estudio comparativo de once aislamientos de virus S de la papa (PVS) de la region Andina. *Fitopatologia* **15,** 42–43.

Silberschmidt, K. M. (1961). The spontaneous occurrence of strains of potato virus X and Y in South America. *Phytopathologische Zeitschrift* **42,** 175–192.

Smith, K. M. (1957). *A Textbook of Plant Virus Diseases*. 2nd Edn. London: Churchill.

Todd, J. M. (1961). Tobacco veinal necrosis on potato in Scotland: control of the outbreak and some characters of the virus. In *Proceedings of the 4th Conference on Potato Virus Diseases, Braunschweig, 1960*, pp. 82–92.

Vanderplank, J. E. (1968). *Disease Resistance in Plants*. London: Academic Press.

Wiersema, H. T. (1972). Breeding for resistance. In *Viruses of Potatoes and Seed-potato Production*, pp. 174–187. J. A. de Bokx. Wageningen: Centre for Agricultural Publishing and Documentation.

Wright, N. S. (1974). Retention of infectious potato virus X on common surfaces. *American Potato Journal* **51,** 251–253.

Wild plants in the ecology of hopper-borne viruses of grasses and cereals

Maurizio Conti
Istituto di Fitovirologia applicata Consiglio Nazionale delle Ricerche, Torino 10135 Italy

Plant and leafhopper vectors

Hoppers are well known plant-sucking insects (taxonomic group: Hemiptera Homoptera Auchenorhyncha) responsible for serious yield losses caused either directly, by infestation of crops which become dried and brown ('hopper-burn'), or indirectly, as vectors of viruses, mycoplasmas and rickettsias. Vector species occur in five different families: Cercopidae, Cicadellidae, Membracidae (sub-family Cicadoidea), Cixiidae and Delphacidae (sub-family Fulgoroidea). The forty or so species which transmit viruses to Gramineae are either cicadellids (=leafhoppers) or delphacids (=planthoppers).

The biology and ecology of hoppers have been reviewed by De Long (1971). Basically, the activity and efficiency of hoppers as virus vectors is influenced by three biological characteristics: overwintering, migration, and feeding behaviour. In cold climates, the means by which hoppers overwinter can be important in sustaining the viruses they transmit from one growing season to another. Leafhoppers overwinter as eggs or as adults, whereas the majority of planthoppers do so as immature forms in diapause. Both adults and immature forms can account for spring transmission of circulative viruses acquired the previous autumn and may thus be responsible for early and heavy infection of crops. By contrast, species which overwinter as eggs do not usually acquire virus until spring. They are therefore less important vectors of circulative viruses, which require a long latent period in the insects. Such species may be initial, though occasional, sources of infection only for viruses which are transmitted transovarially.

The introduction and greatest spread of viruses in crops occur during hopper migrations, when the insects move from their overwintering or breeding sites and infest the surrounding crops, in which they feed actively. The cause of migrations is not known and may differ between species. Several causal factors such as hunger, overcrowding, desiccation of the host-plant and seed maturation have been suggested. Migration mainly involves macropterous forms of both leafhoppers and planthoppers, which are active fliers and can be transported long distances by air currents. Migrant hoppers have been caught at altitudes of 60–4200 m, and can travel hundreds of kilometres (Kuno, 1973; Raatikainen and Vasarainen, 1973; Thresh, *1981*). This emphasizes the magnitude of the phytopathological problems due to hopper-borne viruses as well as the difficulties of forecasting and control.

Feeding behaviour is a specific characteristic of hoppers that determines their ability to transmit viruses (Sôgawa, 1973). It suffices here to stress that vector species feed predominantly in the phloem and occasionally in the xylem, although there is considerable variation between species.

Hopper-borne viruses

The hopper-borne viruses of Gramineae fall into six categories (Table 1) of which the first four are officially recognized taxonomic groups (Matthews, 1979).

Geminiviruses

These have paired isometric particles 18–20 nm in diameter that contain single-stranded circular DNA. Only two members are known to infect Gramineae but wheat dwarf virus may be a third, and the first to be found outside warm climates (Lindsten, 1979a).

Phytoreoviruses

These have icosahedral particles about 70 nm in diameter possessing a double protein coat and containing linear double-stranded RNA, which consists of twelve pieces. The only member that infects Gramineae is rice dwarf virus, which is transmitted by cicadellids.

Fijiviruses

Their particles somewhat resemble phytoreoviruses but the ds-RNA they contain consists of only ten pieces. The eight members presently known to infect Gramineae are all transmitted by delphacid hoppers and not by cicadellids.

Plant rhabdoviruses

These have bacilliform particles 45–60 nm wide and up to 350 nm long, with a single-stranded RNA genome. The geographic distribution is world-wide and maize mosaic and rice transitory yellowing have recently attracted considerable attention.

Small isometric viruses

The viruses in this group of convenience have in general been poorly characterized. Oat blue dwarf is probably the best-known and seems to be unique in infecting both monocotyledonous and dicotyledonous plants. Rice tungro virus may be misplaced within the group, because bacilliform particles have also been found associated with the disease (*cf.* Saito, 1977).

Table 1 Some representative hopper-borne viruses of Gramineae and their vectors

Virus group		*Vectors*
Geminivirus		
Maize streak (133)*	Cic.†	*Cicadulina* spp.
Phytoreovirus		
Rice dwarf (102)	Cic.	*Nephotettix* spp.
		Recilia dorsalis
Fijiviruses		
Cereal tillering disease	Del.	*Dicranotropis hamata*
		Laodelphax striatellus
Sugarcane Fiji (119)	Del.	*Perkinsiella* spp.
Maize rough dwarf (72)	Del.	*Delphacodes propinqua*
		Javesella pellucida
		Laodelphax striatellus
		Sogatella vibix
Oat sterile dwarf (217)	Del.	*Dicranotropis hamata*
		Javesella discolor
		J. dubia
		J. obscurella
		J. pellucida
Rice black-streaked dwarf (135)	Del.	*Laodelphax striatellus*
		Ribautodelphax albifascia
		Unkanodes sapporonus
Plant rhabdovirus		
Barley yellow striate mosaic	Del.	*Javesella pellucida*
		Laodelphax striatellus
Small isometric virus		
Rice tungro (67)	Cic.	*Nephotettix* spp.
		Recilia dorsalis
Others		
Rice hoja blanca	Del.	*Sogatodes oryzicola*
		S. cubanus
Rice stripe	Del.	*Laodelphax striatellus*
		Ribautodelphaux albifascia
		Unkanodes sapporonus

* Numbers refer to CMI/AAB *Descriptions of Plant Viruses* which provide further details of the viruses listed.

† Cic. = Cicadellidae (leafhoppers); Del. = Delphacidae (planthoppers).

Other virus and virus-like agents

The aetiology of rice hoja blanca, rice grassy stunt, and several other important diseases of Gramineae is still uncertain. Some appear to be due to viruses as yet inadequately characterized, whereas others may be due to mycoplasmas or other agents.

Transmission

With only three exceptions (maize chlorotic dwarf, rice tungro, and rice waika viruses), all the hopper-borne viruses of Gramineae are transmitted circulatively and many multiply in the vector and are therefore referred to as propagative. Hoppers infected with virus can transmit only after a latent period of some days (or hours, as in the case of maize streak virus), but, once infective, they retain infectivity through the moult and for several days. By contrast the non-circulative type of transmission is characterized by the absence of a detectable latent period and infectivity is lost at moulting (Harris, 1979).

The virus–vector relationships of both types of transmission are highly specific. Generally, each virus has only one vector or a few that are closely related taxonomically. A generally recognized concept is that the viruses are transmitted by either planthoppers or leafhoppers, but not both (Francki, 1973; Slykhuis, 1977). However, a rhabdovirus, eleusine mosaic virus, is exceptional in that it is transmitted by hoppers of each taxonomic group (Maramorosch *et al.*, 1977). Some hopper species transmit several different viruses (*see* Table 1).

Some propagative viruses are transmitted transovarially (*cf.* Ling, 1972; Milne and Lovisolo, 1977; Brčák, 1979; Harris, 1979). The phenomenon is particularly important epidemiologically in maintaining virus away from host plants and as a means of disseminating viruses over very long distances, in the eggs present in infested plant material.

The ecology of hopper-borne viruses of Gramineae

Many important diseases of maize, rice and other widely grown cereal crops are caused by hopper-borne viruses. Many such viruses may originally have been pathogens of wild plants, becoming economically important when crops were introduced. Evidence for this may be the fact that the symptoms tend to be severe in cultivated plants and generally mild in wild ones in which very aggressive strains can be self-eliminating (Duffus, 1971). Another fascinating theory concerns the propagative viruses that may be 'bridging viruses' between those of plants and animals. It has been suggested that they might be insect viruses which have become adapted to multiply in plant cells in the course of their evolution. An interesting case is that of maize wallaby ear disease (MWED) in Australia: the reo-like virus particles, which appear to be involved, have been detected only in disease-inciting leafhoppers (*Cicadulina bimaculata*) and not in diseased maize plants. The most plausible explanation is that the virus induces the production of toxins in the insect saliva and that these incite MWED symptoms when introduced into plants (Boccardo *et al.*, 1980). It will be interesting to see whether the MWED-associated particles eventually become adapted to multiply in maize.

Whatever the origin of hopper-borne viruses, many wild grasses continue to play an essential role in the epidemiology of several viruses of Gramineae in different parts of the world. Depending on the virus concerned, grasses may be important from one or more of the following epidemiological aspects:

(1) as breeding or feeding hosts of the hopper vectors;
(2) as perpetuating hosts of virus and/or vector between growing seasons;
(3) in selecting new virus strains or specific vector lines.

No seed transmission of hopper-borne viruses has so far been reported in graminaceous or other plants, but perennial grasses may lead to virus dissemination over great distances when moved between countries as vegetative material. Indeed, several viruses have been discovered for the first time in a country near agricultural research stations, e.g. oat sterile dwarf in the UK.

The economic importance of a given hopper-borne virus depends largely on its natural hosts. The number and density of host-plant species, their location near or within susceptible crops, whether such species are perennial and/or hosts of the hopper vectors, etc., are important factors determining virus incidence in crops. Some of these aspects are illustrated in the following examples, with emphasis on personal experience with maize rough dwarf and other viruses of Gramineae in Europe.

Maize rough dwarf virus (MRDV)

This propagative virus infects maize in central and southern Europe and in Israel. In nothern Italy, its natural vector, *Laodelphax striatellus*, has two or three generations per year and overwinters as a nymph in diapause. It does not breed on maize and feeds only occasionally on this species. Maize is grown from April to October and is attractive for the insect only in the early stages of growth, when the overwintered generation of *L. striatellus* migrates. Almost all the infection of maize occurs at this time because plants later become more resistant to MRDV, and high summer temperatures make the insect a less efficient vector.

The host-range of MRDV includes many cereals and grasses including *Digitaria sanguinalis* and *Echinochloa crus-galli*, which seem to play a crucial role in northern Italy. During winter, the diapausing hoppers settle in the crowns of perennial weeds and do not move until spring, when they mature and move to crops. The most common winter hosts of the insect are *Agropyron repens* and *Cynodon dactylon*, but the former is not susceptible to MRDV and the latter has been found occasionally infected in Israel and not in Italy. Other overwintering plant hosts of MRDV have only been detected sporadically, and seem to be poor virus sources for the vector (Conti, 1976; Conti and Milne, 1977). Consequently, MRDV mainly overwinters in planthoppers which acquire virus from the summer weed hosts *D. sanguinalis* and *E. crus-galli*. These are the only available virus sources for the last generation of vectors because infected maize is by then dead, or at least no longer attractive to the hoppers, and other cereal hosts of MRDV have all been harvested. *D. sanguinalis* and *E. crus-galli*, on the other hand, are growing in the maize fields from early summer to late autumn, and are good hosts of both the virus and the vector. Infected plants become particularly attractive to the hopper because of their bushy form and altered chemical composition, as shown in Israel for *C. dactylon* (*cf.* Harpaz, 1972).

Growing alongside maize, these weeds provide the most suitable seasonal habitat for *L. striatellus*, which concentrates on them from its first generation from egg, in June–July. The planthoppers of the following generation hatch in this habitat, acquire MRDV from weeds infected earlier and this leads to further spread. When the weather becomes unfavourable, the nymphs pass into diapause and those infective retain MRDV until the next spring.

The above data refer to maize fields which, as is common practice in northern Italy, are treated with herbicides at sowing time. Normally, the routine use of herbicides delays the development of weeds for about 1 month. Observations in occasional fields where herbicide treatment had been omitted emphasize the importance of maize weeds in the epidemiology of MRDV. Such fields are infested by *D. sanguinalis* and *E. crus-galli* and other weeds from early spring, so that *L. striatellus* can become established in them from the time of the earliest spring migrations (Fig. 1). Consequently, they show much higher than normal MRDV incidences and become centres for secondary spread (Conti, 1976; Milne and Lovisolo, 1977).

Barley yellow striate mosaic virus (BYSMV)

This rhabdovirus has the same vector as MRDV and the two viruses have been studied together in northern Italy. BYSMV is of economic importance only in wheat, although its host-range includes several other widely cultivated cereals and grasses (Conti, 1980). BYSMV resemble MRDV in overwintering in the vector but winter host-plants are also relatively common, and this would suggest a potentially higher incidence of BYSMV than MRDV in susceptible crops. However, wheat is sown in late autumn and emerges when most of the planthoppers have entered diapause and are almost inactive. It becomes

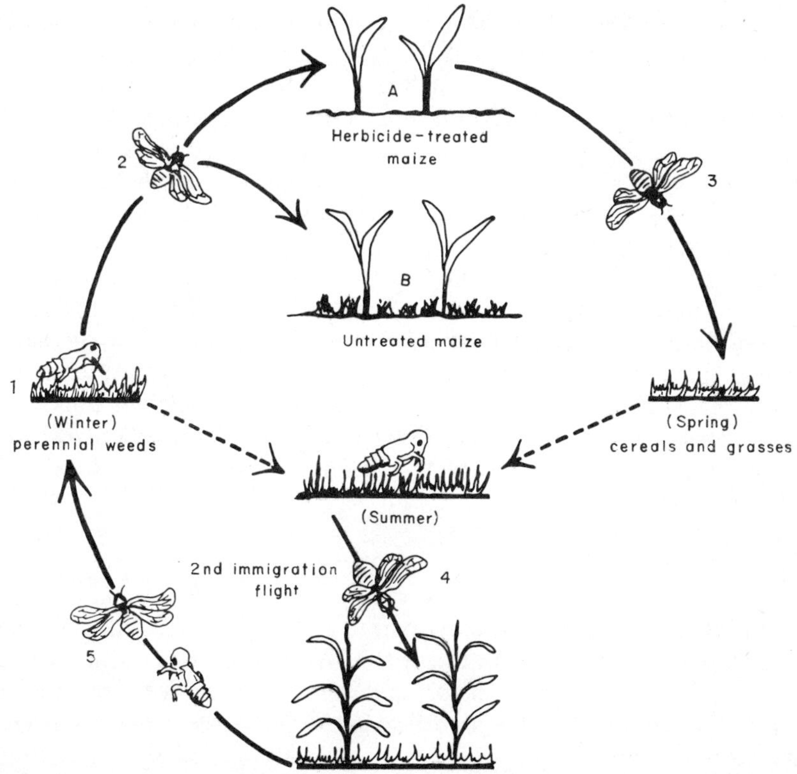

Figure 1 Annual cycle of *Laodelphax striatellus*, vector of maize rough dwarf virus, in northern Italy: (1) overwintering nymph; (2–3) spring migration to grass and cereal fields, during which insects feed also in maize and introduce MRDV. Hoppers do not breed on maize, so do not settle in herbicide-treated fields (A), but do so in untreated ones with grass weeds (B), which become important centres of virus infection and spread; and (4) summer migration to maize-infesting weeds. *Digitaria sanguinalis* and *Echinochloa crus-galli* can be infected with MRDV and become good virus sources to the overwintering generation of hoppers which retain infectivity until the next spring; (5) autumn migration to over-wintering sites.

exposed to BYSMV inoculation in spring when at a more advanced stage of growth and more resistant than the emerging maize contemporaneously exposed to MRDV inoculation. Consequently disease incidence is usually much lower for BYSMV in wheat than for MRDV in maize. BYSMV is an example of a virus having an efficient hopper vector and many natural plant reservoirs, yet spread in crops is limited by cropping practices.

Oat sterile dwarf (OSDV) and cereal tillering disease (CTDV) viruses in Sweden

In Sweden, the planthoppers *Javesella pellucida*, vector of OSDV, and

L. striatellus, vector of CTDV, have one generation per year and overwinter as nymphs in diapause in unploughed stubbles and in fields undersown with grasses. Both viruses overwinter in their vectors and, to a less extent, in perennial grasses, mainly *Lolium perenne*. However, grasses play an important role as perennial hosts of the viruses. Mild and severe strains of both OSDV and CTDV are known, and the mild strains may be those maintained mainly in insects, while severe strains are those which are spread annually from overwintering grasses.

Severe spring outbreaks of OSDV in cereal fields depend largely on the presence of undersown grass leys, as these, in the first year of growth, are very good winter hosts for the hopper vectors that become infective by feeding previously on oats. Replacing oats with barley as a cover crop for undersown leys is the most effective means of controlling OSDV. The epidemiology of CTDV is likely to resemble that of OSDV but it has been little investigated, possibly because this virus is of less economic importance than OSDV (Lindsten, 1979b).

Rice viruses

In Japan, after World War II, early planting of rice became possible due to the widespread use of polyfilm to protect rice nurseries. This practice also provides useful protection from typhoon damage in the western provinces. The introduction of early rice into previously mid-season rice areas led to mixed cropping and changed the density and dynamics of the hopper populations. In the absence of rice, the first generation of *Nephotettix cincticeps*, vector of rice dwarf virus (RDV), develops mainly on a grass weed (*Alopecurus aequalis*), which is not a host of the virus. When the hooper develops on early rice, its rate of multiplication increases five-fold, and the percentage of infected hoppers increases (Nakasuji, 1974). The transmission of RDV to rice is due to adult *N. cincticeps* of the overwintered generation, which invades the nurseries of early- and mid-season rice. By contrast, rice stripe virus (RSV) is seldom found in rice nurseries, due to the different habits of its vector *L. striatellus* (Yashuo *et al.*, 1969). This overwinters as a nymph, and the adults which emerge in spring move to wheat and barley and tend to stay there, so that the rice in the nurseries largely escapes infection. However, extensive spread occurs later in paddies where rice has been transplanted, due to massive immigration of planthoppers of the first generation from eggs.

Rice black-streaked dwarf virus (RBSDV) is also transmitted by *L. striatellus* and is introduced annually to Japanese rice crops as for RSV, although its incidence is usually less. This may be because RSV is transmitted more efficiently than RBSDV and, having a wide natural host-range and being transmitted transovarially in the vector, multiplies more rapidly in the first hopper generation (Ling, 1972).

All the previously mentioned rice-infecting viruses propagate in their vectors and overwinter in them, in winter-sown cereals and in perennial grasses. The non-circulative rice tungro virus (RTV) also infects other graminaceous plants but no clear evidence has been obtained for the existence of any overwintering weed hosts. In India, during the 'off' season, RTV has been detected only in rice stubbles which act as virus sources by producing fresh tillers at the beginning of the new growing season. However, some weeds — *Paspalum distichum* and *Echinochloa colonum* — are important epidemiologically as feeding and breeding hosts of the main local vector, *Nephotettix virescens*, when rice is absent (Rao and John, 1974).

Rice hoja blanca is one of the most destructive diseases of rice in the western hemisphere but its viral aetiology is still uncertain. The vectors are the planthoppers *Sogatodes oryzicola* and *S. cubanus*. Two epidemiological aspects are of particular interest:

(1) that infected insects show a reduction in their fertility and longevity; and
(2) that the disease is likely to have been introduced into rice crops from wild grasses.

The most important hosts are: *Echinochloa walteri, Brachiaria plantaginea, Panicum capillare* and *Sacciolepis striata* (Ling, 1972; Harris, 1979).

Maize streak virus (MSV)

The epidemiology of this virus has been studied extensively by Rose (1979) in Zimbabwe, where there is a cold dry season (May–August), a hot dry season (September–November) and a wet warm season (November–April). The fate of *Cicadulina* spp., vectors of MSV, depends strictly upon rainfall patterns which regulate the grass hosts. Leafhopper populations show a distinct peak of density in August with smaller peaks in October, December, February, and May. These peaks are evident in irrigated and low-lying moist land harbouring the several grasses on which the hoppers concentrate in the hot dry season. During the winter flight period of *Cicadulina*, the insect populations in cultivated fields are highly mobile and change constantly. Populations in different cereal fields later vary widely, showing clear gradients of density which start from the insect breeding sites. This is because breeding areas are limited, non-migrant insects predominate, and only crops close to infested vegetation are invaded.

Another important role of weeds in the epidemiology of MSV is that of virus reservoirs which may select 'new' virus strains. One from *Eleusine* and another from *Sporobulus*, in particular, became highly pathogenic to exotic maize lines when these were first introduced to Africa.

Conclusions

The examples discussed in this paper give some idea of the interactions between plants, viruses, vectors and environment, which determine the epidemiological cycle of hopper-borne viruses of Gramineae. Many such viruses, although economically important, have received scant attention in the field. The continuing changes in cropping techniques, with the introduction of new plant varieties and new agricultural practices, lead to a continually changing epidemiological situation. Therefore, field studies should be encouraged to obtain a better understanding of virus epidemiology and to achieve greater sophistication in manipulating different ecological situations to benefit crops.

References

Boccardo, G., Hatta, T., Francki, R. I. B. & Grivell, C. J. (1980). Purification and some properties of reovirus-like particles from leafhoppers and their possible involvement in wallaby ear disease of maize. *Virology* **100,** 300–315.

Brčák, J. (1979). Leafhopper and planthopper vectors of plant disease agents in Central and Southern Europe. In *Leafhopper Vectors and Plant Disease Agents*, pp. 97–104. K. Maramorosch & K. F. Harris. San Francisco & London: Academic Press.

Conti, M. (1976). Epidemiology of maize rough dwarf virus. III. Field symptoms, incidence and control. *Maydica* **21,** 165–175.

Conti, M. (1980). Vector relationships and other characteristics of barley yellow striate mosaic virus (BYSMV). *Annals of Applied Biology* **95,** 83–92.

Conti, M. & Milne, R. G. (1977). Some new natural hosts of maize rough dwarf virus. *Annales de Phytopathologie* **9,** 255–259.

De Long, D. M. (1971). The bionomics of leafhoppers. *Annual Review of Entomology* **16,** 179–210.

Duffus, J. E. (1971). Role of weeds in the incidence of virus diseases. *Annual Review of Phytopathology* **9,** 319–340.

Francki, R. I. B. (1973). Plant Rhabdoviruses. *Advances in Virus Research* **18,** 257–345.

Harpaz, I. (1972). *Maize Rough Dwarf Virus.* Jerusalem: Israel University Press.

Harris, K. F. (1979). Leafhoppers and aphids as biological vectors: vector–virus relationships. In *Leafhopper Vectors and Plant Disease Agents*, pp. 217–308. K. Maramorosch & K. F. Harris. San Francisco & London: Academic Press.

Kuno, E. (1973). Population ecology of rice leafhoppers in Japan. *Review Plant Protection Research* **6,** 1–16.

Lindsten, K. (1979a). Leaf- and planthoppers as virus vectors in Fennoscandia and measures to prevent their spread. *Entomologisk Tidskrift* **100,** 159–161.

Lindsten, K. (1979b). Planthopper vectors and plant disease agents in Fennoscandia. In *Leafhopper Vectors and Plant Disease Agents*, pp. 155–178. K. Maramorosch & K. F. Harris. San Francisco & London: Academic Press.

Ling, K. C. (1972). *Rice Virus Diseases.* Los Banos Philippines: International Rice Research Institute. 142 pp.

Maramorosch, K., Govindu, H. C. & Kondo, F. (1977). Rhabdovirus particles associated with a mosiac disease of naturally infected *Eleusine coracana* (finger millet) in Karnataka State (Mysore), South India. *Plant Disease Reporter* **61,** 1029–1031.

Matthews, R. E. F. (1979). Classification and nomenclature of viruses. *Intervirology* **12,** 129–296.

Milne, R. G. & Lovisolo, P. (1977). Maize rough dwarf and related viruses. *Advances in Virus Research* **21,** 267–341.

Nakasuji, F. (1974). Epidemiological study of rice dwarf virus transmitted by the green rice leafhopper *Nephotettix cincticeps. Japan Agricultural Research Quarterly* **8,** 84–91.

Raatikainen, M. & Vasarainen, A. (1973). Early- and high-summer flight periods of leafhoppers. *Annales Agricultural Fenniae* **12,** 77–94.

Rao Prasada, R. D. V. J. & John, V. T. (1974). Alternate hosts of rice tungro virus and its vector. *Plant Disease Reporter* **58,** 856–860.

Rose, D. J. W. (1978). Epidemiology of maize streak disease. *Annual Review of Entomology* **23,** 259–282.

Saito, Y. (1977). Rice viruses, with special reference to particle morphology and relationships with cells and tissues. *Review Plant Protection Research* **10,** 83–90.

Slykhuis, J. T. (1977). Virus and virus-like diseases of cereal crops. *Annual Review of Phytopathology* **14,** 189–210.

Sogawa, K. (1973). Feeding of the rice plant- and leafhoppers. *Review Plant Protection Research* **6,** 31–43.

Thresh, J. M. (*1981*). The role of weeds and wild plants in the epidemiology of plant virus diseases. In *Pests, Pathogens and Vegetation*, pp. 53–70. J. M. Thresh. London: Pitman.

Yashuo, S., Ishii, M. & Yamaguchi, T. (1969). Studies on rice stripe disease. *Review Plant Protection Research* **2,** 96–104.

Hawthorn as a source of the fireblight bacterium for pear, apple and ornamental hosts

Eve Billing
East Malling Research Station, Maidstone, Kent ME19 6BJ

Introduction

Fireblight is a bacterial disease of rosaceous plants in the sub-family Maloideae. Besides hawthorn, it affects pears, apples and ornamental species, e.g. *Sorbus* (mountain ash and whitebeam), *Cotoneaster* and *Pyracantha*. The pathogen *Erwinia amylovora* shows no race specificity and cross-infection between hosts is common.

Fireblight probably originated in North America affecting wild rosaceous hosts before cultivated pears and apples were introduced. Pear growing is severely restricted in parts of the USA because of fireblight. There, *Crataegus* spp. are seldom cited as a reservoir of inoculum; most native species are resistant or less susceptible than European hawthorns (Pierstorff, 1931; Thomas and Parker, 1933), presumably because they co-evolved with the pathogen.

In New Zealand, fireblight was found in 1919 and hedgerow hawthorns were commonly affected at first. In the early years when the disease was severe in pears, attempts were made to limit it by removing hawthorns but the regulations were not enforced because hawthorns were important as windbreaks for cattle (Phillips, 1968).

The natural history of hawthorn in English hedges and the epidemiology of fireblight on hawthorn are described elsewhere (Pollard, *et al.*, 1974; Meijneke, 1973; Billing, *et al.*, 1974; Billing, 1978). Emphasis here will be on hawthorns as sources and reservoirs of infection and on the distribution and spread of infection on hawthorn in England and northern Europe.

Distribution and properties of European hawthorns

Crataegus monogyna, C. laevigata (syn. *oxycantha*) and their hybrids are widely distributed in hedgerows and woodlands, along roads, dikes and railway embankments, on waste land and abandoned quarries and on coastal dunes and marshes. No other tree or shrub provides such a cheap, rapid-growing, efficient, impenetrable thorny windbreak which is resistant to salt-laden winds (Meijneke, 1979). Apart from sheltering crops and livestock, they protect and supply food for wild birds and insects and contribute to the landscape. Consequently, removing hawthorns as a fireblight control measure is often unpopular and expensive. Flowering hawthorns are still numerous in many areas despite current farming trends involving removal of hedges and mechanical trimming (Trask, *1981*).

Fireblight hosts are most vulnerable during their flowering period. The later they flower, the greater the risk as the weather becomes warmer. The usual sequence from April to July is: pear, apple, hawthorn and *Sorbus* spp., *Pyracantha* and *Cotoneaster*, with most hawthorns flowering about mid-May to early June. The amount of overlap depends on climate and season. Hawthorns make considerable shoot growth before flowering and shoot infection can occur before bloom.

European hawthorns are highly susceptible to fireblight, despite wide morphological variation (Maas Geesteranus, 1978; Zeller, 1979). Fortunately, the weather at blossom time in northern Europe is often too cool or dry for much blight to occur (Billing, 1980 and unpublished).

Dissemination of inoculum from hawthorns

Bacteria emerge from infected plants embedded in a polysaccharide matrix (Plate 1 (top left)) which becomes sticky on drying and aids survival and dissemination. The bacteria may be disseminated by rain, wind, insects or possibly birds. Once disseminated, they are readily dispersed by rain, mist or dew, which also facilitate infection (Beer, 1979). Patterns of spread of disease from hawthorn are rarely documented but Bech-Andersen (1971) found infected pear trees up to 900 m from affected hedges though 60% were within 100 m.

Wind-blown rain

Inoculum may be carried more than 100 m by driving rain. This occurred in Kent apple orchards with shoot infections in 1969 during a storm with north-westerly winds exceeding 15 ms^{-1} (Glasscock, 1971; Billing, 1974, 1975) where hawthorn was the source of inoculum (Fig. 1).

Wind

In theory inoculum might be spread by wind alone in fine strands (Plate 1 (top right and bottom left)), which are extruded under certain conditions and contain bacteria embedded in a polysaccharide matrix (Eden-Green and Billing, 1972; Billing and Eden-Green, 1973). Contaminated pollen, plant fragments or insects might also be wind-borne, but rain or heavy dew would be necessary after deposition to ensure infection.

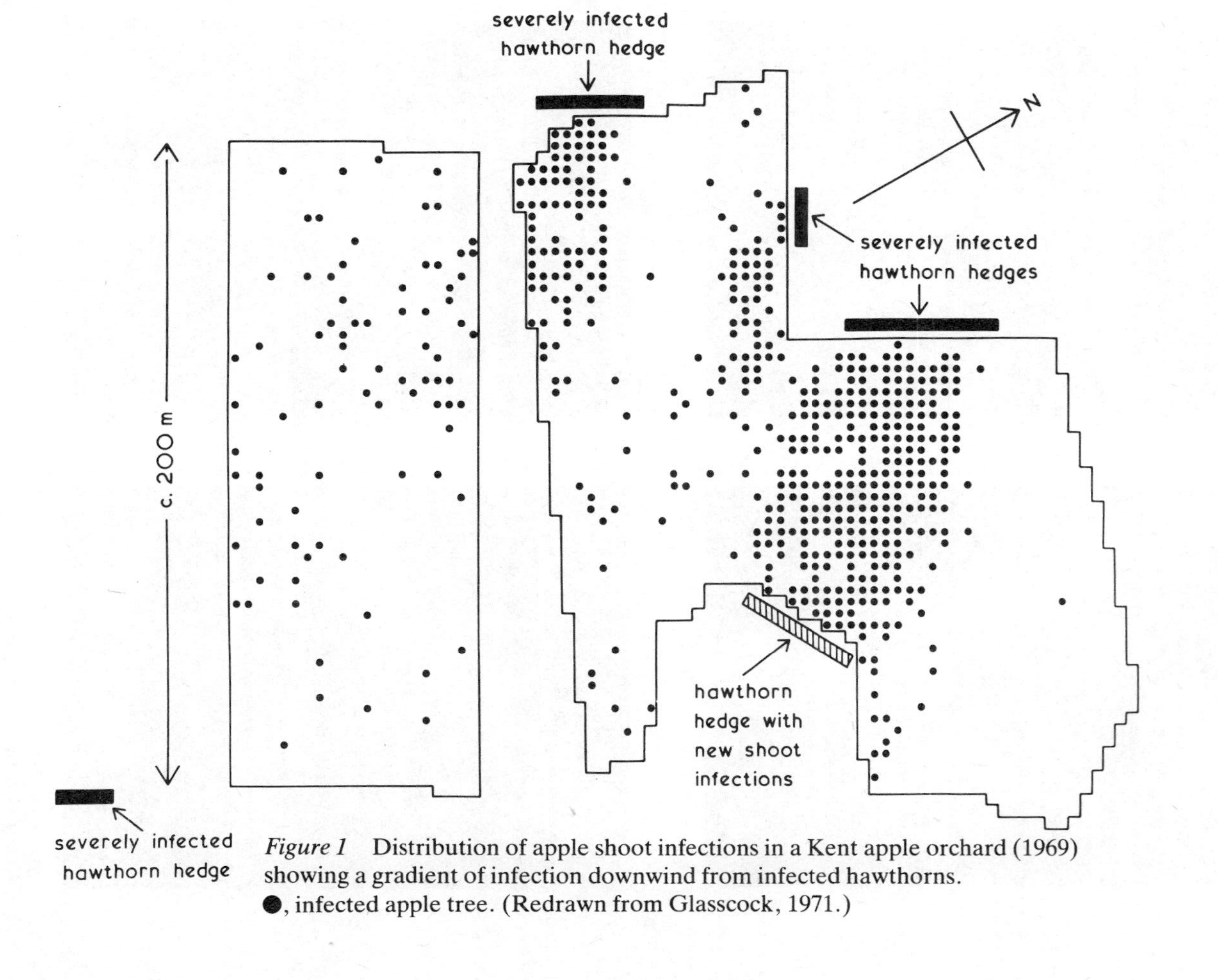

Figure 1 Distribution of apple shoot infections in a Kent apple orchard (1969) showing a gradient of infection downwind from infected hawthorns. ●, infected apple tree. (Redrawn from Glasscock, 1971.)

Plate 1 Hawthorn infections
(Top left) Droplet of bacterial ooze (×5)
(Top right) Ooze strands on shoots (greenhouse specimen, ×5)
(Bottom left) Ooze strand on a pedicel (field specimen, ×20)
(Bottom right) Insect damage on an infected blossom (field specimen, ×10)

Insects

Insects of more than 70 genera have been associated with spread of fireblight on pear and apple (van der Zwet and Keil, 1979). Many of those on pear also infest hawthorn (Emmett and Baker, 1971; Thygesen, *et al.*, 1973) and may play an equivalent role.

Ants and flies are attracted to bacterial ooze and can carry inoculum from overwintering cankers to blossoms (Thomas and Ark, 1934); chewing insects may also be involved (Plate 1 (bottom right)). Insects most commonly associated with spread of infection between blossoms are dipterous flies, aphids and bees. The speed with which infection might be spread between blossoms is shown by observations on the activities of bees and other insects associated with pear blossom (Emmett, 1971).

For spread by insects moving between hosts, coincidence of flowering is important (p. 122). Honey-bees are unlikely to work more than one host at a time (Free, 1970) which may limit the risk, but this may not be true of other flower-visiting insects. Some pear cultivars are more vulnerable than other hosts because from June they produce additional blossom that is mainly visited by syrphids and other flies (Emmett and Baker, 1971).

The importance of post-blossom spread by insects is less certain. Those that have been associated with shoot or fruit infections of apple or pear are chewing or sucking insects such as caterpillars, capsids (e.g. *Lygus* spp.), aphids, leafhoppers (*Empoasca* spp.) and pear psylla (*Psylla pyricola*); the shot-hole borer (*Scolytus rugulosus*) has been implicated in branch infections (van der Zwet and Keil, 1979). Larger damaging insects seem to transmit fireblight most readily (Thygesen *et al.*, 1973). On hawthorns, gall midge larvae (*Dasyneura crataegi*) induce leafy galls on shoots in July which are sometimes associated with fireblight infections (Thygesen *et al.*, 1973; Billing *et al.*, 1974).

Birds

Paradoxically, whilst birds might be concerned in long distance spread of fireblight (*see* below), there is no good evidence that they are important in medium or short distance spread. However, even if they are not very effective, they might sometimes introduce inoculum to new areas beyond the range of spread by insects or wind.

Diverse birds use hawthorns or other hosts for roosting, cover or feeding and any damage they cause increases the chances of inoculum causing infection. The starling (*Sturnus vulgaris*) is a major suspect in terms of damaging potential, particularly during migration to and from the continent. The thrush family (*Turdus* spp.), including blackbirds, redwings and fieldfares, migrate in smaller flocks but will eat fruits of most fireblight hosts. Smaller birds using hawthorn as cover include the bullfinch (*Pyrrhula*

pyrrhula), which attacks both hawthorn and fruit buds in early spring (Matthews and Flegg, *1981*; Summers, *1981*).

Long distance dissemination

Circumstantial and anecdotal evidence suggests that spring migrant birds might have spread fireblight from England to hawthorns in continental Europe (Bech-Andersen, 1971, 1973, 1974; Meijneke, 1972, 1974; Billing, 1974). The spring migration period is from February to early May (Eastwood, 1967), not later as Beer (1979) has suggested. This precedes hawthorn flowering, but infection is possible via scars from swelling buds which are easily detached (Billing, 1978). Delayed migration (Davis, 1964), numerous infected hawthorns in England and favourable weather make 1964 the most likely year. However, this was also a year when, judging by wind trajectories for 7–8 June (J. Cochrane, unpublished) and favourable weather, wind dissemination of inoculum or infected insects was another possibility (Billing, 1974 and unpublished).

Spread of disease between hosts

Hawthorns have frequently been important sources of infection around orchards (Lelliott, 1968a; Glasscock, 1971; Bech-Andersen, 1971), but ornamental hosts such as cotoneaster are sometimes the reservoir of inoculum.

Spread of fireblight between different species is often slow. Infection of hawthorn hedges round the first infected English pear orchards was not seen for several years (Lelliott, 1968b) and in the Netherlands hawthorns remained healthy in areas where infected cotoneasters were common (Meijneke, 1974 and personal communication). Conversely, severely infected hawthorns have been seen near apparently healthy pear and apple trees or cotoneasters (Bech-Andersen, 1971; Billing, personal observations). Possible reasons for this include non-coincidence of flowering, or food or host preferences of insects (Lelliott, 1968b). Cryptic cross-infections may occur (Billing, 1974) with overt disease only developing in a particularly favourable season; weather during the flowering period of a particular host greatly affects the level of disease and hence its chances of detection (Billing, 1980). Damaging storms which encourage shoot infection could also be an important means of spread between hosts as in the case of apple shoot infections (p. 122). Thus hawthorns may be infected for several years without obviously affecting nearby hosts.

Occurrence of fireblight on hawthorns in Europe

In England, the first hawthorn infections were found the year after the first discovery of fireblight on pears in 1957. Between 1958 and 1969, 20 000 infected hawthorns were destroyed in the area east of a line from the Wash through Oxford to Southampton (Anon, 1969). Many were in coastal regions of Kent, Essex and the Thames estuary, but others were found inland in orchard and urban areas including London. The number destroyed, though high, was a minute proportion of the total hawthorn population. In 1971, hawthorn infections were found in south-west England. From 1971 to 1978 the number of affected hawthorns appeared to decline markedly but there was further spread both north and south. Detailed surveys have not been made recently, but the weather was generally less favourable for hawthorn blossom infection in the 1970s than previously (Billing, 1980). In 1979 the weather was again favourable and additional hawthorn infection was seen (Billing, personal observations).

Apart from a few pockets of infection, fireblight is currently of little economic importance in England, although it is widely distributed in many parts of the south (Lelliott, 1979). There is no reason to believe that many trees are infected but the situation could change with a sequence of favourable seasons.

In the Netherlands, fireblight on hawthorn, pear and apple was first seen in 1966 and eradication was attempted as the disease appeared to be restricted to a small area (Meijneke, 1968). In 1971, however, coastal hawthorns in an adjacent area and also to the north were found to be infected (Meijneke, 1974). The disease has since spread widely to inland areas of the Netherlands, mostly on cotoneasters. Inland hawthorns were not severely infected until 1979 (C. A. R. Meijneke, personal communication).

In southern Denmark, fireblight was first found in pear orchards in 1968, the disease at first appeared to spread primarily on hawthorns (Bech-Andersen, 1973) and later on cotoneaster (Hockenhull, 1979). By 1972, infections on coastal hawthorns had been reported from Poland, through north Germany, Jutland and the Netherlands to the Franco–Belgian border (Burkowicz, 1972; EPPO, 1971, 1972, 1973). Infected pear orchards were found in 1978, in south-west France 1000 km from the nearest known focus (Paulin and Lachaud, 1978); hawthorn infections were not seen there until 1979. As in the Netherlands, inland spread on hawthorns appears to have been slow, but once established, the climate is likely to be more favourable for infection inland than on the North Sea coast (Billing, unpublished).

The proportion of hawthorns infected in any one country is very low. Severe outbreaks have mostly remained localized and confined to a few seasons. The disease can destroy whole trees within 1–2 years, but often it fails to progress far and can easily be overlooked. Some trees appear to recover but once a hedgerow is infected there is a risk of subsequent resurgence in favourable seasons.

Prevention and control of hawthorn infections

Ideally, hawthorns should not be planted near orchards or nurseries and existing wind-breaks should be removed within about 200 m. Where hawthorns become infected, uprooting and burning is the best measure or cutting down to ground level and killing the remaining stumps with herbicides. In uprooted hawthorn, survival of the pathogen in tissue appears greatest under dry conditions (Zeller, 1978).

In practice, such measures are often too costly or impracticable and eradication is rarely attempted or achieved. Compromise solutions include trimming hedges annually to reduce blossoming and cutting down infected hedges but allowing regrowth even though reinfection may eventually occur (Anon, 1969; Meijneke, 1974, 1979; Jørgensen, 1978). Spraying hawthorns with insecticide may only reduce insect populations by 50% and valuable predators are killed (Thygesen *et al.*, 1973).

Conclusions

It seems inevitable that fireblight will continue to spread southward in Europe and to a lesser extent northward. Though other hosts may be affected first, hawthorns will ultimately be important reservoirs of inoculum in areas where they are abundant. Their potency will fluctuate according to climatic factors.

References

Anon, (1969). Fireblight of apple and pear. *Ministry of Agriculture, Fisheries and Food, UK, Advisory Leaflet* No. 571.

Bech-Andersen, J. A. (1971). [Fireblight in Denmark caused by *Erwinia amylovora*.] *Ugeskrift for Agronomer* **33,** 677–680.

Bech-Andersen, J. A. (1973). The relationship between fireblight in fruit trees and hawthorn in Denmark. *2nd International Congress of Plant Pathology, Minneapolis.* Abstracts of papers, No. 0749.

Bech-Andersen, J. A. (1974). Spread of fireblight in Europe. In *Proceedings of the XIXth International Horticultural Congress, Warsaw, 1974.* (Abstract) p. 202.

Beer, S. V. (1979). Fireblight inoculum: sources and dissemination. *EPPO Bulletin* **9,** 13–25.

Billing, E. (1974). Environmental factors in relation to fireblight. In *Proceedings of the XIXth International Horticultural Congress, Warsaw, 1974*, pp. 365–372.

Billing, E. (1975). Fireblight (*Erwinia amylovora*). In *Climate and the Orchard* pp. 130–133. H. C. Pereira. *CAB Research Review* No. 5.

Billing, E. (1978). The epidemiology of fireblight on hawthorn in England. In *Proceedings of the IVth International Conference on Plant Pathogenic Bacteria, Angers, France, 1978*, pp. 487–492.

Billing, E. (1980). Fireblight in Kent, England in relation to weather (1955–1976). *Annals of Applied Biology* **95,** 341–364.

Billing, E. & Eden-Green, S. J. (1973). Bacterial strands on hawthorn: glasshouse and field observations. *Report of the East Malling Research Station for 1972*, p. 166.

Billing, E., Bech-Andersen, J. & Lelliott, R. A. (1974). Fireblight in hawthorn in England and Denmark. *Plant Pathology* **23,** 141–143.

Burkowicz, A. (1972). The appearance and current situation of fireblight in Poland. *Plant Health Newsletter, EPPO Publications Series B* No. 72E.

Davis, P. (1964). Aspects of spring migration at the bird observatories, 1964. *Bird Study* **11,** 198–223.

Eastwood, E. (1967). *Radar Ornithology*. London: Methuen.

Eden-Green, S. J. & Billing, E. (1972). Fireblight: occurrence of bacterial strands on various hosts under glasshouse conditions. *Plant Pathology* **21,** 121–123.

Emmett, B. J. (1971). Insect visitors to pear blossom. *Plant Pathology* **20,** 36–40.

Emmett, B. J. and Baker, L. A. E. (1971). Insect transmission of fireblight. *Plant Pathology* **20,** 41–45.

EPPO (1971). Recent changes in the fireblight situation in north-western Europe. *EPPO Report* 71/9 — 354 RSE.

EPPO (1972). Fireblight in France. *EPPO Report* 72/8 — 366 RSE.

EPPO (1973). Fireblight in Belgium. *EPPO Report* 73/1 — 369 RSE.

Free, J. B. (1970). *Insect Pollination of Crops*. London: Academic Press.

Glasscock, H. H. (1971). Fireblight epidemic among Kentish apple orchards in 1969. *Annals of Applied Biology* **69,** 137–145.

Hockenhull, J. (1979). Fireblight in Denmark from 1968 to 1977. EPPO Bulletin **9,** 64–65.

Jørgensen, H. A. (1978). Fireblight control in *Crataegus* hedges by clipping. *Acta Horticulturae* **86,** 69–70.

Lelliott, R. A. (1968a). Fireblight on apple in England. *Plant Pathology* **17,** 48.

Lelliott, R. A. (1968b). Fireblight in England. Its nature and its attempted eradication. *EPPO Publications Series A* No. 45E, pp. 10–14.

Lelliott, R. A. (1979). Present status of fireblight in United Kingdom. *EPPO Bulletin* **9,** 70–71.

Maas Geesteranus, H. P. (1978). Studies on the susceptibility of *Crataegus* species to *Erwinia amylovora* (Burr.) Winsl. *et al.* In *Proceedings of the IVth International Congress on Plant Pathogenic Bacteria, Angers, France*, pp. 499–503.

Matthews, N. J. & Flegg, J. J. M. (*1981*). Seeds, buds and bullfinches. In *Pests, Pathogens and Vegetation*, pp. 375–383. J. M. Thresh. London: Pitman.

Meijneke, C. A. R. (1968). Fireblight: an isolated outbreak in the Netherlands. *EPPO Publications Series A* No. 45E, pp. 17–19.

Meijneke, C. A. R. (1972). Pervuur en zijn verspreiding. [Fireblight and its spread.] *Overdruk Gewasbescherming* **3,** 128–136.

Meijneke, C. A. R. (1973). Meidoorn en perevuur. [Hawthorn and fireblight.] *Nederlands Bosbouw Tijdschrift* **45,** 305–309.

Meijneke, C. A. R. (1974). The 1971 outbreak of fireblight in the Netherlands. In *Proceedings of the XIXth International Horticultural Congress*, pp. 373–382.

Meijneke, C. A. R. (1979). Prevention and control of fireblight. *EPPO Bulletin* **9,** 53–62.

Paulin, J. P. & Lachaud, G. (1978). Fireblight situation in France: August 1978. In *Proceedings of the 4th International Congress on Plant Pathogenic Bacteria, Angers, France*, pp. 519–521.

Phillips, R. (1968). Fireblight in New Zealand. *EPPO Publications Series A* No. 45E, pp. 20–22.

Pierstorff, A. L. (1931). Studies on the fireblight organism, *Bacillus amylovorus. New York Agricultural Experimental Station, Memoir* 136.

Pollard, E., Hooper, M. D. & Moore, N. W. (1974). *Hedges.* London: Collins.

Summers, D. D. B. (*1981*). Bullfinch (*Pyrrhula pyrrhula*) damage in orchards in relation to woodland bud and seed feeding. In *Pests, Pathogens and Vegetation*, pp. 385–391. J. M. Thresh. London: Pitman.

Thomas, H. E. & Ark, P. A. (1934). Fireblight of pears and related plants. *California Agricultural Experimental Station Bulletin* No. 586.

Thomas, H. E. and Parker, K. G. (1933). Fireblight of pear and apple. *New York Agricultural Experimental Station Bulletin* No. 557.

Thygesen, Th., Esbjerg, P. & Eiberg, H. (1973). [Fireblight transmission by insects]. *Saertyrk af Tidsskrift for Planteaul* **77,** 324–336.

Trask, A. B. (*1981*). Changing patterns of land use in England and Wales. In *Pests, Pathogens and Vegetation* pp. 39–49. J. M. Thresh. London: Pitman.

Zeller, W. (1978). Untersuchungen zur feuerbrandkrankheit in der Bundesrepublik Deutschland. 4. Uberlebendsdauer von *Erwinia amylovora* an gerodetem weissdorn (*Crataegus monogyna*). *Nachrichtenblatt des Deutschen Pflanzenschutzdienste* **30,** 186–188.

Zeller, W. (1979). Resistance and resistance breeding in ornamentals, *EPPO Bulletin* **9,** 35–44.

van der Zwet, T. & Keil, H. (1979). Fireblight. A bacterial disease of rosaceous plants. *USA Department of Agriculture, Handbook* No. 510.

The ecology of plant parasitic fungi

B E J Wheeler
Imperial College at Silwood Park, Ascot, Berkshire SL5 7PY

Introduction

Fossil records indicate that millenia ago some plants had fungi growing on them (Pirozynski, 1976), so it is perhaps not surprising that when man started to cultivate plants he found that as his crops grew so did the fungi and that many of these caused damage. However such awareness of the relationship between this damage and fungi developed over many centuries. To the early civilizations these afflictions were yet another manifestation of a godly displeasure at man's shortcomings — and on good authority 'I smote you with blight and mildew . . . I laid waste your gardens and your vineyards . . . yet you did not return to me, says the Lord.' So thundered the prophet Amos at the unfortunate Israelites! Not until the beginning of the ninteenth century and the classical work of Anton de Bary, was it fully appreciated that some fungi living on plants as parasites could also induce disease and thus be termed pathogens (Large, 1940).

The subsequent activities of mycologists and plant pathologists have enlarged our knowledge of fungi both as parasites and pathogens. Amongst other things, they have provided check lists of fungi for many territories, which indicate both the range and diversity of fungal species on wild plants and also record the range of pathogenic fungi on crops (e.g. *see* British Mycological Society, 1962; Holden, 1975). These lists reveal that in most areas wild plants and crops have certain fungal species in common. The basic problem, and the theme of this paper, is the relationship between the populations of these fungi in the two situations, which centres on the contribution of the wild plants in providing infective fungal propagules (inoculum) to the crop. This concept is often crystallized by the farmer in the very practical question 'Where does the disease come from?' which the professional plant pathologist rephrases as 'What is the source of fungal inoculum?'.

Relationships between fungi and plants

Before exploring this concept in detail it is necessary to examine briefly the relationships between fungi and the plants that harbour them, that is their hosts. There are three main points:

(1) Fungi exploit different ecological niches on or within plants (Harper, 1977). Some are confined to the root systems, either living within the root cortex and progressively rotting it (e.g. *Phytophthora cinnamomi*), or so affecting the processes of cell division and enlargement that club roots develop (e.g. *Plasmodiophora brassicae*). Others, similarly exploit young or old leaves, producing necrotic lesions (e.g. *Phytophthora infestans*), leaf curling (e.g. *Taphrina deformans*) or conspicuous masses of fungal growth with or without marked changes in the adjacent host tissue (e.g. *Diplocarpon rosae, Erysiphe graminis*). Yet others exploit the flowers, living in diverse ways within the petals (*Itersonilia perplexans*), anthers (*Ustilago violaceae*) or ovaries (*Ustilago nuda, Tilletia caries*) (Agrios, 1978). With such a diversity of ecological niches, it is not surprising to find that any one plant species is often host to several different fungal species. It is interesting to speculate on what determines the total number of parasitic species colonizing a particular plant species. The limited data available have recently been reviewed (May, 1979). The two factors which most influence abundance appear to be the geographical range and structural complexity of the host. Thus the greater the range of the host the more parasites it is likely to have, with trees having more parasitic species than shrubs and many more than herbs.

(2) Fungi parasitize their hosts in diverse ways and the precise mechanisms by which they obtain their food materials are imperfectly understood for many fungus–host associations. The extremes are represented by the so-called necrotrophs (e.g. *Botrytis cinerea*) and the biotrophs (e.g. the rust and powdery mildew fungi). The necrotrophs rapidly kill the host tissues they invade and live on the disorganized cell contents. By contrast the biotrophs spend much of their life in a balanced relationship with their hosts, directing the products of photosynthesis to the fungal colony but not immediately causing death of cells, an event that would curtail their own activities.

(3) Parasitic fungi have various devices which ensure their distribution spatially or temporally and these are associated with their parasitic habit. The powdery mildews live for many days on the leaves of their hosts producing chains of spores ideally adapted for dissemination by wind. The rust fungi possibly have the additional advantage that the pigments of their uredospores partially protect against UV radiation and thus enable them to retain their viability during long distance transport in the atmosphere (Maddison and Manners, 1972). By contrast, the *Pythium* spp. that rapidly kill the root and hypocotyl of their seedling hosts compensate for their inability to become so widely distributed in space by forming persistent resting spores that ensure long-term survival.

Any generalizations that are made about relationships between populations of fungi in wild plants and those of crops must be tempered by an appreciation of the diversity of the relationships between individual parasitic fungi and their respective hosts which these three points broadly illustrate. It is also true that our knowledge of the parasitic activities of fungi within crops is far greater than that available for wild plants.

The development of fungal populations in relation to the structure of the host population

Since we are concerned with populations of fungi and specifically with plant pathogenic fungi, it is necessary to compare in a general way the development of a population of pathogens in the two host populations, wild plants and crops. In particular, attention must be focussed on the effects of host population structure on the development of the fungal pathogen. Figure 1 shows diagramatically two host populations, composed of individuals assumed equally susceptible to a fungal pathogen, but reflecting the density in stand which might be associated with a crop and a wild plant. Unfortunately, reliance must be placed mainly on data from crop stands to indicate effects on fungal spread and much of it relates to movement between individual plants rather than between clumps of plants as indicated in Fig. 1.

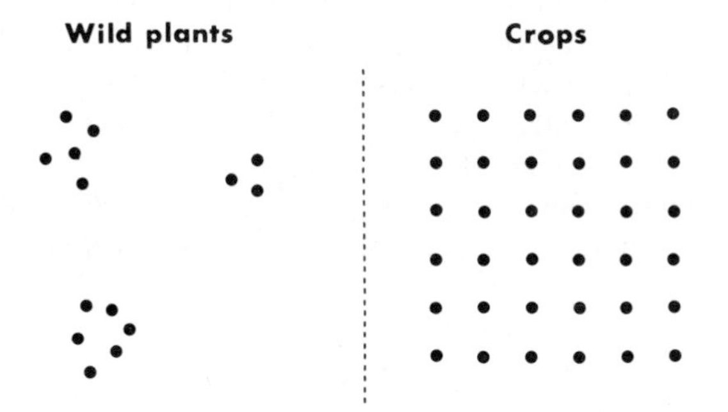

Figure 1 Schematic distribution of host plants, in the wild and in crops.

Early studies by Scott (1956) examined the effect of spacing on the spread of the onion white rot fungus (*Sclerotium cepivorum*) both in seed trays and along rows of onion in the field. In both instances the rate of spread, as indicated by percentage infection after a given time, was much reduced with increased dispersion (Table 1). Similar results were obtained in comparable studies with *Pythium irregulare* using seed boxes of cress (*Lepidium sativum*) planted at different densities (Burdon and Chilvers, 1975). The rate at which the fungus grew through soil and damped-off the seedlings decreased markedly with planting distance (Fig. 2). The ability of such fungi to grow through soil depends largely on how well they conduct essential nutrients

Table 1 Infection by *Sclerotium cepivorum* of onions at different spacings (from Scott, 1956).

Seed trays		*Field*	
Distance between plants (cm)	*Percentage infection*	*Distance between plants (cm)*	*Percentage infection*
2.5 × 2.5	65	5.1	23
5.1 × 5.1	51	7.6	13
7.6 × 7.6	43	12.7	6
10.2 × 10.2	15		

along their mycelia and cope with the competition of other soil micro-organisms. However, even for those fungi which are most successful in these respects (such as *Armillariella mellea* whose aggregated hyphae form rhizomorphs extending many metres through soil), there comes a point when development within the host population ceases. This occurs when the reserves of the food base are exhausted or the hyphae lack the potential for infection when they contact a new host root. Such factors are discussed by Garrett (1970).

It might be thought that such limitations do not apply to spore-producing pathogens of the shoot system but this is not necessarily so. Experiments in Florida with the leaf-spot fungus of celery (*Cerocospora apii*) indicated that the rate of disease increase was reduced in widely-spaced plantings of the host (Table 2). Several factors may be involved of which differences in the micro-climate of the plots were considered to be the most important. With wider separation of susceptible hosts there is opportunity for dilution of inocula in air with decreased disease severity as shown with powdery mildew in plots of barley at different distances from a source of *Erysiphe graminis* (Fig. 3).

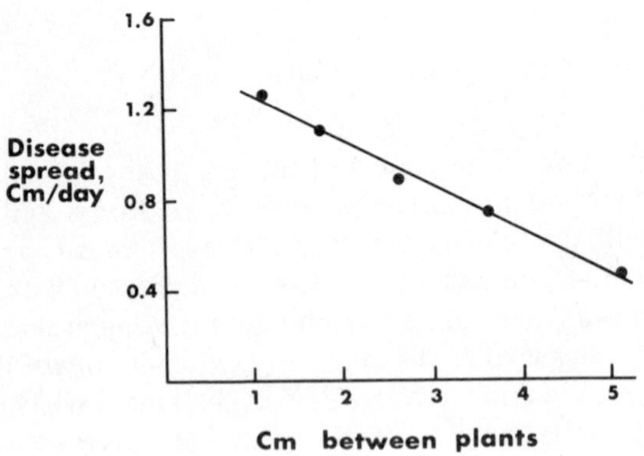

Figure 2 Spread of damping-off by *Pythium irregulare* in relation to distance between seedlings of *Lepidium sativum* (from Burdon and Chilvers, 1975).

Table 2 Spread of *Cercospora apii* amongst 100 celery plants at different spacings (from Berger, 1975).

Area occupied by 100 plants (m^2)	*Percentage disease 26 November, (1973)*	*Infection rate (r) 24 October–26 November (1973)*
6.8	42.1	0.12
7.3	45.3	0.12
10.6	40.1	0.14
15.2	39.1	0.12
16.4	33.6	0.10
20.5	27.3	0.09
29.2	21.1	0.09

Our present model of the two host populations (Fig. 1) supposes that the individuals are equally susceptible to the fungal pathogen. Crops approach this situation as the demands of large scale monoculture have led to the introduction of cultivars with remarkably uniform populations. The individuals develop synchronously and this facilitates harvesting. They also react similarly to a compatible fungal pathogen, thus favouring its epidemic development as with the brown-rust pathogen (*Puccinia hordei*) in a crop of barley (Table 3). This is not true of the wild hosts. Some data obtained at Silwood Park for three rust species suggest that they develop only to a limited extent on their respective hosts, except possibly in recently established sites (Table 4).

Wild plants are also extremely variable in their reactions to a fungal pathogen. Figure 4, for example, indicates the development of powdery mildew on a sample population of a wild barley (*Hordeum spontaneum*), both

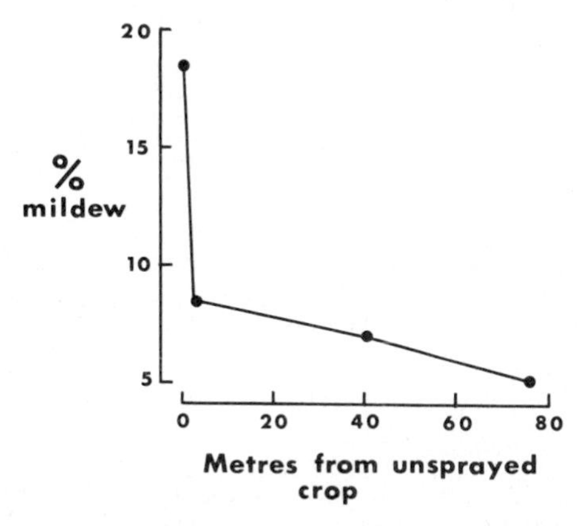

Figure 3 Development of powdery mildew in barley crops at different distances from a source of *Erysiphe graminis* f. sp. *hordei* (from Jenkyn and Bainbridge, 1974).

Table 3 Development of *Puccinia hordei* on barley, cv. Deba Abed 1973 (from Onursal, 1975).

July	*Percentage brown rust*
3	0.2
9	1.2
17	5.4
24	18.2
31	39.4

Table 4 Incidence of rusts in various plant communities.*

		Percentage leaves with rust					
		1977		*1978*			
Site	*Host/rust species*	*July*	*Oct.*	*May*	*June*	*July*	*Oct.*
'Young' field	*Holcus lanatus/ Puccinia coronata*	–	–	1.1	4.4	2.5	22.0
'Old' field	*Lotus corniculatus/ Uromyces pisi*	0	0.6	0	–	0	2.8
Woodland	*Rubus fruticosus/ Phragmidium violaceum*	0	0	0.3	–	–	0

* Communities as described by Southwood, Brown and Reader (1979).
– No record.

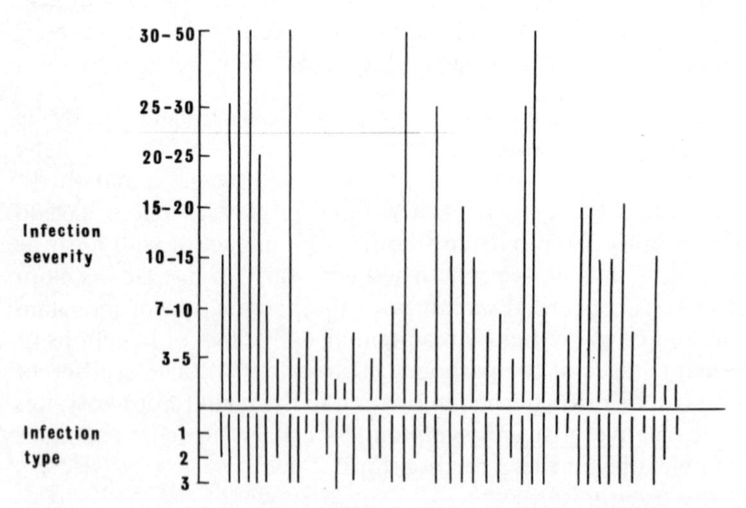

Figure 4 Development of powdery mildew on plants of *Hordeum spontaneum* randomly sampled along transect 217 (from Wahl *et al.*, 1978). *See* p. 168 for comparable details of transect 205.

in terms of the degree to which individual pustules developed (infection type) and the percentage leaf area affected (infection severity). In this instance, plants were sampled at random along a transect and their susceptibility tested by inoculation with *E. graminis* (Wahl *et al.*, 1978). The diversity is striking and the mixture of susceptible individuals with those of considerable if not complete resistance requires an appraisal of the distribution of fungal inoculum in such situations. As we have seen, air-borne inoculum is diluted whilst moving between widely-spaced susceptible plants. This inoculum is reduced further when it also has to move through interspersed resistant ones. The result is a much reduced rate of disease spread in the susceptible population, as with *Cerocospora apii* in celery (Table 5). The adaptation of this principle to the crop situation is discussed by Burdon (1978) and Dinoor (*1981*).

Table 5 Infection rates (*r*) of *Cercospora apii* on susceptible celery interplanted with a tolerant cultivar (from Berger, 1973).

Tolerant plants in population (%)	*Infection rate (15 Feb.–13 Apr., 1972)*
0	0.093
10	0.090
25	0.083
50	0.083
75	0.080
90	0.062
95	0.061

Wild host-fungal pathogen-crop host relationships

We have now reached the point where one generalization can usefully be made concerning fungal inoculum reaching crops from wild hosts. The potential for inoculum production within a crop is usually so great that once a pathogen is established the amount of inoculum is likely soon to exceed considerably that derived from outside sources. The impact of wild hosts as sources of inoculum is thus most pronounced very early in disease development and their role is concerned mainly with the perennation of inoculum between seasons and crops, either by maintaining active foci of infections or harbouring dormant stages of the pathogen (Dinoor, 1974). Three different types of wild host can be distinguished in relation to the main (crop) host, i.e. alternate host, alternative host and symptomless carrier. Specific examples illustrate the complexities of these relationships.

Alternate hosts

Heteroecious rust fungi are unique amongst parasitic fungi in requiring two hosts to complete their life cycle. The detailed elucidation of the alternation of the black-stem rust fungus (*Puccinia graminis*) between its cereal hosts and species of barberry (*Berberis*) was a milestone in the history of plant pathology (Large, 1940). The role of the barberry in providing aecidial inoculum for the infection of adjacent wheat crops was soon established and led to schemes for eradicating it in many states of central USA. The impact of barberry eradication in reducing local inoculum of *P. graminis* can clearly be seen in successive five year periods from 1915 to 1929 (Table 6). To this benefit must be added the limitation in the number of new genetic races that would otherwise have arisen from the sexual reproduction of the fungus on barberry. Unfortunately in other years the influx of massive amounts of uredospore inoculum from the southern limits of the wheat belt undermined this method of control.

Table 6 Yield losses of wheat in relation to barberry eradication (based on Vanderplank, 1963).

Period	*Bushes destroyed: cumulative total (millions)*	*Average annual wheat losses (million bushels)*
1915–1919	<4	50
1920–1924	4–11	26
1925–1929	12–18	11

Puccinia hordei on barleys and its alternate hosts within the genus *Ornithogalum* provide further insights into these relationships. Despite the widespread occurrence of *Ornithogalum* spp. in northern Europe they appear to be unimportant epidemiologically. For example, the aecidial state has been recorded only once in Britain on *O. pyrenaicum* at a site in Berkshire where subsequent systemic searches were unsuccessful (Dennis and Sandwith, 1948). Indeed, there is considerable evidence that *P. hordei* perennates on winter barley as a dormant uredo-mycelium (Simkin and Wheeler, 1974; Parleviet and van Ommeren, 1976). In contrast, in Israel the involvement of *Ornithogalum* spp. is established (Anikster and Wahl, 1979). There the teleutospores of the fungus remain dormant during the hot summers when the barleys dry up. As the rain begins in November they germinate, giving rise to basidiospores which infect the emerging foliage of *Ornithogalum* spp. The aecidiospores on these hosts infect wild barleys (particularly *Hordeum spontaneum*) and on these the uredospores form, some of which are able to initiate epidemics of rust on barley crops. These studies in Israel also emphasized the role of the sexual reproduction of the fungus on *Ornithogalum* in generating new pathogenic forms. Some of the aecidial isolates were virulent on barleys with the Pa7 gene that were resistant to the forms normally found within the cultivated crop. Similar pathogenic forms were isolated subsequently from the wild *H. spontaneum*.

Alternative hosts

Most other parasitic fungi are less complex and the same type of spore can infect the crop and also several other plant species. In many instances the range of alternative hosts and their role in providing inoculum is not clear. For example, powdery mildew occurs each year in the UK on late-planted peas but the fungus, *Erysiphe polygoni* sensu Salmon (now referred to *E. pisi*) does not form cleistothecia on this host and the annual nature of the crop precludes perennation as dormant mycelium. Smith (1969) examined the possibilities that conidia or ascospores discharged from cleistothecia on other supposed wild hosts of the fungus could infect pea. In no instance was a sporulating mildew colony obtained, though the inocula showed varying abilities to develop mycelia on pea (Table 7). Work in Israel on determining alternative hosts of the powdery mildew of cucurbits (*Sphaerotheca fuliginea*) was more successful. This had the added elegance that cross-inoculations were performed between hosts with mildew whose conidia germinate similarly. Mildew colonies were established on cucurbits from conidia on species of *Xanthium, Senecio* and *Papaver*, some of which grow throughout the winter (Dinoor, 1974).

Table 7 Transfers of spores to leaf disks of pea from other hosts of *Erysiphe polygoni* (from Smith, 1969).

Sources of spores	*Max. development on pea**	
	Conidia	*Ascospores*
Ranunculus acris	7	8
Ranunculus repens	7	3
Lathyrus pratensis	7	2
Lupinus sp.	3	3
Delphinium sp.	4	3
Heracleum sphondylium	5	3
Polygonum aviculare	2	3

* Indicated by a numerical scale: 2 = primary appressorium and haustorium; 10 = colony with conidiophores.

A comparable situation exists in the two countries with cereals and the perennation of the mildew, *Erysiphe graminis*. In England, infection of autumn-sown cereals is considered to be mainly by conidial inoculum derived from 'volunteer' plants, though ascospores from cleistothecia on stubble of crops harvested in the summer may supplement this indirectly by infecting such volunteers. Wild hosts are thought to play no part. However, in Israel, a few, widely-distributed, wild grass species serve as alternative hosts. These are infected in the autumn by ascospores that survive the summer on their stubble and the conidia formed subsequently disseminate to the cultivated cereals (Wahl *et al.*, 1978).

Symptomless carriers

Other examples of crop pathogens and their alternative hosts could be quoted. However, there are relatively few studies in which careful cross-inoculations have established the role of the alternative host. In far too many instances this role is assumed from the presence of a morphologically-similar fungus, which merely indicates a possibility. In other situations, soil-borne pathogens of crops have been found associated with the roots of wild species only because attention has been directed to this possibility by the failure to achieve control by allowing long periods between crops known to be susceptible. There is evidence that such symptomless carriers sometimes facilitate the survival of *Gaeumannomyces graminis* (Walker, 1975), *Verticillium dahliae* and *V. albo-atrum* (Pegg, 1974) and *Fusarium solani* f. sp. *phaseoli* (Schroth and Hildebrand, 1964). These can be important, therefore, within cultivated areas in maintaining inoculum. It is doubtful whether they contribute to a dangerous inoculum build-up, though they may promote limited increases of existing inoculum as, for example, do some symptomless hosts of *Furasium solani* f. *phaseoli.* However, where susceptible alternative hosts act as windbreaks at the periphery of crops they can contribute dangerous amounts of fungal inoculum. An example is the spread of *Rosellinia necatrix* from *Pittosporum* hedges into plantings of *Narcissus* in the Scilly Isles (Mantell and Wheeler, 1973).

Concluding remarks

It is worth emphasizing that many fungal pathogens survive on crops either because these are perennials (e.g. the rust, *Hemileia vastatrix* on coffee) or there is some overlap between successive plantings (e.g. *Erysiphe graminis* from autumn-sown to spring-sown cereals in the UK). Many also survive abundantly on crop debris (e.g. *Gaeumannomyces graminis* on cereal roots). For many diseases, the pathogens either do not exist on wild hosts or if they do, they are apparently quite unimportant. From a strictly crop-orientated viewpoint the role of wild hosts in fungal disease epidemiology is a minor one except for relatively few examples, some of which are quoted above.

It would not only be a pity but entirely wrong, however, to conclude that studies of plant parasitic fungi in natural ecosystems are of little relevance to crop plant pathology. As indicated elsewhere in this volume such studies are likely to provide much useful information for disease management in agro-ecosystems. Moreover, there are some areas where relationships between fungal populations on wild hosts and cultivated plants are completely unknown. Plant pathologists might well question how the powdery mildew fungi that form no cleistothecia survive between successive plantings of annual ornamentals. Moreover, in 1976 there were reports of powdery mildew on potatoes in the UK (Lawrence, 1978) yet the disease is otherwise virtually unknown in this country. Was some species on a wild host able to extend its range during the unusually hot summer? These are but two examples that await further studies of plant parasitic fungi in natural ecosystems.

References

Agrios, G. N. (1978). *Plant Pathology*. 2nd Edn. New York: Academic Press.

Anikster, Y. & Wahl, I. (1979). Coevolution of the rust fungi on Graminae and Liliaceae and their hosts. *Annual Review of Phytopathology* **17,** 367–403.

Berger, R. D. (1973). Infection rates of *Cercospora apii* in mixed populations of susceptible and tolerant celery. *Phytopathology* **63,** 535–537.

Berger, R. D. (1975). Disease incidence and infection rates of *Cercospora apii* in plant spacing plots. *Phytopathology* **65,** 485–487.

British Mycological Society (1962). *List of Common British Plant Diseases*. Cambridge: Cambridge University Press.

Burdon, J. J. (1978). Mechanisms of disease control in heterogeneous plant populations — an ecologist's view. In *Plant Disease Epidemiology*, pp. 193–200. P. R. Scott & A. Bainbridge. Oxford: Blackwell Scientific Publications.

Burdon, J. J. & Chilvers, G. A. (1975). Epidemiology of damping-off disease (*Pythium irregulare*) in relation to density of *Lepidium sativum* seedlings. *Annals of Applied Biology* **81,** 135–143.

Dennis, R. G. & Sandwith, N. Y. (1948). Aecidia of barley rust in Britain. *Nature* (London) **162,** 461.

Dinoor, A. (1974). Role of wild and cultivated plants in the epidemiology of plant diseases in Israel. *Annual Review of Phytopathology* **12,** 413–436.

Dinoor, A. (*1981*). Epidemics caused by fungal pathogens in wild and crop plants. In *Pests, Pathogens and Vegetation*, pp. 143–158. J. M. Thresh. London: Pitman.

Garrett, S. D. (1970). *Pathogenic Root-Infecting Fungi*. Cambridge: Cambridge University Press.

Harper, J. L. (1977). *Population Biology of Plants*. London: Academic Press.

Holden, M. (1975). Guide to the literature for the identification of British fungi. *Bulletin of the British Mycological Society* **9,** 67–106.

Jenkyn, J. F. & Bainbridge, A. (1974). Disease gradients and small plot experiments on barley mildew. *Annals of Applied Biology* **76,** 269–279.

Large, E. C. (1940). *The Advance of the Fungi*. London: Jonathan Cape.

Lawrence, N. J. (1978). Powdery mildew of potato. *Transactions of the British Mycological Society* **70,** 161–162.

Maddison, A. C. & Manners, J. G. (1972). Sunlight and viability of cereal rust uredospores. *Transactions of the British Mycological Society* **59,** 429–443.

Mantell, S. H. & Wheeler, B. E. J. (1973). *Rosellinia* and white root rot of *Narcissus* in the Scilly Isles. *Transactions of the British Mycological Society* **60,** 23–35.

May, R. M. (1979). Patterns in the abundance of parasites on plants. *Nature* (London) **281,** 425–426.

Onursal, N. F. (1975). The development of *Puccinia hordei* on different barley cultivars. *PhD Thesis*, University of London.

Parlevliet, J. E. and Ommeren, A. van (1976). Overwintering of *Puccinia hordei* in the Netherlands. *Cereal Rusts Bulletin* **4,** 1–4.

Pegg, G. F. (1974). *Verticillium* diseases. *Review of Plant Pathology* **53,** 157–182.

Pirozynski, K. A. (1976). Fossil fungi. *Annual Review of Phytopathology* **14,** 237–246.

Schroth, M. N. & Hildebrand, D. C. (1964). Influence of plant exudates on root-infecting fungi. *Annual Review of Phytopathology* **2,** 101–132.

Scott, M. R. (1956). Studies of the biology of *Sclerotium cepivorum* Berk. II. The spread of white rot from plant to plant. *Annals of Applied Biology* **44,** 584–589.

Simkin, M. B. & Wheeler, B. E. J. (1974). Overwintering of *Puccinia hordei* in England. *Cereal Rusts Bulletin* **2,** 2–4.

Smith, C. G. (1969). Cross-inoculation experiments with conidia and ascospores of *Erysiphe polygoni* on pea and other hosts. *Transactions of the British Mycological Society* **53,** 69–76.

Southwood, T. R. E., Brown, V. K. & Reader, P. M. (1979). The relationships of plant and insect diversities in succession. *Biological Journal of the Linnean Society* **12,** 327–348.

Vanderplank, J. E. (1963). *Plant Diseases: Epidemics and Control.* New York: Academic Press.

Wahl, I., Eshed, Nava, Segal, A. & Sobel, Z. (1978). Significance of wild relatives of small grains and other wild grasses in cereal powdery mildews. In *The Powdery Mildews*, pp. 84–100. D. M. Spencer. London: Academic Press.

Walker, J. (1975). Take-all diseases of Graminae: a review of recent work. *Review of Plant Pathology* **54,** 113–144.

Epidemics caused by fungal pathogens in wild and crop plants

A Dinoor
Faculty of Agriculture, Hebrew University of Jerusalem, Rehovot, Israel

Introduction

Natural populations of wild plants survive and flourish despite the existence and even epidemic development of numerous diseases (Dinoor, 1974; Browning, 1974). Wild plants have evolved with diseases to develop balanced populations that are adaptive and dynamic in responding to pathogens. Cultivated crops have evolved quite differently and are completely dependent on human activities. It has been envisaged and postulated that a thorough knowledge of the composition of natural populations and the processes operating within them will contribute appreciably to the development of more durable crops.

In this paper epidemic development in the wild is contrasted with the situation in crops. Some of the factors operating in the wild are described and discussed, considering the impact of diseases in the two very different systems. The interactions between these systems are also considered and their importance for the whole environment.

Complete incompatibility* between plants and pathogenic agents is the usual situation. The relatively few cases in which plants are compatible with pathogens are the only ones to cause concern. A basic difference between natural situations and agricultural ones is man's tendency to impose complete incompatibility with some pathogens on plants which are initially compatible with those pathogens. In the wild, different adaptations between hosts and pathogens result in a dynamic co-existence, whereas in agriculture a continuing struggle results in cycles of successes and disasters (e.g. Browning and Frey, 1969).

Comparative epidemiology in crops and in the wild

The life cycles of hosts and their pathogens under natural conditions have resulted from long-term adaptations to each other and to the environment.

* Incompatibility in this context means that the pathogen cannot develop on the host or develops abnormally as a consequence of the interaction between pathogen avirulence and host resistance.

Some of these have been described elsewhere in more detail (Dinoor, 1974) and there are no fundamental differences between life cycles in crops and in wild vegetation.

Sources of inoculum

An important difference between diseases in agriculture and in the wild is in the accessibility and availability of the dormant, over-seasoning stage to the new host. In the wild, dormant material remains at the sites where it was produced in the previous growing season and where subsequent plants will appear. In many agro-ecosystems crop residues are removed from fields, so preventing or at least reducing contact between the new crop and dormant stages of the pathogen. The primary inoculum, which is a crucial factor in initiating epidemics, is often of a completely different order of magnitude in the wild compared with agriculture. Other conditions being equal, the epidemic in the crop will be delayed and the whole ecological situation becomes radically different. For instance, in cultivated winter grasses and cereals in Israel, high levels of primary inoculum promote the early development of epidemics in the warm autumn conditions. Diseases such as barley net blotch (*Pyrenophora teres*) then progress to higher leaves, and inoculum is distributed very effectively. With the lower levels of primary inoculum occurring in crop stands, the propagation of the pathogen is delayed whilst temperatures fall, the host plants develop faster than the pathogen and the 'canopy effect' ensues (Kenneth, 1960). The spread of inoculum is decreased by the lower temperatures and also because diseased leaves are shielded by healthy ones that restrict dispersal by splash and by wind. Early outbreaks sometimes occur in crops due to over-seasoning inoculum from nearby uncontrolled populations in the wild. However, spread into crops is less efficient than spread within them.

The heteroecious rusts represent a unique situation in that they must begin the new season by infecting an alternate host, not related phylogenetically to the main one. In the wild the two hosts are often intermixed or contiguous and inoculum is transferred very effectively. The alternate host can be remote in agricultural areas although the situation is sometimes little different from the natural. For example, until it was declared a noxious weed buckthorn (*Rhamnus cathartica*) was used for hedging fields in Iowa and contributed greatly to outbreaks of crown rust in oats due to *Puccinia coronata* (Dietz, 1923). In Israel, orchards of stone fruits are planted in mountainous or hilly areas on terraces or narrow valleys surrounded by thriving populations of wild anemone (*Anemone coronaria*) that is now protected and serves as alternate host for stone-fruit rust (*Tranzschelia pruni-spinosae*).

Another aspect of disease development concerns the host both as a source of inoculum and as a trap for incoming wind-borne spores. Prolific production of spores by large continuous crop stands results in the mass dissemination of inoculum. Few wild plants occur in dense stands akin to crops, but even then, the production of inoculum is more restricted because of several different adaptions of the wild plants. Incoming spores are trapped more effectively by large continuous areas of crop than by discrete, patchy stands. Long-distance dissemination of pathogens has been described and documented mainly between crops (e.g. Stakman and Harrar, 1957). Wild plants as traps additional to crops and as potential sources of inoculum have rarely been studied (Dinoor, 1967).

Spread of disease

Several factors affect the spread of disease after the establishment of primary inoculum and some will be discussed later. At this stage only the physical arrangement of hosts in the field is considered. Continuous stands of a single species are typical of crops and inoculum efficiency is high in such conditions. In the wild, plants usually grow in mixtures of species that present physical barriers impeding the spread of pathogens between their hosts. Burdon (1978) stressed the importance of spacing between similarly susceptible individuals in the spread of pathogens. The reduced spore concentration away from infected plants at wide spacings was considered to be the main factor in decreasing disease spread. However, other factors can be involved and the microclimate changes with increase in plant density so facilitating infection with *Cercospora apii* (Berger, 1975).

Different conclusions have been reached in our work with leaf rust (*Puccinia recondita*) and *Septoria tritici* of wheat. The spread of these pathogens, especially of *Septoria*, is very limited in erect, dense stands. This is because when infection starts from the lower leaves any water droplets penetrating into the comparatively dense canopy lose much of their impact and then tend to be trapped on adjacent leaves. Splash dispersal is much more effective at wider spacings or after lodging and disease development is faster. The topography of the plant, the type of dispersal and the microclimatic conditions all influence the effects of spacing on disease dispersal.

In those host–pathogen combinations where physical barrier effects occur, this is the main mechanism reducing disease development in a mixed crop and not the separation caused by the incompatible plants present. Similarly, in dense stands, any death of leaves or whole plants, due to disease, breaks the canopy and spread of the pathogen is facilitated provided other conditions permit. Wild species growing in clumps create a discrete pattern in the field. Each clump represents a separate population in which any epidemic develops independently and spreads at a rate influenced by the prevailing conditions.

Adjustments exhibited by plants and pathogens in the wild

Seasonal adjustments

Successful plants and their parasites in the wild have undergone a series of seasonal adjustments that facilitate optimal development and resilience in variable adverse situations. Data on oats, wild oats, crown rust and the alternate host buckthorn (*Rhamnus palaestina*) illustrate some of the adaptations. Crown rust is a winter–spring disease of oats and the limits of its distribution in Israel have been determined by monthly observations throughout the year at 35 locations covering most geographical regions (Fig. 1). The limiting factor seems to be the availability of growing plants and in most regions the oat season terminates in May–June, which is virtually the end of the rust season. However, the disease may persist in a few localities on a few clumps or small populations of wild oats around irrigated plots of summer or perennial crops and near water outflows from reservoirs or poultry houses. Persistence in this way has been recorded in warm regions of the country (e.g. localities 16 and 22 of Fig. 1) and in relatively cool areas (localities 4, 5 and 26).

In many places oats reappear in September in agricultural plots or in the wild after the first rains. Nevertheless, rust does not appear before November, except in the restricted localities where it persisted throughout the summer with adaptation to a wide range of temperatures. However, an obligate parasite cannot remain active without hosts, and when oats reappear in most regions rust inoculum is not available. As a heteroecious species, crown rust may initiate a new cycle of disease only after germination of the dormant spores (teliospores) and the production of basidiospores. These in turn infect buckthorns which are susceptible only at an early stage of their annual growth cycle. The role of the alternate hosts in bridging between seasons provides a striking example of adaptation to prevailing conditions (Fig. 2).

Leaf rust of barley (*Puccinia hordei*) and leaf rust of *Aegilops* (*P. recondita*) are dormant throughout the hot dry summer, when their main hosts and alternate hosts (*Ornithogalum* spp. and *Anchusa* spp., respectively) are also dormant. There is a good synchrony between the germination of the two hosts and of rust teliospores (Anikster and Wahl, 1979); the alternate host serving as a perfect bridge for the rust. Similarly with the stone-fruit rust, although there is a change of phase. This rust is a summer one that develops on the alternate host after a period of winter dormancy.

Perfect adaptation also exists for crown rust in north European and west Mediterranean countries during the transition from the period of summer or winter dormancy to the active season, which depends on region. By contrast, the alternate host for crown rust in Israel seems out of phase in that

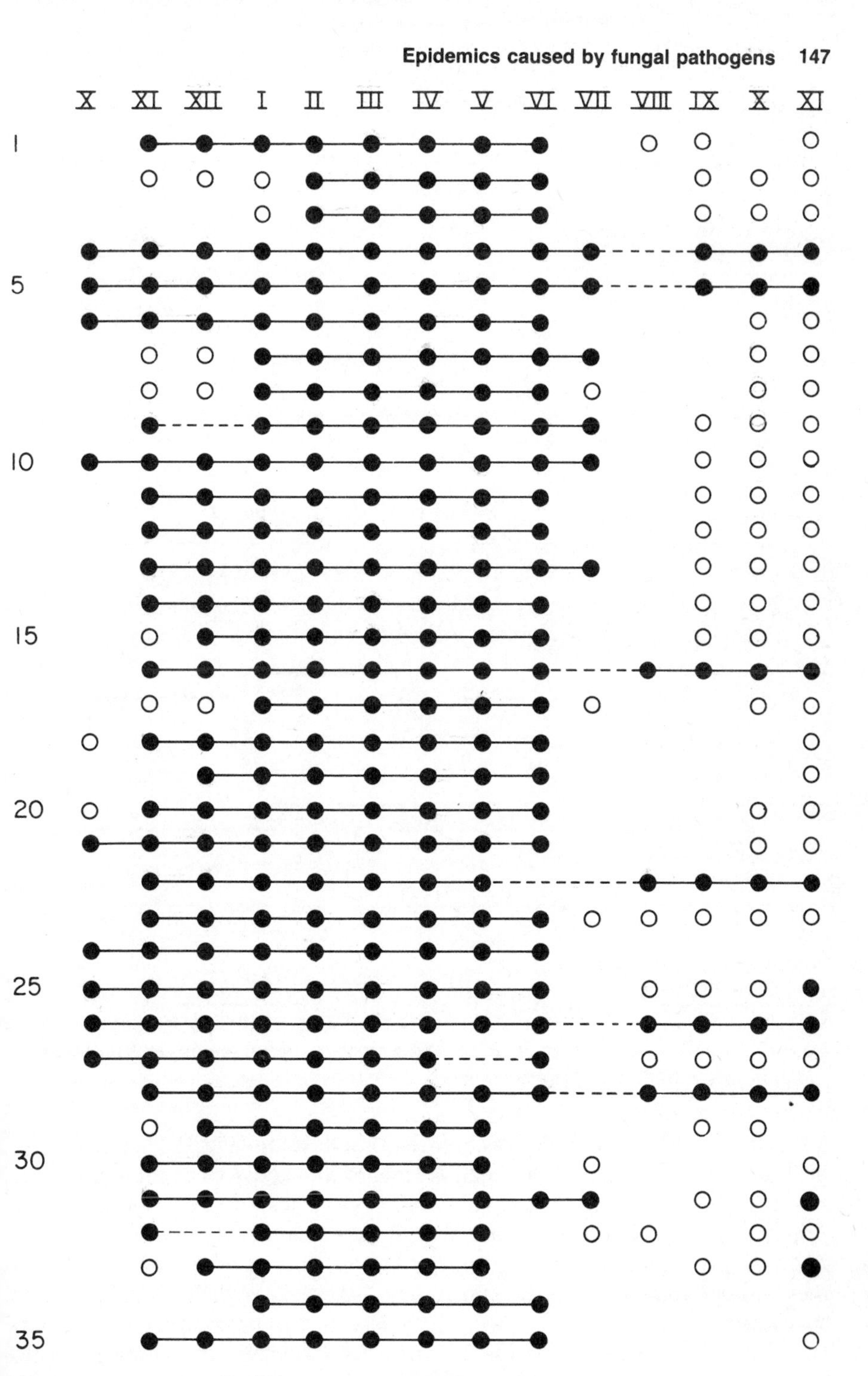

Figure 1 Crown rust activity on oats over the year at thirty-five sites in Israel.
○ non-infected hosts ● infected hosts
●——● continuity of crown rust ●----● assumed continuity.

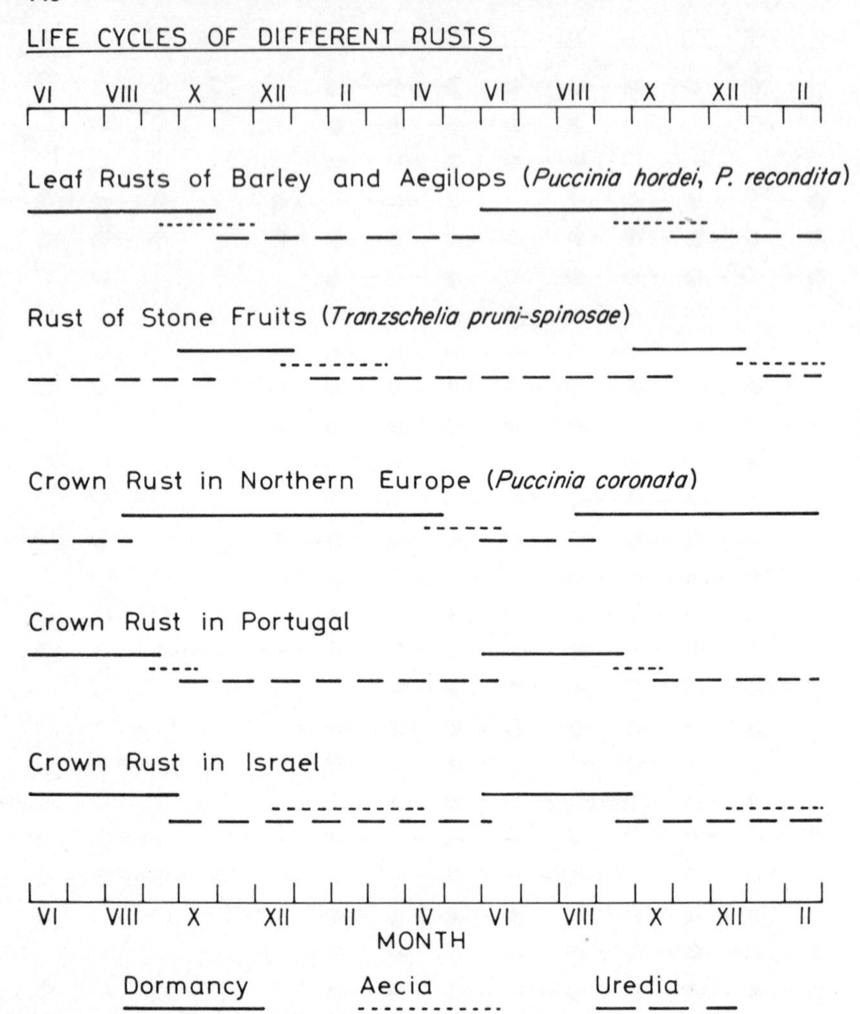

Figure 2 Seasonal adaptation of heteroecious rusts in several regions in relation to compatibility with their alternate hosts.

uredospores occur some months before aecia develop (Fig. 2). The alternate host may provide a bridge between dormancy and activity but there are a few months when the main host is vulnerable yet aeciospores are not produced. Other possible adaptations were sought to explain this situation. Survival of uredospores throughout the summer was discounted by demonstrating that they did not remain viable for more than 2–3 weeks, which is inadequate to span the gap between oat growing seasons. No earlier compatible alternate host could be identified, yet the crown rust season started at least 2 months before rust activity on buckthorn. Small uredial foci were found on wild oats

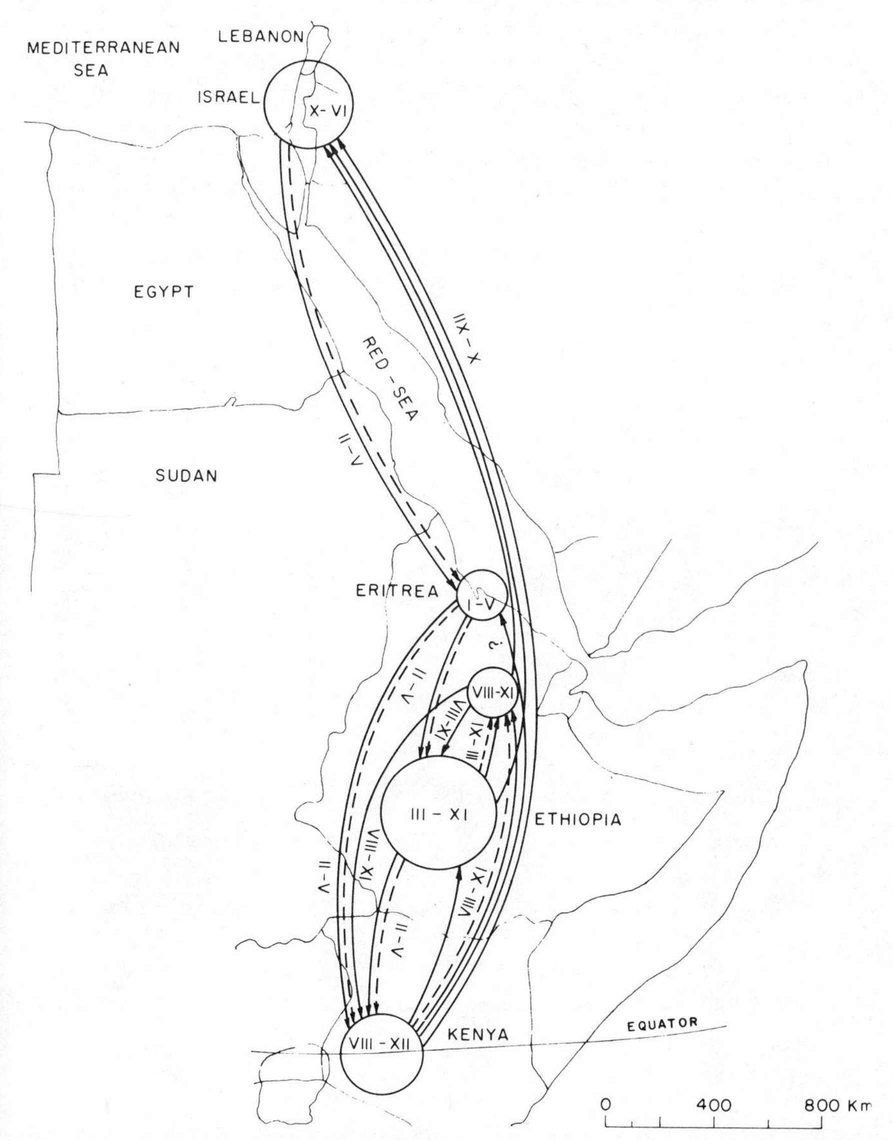

Figure 3 Postulated routes of long-distance dissemination of oat crown rust between the east Mediterranean and east Africa in different months (Roman numerals).

growing throughout the summer in isolated places. However, these foci could not have accounted for the widespread occurrence of crown rust throughout whole areas and systematic studies began on possible distant sources of uredospores.

Information obtained from twenty-four countries in Europe, part of Africa and part of Asia showed that potential sources of crown rust occurred in at least nine countries up to 3500 km away (Norway, Sweden, Denmark, England, Poland, Yugoslavia, Kenya, Ethiopia and Eritrea). However, east Africa was considered to be the most likely source of inoculum after examining wind trajectories and after comparing the virulence of cultures from the various countries with those from Israel. Spores may also on occasions be swept southwards from Israel into Africa to give the cyclical movement illustrated in Figs. 3 and 7. Wind trajectories to the south are shown in naval charts, but they have not been reconstructed from synoptic maps as for the north-bound trajectories and for the long-distance movement of wind-borne locusts (Rainey, *1981*).

Figure 4 Phenological stages in buckthorn *Rhamnus palaestina*, from just before leaf abscission (stage 2) through dormancy and bud burst (4) to the juvenile stage (7). The relative susceptibility to crown rust at each of the different stages can be represented by an index of compatibility on a scale of 0–1.0. Stage 1 = 0.1, S2 = 0.5, S3–S5 = 1.0, S6 = 0.9, S7 = 0.8, S8 = 0.4, S9 = 0.

Apart from possible external sources of inoculum in Africa the alternate host in Israel seems to be a well-adapted link in a complete life cycle, which parallels the asexual cycle. The two critical features of the system are teliospore germination and bud burst of the alternate host. Rust teliospores germinate over a period of several months extending beyond one season, and populations of the deciduous alternate host leaf-out over several weeks. Adequate synchronization is achieved even though each individual bush is fully compatible with crown rust for only a short phase of development (stages 3–5 in Fig. 4).

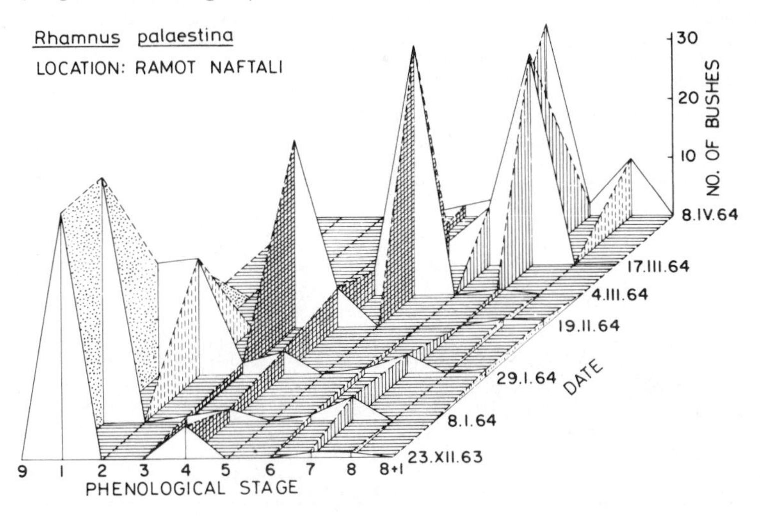

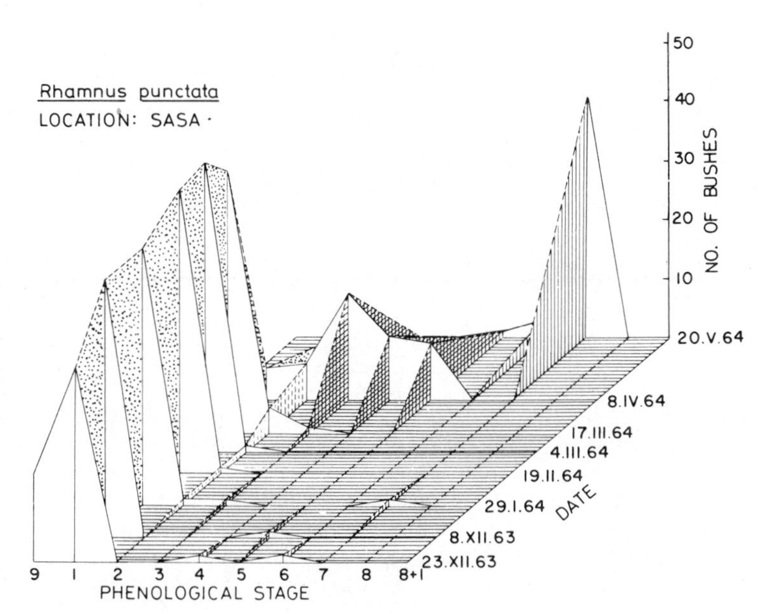

Figure 5 Frequency distribution of individual bushes of representative populations of *Rhamnus palaestina* (top) and *Rhamnus punctata* (bottom) according to phenological stages throughout the season.

The phenological stages of individual buckthorn bushes were followed throughout the season and detailed descriptions of representative populations of two species are shown in Fig. 5. These show frequency distributions of types in the populations according to date and growth stage with the whole population passing through a compatible phenological stage during the season. A general coefficient of compatibility was calculated for the entire population and related to teliospore germination and rust incidence (Fig. 6), ignoring factors such as temperature, humidity and rain. Due to the extended germination of teliospores and the prolonged sprouting of the host population there are many chances for infection and for spread of rust to the main host.

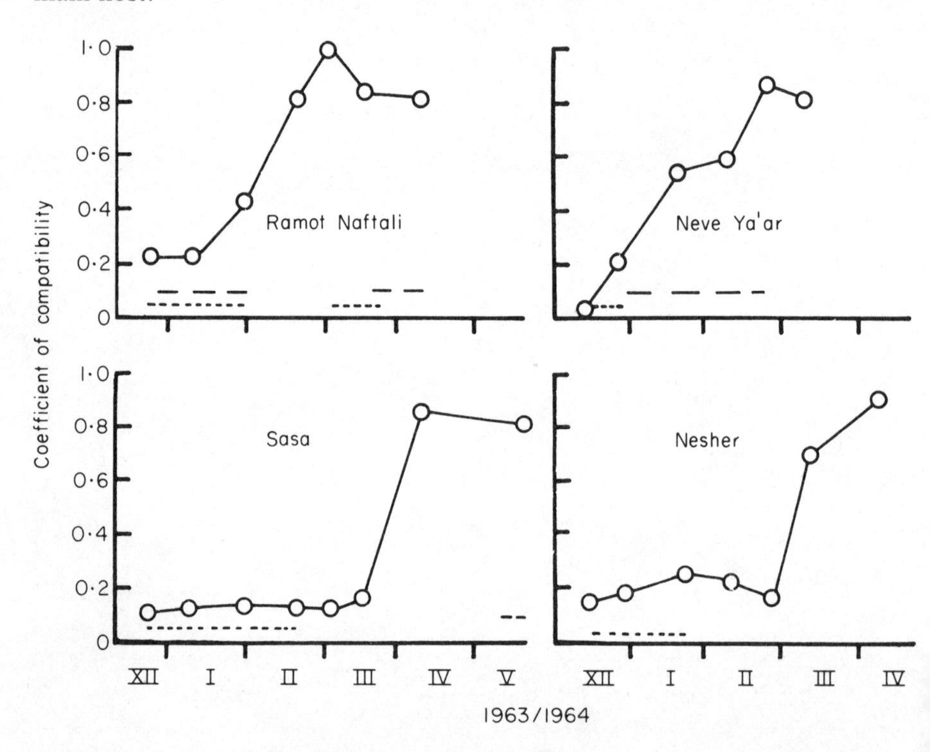

Figure 6 Coefficient of compatibility with rust of buckthorn populations throughout the season (—O—), incidence of teliospore germination (----) and rust infection of the alternate host (——). The two upper populations are of *R. palaestina* and the lower two are of *R. punctata.* A coefficient of 1 means that the entire population of buckthorn is at the compatible growth stages 3–5, whereas a value of 0 means that all plants are at incompatible stage 9 (Dinoor, 1967).

Rhamnus palaestina shoots earlier than *R. punctata* (Fig. 5) and its population is more variable phenologically, leading to greater compatibility with rust under natural conditions. *R. punctata* was found to be susceptible to rust in artificial inoculations and when favourable climatic conditions occurred exceptionally late. Usually it shoots too late to be infected by rust and seems incompatible under field conditions. Aecial production by rusts is a monocyclic process and so there are no further generations of rust on the alternate host to compensate for low infection frequency or inefficiency. A prolonged period of teliospore germination and bud burst ensures more opportunities for successful primary infection. Moreover, there are occasional cases of individual infections becoming systemic with the production of more than one aecial cycle. This arrangement further extends the chances of rust infection from the alternate to the main hosts.

Physiological adjustments

Natural host populations are not free from diseases and may become heavily infected. It is intriguing to try and understand the adjustments by plants to survive attacks by pathogens, and by pathogens so as not to eliminate the host. Browning (1974, *1981*), suggests that studies on natural ecosystems might lead to new and improved strategies of crop protection. Our experience with wheat leaf rust (Dinoor *et al.*, 1979) and septoria leaf blotch (Dinoor, unpublished) indicate the converse situation. These studies on chemical control of foliar diseases of wheat crops have revealed an important principle which probably also operates in the wild. This is concerned with the quantity, quality and distribution of the active photosynthetic area per plant. Spraying experiments have shown that wheat plants have redundant foliage towards the end of the season, at least from the early dough stage and protecting such foliage by chemicals decreases yield. Leaf rust and septoria do not attack the sheaths, stems or spikes at this late stage under Israeli conditions. These organs are more xerophytic than leaves and still contribute appreciably to yield. The effect of diseases in decreasing the amount of green foliage is therefore beneficial. Observations in natural populations have shown that despite severe destruction of foliage by disease the wild plants reproduce adequately. The shift in importance from leaves to stems is also enhanced because more light penetrates into the canopy to reach lower parts of the stems. Another feature of diseased foliage resembles certain forms of mimicry. Wild plants are grazed by livestock or harvested by man, sometimes quite heavily. Diseased clumps are selectively avoided as noted for barley with net blotch and they produce seed and survive for subsequent seasons.

Mixed stands of wild plants are probably protected physiologically by competing avirulent pathogens. This principle was postulated and then demonstrated in agro-ecosystems for multilines and mixed varieties (Johnson and Taylor, 1976; Chin, 1979). It has yet to be demonstrated in the wild and not only for interactions between biotypes of one pathogen, but also between different pathogens. Further speculation is possible along lines suggested by

Burdon (1978), i.e. that certain species in a mixed stand will benefit from a diseased neighbour and not only because the neighbour has been weakened by disease. These species may be partially cross-protected by pathogens moving onto them from susceptible neighbours because they are incompatible with these pathogens and thus become less compatible with their own.

Additional physiological adjustment might occur in wild plants by a reduction in the critical seed size needed for survival. The effect of disease on the quantity and quality of propagules has not yet been estimated for wild plants.

Genetical adjustments

These concern any inherited trait affecting or operating in the host–parasite interaction which will contribute to the ability of either organism to survive and which is being manipulated in the populations. Mechanisms of defence or avoidance in the host and mechanisms of attack or reproduction by the parasite are all included. The balanced populations have gone through a long process of natural adjustment so that various traits have become useless and disappeared. Others are difficult to distinguish because of interactions with ecological and epidemiological factors, although they could be recovered with adequate techniques to identify them. Genes providing different types of resistance may be locally ineffective but could be of great value elsewhere if they can be identified.

Two of the main purposes of studying wild populations are to extract resistance genes and to understand natural and successful strategies of disease resistance (Dinoor, 1977; Browning, *1981*). Methods of evaluation are crucial in such studies, but unfortunately, many investigators still tend to bulk together different or inappropriate cultures despite clear evidence of the need to use single relevant cultures (Dinoor, 1975). A claim that for quantitative resistance any culture will do because it is non-specific is not valid since even quantitative resistance can be differential and therefore specific (e.g. Johnson and Bowyer, 1974). If natural populations are analysed to determine the different types and frequency of resistance (as described by Browning, 1974), the results have severe limitations if bulked or irrelevant cultures are used.

The gene-for-gene concept (Person, 1959), later extended to the one-for-oneness concept, should be the basic one in analysing natural populations. Another important question is the size and limit of a population, how far it is extended and how much genetic drift is being introduced into the system. It clearly depends on the type of plant and especially on the type of pathogen involved. Soil-borne pathogens differ in many respects from those above ground (Sewell, *1981*) and wind-borne pathogens differ from splash-dispersed ones. For some host–parasite combinations the populations will be small, closed and isolated. Others will be large, disseminating systems that are frequently invaded by inoculum from outside sources. The most complex situations occur where rust is transported to and from localities over great distances. The genetic interactions might be complicated, even to the extent that local systems are invaded mainly by incoming parasites originating elsewhere from a totally different region. There could be a sequence of interactions extending in one direction or in a cycle (Fig. 7). Once an appropriate analytical method is established for closed systems it will be possible to proceed to more complex situations.

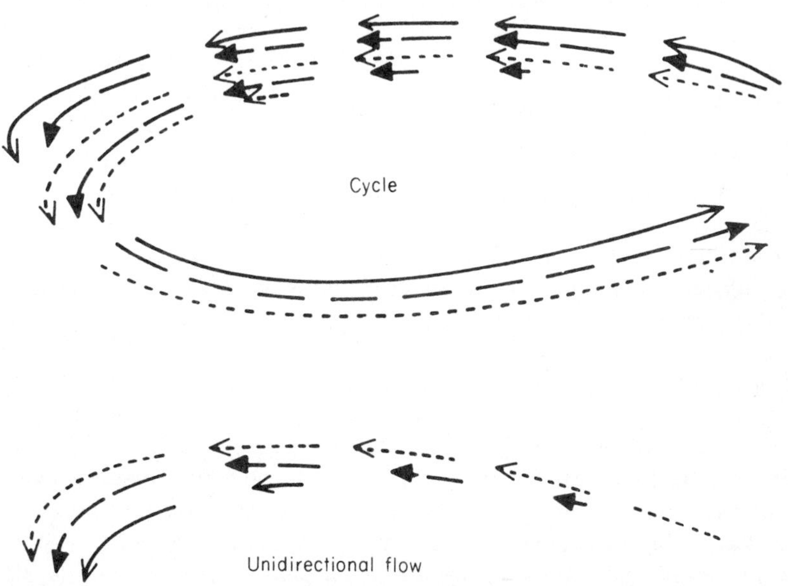

Figure 7 Diagrammatic representation of the long-distance dissemination of rusts by unidirectional (bottom) or cyclical flow (top) within or between continents. Local adjustments are fed into the large-scale system.

The genetic adjustments may use available mechanisms and some will be only partially restrictive. The delicate balance of the system can be judged from the results of any change from the normal situation. Even a well-balanced host population may collapse if the conditions are changed in favour of the pathogen or by prolonging the period over which the host is exposed to infection. Populations of wild oats flourish throughout Israel, with or without crown rust, although some populations are completely destroyed when grown out of season and/or out of the usual habitat.

Mutual relationships between crops and wild vegetation

Wild and weedy plants are notorious in harbouring and spreading pests and diseases into crops and this topic is considered by many contributors to this volume. However, wild plants also provide important genetic reservoirs, breeding grounds and shelter for pathogens (Dinoor, 1974). On wild plants at centres of origin of cultivated plants, there is tremendous variability in pathogen virulence towards resistance genes that continue to be effective elsewhere in the world, and even towards genes that have not been identified or have not yet been introduced into crops (Jones, *1981*). These pathogen reservoirs pose a potential threat to newly-developed cultivars introduced to centres of origin (Harlan, *1981*). It is, therefore, advisable that monitoring of virulences and advance planning should precede the introduction of new varieties. Continually changing pathogenic reservoirs serve as sources of 'selector' cultures to use in resistance breeding.

There is yet another aspect of the mutual relationships between crops and the wild, i.e. the way in which crops growing in uniform masses selectively propagate components of the pathogen population that eventually invade the wild and disrupt the dynamic equilibrium previously existing there. This hypothetical sequence of events has yet to be demonstrated and the full extent and impact is unknown. The buffering capacity of natural populations is not unlimited as becomes apparent when populations are devastated by pathogens when grown beyond the regular season or outside normal habitats (p. 155).

An example of possible spread of a disease from a crop to the wild may be found in the rust of anemones in Israel. Intensive plantings of stone fruits in different regions of the country have greatly augmented the tree hosts available for the rust and the law protecting wild flowers brought about a large increase in the number of anemones. The increase of both hosts has probably led to the subsequent increase in amount of disease. In the wild the disease is systemic and causes sterility of the perennial host that continues to produce rust for years. In cultivation the disease is not perennial and control measures are taken, but the fungus is well-established in wild plants that contribute to the spread of the disease.

A striking case of a pathogen invading a wild habitat is that of *Phytophthora cinnamomi* in the eucalypt forests of Australia (Newhook and Podger, 1972). The spread of the disease into the forests is associated with human activity, forestry, road-building and vehicle movement. There is controversy as to whether the spread of the disease was caused by its introduction to new areas or by ecological changes in areas already infested. Certainly the speed and extent of the destruction is remarkable, first of the under-storey components of the forest and then the canopy trees. There has been a complete change in the flora and effects on the water table and even the climate. Whole plant communities were devastated due to the very wide host range of the fungus and the sequential process was due to differences in the susceptibility of the species involved. Very susceptible species like *Xanthorrhoea* suffer first and are prime indicators of the disease. When such plants are attacked they provide a food base for increased levels of inoculum and spread to relatively less susceptible plants.

Epilogue: the impact of diseases in the wild, and the relevance of studies in natural and agro-ecosystems to the development of disease control strategies

Natural populations of plants that have co-evolved with their parasites seem to be 'successful' communities. It is very tempting to try and elucidate the mechanisms involved in this successful co-existence and to adopt and/or adapt some of them to agro-ecosystems. However, there are several questions to be answered before transposing features of successful natural systems into completely artificial ones. Firstly, how is success measured in natural vs. agricultural systems? In nature success means survival, continuity and competitive ability, but how relevant are these in agriculture? What are the criteria for existence and are they the only relevant ones for success in agriculture? A further question is related to the interactions between hosts, pathogens and environment. The current emphasis is solely on host–pathogen interactions largely ignoring environmental factors, using very crude techniques involving bulked or non-relevant cultures. However, natural populations have co-evolved not only with their pathogens but also with their environment. The adaptations of a patchily distributed population may differ from those of continuous large-scale stands. The final question relates to the intensity and magnitude of the selection pressures imposed by pathogens. Agricultural activities can greatly influence these as by adjusting the timing and length of the growing season or by decreasing the amount of primary inoculum and the efficiency of later inoculum. The genetical structure of a population and integrated measures can also be used and adjusted in agriculture.

Important mechanisms of protection that were proposed from observations on natural populations were successfully adapted to agro-ecosystems. Other mechanisms first encountered in agro-ecosystems may occur also in the wild but have still to be demonstrated there.

References

Anikster, Y. & Wahl, I. (1979). Coevolution of the rust fungi on Gramineae and Liliaceae and their hosts. *Annual Review of Phytopathology* **17,** 367–403.

Berger, R. D. (1975). Disease incidence and infection rates of *Cercospora apii* in plant spacing plots. *Phytopathology* **65,** 485–487.

Browning, J. A. (1974). Relevance of knowledge about natural ecosystems to development of pest management programs for agro-ecosystems. *Proceedings American Phytopathological Society* **1,** 191–199.

Browning, J. A. (*1981*). The agro-ecosystem–natural ecosystem dichotomy and its impact on phytopathological concepts. In *Pests, Pathogens and Vegetation*, pp. 159–172. J. M. Thresh. London: Pitman.

Browning, J. A. & Frey, K. J. (1969). Multiline cultivars as a means of disease control. *Annual Review of Phytopathology* **7,** 355–382.

Burdon, J. J. (1978). Mechanisms of disease control in heterogeneous plant populations — an ecologist's view. In *Plant Disease Epidemiology*, pp. 193–199 P. R. Scott & A. Bainbridge. London: Academic Press.

Chin, K. M. (1979). Aspects of the epidemiology and genetics of the foliar pathogen *Erysiphe graminis* f.sp. *hordei* in relation to infection of homogeneous and heterogeneous populations of the barley host, *Hordeum vulgare. PhD Thesis*, University of Cambridge, UK

Dietz, S. M. (1923). The role of the genus *Rhamnus* in the dissemination of crown rust. *United States Department of Agriculture Bulletin* No. 1162.

Dinoor, A. (1967). The role of cultivated and wild plants in the life cycle of *Puccinia coronata* Cda. var. *avenae* F & L and in the disease cycle of oat crown rust in Israel. *PhD Thesis*, Hebrew University, Jerusalem, Israel, 373 pp. (In Hebrew with English summary.)

Dinoor, A. (1974). Role of wild and cultivated plants in the epidemiology of plant diseases in Israel. *Annual Review of Phytopathology* **12,** 413–436.

Dinoor, A. (1975). Evaluation of sources of disease resistance. In *Crop Genetic Resources for To-day and Tomorrow*, pp. 201–210. O. H. Frankel & J. G. Hawkes. Cambridge: Cambridge University Press.

Dinoor, A. (1977). Oat crown rust resistance in Israel. *Annals New York Academy of Sciences* **287,** 357–366.

Dinoor, A., *et al.* (1979). Optimization of chemical control of wheat leaf rust. *Hassadeh* **59,** 1993–1999. (In Hebrew.)

Harlan, J. R. (*1981*). Ecological settings for the emergence of agriculture. In *Pests, Pathogens and Vegetation*, pp. 3–22. J. M. Thresh. London: Pitman.

Johnson, R. & Bowyer, D. E. (1974). A rapid method for measuring production of yellow rust spores on single seedlings to assess differential interactions of wheat cultivars with *Puccinia striiformis. Annals of Applied Biology* **77,** 251–258.

Johnson, R. & Taylor, A. J. (1976). Effects of resistance induced by non-virulent races of *Puccinia striiformis. Proceedings of the 4th European and Mediterranean Cereal Rusts Conferences*, pp. 49–51.

Jones, R. A. C. (*1981*). The ecology of viruses infecting wild and cultivated potatoes in the Andean region of South America. In *Pests, Pathogens and Vegetation*, pp. 89–107. J. M. Thresh. London: Pitman.

Kenneth, R. (1960). Aspects of the taxonomy, biology and epidemiology of *Pyrenophora teres* Dreschsl., the causal organism of net blotch disease of barley. *PhD Thesis*, Hebrew University of Jerusalem, Israel, 226 pp. (In Hebrew with English summary.)

Newhook, F. J. & Podger, F. D. (1972). The role of *Phytophthora cinnamomi* in Australian and New Zealand Forests. *Annual Review of Phytopathology* **10,** 299–326.

Person, C. (1959). Gene-for-gene relationships in host–parasite systems. *Canadian Journal of Botany* **37,** 1101–1130.

Rainey, R. C. (*1981*). Wild plants in the ecology of the Desert Locust. In *Pests, Pathogens and Vegetation*, pp. 327–337. J. M. Thresh. London: Pitman.

Sewell, G. W. F. (*1981*). Soil-borne fungal pathogens in natural vegetation and weeds of cultivation. In *Pests, Pathogens and Vegetation*, pp. 175–190. J. M. Thresh. London: Pitman.

Stakman, E. C. & Harrar, J. G. (1957). *Principles of Plant Pathology.* 1st Edn. New York: Ronald Press Co.

The agro-ecosystem—natural ecosystem dichotomy and its impact on phytopathological concepts

J Artie Browning
Iowa State University, Ames, Iowa 50011

Introduction

Over three centuries have elapsed since Antony van Leeuwenhoek wrote to the Royal Society on 27 September 1678 on why walkers' shoes can become red when traversing a grassy meadow. The study began as an inquiry into the cause of a widespread fever, probably malaria, because 'the common man' sensed connections among the redness and heat of fire, fever, and the redness of the shoes and concluded that 'the air is therefore infected and very firey'. After confirming that his shoes also became red, Leeuwenhoek turned his 'attention to the grass itself, and saw that some of it was studded with reddish dots. Bringing these before my microscope, I saw that they consisted of small globules, whereof upwards of a thousand did not equal the bigness of a small sand-grain. I find there are various kinds of grass: and among others, one sort that was very rough, which was not contaminated' (Dobell, 1960).

This was probably the first observation of uredospores with a microscope and the first record of pathogenic specialization. It also expressed a phenomenon that continues to this day, the dichotomy between 'contaminated' (diseased) and healthy plants.

Countless other scientists have followed Leeuwenhoek into relatively undisturbed ecosystems, usually faithfully maintaining the dichotomy. Taxonomists named and classified higher plants and, much later, ecologists counted them and studied successional changes but largely ignored any diseases present. Mycologists, on the other hand, embraced the 'contaminated' and determined the life cycles and host ranges of the fungi they named and classified. Plant breeders also became aware of disease and selected 'uncontaminated' plants for use by farmers or as parents in breeding for resistance.

As the plant sciences developed, another, more serious dichotomy emerged — that between 'classical' botanists and agricultural scientists. Indeed, at a meeting marking 'three centuries of the development of botanical thought' Bennett (1971) emphasized that 'agree we must that botanists have long ignored the cultivated flora in favor of the "natural" or the wild'. It is equally true, however, that 'we plant pathologists have ignored diseases in "natural" floras in favor of those in cultivated' (Browning, 1974).

I contend that this last dichotomy is part of a largely unconscious agricentric perspective that permeates agricultural science and that this has obscured basic biological truths that could have come from studies of natural, wild ecosystems. At the very least, it has led to spurious or distorted conclusions of pathogen-crop interactions and to inadequate systems of disease management. In this paper, I attempt to show that several generally accepted concepts of phytopathology can be interpreted differently if they are reconciled with similar phenomena from natural ecosystems and are not interpreted exclusively agricentrically.

Indigenous ecosystems and how plants in them protect themselves

Fortunately, some plant pathologists did eventually take an ecological approach and follow Leeuwenhoek into relatively undisturbed ecosystems. Several have been studied recently to determine how plants in natural communities protect themselves. This change of outlook occurred after a long series of pandemics (e.g. those of cereals caused by their rust fungi, that of maize caused by *Helminthosporium maydis* race T, and fusiform rust of pine caused by *Cronartium fusiforme*) had emphasized the way man's management has sometimes worsened disease problems compared with those occurring in nature.

Strong (1977) eloquently described a natural ecosystem that any standard plant pathology textbook might be expected to predict as a prime candidate for an epidemic. Reviewing a new plant pathology tome, he wrote:

> I am reading this volume in a lowland tropical rain forest, an appropriate place to contemplate plant-parasite interactions. It is almost constantly wet here and never very cold; microbial growth is not restrained by weather. Every plant bears fungal colonies or epiphyllae, and the decay from bacterial digestion begins in foliage within hours after it is cut or falls from plants. Yet the forest is healthy. None of the 500 or so plant species within walking distance is apparently free of microbial activity. The burns, blazes, spots, and lesions of infection are tiny and rare on all but the oldest leaves, even on plants that retain leaves for many months. This is the ultimate enigma of host–parasite interactions. What are the mechanisms and mechanics of the equilibrium between hosts and parasites? That the balance is unstable we see in the case of plagues, pest outbreaks, and disease epidemics in agricultural plants; that it is not simple we see in the plethora of diseases suffered by every plant species. Obviously, the questions of plant pathology are ecological, epidemiological, and morphological as well as biochemical.

Plants in such tropical rain forests reach densities at which individuals interfere with each other as typical K-selectionists (Harper, 1977). They are protected by population buffering, population resistance (Browning and Frey, 1969), or ecosystem resistance (Schmidt, 1978). These plants possess the genetic diversity necessary for population resistance and also diversity in vertical and horizontal structure in space and time, all being components of the *functional diversity* (Schmidt, 1978) of ecosystem resistance. Way (1979) also emphasized that 'Increased diversity *per se* cannot be assumed to decrease pest and disease damage. . . . The quality, not the amount, of diversity matters'.

The tropical rain forest has abundant functional diversity and severe disease fails to develop despite the presence of all components of the disease pyramid: susceptible hosts, virulent pathogens, favourable environment and adequate time. These are all features that from experience with crops might be expected to facilitate spread and cause damaging epidemics. Almost any natural community carries the same message. Components of few such ecosystems have been studied genetically or characterized structurally, however, and thus they reveal few of the 'mechanisms and mechanics of the equilibrium between hosts and parasites' (Strong, 1977).

Fortunately, several indigenous ecosystems have now been analysed epidemiologically and/or genetically for natural protective mechanisms. Segal *et al.* (1980) recently reviewed these, and discussed the 'defense of plant populations against diseases in natural ecosystems undisturbed by man in Israel'. The latter is especially germane, for Israel as part of 'The Fertile Crescent' includes wild species that are progenitors of cultivated barley, oats, and wheat. Each progenitor species is attacked by an array of pathogens and the present populations are the result of aeons of co-evolution. In many cases, the interacting hosts and pathogens are the same species as those in cultivated cereals (e.g. *Avena sterilis* and *A. sativa*; *Hordeum spontaneum* and *H. vulgare*). Thus, results of Israeli studies can be interpreted immediately in relation to a century of research on cereal diseases and they can be applied immediately in developing systems of improved disease-management (Browning, 1974).

Comparative consideration of concepts

It is not necessary here to describe indigenous ecosystems as biological models and it suffices to compare them to agricultural systems and assess their meaning. Few attempts have been made to compare disease development in agricultural and natural ecosystems. It was done recently in general terms (Burdon, 1978; Harper, 1977; Zadoks and Schein, 1979) for land races of maize in Central America and the USA Corn Belt (Borlaug, 1972), for wild grasses in stands contiguous to wheat in the North American 'Puccinia Path' (Harlan, 1976), for forest ecosystems (Schmidt, 1978), and especially in studies on the indigenous ecosystems of Israel and their agricultural counterparts (Browning, 1974; Dinoor, *1981*).

Are formae speciales and races artifacts of agricultural systems?

This is not simply a rhetorical question for many plant pathology and plant breeding resources have been devoted, with only limited success, to controlling races of formae speciales of many highly variable pathogens.

Possibly the most definitive answer comes from Israel, from detailed work on *Puccinia graminis* by Gerechter-Amitai (1973) reviewed by Browning, (1980). From our agro-ecosystem perspective, we normally think of this fungus as showing only great specificity, with race 1 virulent on, say, cultivar A but not B, and race 2 the opposite. Gerechter-Amitai, however, found that 89 grass species in 40 genera were 'susceptible' to infection and also that *P. graminis* cultures from wild grasses in Israel fell into four formae speciales: *avenae*, *tritici*, *secalis*, and *lolii*. The f. sp. *avenae* was most widely distributed and attacked 71 species in 36 genera, whereas *only* 27 species in 8 genera were susceptible to f. sp. *tritici*. In the greenhouse, 39 species (in 16 genera) were susceptible to two formae speciales; 29 species (8 genera) supported three; and 20 species (13 genera) all four. Only one representative example from Gerechter-Amitai's study will be considered here at the pathogenic race level, which is the agriculturalist's main concern. *P. graminis avenae* race 2, which is avirulent on all the standard oat stem rust differential cultivars except Jostrain that carries gene *Pg*. 3, was isolated from 41 grass species in Israel. Even a single-pustule isolate of race 2 sporulated on 80 species of wild grasses in the glasshouse, 69 being rated susceptible. Studies by others with cereal rusts and powdery mildews show a similar pattern.

Early studies of the inheritance of the traits 'formae speciales' and 'race' showed mainly that these are unsuitable units for genetic analyses. However, more recent studies (Hiura, 1978) interpreted in terms of gene-for-gene theory (Flor, 1956) have shown that forma specialis is a level of specialization conditioned by several pairs of corresponding genes in host and pathogen. Similarly, race is a level of specialization based on corresponding gene pairs and would be better described by 'pathogenicity formulae' (Browder *et al.*, 1980). The corresponding gene pairs are clearly the units for genetic studies of pathogenicity and resistance and also for characterizing, classifying, and 'naming' cultures as to their avirulence/virulence capability. Some workers use a 'Formula Number' that serves as a 'nickname' and makes cumbersome pathogenicity formulae more acceptable to plant breeders.

The concepts of formae speciales and race reflect specialization on crop species and cultivars, respectively, and these are of prime interest to man because of his agriculture. They were at one time very useful concepts but now they unnecessarily limit a fully comprehensive assessment of the many highly plastic micro-organisms that have great specificity and also a far greater host range than, for example, *avenae* or *tritici* imply. These biotrophic pathogens can be biological indicators of possible genetic affinities among higher plants that cannot otherwise be crossed and studied genetically, but such phylogenetic relationships among grasses tend to be obscured by the simplistic concepts of formae speciales and race.

Are gene-for-gene host-pathogen associations artifacts of modern plant breeding and of modern agriculture?

Flor's (1956) gene-for-gene hypothesis is a landmark in genetics, plant pathology and plant breeding. Originally elucidated for flax and flax rust, it has been studied extensively and shown to explain 'over 90% of the variability reported in the literature. . . . The same basic pattern seems to hold for fungi, bacteria, nematode, insect, and possibly viral, interactions with plants. The same pattern may also hold for interactions between animals and disease-causing non-antigenic entities' (Ellingboe, 1979). Is it possible that such a broadly applicable biological phenomenon could be in any sense unnatural?

Day (1974), among others, emphasized that 'an important feature of all these (gene-for-gene) systems is that the host is an economic crop and that resistance is oligogenic. None are in any sense "natural" . . .' (p. 96). 'One somewhat ironical view of the gene-for-gene concept is that it is an artifact of breeding for resistance. Plant breeders stripped oligogenic resistance of associated protective polygenic effects and exposed the genes one at a time in agriculture on a tremendous scale, forcing the parasite to respond in kind . . . The gene-for-gene system has not been demonstrated in nature . . .' (p. 110).

Plant breeders have done many great things — witness the Green Revolution with corn, wheat, rice, and other crops — but I know none who claims to have created a single gene that did not already occur in nature, much less a gene that was genetically interlocked with one in another organism and with specificity comparable to that of antigen–antibody reactions. Breeders and pathologists have taken genes from plants in natural systems for study and use. They have concentrated many useful ones in crop plants, made the crop cultivars unnaturally homogeneous, and grown them in unnaturally homogeneous and genetically vulnerable cultural systems. However, this does not make the initial product, the corresponding gene pair, in any sense unnatural.

At least 25 genes for resistance to *Puccinia coronata* and one for *P. graminis* have been incorporated into cultivated *Avena sativa* from wild oats, *A. sterilis* (Simons *et al.*, 1978). Similarly, many genes for resistance to *P. hordei* and *Erysiphe graminis* have been transferred from wild *Hordeum spontaneum* to *H. vulgare* (Wahl *et al.*, 1978). Predictably, these behave as if they had originated in cultivated oats or barley and fit a gene-for-gene model (Person, 1959). These major-effect genes are unlikely to have behaved very differently in their previous genetic background, although they would have been much more difficult to study.

There is no reason to believe that all host–pathogen and host–pest associations are under gene-for-gene control, however. For example, interactions caused by facultative parasites, many insect herbivores, or plant viruses (which cannot readily be crossed to test for such an association) may not be. But those working on such organisms should be aware that gene-for-gene associations are likely to govern interactions between facultative parasites or insects (including virus vectors) and *their* respective pathogens.

How did gene-for-gene systems arise? Was the original progenitor host resistant or susceptible? Was the original progenitor pathogen avirulent or virulent?

These questions are perhaps academic, but they certainly show our agricentric perspective. Diagrams of host-parasite coevolution normally show:

susceptibility (r) → resistance (R) for the host
avirulence (A) → virulence (a) for the pathogen.

This emphasizes the resistance and virulence traits that are of most interest in agriculture without regard to their frequency, mode of inheritance, or direction of evolution in nature.

In a model of coevolution, Browning (1980) took a different view and reasoned:

(1) that progenitor host and pathogen were resistant (R) and avirulent (A), respectively;
(2) that biotrophes probably originated from autotrophic epiphytes when plants moved from sea to land and similar basic cellular processes in these primitive organisms made establishment of 'compatibility' less unlikely;
(3) that host incompatibility genes (genes for host immunity and vertical resistance) are 'carry-over' genes indistinguishable in nature and agriculture from genes that condition non-host immunity — the most common protective mechanism of host plant populations.

Only when the incompatibility barrier (which this model considers to be in a direct lineage from that at the outset of coevolution) is eliminated or has been overcome by the pathogen, is the host-pathogen interaction governed by another system — a rate-controlling one that is the host's basic system of resistance/susceptibility to the pathogen's basic pathogenicity system (Parlevliet, 1981; Browning, 1980). Several authorities (e.g., Ellingboe, 1975) have suggested that this second system (roughly the same as horizontal resistance) is also under gene-for-gene control.

What is the role of resistance genes in nature? Are vertical resistance (VR) and horizontal resistance (HR) genes and concepts artifacts of agriculture?

Vanderplank (1968) stated that VR reduces initial disease (X_0) of some but not all races, while HR reduces the apparent infection rate (r) for all races. The frequent failure of VR when conditioned by oligogenes used uniformly over vast areas, has led to some of the great pandemics. The epidemiological effect attributed to VR is normally impossible in nature, however, for there is neither a source of large quantities of initial inoculum of uniform genotype nor the necessarily large areas of a genetically-uniform target host. Thus, the epidemiological effect attributed by Vanderplank (1968) to VR is probably an artifact of modern agriculture, although VR, or incompatibility, genes are normal components of wild systems (Segal *et al.*, 1980). What then is the role of incompatibility genes in nature? Parlevliet (1981) considered that their role is not to protect the host (that is the role of the basic system of resistance/susceptibility [HR] to basic pathogenicity), but to prevent the pathogen from becoming too aggressive.

The host's basic resistance/susceptibility system is also a natural part of indigenous ecosystems (Segal *et al.*, 1980). Furthermore, its epidemiological effect of reducing r (which Browning *et al.* [1977] called 'dilatory' resistance — the resistance that delays — after separating it from the confusing *genetic* concepts of HR) is the one measured in nature or in crop stands of cultivars protected by this second basic system.

Vanderplank (1968) considered that pathogen aggressiveness is the counterpart of host HR; one is relative to the other and they are measured similarly. Incompatibility (VR) genes in a naturally diverse population work together to condition pathogen aggressiveness (Parlevliet, 1981), which is measured as for HR. Consequently VR and HR in a diverse, natural population *both* manifest themselves by conditioning different degrees of dilatory resistance. (*See* fig. 2 of Browning, [1980], for a graphic illustration of this.) Thus, an indigenous ecosystem 'uses primarily a single type of epidemiologic resistance — dilatory resistance — to protect populations, but it uses many different genetic systems and population structures to achieve it' (Browning, 1980). Man must do likewise in devising ecologically sound and epidemiologically safe methods of crop protection.

Do plant pathologists use the concept of fitness correctly? Is sporulation a valid measure of fitness in a natural ecosystem?*

Thoday (1953) defined fitness as the probability that a population will leave descendents. Similarly, 'The fitness of an organism is measured by the descendents that it leaves, measured over a number of generations' (Harper, 1977). Plant pathologists appropriated the fitness term and concept from this population biology background and used it differently (Vanderplank, 1968; MacKenzie, 1978). Also, they frequently equate fitness with aggressiveness, and spore yield is considered its best single measure (Johnson and Taylor, 1976). The strain that sporulates most, or reproduces best is considered fittest, or most aggressive, on a given host in a given environment. But are these concepts true in the same way, and to the same degree in host-parasite interactions in wild ecosystems?

Obviously, the most 'fit' pathogen cannot necessarily be the one that sporulates best *if the amount of sporulation is limiting* because, in the long

* The term 'fitness' has been introduced to plant pathology from other related disciplines and then used differently (MacKenzie, 1978; Vanderplank, 1968). This is misleading and ideally a separate term should be coined to describe the phytopathological concepts being considered.

term, this would tend to destroy the host in a truly natural system (where the host is not maintained by man) and minimize the probability of *either* pathogen *or* host leaving descendents. In fact, 'balance' in an indigenous ecosystem implies that resistance and environment combined 'keep the average number of unrestricted progeny per parent (pathogen) lesion down to about one' (Vanderplank, 1975); i.e. $iRc = 1$. With 1.01 unrestricted progeny per parent lesion ($iRc = 1.01$), the pathogen increases with time until it levels off at about 2% of susceptible tissue infected; with 1.1, 18%; 1.2, 31% (common in Israel; *see* next section); 2.0, 80% — an epidemic; and with 5.0 progeny per parent lesion, 99.3% (Vanderplank, 1975, table 5.1).

An obligate or near-obligate parasite is on an evolutionary 'knife edge' over the problem of existing without eliminating its host. 'It is not good evolutionary strategy for a pathogen to kill off its host; indeed, the best strategy for the pathogen is one in which the host flourishes' (Harlan, 1976). Fungal pathogens have responded by evolving rates of sporulation that enable them to persist (iRc = about 1) without risk of domination or elimination. A pathogen transferred to an agro-ecosystem with a homogeneous stand of a compatible cultivar, however, will find that the *same* rate of sporulation leads to explosive spread (iRc = 2–5 or more). The changed spatial distribution of the host with crowding of like genotypes unnaturally maximizes the efficiency of spore dispersal. Ecologically, the rate of transfer of energy or biomass from primary producer (the host) to primary consumer (the pathogen) is greatly increased. This highly unstable situation would lead to the cultivar's destruction (were it not maintained by man) and to the self-destruction of an obligate biotroph. This is without evolutionary advantage to the pathogen, of course, but it has no genetic information from the wild ecosystem to behave otherwise (Browning *et al.*, 1979). Fortunately, surprisingly little diversity suffices to restabilize the system, at least in the short term (*see* next section).

Spore yield is a measure of the ability of a given host–pathogen–environment interaction to produce spores. This may be helpful in predicting the success of a given cultivar in an agro-ecosystem but it is no measure of the true fitness of the pathogen with time in either natural or agro-ecosystems. 'Success is measured by continuing to play the game rather than by winning' (Harper, 1977). Thus, the way fitness is commonly used in plant pathology is misleading and a separate term should be coined to describe the phytopathological concepts being considered.

Are susceptible plants and avirulent pathogens eliminated from wild populations? Do super races arise in wild populations? Does a resistant host genotype dominate the wild population?

Many authors have assumed an affirmative answer to the first question, probably because their views have been influenced by observing pathogens kill susceptible entries in breeding nurseries. 'It is a well known hazard in the life of the experimental ecologist, that plant species which appear to be free from disease in nature develop remarkable collections of epidemic diseases as soon as they are grown in pure stands in experimental gardens!' (Harper, 1977).

However, pathologists studying wild systems have expressed different views and Segal *et al.* (1980) wrote: 'We have observed in natural ecosystems that disease incidence, though never ravaging, occasionally becomes intensive in restricted areas, killing vulnerable plants of *A. sterilis* and *H. spontaneum*.' Similarly Dinoor (*1981*) comments that 'some [wild oat] populations are completely destroyed [by crown rust] when grown out of season and/or out of the usual habitat.' Thus, native plants killed by disease are unusual or even rare. Harlan (1976) contrasted native grasses and wheat in the North American 'Puccinia Path', famous for wheat rust pandemics and stated: 'Selection, however, is minimal, and I have never seen a (native grass) plant killed by rust'. Harlan (1976) emphasized, however, that 'selection pressures due to disease are present even if the disease is not causing much damage. Stabilizing selection does not require a raging epidemic to operate'.

Wahl *et al.* (1978) and Segal *et al.* (1980) emphasize genetic features protecting wild *A. sterilis* and *H. spontaneum* populations in Israel. Diverse infection types occurred in transects from areas where environment favoured disease. Segal *et al.* (1980, fig. 2) indicated that for 17 transect samples of *H. spontaneum* from all of Israel, *c.* 60% of the plants had some kind of resistance to powdery mildew and *c.* 33% or less had low infection types. Obviously, a sizeable percentage of each population was rated susceptible and two transects are of particular interest. Transect 205 was from an area in Israel rich in genes for slow mildewing (HR) with some VR and some fully susceptible plants, whereas transect 217 was from an area rich in VR — about a third of the population — and with many intermediate and susceptible plants. (*See* Fig. 1, overleaf and Fig. 4 of Wheeler (*1981*).)

Breeders are disenchanted with VR in several crops because of its supposed failure. With current interest in selection for HR (Browning, 1974), it is instructive to focus on transects 205 and 217 for the insight they give into nature's management of VR genes. Clearly, HR may have protected the source area of transect 205. But where HR was minimal and VR common (transect 217), the diverse population was amply protected mainly by VR, even though this was carried by only a third of the population! Moreover, with present knowledge, it seems that the ecosystem was stabilized at that level! 'Samplings in successive years showed that the defense structures appear to be stable' (Segal *et al.*, 1980). There have been similar findings from Israel on crown and stem rust of *A. sterilis* (Browning, 1974; Segal *et al.*, 1980).

From these data, we may conclude that susceptible plants and avirulent pathogens coexist naturally in wild ecosystems and are not normally eliminated as many authors predict. Super races do not arise to dominate pathogen populations and resistant genotypes do not dominate host populations. This is to be expected from the concepts of fitness and balance in natural ecosystems discussed in the previous section.

It is significant that the vast majority of plants of *H. spontaneum* attacked by *Erysiphe graminis*, and of *A. sterilis* attacked by *Puccinia coronata* or *P. graminis*, had infection severities ⩽ 30% in the transects diagrammed by Wahl *et al.* (1978) and Segal *et al.* (1980). This means (p. 166) that *iRc* ⩽ 1.2 whereas, in genetically homogeneous crop stands of the same host *iRc* frequently exceeds 2.5 and epidemics result.

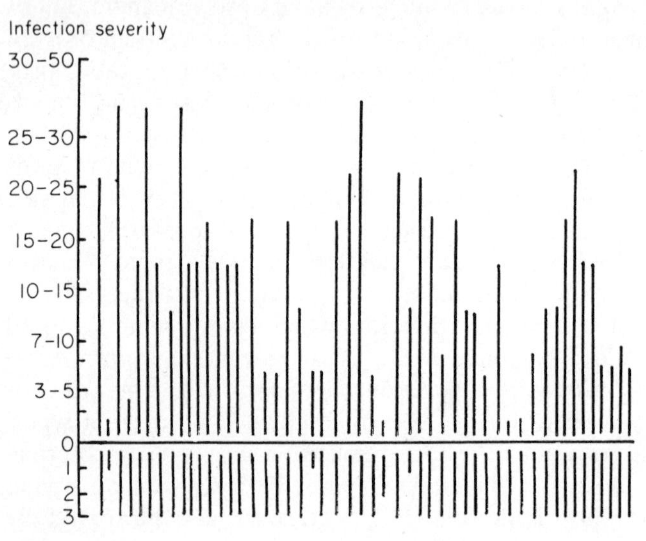

Figure 1 The performance of powdery mildew on plants of wild barley (*Hordeum spontaneum*) in Israel. Infection severity denotes the percentage of leaf area covered with pustules. Designations of infection type are those of Moseman (1968). The data presented here are for alternate plants randomly selected along transect 205 of Wahl *et al.* (1978). Comparable data for transect 217 are presented by Wheeler (*1981*) on p. 136.

In view of the importance of yield and quality, agriculturists need to know how far they must move towards the complete diversity of nature to achieve the minimum necessary level of functional diversity. The answer is, 'Not far'. Blends of 2–10 cultivars or isolines (= multilines; the number actually needed depends on environmental conditions) of cultivated oats against crown rust, and 3-way barley cultivar blends against powdery mildew, have given 'dramatic' protection against these highly epidemic diseases (Browning *et al.*, 1979; Wolfe and Barrett, 1980). The pathogen strains found in these diverse host populations have been more varied than those in the respective pure-line controls, but protection against them has been so effective that they have been unable to increase and become destructive. There seems to be no more reason to fear a 'super race' in these diversified crops than in well-buffered indigenous ecosystems. When a single-pustule culture of 'avirulent' *Puccinia graminis avenae* race 2 is virulent on 69 grass species (Gerechter-Amitai, 1973), the super-race concept is placed in perspective.

What is a super race? It is any race that can increase and damage its host population. With a genetically homogeneous cultivar any race that can attack any given plant in the population is a super race virulent on the entire population. However, given minimum diversity the pathogen seems to retain adequate capacity for its own regulation and the super race is nothing to fear. This contradicts predictions by several recent theoretical models, but 'conclusions ferreted out from studies in defense mechanisms and strategies in natural ecosystems may sometimes be at variance with concepts based on theoretical considerations. However "*validuis est testimonium natural quam argumentum doctrinae*" ' (Segal *et al*., 1980).

Conclusions

The dichotomy between natural and agroecosystems that is analysed here is very real but largely unappreciated. One result is that many host–pathogen phenomena have been interpreted exclusively from the agricultural standpoint and this has led to spurious, or, at least, skewed concepts of plant pathology. Some examples are:

(1) Having all components of the disease pyramid present and favourable does *not* necessarily result in severe disease development in a *natural* ecosystem.
(2) The concepts of formae speciales and race are simplistic and reflect specialization on crop species and cultivars of interest to man. They severely limit man's thinking relative to many highly plastic fungi that have extensive host ranges and yet great specificity.
(3) Gene-for-gene systems involving incompatibility genes probably coevolved from a resistant progenitor host and an avirulent progenitor pathogen. These corresponding gene pairs are perfectly natural and not products of plant breeding. However, the epidemiological effect that is commonly attributed to them of selectively filtering incoming inoculum is probably an artifact of agriculture.
(4) In nature, incompatibility genes do not protect the individual plant; they protect the population by reducing pathogen aggressiveness. Only when the incompatibility barrier is overcome does the second (rate-reducing) system operate to protect the plant *and* the population.
(5) Fitness is a term that should be used in plant pathology consistently with its use in population genetics. The fittest strain to survive in a natural system is *not* necessarily the one that sporulates most, but the one that produces

about one unrestricted progeny per parent lesion, so as not to endanger its host.
(6) Pathogens 'belong' in natural systems as part of the natural order. Susceptible plants and avirulent pathogens are not eliminated and super races and resistant plants seem not to dominate the population.
(7) Emulating Nature's model can improve the effectiveness of crop protection systems with minimum loss of yield and quality traits. Thus, epidemics do not usually occur if an agroecosystem has minimum diversity so that the pathogen retains some of its natural self-regulating ability. There is then no reason to fear a super race that, by definition, could destroy its host unnaturally.

This publication is Journal Paper No. J-9980 of the Iowa Agricultural and Home Economics Experiment Station, Ames, Iowa. (Project 2447).

References

Bennett, E. (1971). The origin and importance of agroecotypes in South-West Asia. In *Plant Life of South-West Asia*, pp. 219–229. P. H. Davis, P. C. Harper & I. C. Hedge. Edinburgh: The Botanical Society.

Borlaug, N. E. (1972). A cereal breeder and ex-forester's evaluation of the progress and problems involved in breeding rust resistant forest trees: Moderator's summary. In *Biology of Rust Resistance in Forest Trees*, pp. 615–642. Washington, D.C.: *United States Department of Agriculture, Forest Service, Miscellaneous Publication* 1221.

Browder, L. E., Lyon, F. L. & Eversmeyer, M. G. (1980). Races, pathogenicity phenotypes, and type cultures of plant pathogens. *Phytopathology* **70,** 581–583.

Browning, J. A. (1974). Relevance of knowledge about natural ecosystems to development of pest management programs for agro-ecosystems. *American Phytopathological Society Proceedings* **1,** 191–199.

Browning, J. A. (1980). Genetic protective mechanisms of plant-pathogen populations: their coevolution and use in breeding for resistance. In *Biology and Breeding for Resistance to Arthropods and Pathogens of Cultivated Plants*, pp. 52–75. M. K. Harris. *Texas Agricultural Experiment Station, Miscellaneous Publication* 1451.

Browning, J. A. & Frey, K. J. (1969). Multiline cultivars as a means of disease control. *Annual Review of Phytopathology* **7,** 355–382.

Browning, J. A., Frey, K. J., McDaniel, M. E., Simons, M. D. & Wahl, I. (1979). The bio-logic of using multilines to buffer pathogen populations and prevent disease loss. *The Indian Journal of Genetics and Plant Breeding* **39,** 3–9.

Browning, J. A., Simons, M. D. & Torres, E. (1977). Managing host genes: epidemiologic and genetic concepts. In *Plant Disease: An Advanced Treatise. I. How Disease is Managed*, pp. 191–212. J. G. Horsfall & E. B. Cowling. New York: Academic Press.

Burdon, J. J. (1978). Mechanisms of disease control in heterogeneous plant populations — an ecologist's view. In *Plant Disease Epidemiology*, pp. 193–200. P. R. Scott & A. Bainbridge. Oxford: Blackwell Scientific Publications.

Day, P. R. (1974). *Genetics of Host–Parasite Interactions*. San Francisco: W. H. Freeman.
Dinoor, A. (*1981*). Epidemics caused by fungal pathogens in wild and crop plants. In *Pests, Pathogens and Vegetation*, pp. 143–158. J. M. Thresh. London: Pitman.
Dobell, C. (1960). *Antony van Leeuwenhoek and His 'Little Animals'*. New York: Dover.
Ellingboe, A. H. (1975). Horizontal resistance: an artifact of experimental procedure? *Australian Plant Pathology Society Newsletter*, pp. 44–46.
Ellingboe, A. H. (1979). Inheritance of specificity: the gene-for-gene hypothesis. In *Recognition and Specificity in Plant Host-Parasite Interactions, pp. 3–17.* J. M. Daly & I. Uritanai. Baltimore: University Park Press.
Flor, H. H. (1956). The complementary genic systems in flax and flax rust. *Advances in Genetics* **8,** 29–54.
Gerechter-Amitai, Z. K. (1973). Stem rust, *Puccinia graminis* Pers. on cultivated and wild grasses in Israel. *PhD Thesis*, Hebrew University, Jerusalem. (Hebrew with English Summary.)
Harlan, J. R. (1976). Disease as a factor in plant evolution. *Annual Review of Phytopathology* **14,** 31–51.
Harper, J. L. (1977). *Population Biology of Plants.* New York: Academic Press.
Hiura, U. (1978). Genetic basis of formae speciales in *Erysiphe graminis* DC. In *The Powdery Mildews*, pp. 101–128. D. M. Spencer. New York: Academic Press.
Johnson, R. & Taylor, A. J. (1976). Spore yield of pathogens in investigations of the race-specificity of host resistance. *Annual Review of Phytopathology* **14,** 97–119.
MacKenzie, D. R. (1978). Estimating parasitic fitness. *Phytopathology* **68,** 9–13.
Moseman, J. G. (1968). Reactions of barley to *Erysiphe graminis* f. sp. *hordei* from North America, England, Ireland and Japan. *Plant Disease Reporter* **52,** 463–467.
Parlevliet, J. E. (1981). Disease resistance in plants and its consequences for plant breeding. In *Plant Breeding II.* K. J. Frey. Ames: Iowa State University Press (in press).
Person, C. (1959). Gene-for-gene relationships in host: parasite systems. *Canadian Journal of Botany* **37,** 1101–1130.
Schmidt, R. A. (1978). Diseases in forest ecosystems: the importance of functional diversity. In *Plant Disease: An Advanced Treatise. II. How Disease Develops in Populations*, pp. 287–315. J. G. Horsfall & E. B. Cowling. New York: Academic Press.
Segal, A., Manisterski, J., Fischbeck, G. & Wahl, I. (1980). How plant populations defend themselves in natural ecosystems. In *Plant Disease: An Advanced Treatise. V. How Plants Defend Themselves*, pp. 75–102. J. G. Horsfall & E. B. Cowling. New York: Academic Press.
Simons, M. D., Martens, J. W., McKenzie, R. I. H., Nishiyama, I., Sadanaga, K., Sebesta, J. & Thomas, H. (1978). Oats: a standardized system of nomenclature for genes and chromosomes and catalog of genes governing characters. *United States Department of Agriculture, Agriculture Handbook* 509.
Strong, D. R., Jr. (1977). Microbial herbivores. *Science* **197,** 1071.
Thoday, J. M. (1953). Components of fitness. *Symposium Society for Experimental Biology* **7,** 96–113.
Vanderplank, J. E. (1968). *Disease Resistance in Plants*. New York: Academic Press.
Vanderplank, J. E. (1975). *Principles of Plant Infection.* New York: Academic Press.

Wahl, I., Eshed, N., Segal, A. & Sobel, Z. (1978). Significance of wild relatives of small grains and other wild grasses in cereal powdery mildews. In *The Powdery Mildews*, pp. 83–100. D. M. Spencer. New York: Academic Press.

Way, M. J. (1979). Significance of diversity in agroecosystems. In *Proceedings: Opening Session and Plenary Session Symposium. IXth International Congress of Plant Protection, Washington D.C.*, pp. 9–12.

Wolfe, M. S. & Barrett, J. A. (1980). Can we lead the pathogen astray? *Plant Disease* **64,** 148–155.

Zadoks, J. C. & Schein, R. D. (1979). *Epidemiology and Plant Disease Management.* Oxford: Oxford University Press.

Section 3 **Soil-borne fungi, nematodes and nematode-borne viruses**

Soil-borne fungal pathogens in natural vegetation and weeds of cultivation

G W F Sewell
East Malling Research Station, Maidstone, Kent ME19 6BJ

Quantitative estimates of soil fungi reflect mainly the fungal 'seed bank' including, as in higher plants, native propagules in various states of dormancy and viability and 'aliens', present by chance and with little likelihood of survival. Most fungus spores in soil are influenced by a general soil fungistasis of biological origin (Dobbs and Hinson, 1953; Watson and Ford, 1972) and germination occurs only when certain stimulatory and/or nutrient materials become available. Some propagules have highly specific stimulatory requirements. While for most species subsequent growth is also restricted to locally enriched zones a minority are able to translocate nutrients and grow relatively freely through soil and some of these fungi form hyphal aggregations of varying complexity (Garrett, 1970). Stimulatory materials are of plant or animal origin. The former occur as cellular debris resulting from plant growth and death, and as exudates from living roots, which may comprise as much as 18% of total photosynthetic products (Barber and Martin, 1976). Fungal activity is high, therefore, in those soil zones influenced by roots, and locally elsewhere on and around plant and animal debris. Successions of species occurring in these microhabitats are influenced by features of the macrohabitat including soil type and horizon and vegetation (Warcup, 1951; Brown, 1958; Sewell, 1959a; Parkinson *et al.*, 1963). For most species activity is brief, localized, and terminated by the production of variably resistant resting propagules. At most times the fungi are subject to antagonism, competition, starvation, parasitism and predation, and the soil environment presents a physical, chemical and biological barrier to dispersal. Inevitably the soil fungal inhabitants have evolved varied methods to accommodate or overcome these problems including partial escape by invading plant roots. Some, the vascular parasites, have adopted systemic invasion, which also permits a degree of aerial dispersal while within the protection of host tissues. The detailed biology and ecology of soil-borne fungi are described by Garrett (1956, 1970) and Griffin (1972) and the effects of plant communities on the pathogenic nature of parasite populations are intriguingly and briefly discussed by Bawden (1957).

The barrier to dispersal provided by the soil appears to have resulted in the evolution of a set of mycelial parasites with wide host ranges and varying, but often considerable, saprophytic abilities. These parasites are widely distributed in natural vegetation and it must be assumed that they are generally non-lethal to their hosts and that there normally exists in natural vegetation a form of homeostasis in host–parasite relationships, as envisaged by Browning (1974), Harlan (1976) and Dinoor (*1981*). Their presence as indigenous pathogens, however, does not preclude some plants being severely debilitated and others killed.

These pathogens are mainly most or all species in the genera *Armillariella, Fusarium, Phytophthora, Pythium, Rhizoctonia, Thielaviopsis* and *Verticillium.* To list their known hosts would be a formidable task and perhaps of value only in indicating those groups of plants escaping detectable infection or utilization. Among British cultivated plants, and 21 years ago, the host list for *Armillariella mellea* included 67 genera of trees, shrubs and ornamental plants, 13 genera of fruit plants and five genera of vegetables, but no pasture or forage crops (Moore, 1959). Nienhaus (1960) considering hosts of *Phytophthora cactorum* listed 222 species in 65 families of Gymnospermae and Angiospermae. He infected 39 of 70 weed species with an isolate of *P. cactorum* from the rotted bark of apple and commented that different results might have been obtained with isolates from other sources. No hosts were recorded in the Gramineae, but Dakwa (1970) showed that severed leaves of *Dactylis* and *Lolium* could be colonized by this pathogen.

Phytophthora cactorum has such an extensive range of hosts, host parts, and host fruits that individual crop species are probably of negligible significance and their occasional destruction little affects its survival. It is doubtful whether there exists a green plant species which, in whole or part, at some stage in its life, could not be exploited by certain Oomycete pathogens. Other soil-borne pathogens may behave similarly and the general applicability of the proposition that 'an efficient parasite does not eliminate its host' (Baker and Cook, 1974) rests with varying strength on the words 'efficient' and 'its host'. It is a plausible and logical statement, but Bawden (1957) warns that logic may be a faulty tool for explaining the reasons underlying ecological situations. If 'efficient' is interpreted in terms of the distribution and size of parasite populations, there are many efficient facultative parasites in the soil to whom the death of one or more of their hosts is quite immaterial and might be ecologically advantageous — as with the wide dispersal of *Verticillium*-infected, wilted leaves from dying hop plants in a weedy plantation.

Dinoor (1974) indicates the importance of genotype in determining the 'strength' of each host-pathogen interaction, and justifiably warns against placing too great an emphasis on the published host list for any pathogen. Dinoor's conclusions however, derive largely from observations on air-borne diseases caused by obligate, host-restricted parasites. Resistance is undoubtedly the normal condition of plants, but it is a variable quality. Genotypic resistance to soil-borne pathogens is more often expressed by restrictions on parasitic activity than by absolute resistance and is not inimical to the survival and multiplication of these parasites. Furthermore, restricted 'benign' infections may result in considerable colonization advantages to facultative parasites on the senescence and death of the host or its infected parts. The soil pathologist therefore should also be wary of placing too great an emphasis on host lists, but for other reasons: where a list is short he should suspect inadequate searches.

With some soil-borne mycelial pathogens, even non-host plant roots may permit 'resting spore' germination and multiplication, as with *Verticillium dahliae* (Evans and Gleeson, 1973) and *Fusarium* spp. (Schroth and Hendrix, 1962). Other pathogens in this group, including species of *Phytophthora* and *Pythium*, produce toxins causing severely diminished plant growth and thus can be pathogenic without necessarily being parasitic (Woltz, 1978). Such non-parasitic pathogenicity may be of wider occurrence and importance than is presently recognized.

Ecological strategies of fungal pathogens

In his chapter on pathogens, Harper (1977) quotes extensively from the epidemiological work of J. E. Vanderplank, and emphasises its relevance to studies on plant population biology. Undoubtedly this is appropriate, but certain restraints may be necessary associated with the characteristics of different groups of pathogens and plants. Vanderplank's concepts are derived almost entirely from studies of diseases caused by host-restricted, obligate, air-borne fungi and vector-borne viruses. Reference to soil-borne mycelial pathogens is made on only 37 of the 702 text-pages in Vanderplank's three books (1963, 1968, 1975). This is not because these pathogens do not cause epidemic disease, but it strongly implies that their omnivorous nature diminishes their value as subjects for the study of epidemics.

Ecologically, the host-restricted, obligate, air-borne fungal pathogens and the omnivorous, facultative, soil-borne pathogens have many similarities with the 'r-species' and 'K-species', respectively, of animals and higher plants; 'r-species being those whose populations spend most of their time in exponential recovery from disasters or successive invasions, and K-species being those whose populations spend a large part of their time under stress from the presence of neighbours' (Harper, 1977). They match closely the 'suites of characteristics' listed for r- and K-species by Southwood (1977), and the perennially colonizing and wasteful (r-selection) genotypes or the endemic, efficient (K-selection) genotypes described by MacArthur and Wilson (1967).

Newman (1978) attempted to distinguish r- and K-types of soil-borne fungi, choosing as examples *Pythium* and *Gaeumannomyces*, respectively. The distinction is not justified and under-estimates the biological range of *Pythium*. The description of *Pythium* as a 'damping-off' fungus which 'quickly spreads through a host plant, probably killing it, and has efficient means for then spreading to another plant' (Newman, 1978) covers only part of its biology. Most *Pythium* species have extensive host ranges. They *are* parasites of young root tissues, but these tissues occur on plants of all ages. Seedlings with one radicle are killed, but older plants with many roots can be severely debilitated, as in certain 'poor growth' or 'replant' diseases (Wilhelm, 1965; Sewell, 1981). Pythiums are also primary invaders of fallen leaves and other fresh plant debris including, importantly, seeds. They can grow and sporulate extensively in soil. Most species invest heavily in elaborate reproductive structures and efficiently convert food into spores (Stanghellini and Hancock, 1971a). They are almost constant and prominent members of soil populations, exploiting many microhabitats and well capable of surviving the subsequent competitive stresses. Strategically, Pythiums are among the stronger K-species of soil-borne pathogens, although tactically, in many of their individual actions they resemble r-species. Similarly Vanderplank's concepts of plant disease epidemiology may be seen as relevant to the *strategies* of population biology with certain types of plant, but perhaps as relevant only to the *tactics* of population biology with others.

If r-species do occur among soil-borne mycelial pathogens, they must be sought among species with the characteristics of obligate, air-borne parasites. Within the genera under consideration *Phytophthora infestans* complies. Its r-selection has been so strong that it has very little (residual?) soil life: ecologically it is more closely related to the downy mildews than to most members of its genus. Possibly *Phytophthora fragariae*, the cause of red-core disease of strawberry, is an example, with little (known) saprophytic ability and with very few (known) alternative hosts among allied genera (Converse and Moore, 1966; Pepin, 1967). The question remains 'how much is not known?' Red-core has been the most severe disease of strawberry for many years but no rotational programme achieves its elimination or control. The *Fusarium formae* are often quoted as examples of 'specialized and host-restricted', soil-borne mycelial pathogens. However, the persistence, and even increases, of soil populations of *F. solani* and *F. oxysporum* in the absence of the *formae* hosts, and in soils where these hosts have never grown (Schroth and Hendrix, 1962), strongly suggests that they are versatile and more omnivorous than is believed. The occurrence of propagules potentially capable of prolonged survival (as those produced by *Fusarium* and *Phytophthora* spp.) may at times have induced complacency among pathologists and subdued interest in the importance of alternative hosts and non-host carriers.

Soil-borne fungal pathogens in natural vegetation

The effects of soil-borne pathogens in determining and modifying the evolution and composition of natural ecosystems have received limited attention from pathologists and ecologists (Newman, 1978). Newman asked 'Do root pathogens often occur in non-crop vegetation?' and 'Do root pathogens have much effect on the plants?' He found few answers in the literature and concluded '. . . although there is little positive evidence on the involvement of pathogens, there is enough to show the possibility that pathogens could have a major influence on species composition and diversity even when no symptoms of pathogen attack are obvious.'

Most root pathologists would answer Newman's first question affirmatively and challenge any doubters to demonstrate freedom from root infection in any field-grown plant. Using one technique, Wilhelm (1956) isolated nine different pathogenic fungi from the roots of *Solanum sarachoides*, a weed of Californian strawberry fields. To the second question the answer again would be affirmative but qualified by reference mainly to pathogen aggressiveness.

The indigenous nature of soil-borne fungal pathogens implies co-existence with native hosts and such co-existence tends to make the *diseases* they cause endemic. With diseases the adjective 'endemic' is used in opposition to 'epidemic' (Vanderplank, 1975). With endemic diseases there exists a precarious balance or equilibrium, with tolerance in the host population, and low aggressiveness in the pathogen population. However, Vanderplank (1975) stresses that host, pathogen and environment never stay in stable equilibrium and, therefore, where disease is mainly endemic there will be local epidemics.

The presence of soil-borne disease in natural vegatation is unlikely to be obvious, but the apparent 'healthiness' (normality) of vegetation is no indication of freedom from soil-borne disease, nor is it evidence that disease is not imperceptibly but continually modifying the patterns of plant populations. Sagar's (1970) observation of increased survival of *Bellis perennis* seedlings after fungicide application to grassland soil provides a field example of pathogen effects. Harper's (1977) observation that wild plant species apparently free from disease in nature frequently develop remarkable collections of epidemic diseases when grown as pure stands in experimental gardens, may partly reflect the relative ease and frequency of recording experimental gardens. The often unexpectedly very large increases in plant growth following soil partial sterilization is further evidence of unsuspected soil pathogens (Wilhelm, 1966). In natural vegetation a dead plant usually disappears quickly; a wild plant severely stunted by *Pythium* or *Verticillium* is simply a small wild plant.

These comments do not deny that some degree of disease protection is afforded by spatial dispersion in diverse communities as discussed by Wheeler (*1981*) and Dinoor (*1981*). However, evidence of protection by mixing plant species or host genotypes derives mainly from observations on air-borne diseases and may not be wholly relevant to the behaviour of soil-borne fungi. The rate of spread of damping-off disease of cress caused by *Pythium* is related to plant spacing (Burdon and Chilvers, 1976a), but even in this simple experimental system, plant clumping effects may be complicated (Burdon and

Chilvers, 1976b). There is field evidence with the Panama disease of banana (*Fusarium oxysporum* f. *cubense*) that rates of spread increased when fewer healthy plants were removed around disease foci (discussed by Vanderplank, 1963). A similar effect is also probably responsible for increased spread of Verticillium wilt of hop in south-east England.

Spatial factors doubtless influence the spread of diseases in general but most soil-borne pathogens have many hosts, albeit varyingly preferred, and in mixed vegetation the spacing of known hosts may not have the expected effects. Experiments by Kerr (1956) suggest that, at times, specific diversity may even result in increased pathogenicity. Kerr showed that root exudates from lettuce and radish stimulated growth of *Pellicularia filamentosa*, whereas those of tomato did not, and the presence of the fungus in soil had relatively little effect on the germination and growth of tomato in monoculture. The emergence of tomato seedlings was greatly diminished, however, when sown in infested soil together with lettuce and radish.

Even allowing that spread is slowed in mixed communities, natural ecosystems are not strongly 'time conscious'. If the rate of spread, or multiplication, of a disease is more or less similar to its rate of disappearance, or decrease, the disease is endemic; if greater the disease is epidemic. Although epidemiology is importantly concerned with rates, the time-scale is unimportant in natural ecosystems and the epidemic or endemic nature of many soil-borne diseases will not readily be apparent.

The richly varied flora of tropical rain forests, with individuals of each species widely separated, has been used to illustrate the strength of the 'dispersal strategy' against diseases (Harlan, 1976). If, however, dispersal reflects past disasters, perhaps involving many and varied epidemics resulting in the fragmentation of not one plant population but many, it could be viewed not as a strategy against disease, but rather as the relic of a series of natural imbalances or routs caused by disease.

In his chapter on plant pathogens, Harper (1977) is relatively dismissive of the soil-borne fungi and treats them as lesser pathogens merely effecting the *coup de grâce* to the vegetation's invalids. In nature such pathogens are indigenous and the diseases they cause are mostly endemic, but they always have the potential for disaster. These are the pathogens that infect young root tissues killing seedlings or causing 'poor growth' diseases. They also infect mature roots and stem or trunk bases, debilitating or killing by girdling, or invade the vascular tissues systemically causing diminished growth or wilt and death. These pathogens are *concealed* agents of plant disease, but are not to be under-estimated. In agriculture they are among the most formidable opponents confronting the crop pathologist. They affect all crops variously and often disastrously; as they have many host species their behaviour is not 'suicidal' if they kill one or more. However, their excesses in crop monocultures are mainly what Bawden (1957) calls 'merry interludes' in their existence.

Soil-borne fungal pathogens and weed plants in crops

The presence of pathogens in natural ecosystems is of direct and obvious relevance to agriculture when virgin land is first planted with crops. It continues to be relevant when native, diseased host plants occur as weeds within crops, or between plantings and growing seasons.

An indigenous soil-borne pathogen may be immediately virulent and cause epidemic disease in an introduced crop, or it may subsequently increase in virulence. Although not all of these fungi possess known sexual stages they have other mechanisms for variation and gene exchange (Webster, 1974). Bawden (1957) comments, 'There is no need to do experiments to show that large-scale changes in the character of plant populations affect the population of parasites; the information is writ large in the history of agriculture by the immense losses from crop diseases'. Continuous monoculture usually results in persistent, severe disease, although occasionally an initially high level of disease may gradually decline (Shipton, 1977). In plantations of perennial fruit crops the initial planting is sometimes apparently free from soil-borne disease but, in the absence of obvious disease, the growth of further plantings of a similar crop is often severely diminished. These so-called 'specific replant diseases' illustrate both the influence of plant species on pathotype enrichment in soil, and the ability of these selected pathogen communities to persist for long periods in the absence of the particular host which originally selected them (Savory, 1966; Sewell and Wilson, 1975; Sewell, 1981). They could be important partial determinants of successions in natural plant communities.

With all crops there is also the possibility that soil-borne pathogens already adapted and aggressive to the host may be introduced with planting material. This is probably rare with crops grown from seed (Baker and Smith, 1966), but is a particular hazard with vegetatively-propagated plants. Such undesirable 'passengers' may occur almost invariably, as with *Thielaviopsis basicola* and layer-bed propagated *Prunus avium* rootstocks (Sewell and Wilson, 1975).

Alternative weed hosts and *Verticillium* wilt of the hop

Whatever the origin of soil-borne pathogens within crops, weed plants affect their spread and persistence and impede eradication or containment policies for disease control. These features are well illustrated by wilt of the hop (*Humulus lupulus*) caused by *Verticillium albo-atrum*.

Verticillium invades the cortical tissue of young roots, but further colonization is impeded by the endodermis (Talboys, 1958). Initial vascular invasion is probably restricted to protoxylem elements. Systemic spread in mature xylem vessels is mainly accomplished by dispersal of conidia in the xylem sap (Sewell and Wilson, 1964). The virulent hop strain of *V. albo-atrum* causes a lethal disease in certain hop cultivars; it has spread from an initial focus in 1930 to most hop farms in south-east England and to some in the West Midlands. It is a statutorily notifiable disease and certain control procedures are enforceable. It provides an example of a major epidemic soil-borne crop plant disease.

The hop plant is a long-lived perennial producing annual bines and is grown at rectangular or square spacing accompanied by one-way or two-way directions of cultivation. Keyworth (1942) showed that spread of disease was influenced by the directions of cultivation. These distinct patterns of spread were attributed to the movement of infected hop debris on cultivators, but in retrospect there can be little doubt that infected weeds were also influential. The most common dicotyledonous weeds in hop plantations were later found to be hosts of the hop-virulent strain of *Verticillium* (Sewell and Wilson, 1958). The virtually complete eradication of weeds throughout the year by the overall application of herbicides resulted in a remarkably constant decrease of disease by, on average, 28% over a wide range of disease incidence levels; a feature which Vanderplank (1963) expected to be a characteristic of control by sanitation in 'simple interest' systemic diseases. Control was attributed primarily to the eradication of alternative weed hosts, although other possible factors were recognized (Sewell and Wilson, 1974). In these experiments the effects of herbicides were compared with areas under traditional tillage, but where field sanitation standards were generally high: in view of the relatively weak saprophytic phase of *V. albo-atrum* (Sewell and Wilson, 1966) the eradication of actively growing and 'available' alternative weed hosts during the period of crop dormancy was regarded as a brief rotation.

The apparent restriction of *V. albo-atrum* to dicotyledonous hosts was utilized in other field trials of measures to eradicate or contain primary disease outbreaks (Sewell and Wilson, 1966). These experiments revealed that, in annually cultivated soil carrying a mixed weed flora, the initial (introduced) level of infectivity steadily declined from 34% over three years, and then appeared to 'stabilize' at 13% after the third year. This effect was attributed to a balance attained between the controlling mechanism of green manuring and the average level of genetic susceptibility of the mixed weed population. In naturally infested soils with weed-free grass cover infectivity was greatly diminished after two years and was apparently nil after 3–5 years. This result provides a unique example of the apparent eradication of a soil-borne pathogen by a strictly controlled form of rotation. Such an approach may not be practicable against other pathogens with greater saprophytic ability and less distinct host preference. Attempts to control by rotation even the closely related *V. dahliae* have been unsuccessful (Nelson, 1950*; Guthrie, 1960*; Huisman and Ashworth, 1976*), probably because this species is able to colonize the surface or invade the cortex of roots of graminaceous as well as dicotyledonous plants (Martinson, 1964; Evans and Gleeson, 1973).

* These authors refer the microsclerotial form of *Verticillium* to *V. albo-atrum*.

In English hop plantations a non-tillage, overall herbicide programme is now generally used whether or not wilt is present. In its absence there is the advantage that newly introduced disease foci are apparent from the incidence of wilt in the crop host plant, without the complication of concealed presence in weed hosts. The successful suppression, or containment, of new disease outbreaks by a 'grubbing and grassing' policy is therefore greatly facilitated.

Verticillium species provide an extreme example of the value of a parasitic, as opposed to an exclusively saprophytic, existence. During the life of a host plant (or its parts) systemic invasion is restricted to the xylem, but as host tissues die the pathogen is released to colonize *from within.* The extent of this colonization advantage has been demonstrated by the observation in soil of vigorous sporulation over the root surfaces of the dying host (Sewell, 1959b); moreover infectivity levels have been demonstrated in the annual debris of hop plants greatly exceeding those expected from the incidence of disease in the crop (Sewell and Wilson, 1961). The principle behind the motto of the House of Marlborough 'Let Curzon hold what Curzon held' is fully exploited. These observations on hop colonization probably apply also to other hosts including weeds, and the total amounts of infected dead host tissue and pathogen biomass are likely to be closely related.

Role of plant residues in survival and pathogenesis

Plant residues may also be invaded and colonized *from without* by certain soil-borne pathogens, and so maintain or increase pathogen populations. The precise threshold of death of residues is not easily defined, and these pathogens are not necessarily the strongest saprophytes. That they are recorded most commonly as being species of *Pythium, Phytophthora* or *Rhizoctonia* (Baker and Cook, 1974) may partly be associated with the relative ease of observation or isolation of these genera, and the superficially greater likelihood that these fungi will behave in this manner: *Verticillium dahliae* a supposed weak saprophyte, was considered a primary colonizer of certain plant residues (Martinson, 1964). Similarly, evidence of colonization comes from observations on crop, rather than weed plant, residues because these are more obvious places to search. Well-documented examples are the colonization of fallen leaves of cotton by *Pythium ultimum* (Hancock, 1977), of papaya by *Pythium aphanidermatum* and *Phytophthora parasitica* (Trujillo and Hine, 1965), of apple and pear by *Phytophthora syringae* (Harris, 1979, *1981*) and of apple fruits by the *Phytophthora* species *cactorum, citricola* and *syringae* (Sewell *et al.*, 1974). These examples are clearly of considerable epidemiological importance, but perhaps more so is the observation by Dakwa (1970) that zoospores of *Phytophthora cactorum* colonized the severed tissues of common weeds on natural soil surfaces and led to sporulation within and from them. These tissues were leaves of *Dactylis glomerata* and *Lolium perenne*, and roots, hypocotyls, cotyledons and leaves of *Galium saxatile, Sinapsis arvensis* and *Stellaria media.* Furthermore, successful colonization was detected when certain tissues (*Dactylis* and *Lolium* leaf; *Stellaria* and *Galium* hypocotyl) were incubated on natural soil surfaces for 5–7 days before inoculation with zoospores.

Fresh green residues of weeds may contribute to the harmful effects of green manuring on crops sown prematurely. This effect has been attributed to invasion of the residues by *Pythium* resulting in increased inoculum for infecting seedlings (Sawada *et al.*, 1964). If initial colonization of the green residues is mainly by *Pythium* spp., which is indeed possible, there may be no need to invoke parasitism, for *Pythium* toxins can suppress germination and seedling growth. This possibility was suggested long ago (Vanterpool, 1933). More recently, strong inhibition of growth of tomato and tobacco has been demonstrated after the incorporation into soil of dead mycelium of *Phytophthora* and *Pythium* spp. (Csinos and Hendrix, 1977, 1978; Csinos, 1979). I have found similar strongly inhibitory effects of the dead mycelium of *Pythium sylvaticum* on the growth of apple roots. The full implications of these discoveries remain to be evaluated and could be considerable. Wilhelm (1965, 1966) has long contended that *Pythium* spp. cause a widespread phytostasis, but his considerations involved parasitism.

Seeds and soil-borne fungal pathogens

The seed bank in soil, and especially in agricultural soils, is usually large. Harper (1977) quotes examples of viable seeds per square metre totalling 39 000 in arable agricultural soil (to 15 cm depth), 23 000 in natural grassland (to 2.5 cm depth) and 12 500 in Callunetum (to 2 cm depth). In uncultivated soils the seed population density rapidly declines with increasing depth, although appreciable populations frequently occur to 15 cm depth. Harper's histograms of seed distributions remarkably resemble those for many soil-borne fungi (e.g. Sewell, 1959a). Harper (1977) considers that most deeply buried seeds will not germinate and will eventually die and decay *in situ*, although this may take a long time. Sagar and Mortimer (1976) reported survivals of 13–53% after six years of species of *Senecio, Stellaria, Urtica, Chenopodium, Fumaria* and *Poa.*

'Numerically the seed bank is often the phase of the life cycle when numbers are highest and it has great importance in the survival of species and, indirectly, in their evolution' (Major and Pyott, 1966). Ecological studies on the mechanisms of population regulation tend to emphasize the role of the more readily observable animal predators (Sagar and Mortimer, 1976; Harper, 1977). With some plant species their influence is great, but with all plant species the successful establishment of the small proportion of seeds that reach suitable germination niches is likely to be determined largely by soil-borne fungal pathogens. The agricultural literature on seedling damping-off diseases and root rots is extensive, and protection of the seeds of many crops with fungicides is a necessary and common practice.

To an ecologist, seeds in soil may be in one of four states: germinating, or in enforced, induced or innate dormancy (Harper, 1957, quoted from Sagar and Mortimer, 1976). The pathologist, with additional interests, would add a fifth state — dead. The genetical control of seed longevity is variable and an important factor in the survival of a species, but, for the individual, it is merely a matter of postponing the hazardous moments of germination. Harper (1977) considers that seed rotting is affected by temperature and moisture and that physical conditions of the environment can determine the probability of seed survival. He was presumably limiting his consideration to survival over the potential dormancy span of the seed, but for most seeds the conditions favouring *germination* are also those favouring the diffusion of seed exudates and fungal activity. During dormancy most seeds are probably relatively safe from fungal attack, but, as they age, diminished viability is accompanied by increased 'leakiness' encouraging fungal activity and invasion. Similarly, the physiological activities of germination are accompanied by exudation of sugars and amino-acids (Parkinson, 1955; Katznelson *et al.*, 1956; Rovira, 1956; Pearson and Parkinson, 1961) and the consequent stimulation of fungal activity (Barton, 1957; Schroth and Snyder, 1961; Schroth *et al.*, 1963; Schroth and Cook, 1964). Prominent among those stimulated are the pathogens *Pythium, Fusarium* and *Rhizoctonia.* Using a soil naturally infected with *Pythium ultimum* and sown with seeds of diverse crops, Singh (1965) found *P. ultimum* populations, in soil adjacent to seeds, increased on average by ×2.4 after 48 hours and by ×7.7 after 96 hours. Stanghellini and Hancock (1971b) recorded 78% germination of *P. ultimum* sporangia in soil adjacent to bean seeds sown only 3–4 hours earlier: germination and growth were stimulated up to 10 mm from seed in 24 hours. Similar effects occurred on the germination and growth of *Fusarium solani* f. sp. *phaseoli.* In my own work, using a quadrat-frequency method and a natural soil amended or not with grass seed (*Festuca* sp.) and assayed when grass seedlings were 6–8 cm tall, the frequency of *Pythium* was 3% in unamended soil, 43% in soil with grass seedlings, and 89% in soil amended with aged non-viable grass seed. In other studies with soil containing a low and relatively ineffectual population of *Pythium sylvaticum*, the addition of non-viable seed of grass or *Antirrhinum* (400 mg litre^{-1} soil) greatly diminished the growth of apple seedlings, on which *P. sylvaticum* is parasitic.

The gravely damaging effects of *Pythium, Fusarium* and *Rhizoctonia* and other pathogens on the germination and growth of *crop* seedlings are well known by pathologists, but the size and possible effects of the *weed* seed and seedling population on the activity of these pathogens perhaps are not widely recognized. With Pythiums, pathogenesis may be of both parasitic and non-parasitic origin; the stimulation of the vigorous growth of these pathogens on and around plant seeds (on their germination or senescence) may provide an important mechanism for the general soil phytostasis described by Wilhelm (1966).

Plant pathology and ecology: concluding remarks

Soil-borne mycelial pathogens are omnivorous facultative parasites, that often possess considerable saprophytic abilities. They are indigenous and important members of natural plant communities. As concealed agents of plant disease they are occasionally lethal but more often important determinants of plant vigour. They are persistent occupiers of soil sites and their effects on species composition and successions in ecosystems are likely to be considerable and perhaps even comparable to the direct effects of competition between plants.

Pathologists, understandably, have been preoccupied with diseases of crop plants rather than those of natural ecosystems, and ecologists, until recently, have interpreted their observations largely in terms of plant competition, physical site factors and predation. The emergence of fungal epidemiology as a mathematically-based science is currently attracting the attention of ecologists because of its probable relevance to studies on the population biology of higher plants: it is also increasing the awareness of ecologists to the effects of fungal pathogens on the development of plant communities. This paper has emphasized that modern fungal epidemiology is largely based on studies of a distinct group of air-borne, obligate and host-restricted pathogens possessing quite different characteristics from their soil-borne counterparts. These two groups of pathogens, the air-borne and the soil-borne, are considered ecologically to be 'r-strategists' and 'K-strategists', respectively. Both groups are undoubtedly influential factors in plant ecology. However, despite the attraction to ecologists of modern epidemiology, and the more readily observable nature of its air-borne subjects, the residential soil-borne fungi are likely to be the more important partial determinants of the structures of natural plant communities.

It is a pleasure to express my sincere thanks to my colleagues D. J. Butt and J. M. Thresh for many helpful discussions and for their critical appraisal of the manuscript of this paper.

References

Baker, K. F. & Cook, R. J. (1974). *Biological Control of Plant Pathogens.* San Francisco: Freeman.

Baker, K. F. & Smith, S. H. (1966). Dynamics of seed transmission of plant pathogens. *Annual Review of Phytopathology* **4,** 311–334.

Barber, D. A. & Martin, J. K. (1976). The release of organic substances by cereal roots into soil. *New Phytologist* **76,** 69–80.

Barton, R. (1957). Germination of oospores of *Pythium mamillatum* in response to exudates from living seedlings. *Nature* (London) **180,** 613–614.

Bawden, F. C. (1957). The role of plant hosts in microbial ecology. In *Microbial Ecology. 7th Symposium of the Society for General Microbiology*, pp. 299–314. Cambridge: Cambridge University Press.

Brown, J. C. (1958). Soil fungi of some British sand dunes in relation to soil type and succession. *Journal of Ecology* **46,** 641–664.

Browning, J. A. (1974). Relevance of knowledge about natural ecosystems to development of pest management programs for agro-ecosystems. *Proceedings of the American Phytopathological Society* **1,** 191–199.

Burdon, J. J. & Chilvers, G. A. (1976a). Epidemiology of *Pythium*-induced damping-off in mixed species seedling stands. *Annals of Applied Biology* **82,** 233–240.

Burdon, J. J. & Chilvers, G. A. (1976b). The effect of clumped planting patterns on epidemics of damping-off disease in cress seedlings. *Oecologia* **23,** 17–29.

Converse, R. H. & Moore, J. N. (1966). Susceptibility of certain *Potentilla* and *Geum* species to infection by various races of *Phytophthora fragariae. Phytopathology* **56,** 637–639.

Csinos, A. S. (1979). Nonparasitic pathogenesis of germinating tomato by *Pythium myriotylum. Canadian Journal of Botany* **57,** 2059–2063.

Csinos, A. S. & Hendrix, J. W. (1977). Nonparasitic stunting of tobacco plants by *Phytophthora cryptogea. Canadian Journal of Botany* **55,** 26–29.

Csinos, A. S. & Hendrix, J. W. (1978). Parasitic and nonparasitic pathogenesis of tomato plants by *Pythium myriotylum. Canadian Journal of Botany* **56,** 2334–2339.

Dakwa, J. T. (1970). Ecological studies of *Phytophthora cactorum* (Leb. and Cohn) Schroet., with reference to the collar rot disease of apple. *PhD Thesis*, University of London, 204 pp.

Dinoor, A. (1974). Role of wild and cultivated plants in the epidemiology of plant disease in Israel. *Annual Review of Phytopathology* **12,** 413–436.

Dinoor, A. (*1981*). Epidemics caused by fungal pathogens in wild and crop plants. In *Pests, Pathogens and Vegetation*, pp. 143–158. J. M. Thresh. London: Pitman.

Dobbs, C. G. & Hinson, W. H. (1953). A widespread fungistasis in soils. *Nature* (London) **172,** 197.

Evans, G. & Gleeson, A. C. (1973). Observations on the origin and nature of *Verticillium dahliae* colonising plant roots. *Australian Journal of Biological Sciences* **26,** 151–161.

Garrett, S. D. (1956). *Biology of Root-Infecting Fungi.* Cambridge: Cambridge University Press.

Garrett, S. D. (1970). *Pathogenic Root-Infecting Fungi.* Cambridge: Cambridge University Press.

Griffin, D. M. (1972). *Ecology of Soil Fungi.* London: Chapman and Hall.

Guthrie, J. W. (1960). Early dying (*Verticillium* wilt) of potatoes in Idaho. *Research Bulletin of the Idaho Agricultural Experiment Station* No. 45.

Hancock, J. G. (1977). Factors affecting soil populations of *Pythium ultimum* in the San Joaquin Valley of California. *Hilgardia* **45,** 107–121.

Harlan, J. R. (1976). Diseases as a factor in plant evolution. *Annual Review of Phytopathology* **14,** 31–51.

Harper, J. L. (1977). *Population Biology of Plants.* London, New York and San Francisco: Academic Press.

Harris, D. C. (1979). The occurrence of *Phytophthora syringae* in fallen apple leaves. *Annals of Applied Biology* **91,** 309–312.

Harris, D. C. (*1981*). Herbicide management in apple orchards and the fruit rot caused by *Phytophthora syringae*. In *Pests, Pathogens and Vegetation*, pp. 429–436. J. M. Thresh. London: Pitman.

Huisman, D. C. & Ashworth, L. J. (1976). Influence of crop rotation on survival of *Verticillium albo-atrum* in soils. *Phytopathology* **66,** 978–981.

Katznelson, H., Rouatt, J. W. & Payne, T. M. B. (1956). The liberation of amino-acids and reducing compounds by plant roots. *Plant and Soil* **7,** 35–48.

Kerr, A. (1956). Some interactions between plant roots and pathogenic fungi. *Australian Journal of Biological Sciences* **9,** 45–52.

Keyworth, W. G. (1942). *Verticillium* wilt of the hop (*Humulus lupulus*). *Annals of Applied Biology* **29,** 346–357.

MacArthur, R. H. & Wilson, E. O. (1967). *The Theory of Island Biogeography.* Princeton, N.J.: Princeton University Press.

Major, J. & Pyott, W. T. (1966). Buried viable seeds in California bunchgrass sites and their bearing on the definition of a flora. *Vegetatio Acta Geobotanica* **13,** 253–282.

Martinson, C. A. (1964). Active survival of *Verticillium dahliae* in soil. *Dissertation Abstracts* **25,** (1), 18.

Moore, W. C. (1959). *British Parasitic Fungi.* Cambridge: Cambridge University Press.

Nelson, R. (1950). *Verticillium* wilt of peppermint. *Technical Bulletin of Michigan (State College) Agricultural Experiment Station* No. 221.

Newman, E. I. (1978). Root microorganisms: their significance in the ecosystem. *Biological Reviews* **53,** 511–554.

Nienhaus, F. (1960). Das Wirtsspektrum von *Phytophthora cactorum* (Leb. et Cohn) Schroet. *Phytopathologische Zeitschrift* **38,** 33–68.

Parkinson, D. (1955). Liberation of amino-acids by oat seedlings. *Nature* (London) **176,** 35.

Parkinson, D., Taylor, G. S. & Pearson, R. (1963). Studies on fungi in the root region, I. *Plant and Soil* **19,** 332–349.

Pearson, R. & Parkinson, D. (1961). The sites of excretion of ninhydrin positive substances from broadbean seedlings. *Plant and Soil* **13,** 391–396.

Pepin, H. S. (1967). Susceptibility of members of the Rosaceae to races of *Phytophthora fragariae. Phytopathology* **57,** 782–784.

Rovira, A. D. (1956). Plant root excretions in relation to the rhizosphere effect, I, II, III. *Plant and Soil* **7,** 178–217.

Sagar, G. R. (1970). Factors controlling the size of plant populations. In *Proceedings of the 10th British Weed Control Conference Brighton*, pp. 965–979.

Sagar, G. R. & Mortimer, A. M. (1976). An approach to the study of the population dynamics of plants with special reference to weeds. *Applied Biology* **1,** 1–48.

Savory, B. M. (1966). Specific replant diseases. *Commonwealth Bureau of Horticulture and Plantation Crops, East Malling, Kent Research Review* No. 1.

Sawada, Y., Nitta, K. & Igarashi, T. (1964). Injury of young plants caused by the decomposition of green manure. *Soil Science and Plant Nutrition* **10,** 163–170.

Schroth, M. N. & Cook, R. J. (1964). Seed exudation and its influence on pre-emergence damping-off of bean. *Phytopathology* **54,** 670–673.

Schroth, M. N. & Hendrix, F. F. (1962). Influence of non-susceptible plants on the survival of *Fusarium solani* f. *phaseoli* in soil. *Phytopathology* **52,** 906–909.

Schroth, M. N. & Snyder, W. C. (1961). Effect of exudates on chlamydospore germination of the bean root rot fungi *Fusarium solani* f. *phaseoli. Phytopathology* **51,** 389–393.

Schroth, M. N., Toussoun, T. A. & Snyder, W. C. (1963). Effect of certain constituents of bean exudate on germination of chlamydospores of *Fusarium solani* f. *phaseoli* in soil. *Phytopathology* **53,** 809–812.

Sewell, G. W. F. (1959a). The ecology of fungi in Calluna-heathland soils. *New Phytologist* **58,** 5–15.

Sewell, G. W. F. (1959b). Direct observation of *Verticillium albo-atrum* in soil. *Transactions of the British Mycological Society* **42,** 312–321.

Sewell, G. W. F. (1981). Effects of *Pythium* species on the growth of apple and their possible causal role in apple replant disease. *Annals of Applied Biology* **97,** 31–42.

Sewell, G. W. F. & Wilson, J. F. (1958). Weed hosts of the 'progressive' hop strain of *Verticillium albo-atrum* Reinke and Berth. *Report of East Malling Research Station for 1957*, pp. 126–128.

Sewell, G. W. F. & Wilson, J. F. (1961). Machine picking in relation to progressive *Verticillium* wilt of the hop, I. *Report of East Malling Research Station for 1960*, pp. 100–104.

Sewell, G. W. F. & Wilson, J. F. (1964). Occurrence and dispersal of *Verticillium* conidia in xylem sap of the hop (*Humulus lupulus*). *Nature* (London) **204,** 901.

Sewell, G. W. F. & Wilson, J. F. (1966). *Verticillium* wilt of the hop: the survival of *V. albo-atrum* in soil. *Annals of Applied Biology* **58,** 241–249.

Sewell, G. W. F. & Wilson, J. F. (1974). The influence of normal tillage and of non-cultivation on *Verticillium* wilt of the hop. *Annals of Applied Biology* **76,** 37–47.

Sewell, G. W. F. & Wilson, J. F. (1975). The role of *Thielaviopsis basicola* in the specific replant disorders of cherry and plum. *Annals of Applied Biology* **79,** 149–169.

Sewell, G. W. F., Wilson, J. F. & Dakwa, J. T. (1974). Seasonal variations in the activity in soil of *Phytophthora cactorum, P. syringae* and *P. citricola* in relation to collar rot disease of apple. *Annals of Applied Biology* **76,** 179–186.

Shipton, P. J. (1977). Monoculture and soilborne pathogens. *Annual Review of Phytopathology* **15,** 387–407.

Singh, R. S. (1965). Development of *Pythium ultimum* in soil in relation to presence and germination of seeds of different crops. *Mycopathologia et Mycologia Applicata* **27,** 155–160.

Southwood, T. R. E. (1977). Habitat, the templet for ecological strategies. *Journal of Animal Ecology* **46,** 337–365.

Stanghellini, M. E. & Hancock, J. G. (1971a). The sporangium of *Pythium ultimum* as a survival structure in soil. *Phytopathology* **61,** 157–164.

Stanghellini, M. E. & Hancock, J. G. (1971b). Radial extent of bean spermosphere and its relation to the behaviour of *Pythium ultimum. Phytopathology* **61,** 165–168.

Talboys, P. W. (1958). Some mechanisms contributing to *Verticillium* resistance in the hop root. *Transactions of the British Mycological Society* **41,** 227–241.

Trujillo, E. E. & Hine, R. B. (1965). The role of papaya residues in papaya root rot caused by *Pythium aphanidermatum* and *Phytophthora parasitica. Phytopathology* **55,** 1293–1298.

Vanderplank, J. E. (1963). *Plant diseases: Epidemics and Control.* New York: Academic Press.

Vanderplank, J. E. (1968). *Disease Resistance in Plants.* New York: Academic Press.

Vanderplank, J. E. (1975). *Principles of Plant Infection.* New York: Academic Press.

Vanterpool, T. C. (1933). Toxin formation by species of *Pythium* parasitic on wheat. In *Proceedings of the World's Grain Exhibition and Conference, Regina, Saskatchewan,* **2,** 294–298.

Warcup, J. H. (1951). The ecology of soil fungi. *Transactions of the British Mycological Society* **34,** 376–399.

Watson, A. G. & Ford, E. J. (1972). Soil fungistasis — a reappraisal. *Annual Review of Phytopathology* **10,** 327–348.

Webster, R. K. (1974). Recent advances in the genetics of plant pathogenic fungi. *Annual Review of Phytopathology* **12,** 331–353.

Wheeler, B. E. J. (*1981*). The ecology of plant parasitic fungi. In *Pests, Pathogens and Vegetation*, pp. 131–142. J. M. Thresh. London: Pitman.

Wilhelm, S. (1956). A sand-culture technique for the isolation of fungi associated with roots. *Phytopathology* **46,** 293–295.

Wilhelm, S. (1965). *Pythium ultimum* and the soil fumigation growth response. *Phytopathology* **55,** 1016–1020.

Wilhelm, S. (1966). Chemical treatments and inoculum potential of soils. *Annual Review of Phytopathology* **4,** 53–78.

Woltz, S. S. (1978). Nonparasitic plant pathogens. *Annual Review of Phytopathology* **16,** 403–430.

The role of grasses in the ecology of take-all fungi

J W Deacon
Microbiology Department, School of Agriculture, West Mains Road, Edinburgh, EH9 3JG

Introduction

Take-all is a major root disease of intensively grown cereals. The fungus, *Gaeumannomyces graminis*,* invades the root cortex from superficially growing runner-hyphae and gains access to the stele, where it causes plugging and discoloration. Attempts to breed resistant wheat and barley cultivars have proved largely unsuccessful, and chemical control is not yet feasible, so control methods have centred mainly around cultural practices and crop rotations. The latter are effective because the fungus is a specialized parasite of the Gramineae and its air-borne ascospores have negligible infection potential in normal agricultural conditions. Disease normally results from mycelial inoculum persisting in previously colonized host residues and which, therefore, can be 'starved out'. These and other aspects of the disease cycle are discussed by Asher and Shipton (1981).

Grasses and grasslands have long been known to harbour the take-all fungus, with potentially serious consequences for subsequent cereal crops. In the last decade, however, it has become clear that grasslands also support populations of related but weakly or non-pathogenic fungi, which are largely confined to the root cortex. Such fungi, like *Phialophora graminicola*,* are effective biological control agents of take-all. Consequently, two contrasting roles can be distinguished for grasses in the ecology of take-all. The balance between them is considered here and placed in the wider context of the natural environment.

Grasses as hosts of the take-all fungi

There are many reports of grasses as hosts of the two highly pathogenic forms

* The taxonomic confusion surrounding *G. graminis* and related fungi is discussed by Walker (1975). The names used here are those recommended by Walker in Asher and Shipton (1981) and are adopted by all contributors to the forthcoming monograph on take-all (Asher and Shipton, 1981). *P. graminicola* = *P. radicicola* var. *graminicola* Deacon. The name *G. graminis* is used collectively for the pathogenic varieties *G. graminis* vars *tritici* and *avenae*, unless otherwise stated.

of *G. graminis* (vars. *tritici* and *avenae*), and four main themes are apparent, as follows.

Infection decreasing the yield or vigour of useful grasses

Few grass crops are reported to be severely damaged by take-all, even though they often support populations of *G. graminis*. This may reflect the widespread occurrence of biological control agents in grasslands, as described later, but a major factor is undoubtedly the general resistance or 'field tolerance' of grasses to infection. Much evidence is available on this point and on the relative susceptibilities of different grass species (Nilsson, 1969).

In contrast to pastures and leys, special attention has been given to grass seed crops because of their high commercial value. Doling and Hepple (1959) surveyed a range of these crops, especially *Dactylis glomerata*, in Lincolnshire and reported substantial infection by 'dark runner-hyphae of *Ophiobolus graminis*' (syn. *G. graminis*). They implied that yield loss of grass seed crops could result from take-all infection, and this has been generally accepted. However, the evidence is unsatisfactory, and it is now known that dark runner-hyphae on grass roots in Britain are just as likely to be of the avirulent parasite *Phialophora graminicola*. Also, *D. glomerata* is usually listed amongst the more resistant grasses to take-all infection.

With these reservations, the only well-documented example of serious take-all damage to grasslands is the ophiobolus patch disease of bent grass turf (*Agrostis* spp.) caused by *G. graminis* var *avenae* (Smith, 1965). This disease is quite common on the acidic soils of north and west Britain, where it characteristically occurs after liming turf. It occurs also in much of Northern Europe and in parts of Australia and the USA, especially in turf sown into fumigated soils. These predisposing factors can be explained in terms of a reduction in the normal resident populations of control agents (Deacon, 1973), but also it is significant that *Agrostis* spp. are amongst the most susceptible grasses, particularly to *G. g. avenae*.

Pastures as reservoirs of *G. graminis*

Field experience in several countries suggests that natural or other long-standing pastures can harbour *G. graminis*, with serious consequences for subsequent cereal crops. The best-documented and most spectacular examples occur in Australia, on light sandy soils conducive to take-all damage. Adam (1951), for example, recorded severe take-all infection of wheat grown after a barley crop in land that had previously been in pasture for at least 25 years. Wheat grown on the same site after a 13 month bare fallow was much less infected, but the least disease occurred in wheat after oats grazed by sheep. Apart from its practical value, this work is interesting in suggesting that the pasture had supported *G. g. tritici* rather than *avenae*. The latter causes severe damage to oats (and wheat) because it can detoxify avenacin, a pre-formed inhibitor in oat roots (Turner, 1961). As explained later, it may be significant that the pasture studied by Adam comprised only 20% grass cover, the remainder being of clover and dicotyledonous weeds. Similar pastures occur in much of Western Australia, where subsequent wheat crops often suffer severe take-all damage.

Carry-over of *G. graminis* by short-term grass leys

Grass or grass-legume leys are commonly used as breaks from cereals and may carry-over substantial amounts of *G. graminis*. In Australia they almost invariably consist of self-sown 'weed grasses' and legumes, and thus resemble in composition and effect the pastures considered previously. In Britain the leys comprise mixtures or pure stands of selected species and interest has focussed on the effects of different grasses in carrying-over *G. graminis*. In artificially infested small plots, Brooks (1965a) showed that sixteen common ley grasses or grass weeds all maintained *G. graminis* at higher levels than in control plots sown to legumes. However, the grasses differed in their overall effects and in their relative effectiveness with respect to *G. graminis tritici* and *avenae*. It is not known if these differences are relevant in agricultural practice; they may be of little consequence when grasses are intersown with legumes (Wehrle and Ogilvie, 1955). Further study of grass-legume mixtures in comparison with pure grass swards may be rewarding because the mixtures are likely to support much lower levels of the control agent, *P. graminicola* (Deacon, 1976).

Infection of grass weeds in and around cereal crops

Most of the foregoing comments apply also to grass weeds, the important feature being that most, if not all, grasses are potential hosts of *G. graminis*. Particular attention has been given to this problem on the Dutch polders. The newly-drained land is rapidly colonized by grass weeds, but then reeds (*Phragmites communis*) are sown to suppress grass growth and to help dry and stabilize the soil. The first crop is usually winter rape, followed by one or more cereals, which are soon infected by *G. graminis*, severely so if they are grown sequentially. Gerlagh (1968) recorded infection of many grass weeds, though the only references are to dark runner-hyphae. His conclusion that air-borne ascospores initiate infection of the colonizing grasses and those persisting around dykes seems reasonable. However, infection of grasses by ascospores has never been demonstrated, and cereal roots are infected by them only where a stable root-surface microflora is absent, as in the polders (Brooks, 1965b).

In Britain, rhizomatous grass weeds such as *Agropyron repens*, *Agrostis* spp. and *Holcus lanatus* seem particularly important as 'bridging hosts' of *G. graminis* where cereals are grown, because they provide living tissues in which the fungus can grow between crops. Edge or boundary effects of grasses have received relatively little attention, however, because the rate of mycelial spread of *G. graminis* through soil is slow, and even unidirectional cultivations do not spread detectable inoculum more than a few metres. Nevertheless, further study of field margins may be useful for two reasons. Firstly, *Arrhenatherum elatius* and *Avena fatua* are common around cereal crops and as they are resistant to *G. graminis tritici* but no *G. g. avenae* they may support populations of the latter in regions where oats are seldom grown. Secondly, and more important, the levels of take-all infection often rise dramatically to damage second and third successive wheat crops, from low or even undetectable levels in first cereal crops. In such cases the pathogen probably occurs throughout the first crop but at a 'sub-clinical' level. It might increase its inoculum levels for early infection of subsequent crops by growing on senescing root tissues of the first crop (Deacon and Henry, 1980). If so, then ascospores may be more important in spread of the disease from the field margins than is currently thought: their role has previously been considered mainly in terms of establishing *progressive*, damaging infections of roots, rather than of establishing foci of infection from which the fungus can grow and damage succeeding crops.

Beneficial effects of grasses

Despite all the evidence outlined in previous sections, grass-cropping has seldom been shown to have detrimental effects on take-all of cereals in Britain. Indeed, grasses can be beneficial, as shown clearly by the survey done in central England by Rosser and Chadburn (1968). Second wheat crops grown sequentially after grass both out-yielded and had less take-all than corresponding second wheat crops in 'arable' rotations. The difference in yield was maintained in third wheat crops, but take-all levels were then similar. Recent experiments at Rothamsted have confirmed these findings, by showing that the onset of severe take-all is delayed by 1 year after grass (*see* Slope *et al.*, 1979).

A satisfactory explanation of this phenomenon became possible with the finding that most grasslands in Britain support large populations of the control agent, *P. graminicola*. These populations persist — albeit at progressively declining levels — in sequences of cereals after grass, whereas *P. graminicola* occurs at much lower levels in other cereal crops. Much of the evidence for the role of this fungus in biological control of take-all is summarized elsewhere (Deacon, 1976). Briefly, *G. graminis* is controlled when *P. graminicola* or similar fungi are added to natural soils in the glasshouse; the controlling effects of grass turf or soils from beneath grass leys are directly proportional to their *Phialophora* contents, and in surveys of cereal crops or soils the amount of infection by *G. graminis* is inversely related to that by *Phialophora*. The evidence suggests that there is a natural, widespread biological control of take-all in British grasslands and that this has been exploited unknowingly in the past in the use of grass leys in cereal-cropping systems. Admittedly, only a temporary control is achieved in the cereal crops, but this is to be expected because wheat is highly susceptible to take-all infection and *P. graminicola* does not naturally occur at high levels in cereals.

The argument that there is a natural control of take-all in British grasslands may seem inconsistent with the occurrence of ophiobolus patch disease in some types of turf. Indeed, in a preliminary study as much *Phialophora* was found in the centres of disease patches as in the surrounding disease-free turf. But these conflicting points were reconciled in subsequent work, which showed that the control agent was absent, or present at only low levels, in turf of very low pH, i.e. turf that would be expected to develop the disease after liming (Deacon, 1973). Thus, the key to this disease in this situation seems to be that turf needing lime to improve its vigour has a low resident population of *Phialophora* and perhaps other control agents, and liming creates conditions ideal for growth by *G. graminis* in the absence of its normal control agents. Several other peculiar features of the disease may similarly reflect the activity of *P. graminicola*. For example, the spontaneous disappearance of disease within a few years may be due to the population of *P. graminicola* returning to a high level.

The mechanism of biological control by *Phialophora* is still debatable. The fungi show no obvious interaction in agar culture and so control has been assumed to be host-mediated. *P. graminicola* and similar fungi elicit lignification of the root endodermis and stele whilst invading the cortex, to the possible detriment of *G. graminis* which must enter the stele to cause disease (Speakman and Lewis, 1978). But the main site of interaction seems to be outside of the endodermis, because *P. graminicola* can control similar weak pathogens that are normally restricted to the root cortex, and vice-versa (Deacon, 1974). Attempts to explain this by induction of phytoalexins have so far proved unsuccessful, although competition for nutrients or infection sites remains a possible explanation. Deacon and Henry (1980) proposed this after concluding that *G. graminis* can benefit from early, natural senescence of the root cortex (*see later*). It was argued that *G. graminis* uses nutrients that are more readily available from senescing than from fully functional cells to increase its inoculum potential, thus enabling progressive infection to take place from a low initial inoculum level. Other parasites, like *P. graminicola*, may compete for these nutrients and in so doing they may reduce take-all infection.

The population balance between *G. graminis* and *P. graminicola*

Many glasshouse experiments have shown that the population balance between *G. graminis* and *P. graminicola* is crucial for control; the population of *P. graminicola* must significantly outweigh that of *G. graminis*, or the control agent must be present before the pathogen. Further development of control programmes in the field thus depends on detailed knowledge of factors affecting the population of *P. graminicola*. Significantly, Holden (1976) showed that cereal root cortices — especially of wheat — have a very limited life-span, even in the absence of *G. graminis*. He concluded from experiments, that *P. graminicola* is a weak parasite, adapted to colonize cells as they senesce. Its large populations in British grasslands may therefore result from the high root density of a closed grass sward, coupled with the periodic production of new roots and the progressive cortical death of existing ones. These comments have obvious relevance to the problems of establishing and maintaining populations of *P. graminicola* in cereal crops; in practice, the manipulation of *P. graminicola* may depend on influencing root cortex death. Encouraging preliminary results have been obtained in a field trial in which first, second and third successive wheat crops showed significantly more cortical death after a two-year grass ley than after a two-year break of non-graminaceous crops. Moreover, differences in the amounts of *P. graminicola* were found between identically treated plots and were matched by differences in amounts of cortical death (C. M. Henry and J. W. Deacon, unpublished). There is no evidence that *P. graminicola* causes or enhances the rate of cortical death; rather, this fungus seems to respond to differences caused by other factors. Moreover, there is no evidence that cortical death *per se* is detrimental to the crop.

Cortical senescence — a unifying theme?

The highly pathogenic varieties of *G. graminis* and the control agent *P. graminicola* represent opposite ends of a range of dark mycelial parasites found on roots of the Gramineae throughout the world. Most of the others are of intermediate pathogenicity towards cereals (*see* Asher and Shipton, 1981). All these fungi seem largely restricted to graminaceous hosts; many of them seem interrelated, and all those tested to date can control take-all in glasshouse conditions. Moreover, all have the same distinctive infection habit: they grow on cereal and grass roots by dark runner-hyphae, and penetrate the root cortex at intervals by narrower hyaline hyphae. This infection habit can be viewed as a response to the progressive death of root cortical cells, which starts in the epidermis behind the zone of living root hairs and proceeds layer by layer back along the root as it ages. In winter wheat in field conditions, dark runner-hyphae of *P. graminicola* occurred most often in the intercellular spaces of the cortex, next to the outermost living host cells (Deacon, 1980). The implication was that the fungus could 'keep pace' with the rate of cortical cell death in the field, so its dark, lysis-resistant runner-hyphae might form the inoculum bases from which it invades each successive cell layer as host resistance declines. It remains uncertain why these parasites differ so markedly in pathogenicity and a solution to this must await more detailed study of their *natural* host-ranges. Meanwhile, there is increasing evidence that *G. graminis* var *tritici*, the aggressive pathogen of wheat and barley, also has non-pathogenic forms (e.g. Asher, 1978) and these, too, must now be sought in natural communities.

References

Adam, D. B. (1951). The control of take-all in wheat after a long period of pasture. *Australian Journal of Agricultural Research* **2,** 273–282.

Asher, M. J. C. (1978). Isolation of *Gaeumannomyces graminis* var. *tritici* from roots. *Transactions of the British Mycological Society* **71,** 322–325.

Asher, M. J. C. & Shipton, P. J. (1981). *The Biology and Control of Take-all.* London: Academic Press.

Brooks, D. H. (1965a). Wild and cultivated grasses as carriers of the take-all fungus (*Ophiobolus graminis*). *Annals of Applied Biology* **55,** 307–316.

Brooks, D. H. (1965b). Root infection by ascospores of *Ophiobolus graminis* as a factor in epidemiology of the take-all disease. *Transactions of the British Mycological Society* **48,** 237–248.

Deacon, J. W. (1973). Factors affecting occurrence of the ophiobolus patch disease of turf and its control by *Phialophora radicicola. Plant Pathology* **22,** 149–155.

Deacon, J. W. (1974). Interactions between varieties of *Gaeumannomyces graminis* and *Phialophora radicicola* on roots, stem bases and rhizomes of the Gramineae. *Plant Pathology* **23,** 85–92.

Deacon, J. W. (1976). Biological control of the take-all fungus. *Gaeumannomyces graminis*, by *Phialophora radicicola* and similar fungi. *Soil Biology and Biochemistry* **8,** 275–283.

Deacon, J. W. (1980). Ectotrophic growth by *Phialophora radicicola* var *graminicola* and other parasites of cereal and grass roots. *Transactions of the British Mycological Society* **75,** 158–160.

Deacon, J. W. & Henry, C. M. (1980). Age of wheat and barley roots and infection by *Gaeumannomyces graminis* var *tritici. Soil Biology and Biochemistry* **12,** 113–118.

Doling, D. A. & Hepple, S. (1959). Occurrence of take-all in cocksfoot and other grasses. *Plant Pathology* **8,** 73–75.

Gerlagh, M. (1968). Introduction of *Ophiobolus graminis* into new polders and its decline. Wageningen: *Centre for Agricultural Publishing and Documentation, Agricultural Research Reports* No. 713.

Holden, J. (1976). Infection of wheat seminal roots by varieties of *Phialophora radicicola* and *Gaeumannomyces graminis. Soil Biology and Biochemistry* **8,** 109–119.

Nilsson, H. E. (1969). Studies of root and root rot diseases of cereals and grasses. I.

On resistance to *Ophiobolus graminis* Sacc. *Annals of the Agricultural College of Sweden* **35,** 275–807.

Rosser, W. R. & Chadburn, B. L. (1968). Cereal diseases and their effects on intensive wheat cropping in the East Midlands Region. *Plant Pathology* **17,** 51–60.

Slope, D. B., Prew, R. D., Gutteridge, R. J. & Etheridge, J. (1979). Take-all, *Gaeumannomyces graminis* var. *tritici*, and yield of wheat grown after ley and arable rotations in relation to the occurrence of *Phialophora radicicola* var. *graminicola. Journal of Agricultural Science, Cambridge* **93,** 377–389.

Smith, J. D. (1965). *Fungal Diseases of Turf Grasses.* 2nd Edn. Bingley: Sports Turf Research Institute.

Speakman, J. B. & Lewis, B. G. (1978). Limitation of *Gaeumannomyces graminis* by wheat root responses to *Phialophora radicicola. New Phytologist* **80,** 373–380.

Turner, E. M. (1961). An enzymic basis for pathogen specificity in *Ophiobolus graminis. Journal of Experimental Botany* **12,** 169–175.

Walker, J. (1975). Take-all diseases of Gramineae: a review of current work. *Review of Plant Pathology* **54,** 113–144.

Wehrle, V. M. & Ogilvie, L. (1955). Effect of ley grasses on the carry-over of take-all. *Plant Pathology* **4,** 111–113.

Role of wild plants and weeds in the ecology of plant-parasitic nematodes

D J Hooper and A R Stone
Rothamsted Experimental Station, Harpenden, Herts AL5 2JQ

Introduction

J. B. Goodey *et al.* (1965) and Bendixen *et al.* (1979) list many weeds and wild plants as hosts of nematodes attacking crops but the large literature has not been reviewed systematically. In this review examples have been selected from temperate, sub-tropical and tropical zones to illustrate the importance of weeds and wild plants in the ecology of the nematode pests of crops.

Wild plants as original hosts of crop nematodes

Crops have been cultivated for generally no more than 9000 years (Hutchinson, 1965; Harlan, *1981*), a time insufficient for much evolution of plant-parasitic nematodes to have occurred. Even the selection of host-specific races or pathotypes within species is likely to have occurred on wild relatives and precursors of cultivated plants (Stone, 1979). Some nematodes have co-evolved with certain plants and remained with them as they were selected for cultivation. Others have transferred from indigenous wild hosts to introduced crops. Probably all plant-parasitic nematodes include wild species in their host ranges and virtually all wild and crop species are attacked.

Paramonov (1962) and Maggenti (1971) thought that feeding on higher plants could have evolved from feeding on fungi. The potato-rot nematode (*Ditylenchus destructor*) and several bud and leaf nematodes (*Aphelenchoides* spp.) feed facultatively on fungi and higher plants. Feeding and parasitism of higher plants was presumably via the algae, liverworts and mosses. Mosses are hosts of several tylenchids and close relationships have evolved in *Tylenchus*-, *Ditylenchus*- and *Anguina*-like species, which reproduce in galls induced on mosses. Many ferns are hosts of *Aphelenchoides* spp., especially *A. fragariae* which also attacks many angiosperms. Maggenti noted that the association between insects and fungal-feeding nematodes may have been important in the evolution of some nematodes parasitizing aerial parts of plants. Red-ring disease of coconut palms caused by *Rhadinaphelenchus cocophilus*, transmitted by the weevil *Rhynchophorus palmorum* (*see* Blair, 1969), and the lethal wilting disease of pines caused by *Bursaphelenchus lignicolus*, transmitted by the beetle *Monochamus alternatus* (*see* Mamiya and Enda, 1972), may have arisen through such close associations.

The ability of many nematodes to withstand desiccation facilitates dispersal, especially with seeds of their hosts. Hence, as wild plant species were gradually domesticated nematodes accompanied them on their seed. This applies particularly to seed-gall nematodes such as *Anguina tritici*, which attacks only wild or domesticated wheats and related grains. *Anguina agrostis* which galls *Agrostis tenuis*, and occurs in the wild, is sometimes troublesome in pastures, decreasing seed yields and, because of an associated bacterium, making the crop toxic to livestock (Galloway, 1961). Similarly *Anguina funesta* galls with an associated bacterium on *Lolium rigidum* are toxic to sheep in Australia (Price *et al.*, 1979). The stem and bulb nematode (*Ditylenchus dipsaci*) attacks many crops in temperate regions (Hooper, 1972) and readily withstands desiccation, the resistant fourth-stage juveniles forming 'eelworm wool'. It is seed-borne by several crops and its races have wide host ranges that include many weeds. *Aphelenchoides besseyi*, which attacks wild and cultivated rice, is seed-borne and occurs wherever rice is grown (Fortuner and Orton Williams, 1975).

Radopholus similis causes severe necrosis of banana roots and corms making them susceptible to fungal attack. Infested plants with poor root systems are often toppled by wind. Infestations are spread mainly in planting material (O'Bannon, 1977). However, the genus *Radopholus* seems to be indigenous to Australia where wild bananas occur (Simmonds, 1976) and may have been the original source of this nematode (Sher, 1968).

Cyst nematodes have a swollen, sedentary female that must establish and maintain a feeding site (transfer cell) for a prolonged period. They are well-adapted to their hosts and often have small host ranges which, for a given species or species group, may be limited to a single plant genus, family, or order (Stone, 1979). There is much evidence for co-evolution between these nematodes and their hosts and the original wild hosts can sometimes be identified.

Punctodera is a cosmopolitan genus with host ranges limited to the Gramineae; *P. chalcoensis* is a pest of maize known only from central Mexico. Its known hosts are maize and teosinte (*Zea mexicana*); various cereals and grasses are non-hosts (Stone *et al.*, 1976). Maize is thought to have originated from a wild precursor in central Mexico where the earliest archaeological record of maize cultivation is c. 5000 years BC (Goodman, 1976; Harlan, *1981*). Teosinte has a similar distribution to maize and is a close relative, possibly a precursor (Wilkes, 1972). It seems probable that *P. chalcoensis,* with its host range limited to maize-like plants and its distribution limited to the centre of origin of maize, is a species which originated on wild relatives of maize and was able to develop on the cultigen as it evolved.

Teosinte is a weed in maize fields and is difficult to control by cultural practices because its growth habit resembles that of the crop.

Wild hosts of cyst nematodes are likely to be most abundant in regions where the nematodes evolved. Several of the *Globodera* complex parasitizing Solanaceae are crop pests with wide host ranges within the family (Stone, 1972). The potato cyst-nematodes, *G. rostochiensis* and *G. pallida*, are naturally distributed in South America (K. Evans *et al.*, 1975) on wild and cultivated *Solanums*. Wild Solanaceae frequently occur in Andean fields, where cultural practices are often primitive (Jones, *1981*). In North America other *Globodera* species (*G. virginiae, G. solanacearum* and *G. tabacum*) occur on the eastern seaboard on indigenous wild Solanaceae and especially in Mexico where the solanaceous flora supports a rich *Globodera* fauna.

There are other examples of cyst nematodes occurring in the centres of origin of their hosts. *Heterodera daverti*, a parasite of clovers, occurs in the eastern Mediterranean (Stone *et al.*, 1980) which is the centre of diversity of *Trifolium* (A. M. Evans, 1976). Similarly *H. medicaginis* is a serious pest of lucerne in the Caspian region, which occurs there as a primitive diploid (Lesins, 1976).

Root-knot nematodes (*Meloidogyne* spp.) also have swollen, sedentary females which feed upon modified cells (giant cells). Many have narrow host ranges and are local in distribution. However, the agriculturally most important species (*M. incognita*, *M. javanica*, *M. arenaria* and *M. hapla*) are widespread and have extensive host ranges (Sasser, 1979). The ancestral stock of the group may be a temperate climate form with limited host range (Triantaphyllou, 1979) and descendants still exist with localized distributions and few known hosts. Tropical and sub-tropical species of *Meloidogyne* have shorter generation times than those of cyst nematodes, most of which are temperate animals. Feeding sites for these *Meloidogyne* have to be maintained for shorter periods than those for cyst nematodes and the interrelationships between host and parasite may not have need to be so finely developed, permitting a wider host range than is possible for temperate Heteroderoidea.

Many papers listing hosts of *Meloidogyne* species include wild plants. Krall (1970) listed 103 wild hosts of *M. hapla* in Estonia and refers to 'vast natural foci' of the nematode. Elgindi and Moussa (1971) found *M. incognita*, *javanica*, *thamesi* and *arenaria* in newly cultivated land and recorded many wild plants as hosts. D. P. Taylor *et al.* (1978) found a *Meloidogyne* species of the *incognita-javanica* group parasitizing the wild baobab tree (*Adansonia digitata*) in Senegal. Natural populations of cyst and root-knot nematodes on wild plants threaten land newly brought into cultivation.

Wild plants as reservoirs of nematodes of crops

A distinction between wild plants outside crops and those within them is difficult to make. Plants in uncultivated land, woodlands and hedgerows alongside cultivated crops may act as reservoirs for pest species. Virgin areas, e.g., the Great Plains of North America (Thorne and Malek, 1968) and the Russian Virgin Lands (Gritsenko, 1968) harbour abundant nematodes known to parasitize crops elsewhere, which demonstrates that wild plants are indeed reservoirs of pest species. Locally, adjacent woodlands and/or uncultivated lands also contain important plant-parasitic nematodes (Crow and Macdonald, 1978; Goheen and Braun, 1956; Gritsenko, 1968; Kiryanova and Shagalina, 1974; Rebois and Golden, 1978; Ruehle, 1968; Sharma, 1976; McNamara and Flegg, *1981*).

Ditylenchus dipsaci with its many weed hosts often occurs in uncultivated land in weeds, e.g. plantains (*Plantago* spp.), dandelion (*Taraxacum officinale*), cat's ear (*Hypochoeris radicata*) and wild strawberry (*Fragaria chiloensis* and *F. vesca*) (Green, *1981*). Its dispersal by weed seeds is well illustrated by its widespread occurrence in *H. radicata* in the western USA. There it and *D. dipsaci* were apparently introduced together but this race of *D. dipsaci* did not transfer to other likely hosts (Godfrey, 1924). In the same areas, a different race of *D. dipsaci* in wild strawberries was apparently the source of infestations in cultivars (Godfrey and McKay, 1924).

Ditylenchus destructor is a pest of potatoes in Europe but not in South America from whence European potatoes originated. This nematode reproduces in weeds such as field mint (*Mentha arvensis*) and creeping sowthistle (*Sonchus arvensis*), upon which it survives between crops (J. B. Goodey, 1952).

Paranguina agropyri, which galls stems of couch grass (*Agropyron repens*) in parts of Russia, also attacks seedlings of rye, wheat and barley (Krall and Krall, 1968). Some gall-forming species have restricted host ranges and attempts are being made to use *Anguina picridis* to control weed grasses (T. S. Ivanova, 1966; Kovalev and Danilov, 1973; Watson, 1976). Similarly *Nothanguina phyllobia* might be used to control *Solanum elaeagnifolium* (Robinson *et al.*, 1978).

Subanguina radicicola galls the roots of *Poa annua* growing wild or as a weed. Lewis and Webley (1966) found it damaged *Poa pratensis* and *P. trivialis* in pasture and also attacked wheat, oats and barley. T. Goodey (1932) noted that British populations on *Elymus arenarius* had different host ranges from continental populations: six races were recognized (Krall and Krall, 1970).

Rotylenchulus reniformis is an obligate, sedentary semi-endoparasite of roots, commonly parasitizing crops and weeds throughout tropical and sub-tropical countries (Ayala and Ramirez, 1964; Siddiqi, 1972). It occurs in uncultivated regions on wild Gramineae and cactus (*Opuntia*) in the French West-Indies (Scotto La Massèse, 1969) and on baobab, common in arid regions of W. Africa (D. P. Taylor *et al.*, 1978).

Hirschmanniella oryzae, a root endoparasite of rice (Babatola and Bridge,

1979; Fortuner and Merny, 1979), parasitizes weeds in uncultivated areas and rice in flooded paddies (Fortuner, 1976). The rice stem nematode (*Ditylenchus angustus*) distorts panicles and decreases yields; although not readily dispersed on rice seed (A. L. Taylor, 1969) wild rice species are also hosts (Miah and Bakr, 1977) and infestations are carried downstream to cultivated paddies (Tin Sein and Kaung Zan, 1977).

Aphelenchoides arachidis infests the pods and testas of groundnuts. Although seed-borne (Bridge *et al.*, 1977) it is known only from a restricted area in Northern Nigeria where other crops and wild grasses are hosts (Bos, 1977) from which it has presumably spread.

Wild plants and particularly those of hedgerows are reservoirs of the root-ectoparasitic longidorid and trichodorid nematodes, some of which are vectors of plant viruses (McNamara and Flegg, *1981*; Murant, *1981*). Besides transmitting viruses these nematodes sometimes severely damage roots (Whitehead and Hooper, 1970).

Wild Solanaceae in the eastern USA and Mexico are reservoirs for *Globodera tabacum*, *virginiae* and *solanacearum*. Cultivated potatoes are not parasitized by these nematodes (Lownsbery and Lownsbery, 1954; L. I. Miller and Gray, 1968; L. I. Miller *et al.*, 1975) but they severely damage tobacco. In Mexico the *Globodera* fauna is rich (*see* Stone, 1979) in wild places and along field margins. Paradoxically, where *G. rostochiensis* and *G. pallida* occur in North America they are introductions.

Records of cyst nematodes in wild sites are few because most attention has been paid to fields where hosts occur as weeds. In a survey of uncultivated land in nature reserves in southern England, Hill and Stone (1980) found 45% of soils from around grasses and 75% of soils from around dicotyledonous plants contained cyst nematodes in the *avenae*, *schachtii* and *goettingiana* sub-groups of the genus *Heterodera*, all of which include agricultural pests.

In tropical and sub-tropical peasant farming systems, where wild plants readily encroach onto cultivated land, reservoirs of *Meloidogyne* spp., especially those with extensive host ranges, are important. 'Shifting cultivation' and the use of previously uncultivated ground provide further opportunity for the incursion into agriculture of *Meloidogyne* on wild hosts. However, plant-parasitic nematodes in cassava crops planted in former Venezuelan savannah were surprisingly few (Stone, unpublished). Crow and Macdonald (1978) provide an example of the northern root-knot nematode, *Meloidogyne hapla*, occurring on native vegetation adjacent to strawberry fields in Minnesota.

Weeds as foci of infestation within crops and as carry-over hosts

Weeds support many nematodes that attack crop plants (Caveness, 1967; Hogger and Bird, 1976; Kasimova, 1969; Pavlyuk, 1974; Shlepetene, 1965; Soloveva, 1965; Zem, 1977; Zem and Lordello, 1976). For example, weeds are important carry-over hosts of the stem nematode *Ditylenchus dipsaci* (Green, *1981*). Seinhorst (1956) noted that numbers increased under a non-host crop of wheat containing the host weeds chickweed (*Stellaria media*), scarlet pimpernel (*Anagallis arvensis*) and cleavers (*Galium* spp.). Wild oat (*Avena sativa*), a troublesome weed (Fryer, *1981*) maintains the race that attacks oats, faba beans, onions and several other crops. There are at least 11 distinct races in Europe with different host ranges on crops and weeds (Seinhorst, 1957). Some can interbreed and their progeny have different host preferences (Sturhan, 1966; Webster, 1967). Thus weeds like *S. media*, a host to several races (T. Goodey, 1947; Southey and Staniland, 1950), is a reservoir in which races may interbreed and possibly produce new ones. *S. media* and other weeds are also important as carry-over hosts that permit reproduction of *D. dipsaci* in the autumn and winter (Gentzsch, 1973). Some karyologically distinct races of *D. dipsaci* have restricted host ranges, including weeds (Barabashova, 1978; Ladygina, 1977, 1978).

Besides being a pest of potatoes, *Ditylenchus destructor* also attacks bulbous iris (*Iris* spp.) and some weeds are important carry-over hosts (J. B. Goodey, 1951; I. V. Ivanova, 1973).

Aphelenchoides fragariae, damaging to strawberry, and *A. ritzemabosi*, which attacks chrysanthemums and strawberry, both have wide host ranges including many common weeds (Burkhardt, 1967; Juhl, 1978; Siddiqi, 1974; Szczygiel, 1977). *A. besseyi* attacks rice and strawberries and has a limited host range (Franklin and Siddiqi, 1972) but some weeds are carry-over hosts (Vuong, 1969). Likewise *Hirschmaniella oryzae* and *D. angustus*, root and stem parasites of rice respectively, have few weed hosts (Mathur and Prasad, 1974; Vuong, 1969).

The burrowing nematode (*Radopholus similis*) on bananas also infests weeds (Edwards and Wehunt, 1971; Koshy and Sosamma, 1977; Maas, 1969) and the race that causes spreading decline of citrus in Florida can survive on many other hosts including weeds (O'Bannon, 1977; Suit *et al.*, 1954).

Wild yams that appear in 'bush fallow' are a main source of infestation of *Scutellonema bradys*, which causes a dry rot of yam tubers (Bridge, 1972); other weeds also act as carry-over hosts (Adesiyan, 1976).

Pratylenchus spp., damaging root endoparasites, generally have weed hosts; notably *P. penetrans*, a pest in temperate regions (Kasimova *et al.*, 1976; Townshend and Davidson, 1960; Willis and Thompson, 1969), and *P. brachyurus*, which attacks many tropical crops (Corbett, 1976; Guerout, 1975; Egunjobi, 1974; Shepherd, 1977).

Many weeds are carry-over hosts for longidorid and trichodorid nematodes (Cohn and Mordechai, 1969; McElroy, 1972; P. M. Miller, 1980; Thomas, 1969, 1970; Winfield and Cooke, 1975) and of the viruses they transmit (Murant, *1981*).

Wild hosts as weeds in crops may be carry-over hosts or foci of infestations where a cyst nematode occurs indigenously. Some of the agriculturally important cyst nematodes in western Europe provide examples. *Heterodera avenae* is indigenous and there are several extensive lists of wild hosts. Duggan (1959) reported that, except for *Agropyron repens* and *Agrostis tenuis*, 80 wild and cultivated cereals, and other Gramineae tested, were hosts. In that study, other then unrecognized, members of the *avenae* group were possibly present. Mowat (1974) recorded many grass hosts for the cereal cyst nematode (*H. avenae*) and the related *H. iri* and *H. mani*. Videgard (1969) found wild oat supports large populations of *H. avenae* but numbers decline under grasses of agronomic importance (Gair, 1968) and most wild graminaceous hosts maintain small population densities. Other members of the *H. avenae* complex, *H. longicaudata* (*H. bifenestra*?), *H. hordecalis* and *H. latipons* are known from wild and cultivated hosts. The beet cyst-nematode (*Heterodera schachtii*) has a wide host range among several botanical families with a great diversity of wild hosts: Vinduska (1972) recorded increases on *Chenopodium album* and *Sinapis arvensis*. The pea cyst-nematode (*H. goettingiana*), occurs on some wild Leguminosae and the clover cyst-nematode (*H. trifolii*) has wild hosts among the Leguminosae and in some other families. Yeates and Visser (1979) concluded that in view of its persistence and wide host range, control of *H. trifolii* in New Zealand pastures is unlikely to be achieved by cultural practices. Among indigenous western European cyst nematodes the carrot cyst-nematode (*H. carotae*) is unusual in having a known host range confined to cultivated and wild carrot (*Daucus carota*).

Similar patterns of hosts can be found for cyst nematodes outside western Europe (e.g., Caveness, 1967; Odihirin, 1975; Tikhonova, 1968). Epps and Chambers (1966) found *H. glycines* reproduced more rapidly on some weed species than on soybean, the main crop host. Stone (1979) lists host families for species of cyst nematodes.

Volunteer potato plants often occur in cereal and other crops in western Europe and may act as carry-over hosts for potato cyst-nematodes.

The role of weeds in maintaining *Meloidogyne* populations is commensurate with the wide host ranges of the economically important species. The literature is large and 47 of 59 papers on the host range of *Meloidogyne* published between 1965 and 1979 mentioned weeds and wild hosts.

Lamberti (1979) and others emphasize the importance of controlling weeds in attempts to reduce numbers of polyphagous *Meloidogyne* spp. by non-host crops. However, because of their wide host ranges such species cannot be eradicated by crop rotation. Hogger (1975) records many weed species as hosts of *Meloidogyne incognita* in Georgia and they decreased the rate nematode populations declined over winter. Dry season decline in the tropics may also be reduced. For instance, *Meloidogyne* spp. infest weeds in more than 20 plant families in Nigeria, some of which permit relatively slow reproduction of the nematodes during the dry season (Odihirin and Adesida, 1975).

Carry-over of root-knot nematodes on weeds has received considerable attention. Clayton *et al.* (1944) found that *Digitaria* species in an oat crop nullified the population decrease expected for *Meloidogyne* infesting tobacco. There is an early record of *Meloidogyne* on weeds in two-year fallows between tobacco crops in Rhodesia (Anon., 1946).

The impact of agricultural practices on nematode populations

Exhortations to control weeds and prevent nematode multiplication abound (Franklin, 1970; Kavanagh, 1974), but papers detailing the effects of weed control on nematodes are few. An example is the annual use of herbicides to control weeds in Georgia cotton fields (Bird and Hogger, 1973). As the primary weeds declined the incidence of nut sedges (*Cyperus esculentus* and *C. rotundus*) increased. As these species are good hosts of *Meloidogyne incognita*, a pest of cotton (and are also hosts of *Hoplolaimus robustus*), an increase in root-knot disease ensued. Nematologists may exploit the changed weed flora by using *Cyperus* as an indicator host for *Meloidogyne incognita*, which can be found in the roots two months before cotton is planted (Hogger and Bird, 1976).

Crop rotation is often a useful method of combating nematodes but, as one species declines, others can increase (Ferris and Bernard, 1971; Furstenberg and Heyns, 1978; Luc *et al.*, 1964; Murphy *et al.*, 1974). Cultural practice influences nematode populations (Good, 1968) and the mixed cropping common in tropical regions tends to maintain large nematode populations. Cover crops have been suggested to decrease nematodes and one of the more successful is pangola grass (*Digitaria decumbens*), used to control pineapple nematodes in Puerto Rico (Ayala *et al.*, 1969). A summer cover crop is often preferable to a weed fallow; *Belonolaimus longicaudatus* and *Meloidogyne incognita* declined under a cover crop of *Indigofera hirsuta* but increased under weeds (Rhoades, 1978). Brodie *et al.* (1970) noted changes in populations under different cover crops and Miller and Aheras (1969) found that some weeds in a cover crop/fumigation study were very good hosts of *Pratylenchus penetrans* and *Tylenchorhynchus claytoni*.

Nematode control via weed control presents greater problems in tropical and sub-tropical agriculture, especially peasant agriculture, than in the temperate zones. This is because of the greater abundance of weeds and the greater host range of the main warm-climate *Meloidogyne* species.

Some cultural practices may increase nematode problems. For example, some plants grown as supports for black pepper (*Piper nigrum*) in India are good hosts for *M. incognita*, providing established foci from which the nematodes spread to pepper (Koshy *et al.*, 1977). The practice of intercropping extends both the duration and range of available host plants.

The rapid loss of fertility of tropical soils in the Americas has been attributed to the build up of nematodes and other pests, rather than the exhaustion of nutrients (Steiner and Buhrer, 1964). Shifting cultivation reduces this problem (Ogbuji, 1979) but each new site contains nematodes left by the original flora and seeds of wild plants, which are potential weeds.

Roberts and Stone (Stone, 1979) found potential differences in the host ranges of European populations of potato cyst-nematodes compared with that of a Bolivian population or those of other *Globodera* species from the Americas. European populations of *G. rostochiensis* developed in only five *Solanum* subgenus *Leptostemonum* spp. compared with the fifteen or more host *Leptostemonum* spp. of American populations. A European *G. pallida* population behaved similarly, although two others had extensive host ranges. Potato cyst-nematodes are thought to have been introduced into Europe with the influx of new potato breeding material, which followed the blight epidemics of the 1840s, and cultivation over 130 years may have resulted in European nematode populations losing some genes conferring ability to develop on other host species, the effect being less marked in later introductions of nematodes. Dasgupta and Seshadri (1971) reported a similar loss of host range in pathotypes of the sedentary endoparasitic nematode *Rotylenchulus reniformis* raised on one host species for 36 generations. These findings support the view of Vanderplank (1968) that there is always selection against redundant genes for virulence.

Nematodes from wild or weed hosts may be confused with related pest species of crops. *Heterodera mani* occurs on grasses in the UK and, although not a pest of cereals, may be confused with *H. avenae*. Similarly *Globodera achilleae*, a parasite of yarrow (*Achillea millefolium*) in England and Wales (Sykes and Webley, 1979) and Scotland (Mrs. M. McKenzie, personal communication) closely resembles *G. pallida*. Cereal and potato cyst-nematodes cannot be identified with certainty on gross morphology alone and misidentification of species on wild hosts can result in inappropriate advice to farmers.

Conclusion

Because wild and weed species are important in the ecology of many agricultural plant-parasitic nematodes, good farming practice can help to control nematode damage to crops, especially where the nematode is indigenous or has an extensive host range.

References

Adesiyan, S. O. (1976). Host range studies of the yam nematode, *Scutellonema bradys*. *Nematropica* **6,** 60–63.

Anon. (1946). Root knot nematode. *Tobacco Research Board Bulletin, Rhodesia*, pp. 39–82.

Ayala, A., Gonzalez-Tejera, E. & Irizarry, H. (1969). Pineapple nematodes and their control. In *Nematodes of Tropical Crops*, pp. 210–224. J. E. Peachey. Farnham Royal: Commonwealth Agricultural Bureaux.

Ayala, A. & Ramirez, C. T. (1964). Host range, distribution and bibliography of the reniform nematode, *Rotylenchulus reniformis*, with special reference to Puerto Rico. *Journal of Agriculture, University of Puerto Rico* **48,** 140–161.

Babatola, J. O. & Bridge, J. (1979). Pathogenicity of *Hirschmanniella oryzae, H. spinicaudata* and *H. imamuri* on rice. *Journal of Nematology* **11,** 128–132.

Barabashova, V. N. (1978). [Karyological investigations of stem nematodes from the *Ditylenchus dipsaci* complex.] *Nauchy e Doklady Vysshei Shkoly, Biologicheskie Nauki* No. 5, 109–114.

Bendixen, L. E., Reynolds, D. A. & Riedel, R. M. (1979). An annotated bibliography of weeds as reservoirs for organisms affecting crops. 1. Nematodes. *Ohio Agricultural Research and Development Center, Research Bulletin* No. 1109, 64 pp.

Bird, G. W. & Hogger, C. (1973). Nutsedges as hosts of plant-parasitic nematodes in Georgia cotton fields. *Plant Disease Reporter* **57,** 402.

Blair, G. P. (1969). The problem of control of red ring disease. In *Nematodes of Tropical Crops*, pp. 99–108. J. E. Peachey. Farnham Royal: Commonwealth Agricultural Bureaux.

Bos, W. S. (1977). A preliminary report on the distribution and host range of the nematode *Aphelenchoides arachidis* Bos, in north Nigeria. *Samaru Agricultural Newsletter* **19,** 21–23.

Bridge, J. (1972). Nematode problems with yams (*Dioscorea* spp.) in Nigeria. *PANS* **18,** 89–91.

Bridge, J., Bos, W. S., Page, L. J. & McDonald, D. (1977). The biology and possible importance of *Aphelenchoides arachidis*, a seed-borne endoparasitic nematode of groundnuts from northern Nigeria. *Nematologica* **23,** 253–259.

Brodie, B. B., Good, J. M. & Jaworski, C. A. (1970). Population dynamics of plant nematodes in cultivated soil: effect of summer cover crops in newly cleared land. *Journal of Nematology* **2,** 217–222.

Burckhardt, F. (1967). [The occurrence of leaf eelworm on weeds and other wild plants.] *Mitteilungen aus der Biologischen Bundesanstalt für Land- und Forstwirtschaft* (Berlin-Dahlem) **121,** 71–75.

Caveness, F. E. (1967). Shadehouse host ranges of some Nigerian nematodes. *Plant Disease Reporter* **51,** 33–37.

Clayton, E. E., Shaw, K. J., Smith, T. E., Gaines, J. G. & Graham, T. W. (1944). Tobacco disease control by crop rotation. *Phytopathology* **34,** 870–883.

Cohn, E. & Mordechai, M. (1969). Investigations on the life cycles and host preference of some species of *Xiphinema* and *Longidorus* under controlled conditions. *Nematologica* **15,** 295–302.

Corbett, D. C. M. (1976). *Pratylenchus brachyurus. CIH Descriptions of Plant-parasitic Nematodes*, Set 6, No. 89.

Crow, R. V. & Macdonald, D. H. (1978). Phytoparasitic nematodes adjacent to established strawberry plantations. *Journal of Nematology* **10,** 204–207.

Dasgupta, D. R. & Seshadri, A. R. (1971). Reproduction, hybridization and host adaptation in physiological races of the reniform nematode, *Rotylenchulus reniformis*. *Indian Journal of Nematology* **1,** 128–144.

Duggan, J. J. (1959). Host range of cereal root eelworm. *Agricultural Record* **13,** 2–8.

Edwards, D. I. & Wehunt, E. J. (1971). Host range of *Radopholus similis* from banana areas of Central America with indications of additional races. *Plant Disease Reporter* **55,** 414–418.

Egunjobi, O. E. (1974). Nematodes and maize growth in Nigeria. 1. Population dynamics of *Pratylenchus brachyurus* in and about the roots of maize and its effects on maize production at Ibadan. *Nematologica* **20,** 181–186.

Elgindi, D. M. & Moussa, F. F. (1971). Root knot nematodes in recently reclaimed sandy areas of U.A.R. II. New host records for root knot nematodes, *Meloidogyne* spp. *Mededelingen van de Faculteit Landbouwwetenschappen Rijksuniversiteit Gent* **36,** 1341–1344.

Epps, J. M. & Chambers, A. Y. (1966). Comparative rates of reproduction of *Heterodera glycines* on 12 host plants. *Plant Disease Reporter* **50,** 608–610.

Evans, A. M. (1976). Clovers. In *Evolution of Crop Plants*, pp. 175–179. N. W. Simmons. London and New York: Longman.

Evans, K., Franco, J. & De Scurrah, M. M. (1975). Distribution of species of potato cyst-nematodes in South America. *Nematologica* **21,** 365–369

Ferris, V. R. & Bernard, R. L. (1971). Crop rotation effects on population densities of ectoparasitic nematodes. *Journal of Nematology* **3,** 119–122.

Fortuner, R. (1976). [Ecological study of nematodes of rice paddies in Senegal.] *Cahiers ORSTOM, Série Biologie, Nématologie* **11,** 179–191.

Fortuner, R. & Merny, G. (1979). Root-parasitic nematodes of rice. *Revue de Nematologie* **2,** 79–102.

Fortuner, R. & Orton Williams, K. (1975). Review of the literature on *Aphelenchoides besseyi* Christie, 1942, the nematode causing 'white tip' disease in rice. *Helminthological Abstracts, Series B* **44,** 1–40.

Franklin, M. T. (1970). Interrelationships of nematodes, weeds, herbicides and crops. In *Proceedings of the 10th British Weed Control Conference, Brighton*, pp. 927–933.

Franklin, M. T. & Siddiqi, M. R. (1972). *Aphelenchoides besseyi. CIH Descriptions of Plant-parasitic Nematodes*, Set 1, No. 4.

Fryer, J. D. (*1981*). Weed control practices and changing weed problems. In *Pests, Pathogens and Vegetation*, pp. 403–414. J. M. Thresh. London: Pitman.

Furstenberg, J. P. & Heyns, J. (1978). The effect of cultivation on nematodes. Part 1. *Rotylenchulus parvus. Phytophylactica* **10,** 77–80.

Gair, R. (1968). Population changes of cereal cyst eelworm under various grass species. *Plant Pathology* **17,** 145–147.

Galloway, J. H. (1961). Grass seed nematode poisoning in livestock. *Journal of the American Veterinary Medical Association* **139,** 1212–1214.

Gentzsch, D. (1973). Die Vermehrung des Zwiebeln schädigenden Stengelnematoden (*Ditylenchus dipsaci*) in Unkräutern und Schlussfolgerungen für den Herbizideinsatz. Berichte 12. Tage *Probleme der Phytonematologie Gross Lüsewitz, 1 Juni, 1973*, pp. 109–120.

Godfrey, G. H. (1924). Dissemination of the stem and bulb-infesting nematode *Tylenchus dipsaci*, in seeds of certain composites. *Journal of Agricultural Research, Washington* **28,** 473–478.

Godfrey, G. H. & McKay, M. B. (1924). The stem nematode *Tylenchus dipsaci* on wild hosts in the Northwest. *United States Department of Agriculture, Bulletin* No. 1229, 9 pp.

Goheen, A. C. & Braun, A. J. (1956). Some parasitic nematodes associated with wild strawberry plants in woodlands in Maryland. *Plant Disease Reporter* **40,** 43.

Good, J. M. (1968). Relation of plant parasitic nematodes to soil management practices. In *Tropical Nematology*, pp. 113–118. G. C. Smart & V. G. Perry. Gainesville: University of Florida Press.

Goodey, J. B. (1951). The potato tuber nematode *Ditylenchus destructor* Thorne, 1945; the cause of eelworm disease in bulbous iris. *Annals of Applied Biology* **38,** 79–90.

Goodey, J. B. (1952). Investigations into the host ranges of *Ditylenchus destructor* and *D. dipsaci. Annals of Applied Biology* **39,** 221–228.

Goodey, J. B., Franklin, M. T. & Hooper, D. J. (1965). *T. Goodey's the Nematode Parasites of Plants Catalogued Under Their Hosts*, 3rd edn. Farnham Royal: Commonwealth Agricultural Bureaux.

Goodey, T. (1932). Some observations on the biology of the root-gall nematode *Anguillulina radicicola* (Greeff, 1872). *Journal of Helminthology* **10,** 33–44.

Goodey, T. (1947). On the stem eelworm, *Anguillulina dipsaci*, attacking oats, onions, field beans, parsnips, rhubarb, and certain weeds. *Journal of Helminthology* **22,** 1–12.

Goodman, M. M. (1976). Maize. In *Evolution of Plant Crops*, pp. 128–136. N. W. Simmons. London and New York: Longman.

Green, C. D. (*1981*). The effects of weeds and wild plants on the reinfestation of land by *Ditylenchus dipsaci*. In *Pests, Pathogens and Vegetation*, pp. 217–224. J. M. Thresh. London: Pitman.

Gritsenko, V. P. (1968). [Some data on the nematode fauna of virgin soils in the Chuisk Valley.] In [*Helminths of animals and plants in Kirgizia.*], pp. 157–163. K. K. Karakeev *et al.*). Izdat. 'Ilim': Frunze.

Guerout, R. (1975). Nematodes of pineapple: a review. *PANS* **21,** 123–140.

Harlan, J. R. (*1981*). Ecological settings for the emergence of agriculture. In *Pests, Pathogens and Vegetation*, pp. 3–22. J. M. Thresh. London: Pitman.

Hill, A. J. & Stone, A. R. (1980). *Annual Report of Rothamsted Experimental Station for 1979*, p. 143.

Hogger, C. H. (1975). Plant-parasitic nematodes assòciated with weeds and agronomic crops in Georgia. *Dissertation*, University of Georgia, Athens.

Hogger, C. H. & Bird, G. W. (1976). Weed and indicator hosts of plant parasitic nematodes in Georgia cotton and soybean fields. *Plant Disease Reporter* **60,** 223–226.

Hooper, D. J. (1972). *Ditylenchus dipsaci. CIH Descriptions of Plant-parasitic Nematodes*, Set 1. No. 14.

Hutchinson, J. B. (1965). Crop plant evolution: a general discussion. In *Essays on Crop Plant Evolution*, pp. 166–181. J. Hutchinson. Cambridge: Cambridge University Press.

Ivanova, I. V. (1973). [The infection rate of weeds with the nematode *Ditylenchus destructor*.] *Byulleten' Vsesoyuznogo Instituta Gel'mintologii im K.I. Skryabina* No. 11, 39–42.

Ivanova, T. S. (1966). [Experiments on biological control of *Acroptilon picris*.]

Izvestiya Akademii Nauk Tadzhikskoi SSR, Otdelenie Biologicheskaya Nauk No. 2, 51–63.

Jones, R. A. C. (*1981*). The ecology of viruses infecting wild and cultivated potatoes in the Andean region of South America. In *Pests, Pathogens and Vegetation*, pp. 89–107. J. M. Thresh. London: Pitman.

Juhl, M. (1978). [Chrysanthemum nematodes: list of hosts for the leaf nematode *Aphelenchoides ritzemabosi.*] *Ugeskrift for Agronomer, Hortonomer, Forstkandidater of Licentiater* **123,** 183–186.

Kasimova, G. A. (1969). [Nematodes of some weeds from vegetable fields in the Kuba-Khachmas zone of Azerbaidzhan.] *Materialy Sessi Zakavkazskogo Soveta po Koordinatsii Nauchno-Issledovatel' skikh Rabot po Zashchite Rastenii, Baku* **4,** 92–93.

Kasimova, G. A. & Atakishieva, Ya. Yu. (1976). [Ecological and faunistic characterization of nematodes of the weeds of vegetable crops on the Apsheron peninsula (Azerbaidzhan).] *Izvestiya Akademii Nauk Azerbaidzhanskoi SSR, Biologicheskie Nauki* No. 5, 45–51.

Kavanagh, T. (1974). The influence of herbicides on plant disease. II. Vegetables, root crops and potatoes. *Scientific Proceedings of the Royal Dublin Society, B* (1971, published 1974), 251–265.

Kiryanova, E. S. & Shagalina, L. M. (1974). [Natural foci of plant parasitic nematodes, possible source of infection for cultivated plants.] *Izvestiya Akademii Nauk Turkmenskoi SSR, Biologicheskie Nauki* No. 2, 73–74.

Koshy, P. K. & Sosamma, V. K. (1977). Host range of the burrowing nematode *Radopholus similis* (Cobb, 1893) Thorne, 1949. *Indian Journal of Nematology*, (1975, published 1977) **5,** 255–257.

Koshy, P. K., Sosamma, V. K. & Sundararaju, P. (1977). Screening of plants used as pepper standards against root-knot nematode. *Indian Phytopathology* **30,** 128–129.

Kovalev, O. V. & Danilov, L. G. (1973). [The introduction of *Paranguina picridis* into the Crimea.] *Zashchita Rastenii* No. 4, p. 43.

Krall, E. (1970). [On natural foci of distribution of root-knot nematodes in Estonia.] *Materialy 7go Pribaltiiskogo Soveshchaniya po Zashchite Rastenii*, Part 1, 9–11.

Krall, E. & Krall, H. (1968). Nematoodikahjustustest teraviljapoldudel. *Sotsialistlik Pollumajandus* **23,** 732–734.

Krall, E. & Krall, H. (1970). [A new biological race of *Anguina radicicola* parasitizing fodder grasses in Estonia.] *Materialy 7go Pribaltiiskogo Soveshchaniya po Zashchite Rastenyii*, Part 1, 57–60.

Ladygina, N. M. (1977). [Species composition and distribution of stem nematodes in the Ukraine.] *Nauchnye Doklady Vysshei Shkoly, Biologicheskie Nauki* No. 7, 48–52.

Ladygina, N. M. (1978). [The genetic and physiological compatibility of different forms of stem nematodes. VI. Cross-breeding of *Ditylenchus* from cultivated plants and from weeds.] *Parazitologiya* **12,** 349–353.

Lamberti, F. (1979). Chemical and cultural methods of control. In *Root-knot Nematodes* (Meloidogyne *species*) *Systematics, Biology and Control*, pp. 405–423. London and New York: Academic Press.

Lamberti, F., Taylor, C. E. & Seinhorst, J. W. (1979). *Nematode Vectors of Plant Viruses*. London: Plenum.

Lesins, K. (1976). Alfalfa, lucerne (*Medicago sativa*) (Leguminosae-Papilionatae). In

Evolution of Crop Plants, pp. 165–168. N. W. Simmons. London and New York: Longman.

Lewis, S. & Webley, D. (1966). Observations on two nematodes infesting grasses. *Plant Pathology* **15,** 184–186.

Lownsbery, B. F. & Lownsbery, J. W. (1954). *Heterodera tabacum* new species, a parasite of solanaceous plants in Connecticut. *Proceedings of the Helminthological Society of Washington* **21,** 42–47.

Luc, M., Merny, G. & Netscher, C. (1964). Enquête sur les nématodes parasites des cultures de la République Centrafricaine et du Congo-Brazzaville. *L'Agronomie Tropicale* **19,** 723–746.

Maas, P. W. T. (1969). Two important cases of nematode infestation in Surinam. In *Nematodes of Tropical Crops*, pp. 149–154. J. E. Peachey. Farnham Royal: Commonwealth Agricultural Bureaux.

McElroy, F. B. (1972). Studies on the host range of *Xiphinema bakeri* and its pathogenicity to raspberry. *Journal of Nematology* **4,** 16–22.

McNamara, D. G. & Flegg, J. J. M. (*1981*). The distribution of virus-vector nematodes in Great Britain in relation to past and present natural vegetation. In *Pests, Pathogens and Vegetation*, pp. 225–235. J. M. Thresh. London: Pitman.

Maggenti, A. R. (1971). Nemic relationships and the origins of plant parasitic nematodes, pp. 68–81. In *Plant Parasitic Nematodes. 1.* B. M. Zuckerman, W. F. Mai & R. A. Rohde. London and New York: Academic Press.

Mamiya, Y. & Enda, N. (1972). Transmission of *Bursaphelenchus lignicolis* (Nematoda: Aphelenchoididae) by *Monochamus alternatus* (Coleoptera: Cerambycidae). *Nematologica* **18,** 159–162.

Mathur, V. K. & Prasad, S. K. (1974). Survival and host range of the rice root nematode, *Hirschmanniella oryzae. Indian Journal of Nematology* **3,** 88–93.

Miah, S. A. & Bakr, M. A. (1977). Sources of resistance to ufra disease of rice in Bangladesh. *International Rice Research Newsletter* **2,** 8.

Miller, L. I. & Gray, B. J. (1968). Horsenettle cyst nematode, *Heterodera virginiae* n.sp., a parasite of solanaceous plants. *Nematologica* **14,** 535–543.

Miller, L. I., Stone, A. R., Spasoff, L. & Evans, D. M. (1975). Resistance of *Solanum tuberosum* to *Heterodera solanacearum. Proceedings of the American Phytopathological Society* **1,** 153.

Miller, P. M. (1980). Reproduction and survival of *Xiphinema americanum* on selected woody plants, crops, and weeds. *Plant Disease* **64,** 174–175.

Miller, P. M. & Aheras, J. F. (1969). Influence of growing marigolds, weeds, two cover crops, and fumigation on subsequent populations of parasitic nematodes on plant growth. *Plant Disease Reporter* **53,** 642–646.

Mowat, D. J. (1974). The host range and pathogenicity of some nematodes occurring in grassland in Northern Ireland. *Record of Agricultural Research* **22,** 51–58.

Murant, A. T. (*1981*). The role of wild plants in the ecology of nematode-borne viruses. In *Pests, Pathogens and Vegetation*, pp. 237–248. J. M. Thresh. London: Pitman.

Murphy, W. S., Brodie, B. B. & Good, J. M. (1974). Population dynamics of plant nematodes in cultivated soil: effects of combinations of cropping systems and nematicides. *Journal of Nematology* **6,** 103–107.

O'Bannon, J. H. (1977). Worldwide dissemination of *Radopholus similis* and its importance in crop production. *Journal of Nematology* **9,** 16–25.

Odihirin, R. A. (1975). Occurrence of *Heterodera* cyst nematode (Nematoda: Heteroderidae) on wild grasses in southern Nigeria. *Nigerian Society for Plant Protection, Occasional publication*, No. 1, pp. 24–25.

Odihirin, R. A. & Adesida, T. O. (1975). Locations and situations in which plant parasitic nematodes survive the dry season in Nigeria. II. The role of weeds in carrying root-knot nematodes over the dry season in southern Nigeria. *Nigerian Society for Plant Protection, Occasional Publication* No. 1, 17.

Ogbuji, R. O. (1979). Shifting cultivation discourages nematodes. *World Crops* (May/June), 113–114.

Paramonov, A. A. (1962). [*Principles of phytonematology. Vol. 1. The origin of nematodes. Ecological and morphological characteristics of phytonematodes. General principles of taxonomy.*] Moscow: Izdatelstvo Akademii Nauk SSSR. (See also: *Plant parasitic nematodes. Vol. 1.* Translation from Russian by Israel Program for Scientific Translations, Jerusalem, 1968.)

Pavlyuk, L. V. (1974). [Quantitative and qualitative characteristics of the nematodes of *Senecio rhombifolius, Adonis vernalis* and *Valeriana officinalis*.] *Trudy Gel'mintologicheskoi Laboratorii (Ekologiya i geografiya gel'mintov)* **24,** 126–130.

Price, P. C., Fisher, J. M. & Kerr, A. (1979). *Anguina funesta* n.sp. and its association with *Corynebacterium* sp., in infecting *Lolium rigidum. Nematologica* **25,** 76–85.

Rebois, R. V. & Golden, A. M. (1978). Nematode occurrences in soybean fields in Mississippi and Louisiana. *Plant Disease Reporter* **62,** 433–437.

Rhoades, H. L. (1978). *Indigofera hirsuta* as a summer cover crop for controlling *Belonolaimus longicaudatus* and *Meloidogyne incognita* in Florida, USA. In *Proceedings of the 3rd International Congress of Plant Pathology, München, 16–23 August, 1978.* Göttingen: Deutsche Phytomedizinische Gesellschaft.

Robinson, A. F., Orr, C. C. & Abernathy, J. R. (1978). Distribution of *Nothanguina phyllobia* and its potential as a biological control agent for silver-leaf nightshade. *Journal of Nematology* **10,** 362–366.

Ruehle, J. L. (1968). Plant parasitic nematodes associated with southern hardwood and coniferous forest trees. *Plant Disease Reporter* **52,** 837–839.

Sasser, J. N. (1979). Pathogenicity, host ranges and variability in *Meloidogyne* species. In *Root-knot Nematodes* (Meloidogyne *species*). *Systematics, biology and control,* pp. 257–268. F. Lamberti & C. E. Taylor. London: Academic Press.

Scotto La Massèse, C. (1969). The principal plant nematodes of crops in the French West Indies. In *Nematodes of Tropical Crops*, pp. 164–183. J. E. Peachey. Farnham Royal: Commonwealth Agricultural Bureaux.

Seinhorst, J. W. (1956). Population studies on stem eelworms (*Ditylenchus dipsaci*). *Nematologica* **1,** 159–164.

Seinhorst, J. W. (1957). Some aspects on the biology and ecology of stem eelworms. *Nematologica* II, Suppl., 355–361 S.

Sharma, R. D. (1976). Nematodes of the cacao region of the State of Espirito Santo, Brazil. II. Nematodes associated with field crops and forest trees. *Revista Theobroma* **6,** 109–117.

Shepherd, J. A. (1977). Hosts of non-gall-forming nematodes associated with tobacco in Rhodesia. *Rhodesian Journal of Agricultural Research* **15,** 95–97.

Sher, S. A. (1968). Revision of the genus *Radopholus* Thorne, 1949 (Nematoda: Tylenchoidea). *Proceedings of the Helminthological Society of Washington* **35,** 219–237.

Shlepetene, Yu. A. (1965). [Nematodes of some weeds.] *Zashchita Rastenii ot Vreditelei i Boleznei, Sornyak., Vilnius*, pp. 74–76.

Siddiqi, M. R. (1972). *Rotylenchulus reniformis. CIH Descriptions of Plant-parasitic Nematodes*, Set 1, No. 5.

Siddiqi, M. R. (1974). *Aphelenchoides ritzemabosi. CIH Descriptions of Plant-parasitic Nematodes*, Set 3, No. 32.

Simmonds, N. W. (1976). *Evolution of Crop Plants*. London and New York: Longman.

Soloveva, G. I. (1965). [Plant nematodes of weeds in the cabbage field.] *Trudy Gel'mintologicheskoi Laboratorii, Akademiya Nauk SSR* **16,** 115–119.

Southey, J. F. & Staniland, L. N. (1950). Observations and experiments on stem eelworm *Ditylenchus dipsaci* (Kühn, 1857) Filipjev, 1936, with special reference to weed hosts. *Journal of Helminthology* **24,** 145–154.

Steiner, G. & Buhrer, E. M. (1964). The plant-nematode problem of the American Tropics. *Journal of Agriculture of the University of Puerto Rico* **48,** 69–100.

Stone, A. R. (1972). The round-cyst species of *Heterodera* as a group. *Annals of Applied Biology* **71,** 280–283.

Stone, A. R. (1979). Co-evolution of nematodes and plants. *Symbolae Botanicae Upsalienses* **22,** 46–61.

Stone, A. R., Moss, C. Sosa & Mulvey, R. H. (1976). *Punctodera chalcoensis* n.sp. (Nematoda: Heteroderidae) a cyst nematode from Mexico parasitising *Zea mays. Nematologica* **22,** 381–389.

Stone, A. R., Bennison, J. & Burrows, P. (1980). *Annual Report of Rothamsted Experimental Station for 1979*, p. 143.

Sturhan, D. (1966). Wirtspflanzenuntersuchungen an Bastardpopulationen von *Ditylenchus dipsaci*-Rassen. *Zeitschrift für Pflanzenkrankheiten Pflanzenpathologie und Pflanzenschutz* **73,** 168–174.

Suit, R. F., DuCharme, E. P. & Brooks, T. L. (1954). Non-citrus plants in relation to spreading decline. *Proceedings of the Soil and Crop Science Society of Florida* **14,** 182–184.

Sykes, G. B. & Webley, D. P. (1979). *Globodera achilleae* (Golden and Klindic) Mulvey and Stone: first records for Wales and England. *Plant Pathology* **28,** 54.

Szczygiel, A. (1977). [Strawberry bud and leaf nematode *Aphelenchoides fragariae* (Ritzema Bos 1890).] *Warsaw, Poland: Polska Akademia Nauk, Konitet Ochrony Roslin*, 39 pp.

Taylor, A. L. (1969). Nematode parasites of rice. In *Nematodes of Tropical Crops*, pp. 264–268. J. E. Peachey. London and New York: Longman.

Taylor, D. P., Netscher, C. & Germani, G. (1978). *Adansonia digitata* (Baobab) a newly discovered host for *Meloidogyne* sp. and *Rotylenchulus reniformis:* agricultural implications. *Plant Disease Reporter* **62,** 276–277.

Thomas, P. R. (1969). Crop and weed plants compared as hosts of viruliferous *Longidorus elongatus* (de Man). *Plant Pathology* **18,** 23–28.

Thomas, P. R. (1970). Host status of some plants for *Xiphinema diversicaudatum* (Micol.) and their susceptibility to viruses transmitted by this species. *Annals of Applied Biology* **65,** 169–178.

Thorne, G. & Malek, R. B. (1968). Nematodes of the Northern Great Plains. Part 1. Tylenchida (Nemata: Secernentea). *South Dakota State University, Agricultural Experiment Station, Technical Bulletin* No. 31, 111 pp.

Tikhonova, L. V. (1968). [*Heterodera avenae.*] *Izdatestvo Akademii Nauk SSSR, Moscow*, pp. 407–413.

Tin Sein & Kaung Zan (1977). Ufra disease spread by water flow. *International Rice Research Newsletter* **2,** (2), 5.

Townshend, J. L. & Davidson, T. T. (1960). Some weed hosts of *Pratylenchus penetrans* in Premier strawberry plantations. *Canadian Journal of Botany* **38,** 267–273.

Triantaphyllou, A. C. (1979). Cytogenetics of root-knot nematodes. In *Root-knot Nematodes* (Meloidogyne *species*) *Systematics, Biology and Control*, pp. 85–109. F. Lamberti & C. E. Taylor. London and New York: Academic Press.

Vanderplank, J. E. (1968). *Disease Resistance in Plants.* New York: Academic Press.

Videgard, G. (1969). Inventering av havrecystnematod 1965–68. *Vaxtskyddanotiser* **33,** (2/3), 23–29.

Vinduska, L. (1972). [Number of generations of *Heterodera schachtii* Schmidt on winter rape and on weeds.] *Ochrana Rostlin* **8,** 207–210.

Vuong, H. H. (1969). The occurrence in Madagascar of the rice nematodes *Aphelenchoides besseyi* and *Ditylenchus angustus.* In *Nematodes of Tropical Crops*, pp. 274–287. J. E. Peachey. Farnham Royal: Commonwealth Agricultural Bureaux.

Watson, A. K. (1976). The biological control of Russian knapweed with a nematode. In *Proceedings of the IV International Symposium on biological control of weeds, 30 August–2 September 1976, University of Florida*, pp. 221–223. T. E. Freeman. Quebec: Institute of Food and Agricultural Sciences, Macdonald College, McGill University.

Webster, J. M. (1967). The significance of biological races of *Ditylenchus dipsaci* and their hybrids. *Annals of Applied Biology* **59,** 77–83.

Whitehead, A. G. & Hooper, D. J. (1970). Needle nematodes (*Longidorus* spp.) and stubby root nematodes (*Trichodorus* spp.) harmful to sugar beet and other field crops in England. *Annals of Applied Biology* **65,** 339–350.

Wilkes, G. H. (1972). Maize and its wild relatives. *Science* **177,** 1071–1077.

Willis, C. B. & Thompson, L. S. (1969). The influence of soil moisture and cutting management on *Pratylenchus penetrans* reproduction in birdsfoot trefoil and the relationship of inoculum levels to yields. *Phytopathology* **59,** 1872–1875.

Winfield, A. L. & Cooke, D. A. (1975). The ecology of *Trichodorus.* In *Nematode Vectors of Plant Viruses*, pp. 309–341. F. Lamberti, C. E. Taylor & J. W. Seinhorst. London and New York: Academic Press.

Yeates, G. W. & Visser, T. A. (1979). Persistence of *Heterodera trifolii* (Nematoda) cysts in the absence of host plants. *New Zealand Journal of Agricultural Research* **22,** 649–651.

Zem, A. C. (1977). Informacoes preliminares sobre os nematoides que se hospedam em plantas invasoras. II. *Reuniao de Nematologia, Sociedade Brasileria de Nematologia, Publicacao* No. 2, pp. 45–48.

Zem, A. C. & Lordello, L. G. E. (1976). [Nematodes associated with weed plants in Brazil.] *Anais de Escola Superior de Agricultura 'Luiz de Queiroz'* **33,** 597–615.

The effects of weeds and wild plants on the reinfestation of land by *Ditylenchus dipsaci* (stem and bulb nematode) and on the stability of its populations

C D Green
Rothamsted Experimental Station, Harpenden, Herts AL5 2JQ

Introduction

The stem and bulb nematode (*Ditylenchus dipsaci*) is an endoparasite of plant stems, leaves and bulbs and is indigenous in Great Britain. It is most often found in large numbers, primarily because it is usually sought in very susceptible cultivated plants in which it causes distortion, necrosis and death. *D. dipsaci* populations decline rapidly after stands of such plants are removed (Seinhorst, 1956; Wilson and French, 1975). The nematode is found in many crops, weeds and in native plant communities which show little or no damage. In them populations are smaller but probably more stable and they are important sources for the infestation of cultivated areas. This paper assesses the role of such populations in relation to the ecology of the species in agricultural land and crops.

Life cycle and multiplication

D. dipsaci is amphimictic and lays its eggs in plant tissues. Development is continuous through four juvenile stages. The first occurs in the egg; the second and third are most often found in plant tissues. Development may be arrested in the fourth stage which can persist in moist soil and withstands drying in air or plant tissues. *D. dipsaci* spreads into the flowers and seeds of hosts which survive to fruiting. The dry fourth stage occurs in seeds of many cultivated plants and in vegetable crops is present most often in seeds of species that are seldom severely damaged (Green and Sime, 1979).

D. dipsaci multiplies rapidly with a generation time of 2–3 weeks at 15°C (Yuksel, 1960) and at a rate that is independent of the numbers inoculated or their density in the tissues of artificially-infested *Vicia faba* or *Pisum sativum* (Green, 1977). Multiplication continues until the entire host or infested shoots die, but some infested seedlings are killed before reproduction occurs

(Griffin, 1975). Such rapid uncontrolled multiplication and exploitation of the host are typical of opportunist parasites with the characteristics of 'r' strategists (Pianka, 1970; Jones, 1980). Such animals are likely to have variable, unstable populations which are liable to become extinct and effective dispersal or survival mechanisms are essential for recolonization. Often *D. dipsaci* only survives long enough to persist between crops in soils with more than 30% clay (Seinhorst, 1956; Green and McNee, 1978; Green and Wilson, 1980).

Host range

D. dipsaci reproduces in many plants (Table 1) and seems to be an unselective and unspecialized parasite, yet it has been sub-divided into many biological races (Hesling, 1966; Sturhan, 1969) based upon their ability to reproduce in selective hosts (Seinhorst, 1957). Populations of *D. dipsaci* from *Vicia faba, Plantago maritima, Taraxacum officinale* and *Falcaria* sp. were found to be tetraploid but most races were diploid (Barabaschova, 1974; Ladygina and Barabaschova, 1976; Sturhan, 1969), morphologically indistinguishable, and produced fertile progeny after hybridization (Eriksson, 1974; Ladygina, 1973; Sturhan, 1964, 1966; Webster, 1967). The host ranges of the hybrid progeny sometimes differed from those of the parents indicating great genetic variation in host-compatibility. Races from *Medicago sativa, Trifolium pratense, T. repens* and *Phlox* sp. were partially isolated reproductively as the progeny of some crosses with other races were infertile (Erikson, 1974;

Table 1 The number and taxonomic grouping of the species recorded as hosts of *Ditylenchus dipsaci* by Goodey, Franklin and Hooper (1965).

*Class**	*Subclass*	*Orders‡*	*Families‡*	*British† flora*	*Cultivated/ foreign*
Liliatae	Alismatidae	0/14	0/14	0	0
(monocotyledonous)	Commelinidae	0/8	2/25	21	13
	Arecidae	0/4	0/5	0	0
	Lilidae	2/2	4/17	10	59
Magnoliatae	Magnolidae	2/6	3/36	10	9
(dicotyledonous)	Hamamelidae	1/9	1/23	0	1
	Caryophyllidae	2/4	4/14	25	9
	Dilleniidae	4/12	6/69	21	50
	Rosidae	9/16	11/108	34	30
	Asteridae	8/9	11/43	61	65

* System of Cronquist (1968).
† Clapham, Tutin and Warburg (1962).
‡ The number of orders or families containing host species as a fraction of the total number.

Ladygina, 1974, 1976) and their host ranges were also more restricted than those of most populations. *D. dipsaci* infestations in *M. sativa* and *T. pratense* crops often start from seed-borne parasites (Brown, 1957; Shipstra, 1952) and it is likely that they have become isolated due to continuous propagation in forage legumes. If most dispersal occurs in such seeds there is little mixing with other races to maintain or introduce genetic variation.

Dispersal and re-colonization

Dispersal of *D. dipsaci* occurs in water running over the soil surface which mainly leads to local spread within crops. On sloping land *D. dipsaci* can also spread between fields or even farms. In one area studied in Somerset most of the strawberry fields infested with *D. dipsaci* were in the drainage zones from three shallow depressions in a permanent grazed pasture containing infested plants (p. 221). Fields below these bowls were readily colonized from the native plant communities. These circumstances are somewhat exceptional and fields are usually separated from neighbouring infested land and are colonized if infested debris or soil is introduced. *D. dipsaci* dispersed in water, debris, or soil is likely to initiate colonies with the genetic variation of the initial population. However, many individuals will fail to establish colonies and some will be founded by one or a few females with limited genetic resources.

Table 2 Mechanisms of seed dispersal of British plant species recorded as hosts for *Ditylenchus dipsaci*.

Seed dispersal mechanism	*Usual plant habit* *Annual*	*Biennial*	*Perennial*
Winged	3	3	7
Pappus	2	6	5
Awned	3	1	1
Capsule or pod	19	6	27
Other	45	5	37

Natural dispersal on weed seeds will spread *D. dipsaci* locally into and/or within fields and for longer distances with seed of the relatively few hosts with winged, awned or feathered seeds (Table 2). Moreover, crop seeds distribute the nematode very widely and frequently lead to the infestation or reinfestation of arable soils. *D. dipsaci* tends to be aggregated in infested samples of seeds of *Vicia faba* and *Allium cepa*. In one detailed study (Green, 1979) a few seeds had many nematodes whereas many had none. Consequently most nematodes in seeds were in groups that could reproduce. The number of potentially infested seedlings from infested samples varied between 1% and

5% and increased little in relation to the numbers of nematodes found. Seed-borne *D. dipsaci* therefore stabilizes the number of colonies initiated from the seed at each host generation. *D. dipsaci* may be carried on weed seed contaminating crop seed and even if there are few weed seeds, nematodes if present will be sufficiently numerous to initiate colonies because of their aggregation. The host compatibility of colonies initiated by seed-dispersed *D. dipsaci* will be restricted and represents only part of the genetic variation present in the initial soil population.

Population dynamics

Models for population changes of parasites (Crofton, 1971; Anderson and May, 1978; May and Anderson, 1978) demonstrate that colonies of opportunist parasites become stabilized if they are aggregated and the parasite is killed by death of the host as the parasite population becomes too great. The host–parasite system of *D. dipsaci* differs from the models in two respects. Firstly, spread of the parasite within the host may be restricted so that *D. dipsaci* kills only part of the host. However, death of most of the parasites within the most densely infested parts of the host stabilizes the parasite population in the same way as host death in the models. Secondly, migration of *D. dipsaci* from hosts occurs readily and is density-dependent (Webster, 1964) so that very efficient invasion of hosts by the parasite after emigration may offset the effects of host death in reducing nematode numbers. Transmission between hosts can be increased by transport of the nematodes in water running over the soil causing widespread, uniform infestations which quickly eliminate both host and parasite (Fig. 1). Unaided, *D. dipsaci* spreads through a clay loam at less than 1 cm day^{-1} (Webster and Greet, 1967) by prolonged random wandering. Movement is quicker in lighter soils (Wallace, 1962) and, near hosts, is directed towards them (Blake, 1962), yet the distances travelled daily in search of hosts can be only a few centimetres. A dense plant stand facilitates transmission between hosts as in fields of *Medicago sativa*, where plants are killed in expanding patches (Brown, 1957). In arable crops, particularly those that are precision-drilled to the final stand required the spacing may restrict transmission, but weeds can more than double the total plant density. A typical arable soil contained 1200–6500 weed seeds m^{-2} in the top 10 cm, from which 60–300 seedlings m^{-2} grew after a cultivation (Roberts and Ricketts, 1979). Such weed populations provide alternative hosts for *D. dipsaci* and the increased plant density facilitates transmission between hosts. *D. dipsaci* numbers were greater in weedy oat crops than in weed-free stands, although the number of attacked oat plants was similar in both (Wilson and French, 1975), suggesting that number of hosts rather than number of invading nematodes was limiting.

In permanent communities of native plants *D. dipsaci* can be widespread but in small numbers, as in *Lamium purpureum* around fields in Warwickshire, in *Plantago major* and nearby naturalized *Narcissus pseudo-narcissus* in a meadow in Yorkshire and in a herbaceous community in Somerset. There *D. dipsaci* was in dried stems of *Eupatorium cannabinum* and *Dipsacus*

Figure 1 An area of onions with most killed by *D. dipsaci* spread by water on the soil surface.

sylvestris in autumn, and in shoots of *Arum maculatum* and *Stachys sylvatica* in spring. Occasional individuals were also found in *Allium ursinum, Rumex crispus, Urtica dioica* and *Plantago major.* The populations were non-destructive and seemed to be in stable equilibrium, although the plants were sufficiently dense for successful emigration and reinvasion.

Discussion

The host range, population-stabilizing and dispersal mechanisms of *D. dipsaci* have evolved in wild plant communities in which the nematode is endemic. The nematodes' rapid uncontrolled reproduction in some crops and effective means of dispersal are attributes of the 'r' strategists, referred to as 'exploiters' by Jones (1980). However, in wild plant communities and other crops the nematode populations appear more stable and similar to those of 'K' strategists or 'persisters'.

Death of invaded host tissues stops reproduction and enforces movement between plants. In sparse vegetation the wide spacings restrict transmission between plants and the occurrence of incompatible species also impedes transmission in mixed plant communities. Lines of *D. dipsaci* that have been selected for host compatibility with different plant species by several generations of reproduction in a plant, become mixed in soil. The number of individuals that can reproduce on a particular plant species depends on their compatibility with that host. Constant changing of hosts in a mixed plant stand will restrict reproduction as there will always be a proportion of the population which is unable to breed. Moreover, seasonal changes in the flora as happens in natural plant communities will periodically restrict reproduction giving longer term cycling of the population. These selection pressures will tend to maintain genetic diversity in *D. dipsaci* populations with regard to host compatibility, and in consequence diminish reproductive capacity.

The dispersal of *D. dipsaci* in seed which, in crops, enables *D. dipsaci* to re-colonize efficiently is likely to stabilize and maintain populations in wild plant communities. Few host species have winged, awned or feathered seeds that facilitate long-distance dispersal (Table 2). Consequently most seeds and any nematodes they carry remain near the parent and infestations spread slowly. They will effectively revitalize populations which have dwindled because transmission between hosts has reduced reproduction too much and will stabilize the proportion of infested plants at the same time.

Weeds in crops increase plant density and the likelihood of migrating *D. dipsaci* reinvading plants. The numbers of the nematodes surviving may be increased but the rate of reproduction in individual plants may be decreased because weeds increase plant diversity in both space and time. The growth of weeds out of phase with the crop and the local distribution of infested weed seed both increase the likelihood of the population surviving between crops and, as in wild plant communities, maintain a wide genetic diversity in the host compatibility of the *D. dipsaci* populations.

Summary

D. dipsaci populations are stabilized by death of heavily infested host tissue, by the limitation of re-invasion after migration by the density of the hosts and because reproduction after invasion is limited by the occurrence of non-compatible hosts. In monospecific and ephemeral or harvested plant communities nematode populations are unstable (r strategy) and persist in seed, soil or in weeds. Reinfestation of land can occur by transport of *D. dipsaci* in soil, water or with the seeds of wild or crop plants. The number of seeds infested by *D. dipsaci* is stabilized by aggregation of the nematodes on the seed and, in the absence of other sources of infestation, this stabilizes the number of colonies initiated. Seed dispersal selects *D. dipsaci* for host compatibility and encourages race specialization. In contrast mixed plant communities encourage genetic diversity for host compatibility of the parasite and enable its populations to stabilize (K strategy).

References

Anderson, R. M. & May, R. M. (1978). Regulation and stability of host–parasite population interactions. 1. Regulatory processes. *Journal of Animal Ecology* **47,** 219–247.

Barabaschova, V. N. (1974). Karyotypical peculiarities of some forms of stem eelworms of the collective species *Ditylenchus dipsaci. Parazitologiya* (Leningrad) **8,** 408–412.

Blake, C. D. (1962). Some observations on the orientation of *Ditylenchus dipsaci* and invasion of oat seedlings. *Nematologica* **8,** 177–192.

Brown, E. B. (1957). Lucerne stem eelworm in Great Britain. *Nematologica* **2** (Supplement), 369–375.

Clapham, A. R., Tutin, T. G. & Warburg, E. F. (1962). *Flora of the British Isles.* Cambridge: Cambridge University Press.

Crofton, H. D. (1971). A model of host–parasite relationships. *Parasitology* **63,** 343–364.

Cronquist, A. (1968). *The Evolution and Classification of Flowering Plants.* London: Nelson.

Eriksson, K. B. (1974). Intra specific variation in *Ditylenchus dipsaci.* 1. Compatibility tests with races. *Nematologica* **20,** 147–162.

Goodey, J. B., Franklin, M. T. & Hooper, D. J. (1965). *The nematode parasites of plants catalogued under their hosts.* Farnham Royal: Commonwealth Agricultural Bureaux.

Green, C. D. (1977). Stem eelworm damage to pulse crops, *Annual Report of the National Vegetable Research Station for 1976*, p. 91.

Green, C. D. (1979). Aggregated distribution of *Ditylenchus dipsaci* on broad bean seeds. *Annals of Applied Biology* **92,** 271–274.

Green, C. D. & McNee, H. (1978). Stem-eelworm persistence. *Annual Report of the National Vegetable Research Station for 1977*, p. 91.

Green, C. D. & Sime, S. (1979). The dispersal of *Ditylenchus dipsaci* with vegetable seeds. *Annals of Applied Biology* **92,** 263–270.

Green, C. D. & Wilson, A. (1980). Persistence of *Ditylenchus dipsaci. Annual Report of the National Vegetable Research Station for 1979*, p. 46.

Griffin, G. D. (1975). Parasitisation of non-host cultivars by *Ditylenchus dipsaci. Journal of Nematology* **7,** 236–238.

Hesling, J. J. (1966). Biological races of stem eelworm. *Annual Report of the Glasshouse Crops Research Institute for 1965*, pp. 132–141.

Jones, F. G. W. (1980). Some aspects of the epidemiology of plant parasitic nematodes. In *Comparative Epidemiology*, pp. 71–92. J. Palti & J. Kranz. Wageningen: Pudoc.

Ladygina, N. M. (1973). Physiological compatibility of different forms of stem eelworms (3) crossing of parsley, parsnip, bulb and strawberry eelworms. *Parazitologiya* (Leningrad) **7,** 67–71.

Ladygina, N. M. (1974). Genetic and physiological compatibility of different forms of stem eelworms (4) crossing the Phlox eelworm with other stem eelworms. *Parazitologiya* (Leningrad) **8,** 63–69.

Ladygina, N. M. (1976). Genetic and physiological compatibility of different forms of stem eelworm (5) crossing of the Red Clover race with other stem eelworms. *Parazitologiya* (Leningrad) **10,** 40–47.

Ladygina, N. M. & Barabaschova, V. N. (1976). Genetic and physiological compatibility and karyotypes of stem eelworms. *Parazitologiya* (Leningrad) **10,** 449–456.

May, R. M. & Anderson, R. M. (1978). Regulation and stability of host–parasite population interactions. 2. Destabilising processes. *Journal of Animal Ecology* **47,** 249–267.

Pianka, E. R. (1970). On r and K selection. *American Naturalist* **104,** 592–597.

Roberts, H. A. & Ricketts, M. E. (1979). Quantitative relationships between the weed flora after cultivation and the seed population in the soil. *Weed Research* **19,** 269–275.

Schipstra, K. (1952). Stem eelworm (*Ditylenchus dipsaci* (Kuhn) Filipjev) in clover seed. *Tijdschrift over plantenziekten* **58,** 87.

Seinhorst, J. W. (1956). Population studies on stem eelworm (*Ditylenchus dipsaci*). *Nematologica* **1,** 159–165.

Seinhorst, J. W. (1957). Some aspects of the biology and ecology of stem eelworms. *Nematologica* **2** (Supplement), 355–361.

Sturhan, D. (1964). Kreuzungsversuche mit biologischen Rassen des Stengelachens (*Ditylenchus dipsaci*). *Nematologica* **10,** 328–334.

Sturhan, D. (1966). Wirtspflanzenuntersuchungen an Bastardpopulationen von *Ditylenchus dipsaci* Rassen. *Zeitschrift fur Pflanzenkrankheiten Pflanzenpathologie und Pflanzenschutz* **73,** 168–174.

Sturhan, D. (1969). Das Rassen problem bei *Ditylenchus dipsaci. Mitteilungen der Biologischen Bundesanstalt fur Land-u. Forstwirschaft* **136,** 87–89.

Wallace, H. R. (1962). Observations on the behaviour of *Ditylenchus dipsaci* in soil. *Nematologica* **7,** 91–101.

Webster, J. M. (1964). Population increase of *Ditylenchus dipsaci* (Kühn) in the narcissus and the spread of the nematode through the soil. *Annals of Applied Biology* **53,** 485–492.

Webster, J. M. (1967). The significance of biological races of *Ditylenchus dipsaci* and their hybrids. *Annals of Applied Biology* **59,** 77–83.

Webster, J. M. & Greet, D. N. (1967). The effect of host crop and cultivation on the rate that *Ditylenchus dipsaci* reinfested a partially sterilized area of land. *Nematologica* **13,** 295–300.

Wilson, W. R. & French, N. (1975). Population studies of stem eelworm (*Ditylenchus dipsaci*) in microplots, field plots and fields in the north of England under different crop rotations. *Plant Pathology* **24,** 176–183.

Yuksel, H. S. (1960). Observations on the life cycle of *Ditylenchus dipsaci* on onion seedlings. *Nematologica* **5,** 289–296.

The distribution of virus-vector nematodes in Great Britain in relation to past and present natural vegetation

D G McNamara and J J M Flegg
East Malling Research Station, Maidstone, Kent ME19 6BJ

Introduction

The soil-inhabiting, ectoparasitic nematodes of the Order Dorylaimida have recently received much attention as vectors of plant viruses. Species from four genera transmit viruses: *Xiphinema* and *Longidorus* transmit the polyhedral nepoviruses, while *Trichodorus* and *Paratrichodorus* transmit the rod-shaped tobraviruses (McElroy, 1977; Murant, *1981*). Since Hewitt *et al.* (1958) demonstrated that *X. index* was responsible for the spread of grapevine fanleaf virus, the biology, ecology and taxonomic relationships of this group of nematodes have been closely studied. About 80% of the two hundred or so known species have been described since 1958. Most interest, however, has centred on their ability to transmit viruses; 32 species have been implicated and much knowledge has been gained on nematode/virus/plant interrelationships.

Just as there are striking differences in the structure of nepo- and tobraviruses, so there are marked differences between their vectors. The trichodorids (*Trichodorus* and *Paratrichodorus*) are about 1 mm long. Their life-cycle may last less than 55 days and they reproduce at any time of year, whenever host-plant roots are available (Pitcher and McNamara, 1970). The longidorids (*Xiphinema* and *Longidorus*) are longer, thin nematodes, some as long as 10 mm but most being 2–5 mm. They have one, sometimes two egg-laying peaks each year and the life-cycle from egg to adult can take several years (Flegg, 1968b; Griffin and Darling, 1964). Trichodorids are biologically and ecologically similar to most other plant-parasitic nematodes whereas the longidorids form a separate group.

This paper deals with some longidorids and certain aspects of their ecology. Four of the ten species found in Great Britain are known to transmit viruses. *L. elongatus* is distributed throughout Great Britain, *X. diversicaudatum* occurs in England, Wales and southern Scotland, whereas *L. macrosoma* and *L. attenuatus* have limited distributions in southern England.

There has been no previous attempt to explain the origins of longidorid nematodes in Britain and efforts to relate their distribution to soil type have not been entirely successful. We here suggest that the distribution of longidorids was initially determined by prehistoric natural vegetation and associated opportunities for colonization, and that the distribution was modified subsequently by agricultural practices. As nematodes leave no fossilized remains, it has been necessary, in developing these views, to use information from diverse sources, including some collected for purposes bearing little relation to our own.

Distribution of longidorids in relation to natural vegetation

Rau (1975) sampled in 'undisturbed' (= recently undisturbed by man) habitats in Lower Saxony, Germany for virus-vector nematodes and was able to associate certain longidorid species with particular vegetation types. For example, *X. diversicaudatum* occurs almost exclusively in oak (*Quercus*) woodland of the types of Stellario-Querco-Carpinetum and *Eichen Auenwald* (wet alluvial oak wood — probably Querco-Fraxinetum). This nematode also occurs locally in beech (*Fagus*) woods on chalk soils. The distribution of *X. diversicaudatum* in Great Britain (Figs. 1 and 2: Taylor and Brown, 1976) is

Figure 1 Sites in Great Britain that have been sampled for longidorid nematodes (Taylor and Brown, 1976).

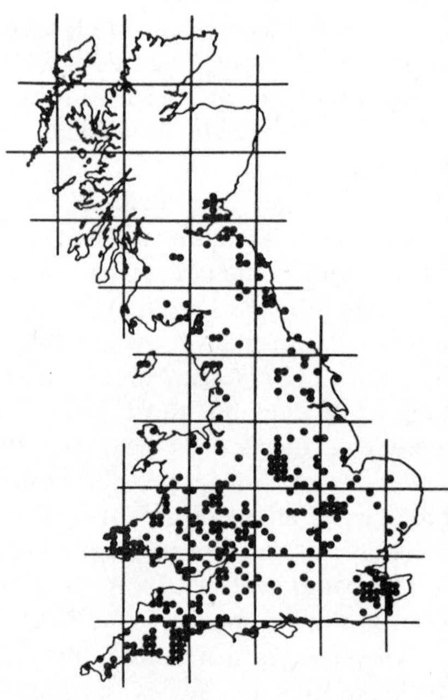

Figure 2 Distribution of *Xiphinema diversicaudatum* in Great Britain (Taylor and Brown, 1976).

consistent with these findings, in that the species is largely confined to regions of Britain that were covered by deciduous forests in the 'Atlantic' post-glacial period, 3000–5000 years ago (Godwin, 1975). The frequency of oak pollen in peat bogs shows that oak forests reached their greatest extent during this period and covered the country as far north as a line joining the Clyde and Forth estuaries (Fig. 3). Small woods of oak occurred in sheltered valleys north of this line, but the colder climate was (and still is) more suitable for pine (*Pinus* sp.) forests. In France, Dalmasso (1970) defined the southern limit of *X. diversicaudatum* by a line from the Alps of Provence in the east, running south of the Massif Central and turning south towards the Atlantic coast; this line is similar to that defining the southern limit of the European deciduous forest. The limited information available suggests that the distribution of *X. diversicaudatum* in western Europe corresponds to areas where deciduous woodland is or was the natural climax vegetation.

In Lower Saxony *L. macrosoma* occurs mainly in beech woods on chalk soil (Rau, 1975). In Britain the nematode is limited to southern England and concentrated on the chalk belt running obliquely north-eastwards towards East Anglia (Fig. 4). Beech is thought to have entered Britain late in post-glacial times and to have colonized initially the shallow chalk soils where

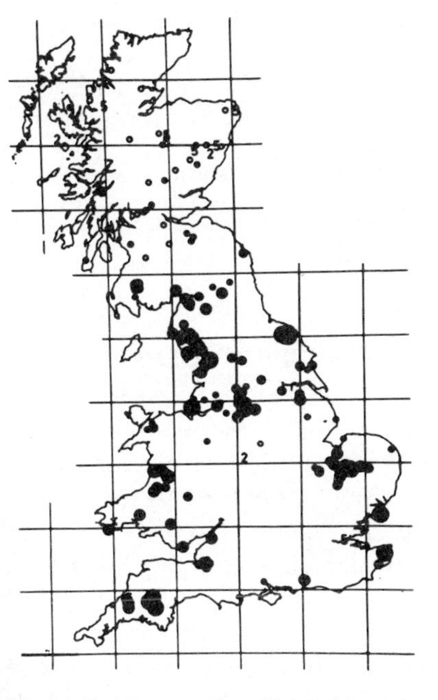

Figure 3 Occurrence of oak (*Quercus* spp.) pollen throughout Great Britain in peat deposited 3000–5000 years ago, during the period of maximum expansion of deciduous forests (Godwin, 1975). The size of the black circles indicates the frequency of occurrence and the open circles represent trace findings.

oak had difficulty in establishing; however, it would eventually have spread to dominate oak forests (Tansley, 1968).

L. elongatus (*sensu lato*) may be a species complex (Koslowska and Seinhorst, 1979) with different forms having different habitat requirements. In Lower Saxony one morphological form is associated with grassland, whereas a second form occurs only in oak woodlands (Rau, 1975). British populations of *L. elongatus* have not been closely studied and taxonomic uncertainty makes it difficult to link distribution to present or past vegetation types.

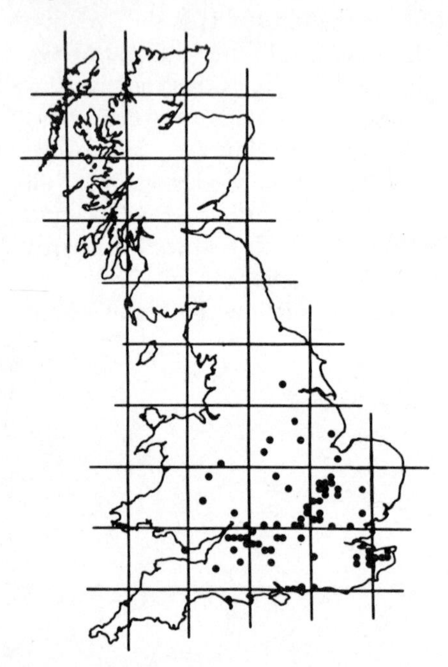

Figure 4 Distribution of *Longidorus macrosoma* in Great Britain (Taylor and Brown, 1976).

L. attenuatus is more typical of most *Longidorus* spp., which are commonly parasites of herbaceous hosts, especially grasses, in contrast to the woody perennial hosts of *Xiphinema* spp. (Cohn, 1975). The limited distribution of *L. attenuatus* reflects the paucity of natural grassland in Britain; this species is concentrated around the Breckland of East Anglia which has been grassland since Neolithic times (Godwin, 1975).

We suggest that the present distribution of longidorid nematodes is related to the distribution of natural climax vegetation and, in particular, that of *X. diversicaudatum* and *L. macrosoma* to the distribution of oak and beech woodland, respectively, in prehistoric Britain. If this is so, then soil characteristics, which have preoccupied those previously attempting to explain nematode distribution, become less dominant. Nevertheless, they remain important, together with climatic and biotic factors, in determining the prevailing vegetation type. In other words, longidorids are found where a particular plant community exists, or has existed; the type of plant community being determined by climatic, edaphic and biotic factors.

The origins of longidorids in Britain

How longidorids have colonized suitable habitats in Britain has never been explained. They cannot always have been present nor could they have spread with their host plants.

The most recent glaciation, the Devensian, which covered the north of Britain and reached southwards to the English midlands, lasted for some 40 millenia, ending about 12 000 years ago (West, 1977). The glaciation scoured the soil from the land surface and it seems virtually impossible that nematodes could have survived beneath the glaciers. Far south of the glaciers the soil was permanently frozen and supported only a tundra-type vegetation. It is inconceivable that denizens of temperate deciduous forest could have persisted for long under these conditions. Therefore, it seems that Britain was recolonized by nematodes from warmer areas to the south after the retreat of the glaciers and after the return of temperate vegetation.

Longidorid nematodes are '*prisonniers de leur milieu*' (Dalmasso, 1970) living beneath the soil surface to a depth of 1 m and sometimes deeper (Flegg, 1968a) with little access to the world above. There are no dispersal stages in their life-cycles, the four juvenile stages being morphologically (apart from smaller size and sexual immaturity), and in terms of behaviour broadly indistinguishable from adults. Their only autonomous means of spread is by slow movement through the soil.

The ability of nematodes to move through soil is determined by a complex of interrelating factors, the most important of which are nematode length, soil particle size, aggregation and water content (Wallace, 1968; Jones *et al.*, 1969). In general, in similar soils at a particular water content, a small nematode can move more rapidly than a larger one, and thus under similar laboratory conditions, *Trichodorus viruliferus* (length 730 μm, diameter 30 μm) travelled 10 mm day^{-1} (Derrett, 1981), whereas *X. diversicaudatum* (length 4.3 mm, diameter 60 μm) covered only 1.4 mm day^{-1} (Fritzsche, 1968). These distances, measured in soil, represent radii of dispersal rather than straight-line movement; on the surface of agar, *T. viruliferus* produced tracks measuring 83 mm day^{-1} (Derrett, 1978). Harrison and Winslow (1961) estimated that a population of *X. diversicaudatum* spread 22.5 m from a hedgerow into land on which cultivation had ceased 75 years previously and oak forest had subsequently regenerated. From these estimates it seems that the maximum rate of dispersal through soil by this species is less than

0.5 m year^{-1}. Therefore, the nematodes could not have moved autonomously more than 4 km during the 7–8000 years since the reappearance of deciduous woodland in Britain after the retreat of the Devensian glaciation (Godwin, 1975). This is obviously insufficient to account for their present distribution and other modes of dispersal must have been involved.

Longidorids can survive for hours or days in small quantities of moist soil and could have been dispersed in soil adhering to the feet, fur or feathers of the mammals or birds of the forests. For long-distance transport, several large birds, with appropriately large feet, inhabit woodland at some times of the year and could carry nematodes very rapidly for hundreds of kilometres. Prominent among these is the woodcock (*Scolopax rusticola*), which feeds throughout the year by probing for invertebrates in moist woodland soils. Most woodcocks migrate in spring and autumn on a north/south or northeast/southwest axis. A flight duration of several hours at *c*. 50 km h^{-1} is normal, and it is likely that, on longer journeys, the birds would stop to feed and recuperate *en route* (Shorten, 1974). The heron (*Ardea cinerea*) and certain ducks (*Anatidae*) also visit woodland, particularly during the summer, and make similar long-distance flights. Although no direct evidence for nematode transport by birds has yet been sought, it is suspected that birds are partly responsible for carrying nematodes to newly-formed polder soils in the Netherlands (Kuiper, 1962).

Medium range dispersal could have been provided by mammals. Obvious possibilities are ungulates, such as wild boar (*Sus scrofa*), red deer (*Cervus elaphus*), roe deer (*Capreolus capreolus*), wild ox (*Bos primigenius*) and elk (*Alces alces*), and the burrowing carnivors, fox (*Vulpes vulpes*) and badger (*Meles meles*). The home ranges of these mammals vary from 7–15 ha for the roe deer (Chapman, 1977) to 250–1500 ha for the fox (Lloyd, 1977). Short-range, local spread of nematodes could have been by small, burrowing mammals such as common shrew (*Sorex araneus*), bank vole (*Clethrionomys glareolus*) or wood mouse (*Apodemus sylvaticus*), which have home ranges generally considerably less than 1 ha (Michielsen, 1966; Brown, 1966). Flooding could also account for short-range dispersal, supplemented by the autonomous movement of the nematodes themselves.

Grassland nematodes could have been dispersed similarly by other species of birds and mammals.

Distribution of longidorids as influenced by agriculture

During the past 2000 years the human inhabitants of Great Britain have progressively cleared the indigenous vegetation to grow crops and to provide timber for fuel, homes and ships. Deciduous forests have decreased over this period from 60% to 6–8% of the land surface (Stamp, 1969) and, although this spread of agriculture with its concomitant increase in the amount of grassland may have favoured some longidorids, it has been harmful to the forest-dwelling species.

Several authors have noted the relatively small numbers of *X. diversicaudatum* in cultivated soils (Pitcher and Jha, 1961; Harrison and Winslow, 1961; Fritzsche, 1968) and the reasons are apparent from the ecological and biological characteristics of this species. Soil disturbance damages these long nematodes (Flegg, 1967) and cultivations during the breeding period of spring and early summer are particularly harmful. McNamara *et al.* (1980) found *Longidorus* spp. in undisturbed grassland surrounding hop plantings in southern Germany but never in frequently cultivated hop soils. In contrast, these nematodes occur commonly (albeit in small numbers) in English hop gardens where the soil is rarely disturbed now that weed control is almost exclusively by means of herbicides.

An additional reason why few longidorids exist under agricultural crops may be the plant species cultivated. Although longidorids feed on (and transmit virus to) many plant species (Fritzsche and Hofferek, 1969; Thomas, 1970; Murant, *1981*), the number from which they can derive adequate nourishment to complete their life cycles seems relatively restricted (Cohn, 1975). Moreover, annual crops, with a short season of root production, may not provide food supplies at appropriate times of year for nematodes that live longer than a year and have restricted breeding seasons (Cotten and Roberts, 1980). However, some crops, including strawberries and newly-planted pastures, are suitable and enable the nematodes to multiply (McNamara and Pitcher, 1977). Between suitable crops, when the soil is fallowed or poor-host crops are grown, the nematodes can survive for several years (McNamara, 1980), although their numbers gradually decline until they eventually disappear.

Authors, who found that agricultural crops were poor hosts for longidorids, also noticed the striking concentrations of *X. diversicaudatum* in hedgerows. Presumably the nematodes thrive in the undisturbed soil containing roots of diverse woody perennial species, many of which also occur in the shrub layer of deciduous forests. Hedgerows certainly act as major sources from which the nematodes can spread into fields where suitable crops are grown and some of the viruses transmitted also have alternative hosts among hedgerow shrubs and trees. Outbreaks of nepovirus diseases tend to occur alongside hedgerows or at former hedgerow sites; the occurrence of arabis mosaic virus (AMV) in hop gardens is well documented (Thresh and Pitcher, 1978). Harrison and Winslow (1961) reported the spread of AMV from hedgerows into adjacent rows of strawberries, and Taylor and Thomas (1968) found a similar spread of strawberry latent ringspot virus from a hedge into a field of raspberries. In both cases there were many *X. diversicaudatum* in the hedges and progressively fewer at distances into the field (Fig. 1 of Murant, *1981*).

Pitcher and Jha (1961) illustrated both the preference of *X. diversicaudatum* for woodland and hedgerows and the hazards of planting a favoured host-crop on land where small numbers of nematodes survived under a poor host-crop (Fig. 5). Few nematodes were present in long-term pasture but more in the surrounding hedge and adjoining woodland. When strawberries replaced part of the pasture the numbers of *X. diversicaudatum* increased significantly, with resulting rapid spread of AMV.

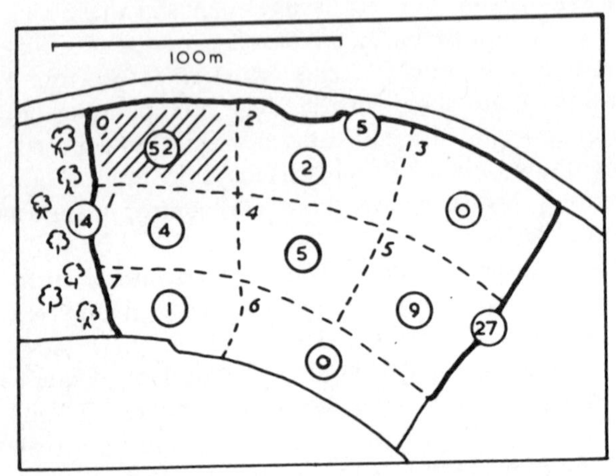

Figure 5 Pasture field surrounded by hedge (thick lines) and adjoining woodland. The field division at top left was planted with strawberries (cross hatching). Encircled numbers are mean numbers of *Xiphinema diversicaudatum* per 200 ml of soil. (Pitcher and Jha, 1961).

Agricultural practices also provide additional means of transport for nematodes in soil adhering to implements or plants. These means of dispersal are thought to be less important for longidorids than for other nematodes (Murant, 1970), but they could account for the occurrence of some species outside their normal geographic range.

Conclusion

We suggest that the distribution of longidorid nematodes in Great Britain was determined initially by that of prehistoric natural vegetation and that, for at least two species, *X. diversicaudatum* and *L. macrosoma*, there were two distinct phases:

(1) From the return of deciduous forests to Britain after the last glaciation until the beginning of the era of man's influence. The nematodes spread northwards, assisted by birds for long-distance dispersal and supplemented by large mammals for shorter distances, to all regions where suitable vegetation existed. There, local spread was by small mammals, flooding and by the nematodes' own locomotion.

(2) From the beginning of deforestation to the present day. The nematodes declined and became largely extinct in cultivated areas, yet survived in woodland and hedgerows, whence they occasionally invade and carry viruses into suitable crops.

References

Brown, L. E. (1966). Home range and movement in small mammals. *Symposium, Zoological Society of London* **18,** 111–142.

Cohn, E. (1975). Relations between *Xiphinema* and *Longidorus* and their host plants. In *Nematode Vectors of Plant Viruses*, pp. 365–386. F. Lamberti, C. E. Taylor & J. W. Seinhorst. London: Plenum Press.

Chapman, D. I. (1977). Roe deer *Capreolus capreolus*. In *The Handbook of British Mammals*, pp. 437–446. G. B. Corbet & H. N. Southern. Oxford: Blackwell Scientific Publications.

Cotten, J. & Roberts, H. (1980). Effect of four crops on the reproduction and development of *Xiphinema diversicaudatum* and on the pore space moisture content and temperature of soil. *Plant Pathology,* **29,** 70–76.

Dalmasso, A. (1970). Influence directe de quelques facteurs écologique sur l'activité biologique et la distribution des espèces françaises de la famille des Longidoridae (Nematoda–Dorylaimida). *Annales de Zoologie, Écologie Animale* **2,** 163–200.

Derrett, K. J. (1978). *Report of East Malling Research Station for 1977*, p. 119.

Derrett, K. J. (1981). *PhD Thesis*, University of London (in preparation).

Flegg, J. J. M. (1967). Studies of the biology and ecology of some plant virus vectors and allied species of the genera *Xiphinema* and *Longidorus* (Nematoda: Dorylaimida). *PhD Thesis*, University of London.

Flegg, J. J. M. (1968a). The occurrence and depth distribution of *Xiphinema* and *Longidorus* species in South-eastern England. *Nematologica* **14,** 189–196.

Flegg, J. J. M. (1968b). Life-cycle studies of some *Xiphinema* and *Longidorus* species in South-eastern England. *Nematologica* **14,** 197–210.

Fritzsche, R. (1968). Beitrag zum Wanderungsverhalten von *Xiphinema diversicaudatum* (Mikoletzky) Thorne, *X. coxi* Tarjan und *Longidorus macrosoma* Hooper sowie der Ausbreitung des Rhabarbermosaik-Virus im Feldbestand. *Biologisches Zentralblatt* **87,** 481–488.

Fritzsche, R. & Hofferek, H. (1969). Beiträge zum Saugverhalten und Nährpflanzenkreis von *Xiphinema diversicaudatum* (Mikoletzky) Thorne. *Archiv für Pflanzenschutz* **5,** 111–118.

Godwin, H. (1975). *The History of the British Flora*. Cambridge: Cambridge University Press.

Griffin, G. D. & Darling, H. M. (1964). An ecological study of *Xiphinema diversicaudatum* in an ornamental spruce nursery. *Nematologica* **10,** 471–479.

Harrison, B. D. & Winslow, R. D. (1961). Laboratory and field studies on the relation of arabis mosaic virus to its nematode vector *Xiphinema diversicaudatum* (Micoletzky). *Annals of Applied Biology* **49,** 621–633.

Hewitt, W. B., Raski, D. J. & Goheen, A. C. (1958). Nematode vector of soil borne fanleaf virus of grapevines. *Phytopathology* **48,** 586–595.

Jones, F. G. W., Larbey, D. W. & Parrott, D. M. (1969). The influence of soil structure and moisture on nematodes, especially *Xiphinema, Longidorus, Trichodorus* and *Heterodera* spp. *Soil Biology and Biochemistry* **1,** 153–165.

Keyworth, W. G. & Hitchcock, M. M. (1948). Aerial surveys of the incidence of nettlehead disease of hop on former hedgerow and pasture sites. *Report of East Malling Research Station for 1947*, pp. 153–156.

Kuiper, K. (1962). Dissemination of plant parasitic nematodes into new polder soil. *Nematologica* **7,** 16.

Kozlowska, J. & Seinhorst, J. W. (1979). *Longidorus elongatus* and closely related species in the Netherlands and Lower Saxony (Germany) with the description of two new species, *L. cylindricaudatus* and *L. intermedius* (Nematoda: Dorylaimida). *Nematologica* **23,** 42–53.

Lloyd, H. G. (1977). Fox *Vulpes vulpes*. In *The Handbook of British Mammals*, pp. 311–320. G. B. Corbet & H. N. Southern. Oxford: Blackwell Scientific Publications.

McElroy, F. D. (1977). Nematodes as vectors of plant viruses — A current review. *Proceedings of the American Phytopathological Society* **4,** 1–10.

McNamara, D. G. (1980). The survival of *Xiphinema diversicaudatum* (Micol.) in plant-free soil. *Nematologica,* **26,** 170–181.

McNamara, D. G. & Pitcher, R. S. (1977). The long-term effects of four monocultural regimes on two populations of the nematodes *Xiphinema diversicaudatum* and *Longidorus* spp. *Annals of Applied Biology* **86,** 405–413.

McNamara, D. G., Sander, E. & Eppler, A. (1980). The distribution of virus-vector nematodes in the hop-growing region of Tettnang, West Germany. *Zeitschrift für Pflanzenkrankheiten und Pflanzenschutz* **87,** 73–82.

Michielsen, N. (1966). Intraspecific and interspecific competition in the shrews *Sorex araneus* L. and *Sorex minutus* L. *Archives néerlandaises de Zoologie* **17,** 73–174.

Murant, A. F. (1970). The importance of wild plants in the ecology of nematode-transmitted plant viruses. *Outlook on Agriculture* **6,** 114–121.

Murant, A. F. (*1981*). The role of wild plants in the ecology of nematode-borne viruses. In *Pests, Pathogens and Vegetation*, pp. 237–248. J. M. Thresh. London: Pitman.

Pitcher, R. S. & Jha, A. (1961). On the distribution and infectivity with arabis mosaic virus of a dagger nematode. *Plant Pathology* **10,** 67–71.

Pitcher, R. S. & McNamara, D. G. (1970). The effect of nutrition and season of year on the reproduction of *Trichodorus viruliferus. Nematologica* **16,** 99–106.

Rau, J. (1975). Das Vorkommen virusübertragender Nematoden in ungestörten Biotopen Niedersachsens. *Dissertation*, Technische Universität Hannover.

Shorten, M. (1974). The European woodcock (*Scolopax rusticola* L.); a search of the literature since 1940. *Game Conservancy Report*, No. 21, pp. 1–93.

Stamp, L. D. (1969). *Man and the Land*. London: Collins.

Tansley, A. G. (1968). *Britain's Green Mantle, Past, Present and Future*. London: George Allen & Unwin.

Taylor, C. E. & Brown, D. J. F. (1976). The geographical distribution of *Xiphinema* and *Longidorus* nematodes in the British Isles and Ireland. *Annals of Applied Biology* **84,** 383–402.

Taylor, C. E. & Thomas, P. R. (1968). The association of *Xiphinema diversicaudatum*

(Micoletsky) with strawberry latent ringspot and arabis mosaic viruses in a raspberry plantation. *Annals of Applied Biology* **62,** 147–157.

Thomas, P. R. (1970). Host status of some plants for *Xiphinema diversicaudatum* (Micol.) and their susceptibility to virus transmitted by this species. *Annals of Applied Biology* **65,** 169–178.

Thresh, J. M. & Pitcher, R. S. (1978). The spread of nettlehead and related virus diseases of hop. In *Plant Disease Epidemiology*, pp. 291–298. P. R. Scott & A. Bainbridge. Oxford: Blackwell Scientific Publications.

Wallace, H. R. (1968). The dynamics of nematode movement. *Annual Review of Phytopathology* **6,** 91–114.

West, R. G. (1977). *Pleistocene Geology and Biology*. London: Longman.

The role of wild plants in the ecology of nematode-borne viruses

A F Murant
Scottish Crop Research Institute, Invergowrie, Dundee, Scotland DD2 5DA

It is now more than 20 years since Hewitt, Raski and Goheen (1958) first implicated a soil-inhabiting nematode (*Xiphinema index*) as the vector of a plant virus (grapevine fanleaf virus). Since then at least twelve viruses have been shown to have nematode vectors and several more are suspected to be transmitted in this way. Nematode-borne viruses cause important diseases in a wide range of crop species throughout the world. The main features of their ecology (Harrison, 1977) and the important part played by wild plants (Murant, 1970) are now well understood by most plant virologists, but the literature is very relevant to the subject of this volume and is summarized here for a wider readership.

Nematode-borne viruses and their vectors

There are two groups of nematode-borne viruses: the *tobraviruses* (Harrison and Robinson, 1978), which have rod-shaped particles of two lengths, *c.* 200 nm and 45–115 nm, and the *nepoviruses* (Harrison and Murant, 1977; Murant, 1981), which have 28–30 nm isometric particles. The tobravirus group has two members, tobacco rattle virus (TRV) and pea early-browning virus (PEBV), which are serologically related and are transmitted by nematodes of the genera *Trichodorus* and *Paratrichodorus*. The nepovirus group comprises more than twenty viruses, of which the best known are arabis mosaic (AMV), cherry leaf roll (CLRV), grapevine fanleaf (GFLV), raspberry ringspot (RRV), strawberry latent ringspot (SLRV), tobacco ringspot (TobRV), tomato ringspot (TomRV) and tomato black ring (TBRV). Many nepoviruses have no serological relationship to each other, but they share many characteristics including particle shape and vector relations. All the known vectors are nematodes of the genera *Longidorus* and *Xiphinema*.

Nematodes of all four vector genera are free-living species, mostly 1–10 mm long, which feed ectoparasitically on plant roots by puncturing the cells with a needle-like stylet. The nematodes and many of the viruses they transmit have wide natural host ranges. This feature ensures survival of the viruses in the wild and also makes them important in a wide range of annual and perennial crops. Nematode-borne viruses are primarily pathogens of wild plants that become important economically when a sensitive crop is grown in infested soil. For this to happen the adaptations that enable the viruses to survive and spread in the wild must also ensure their survival in arable land.

Wild plants are important in the ecology of nematode-borne viruses in four ways:

(1) as hosts of the nematodes,
(2) as hosts of the viruses,
(3) as a means of dissemination of the viruses, and
(4) as a means by which the viruses persist in soils.

Wild plants as hosts of vector nematodes

Some aspects of the role of wild plants in the ecology of virus vector nematodes are considered elsewhere in this volume (McNamara and Flegg, *1981*), but a few additional observations are relevant here.

Nepoviruses exhibit considerable vector specificity: each is transmitted usually by only one, or rarely two, nematode species and major serological variants of the same virus are usually associated with different specific nematode vectors. Thus the Scottish form of RRV is transmitted by *Longidorus elongatus* but rarely by *L. macrosoma*, whereas the English form, which is serologically somewhat distantly related, is transmitted by *L. macrosoma* more efficiently than by *L. elongatus*. Similarly, forms of TBRV that occur in England are transmitted best by *L. attenuatus* whereas the serologically distinctive Scottish forms have *L. elongatus* as vector (Harrison, 1964; Taylor and Murant, 1969). The distribution of the viruses is correlated with that of the most efficient of their nematode vectors. The geographical distribution of the nematodes is governed by climatic and edaphic factors, but the local distribution and the size of the nematode population (and hence the incidence of nematode-borne viruses at a site and the characteristic patchy shape of disease outbreaks) are determined by various factors of which one of the most important is the cropping or vegetational history of the site.

Although most vector nematodes feed on and transmit viruses to a wide range of plant species, there is evidence that their breeding host range (the range of plants from which the nematode can derive sufficient food to complete its life cycle) is much more restricted. In general, *Xiphinema* spp. thrive best on woody perennial plants, whereas such plants form a smaller part of the host spectrum of *Longidorus* spp. (Cohn, 1975). *X. index*, the vector of GFLV, seems to have a very narrow host range, being confined to grapevine, mulberry and fig. *X. diversicaudatum* is particularly associated with woody plants, and viruses transmitted by this nematode (AMV and SLRV) are often prevalent alongside hedges (Fig. 1; Harrison and Winslow,

1961; Pitcher and Jha, 1961; Taylor and Thomas, 1968) or in parts of fields from which woody plants have recently been removed. In south-western England, however, this nematode seems more generally distributed in grassland, possibly in association with white clover (Harrison and Winslow, 1961). *X. americanum*, the vector of TobRV and TomRV, is also especially

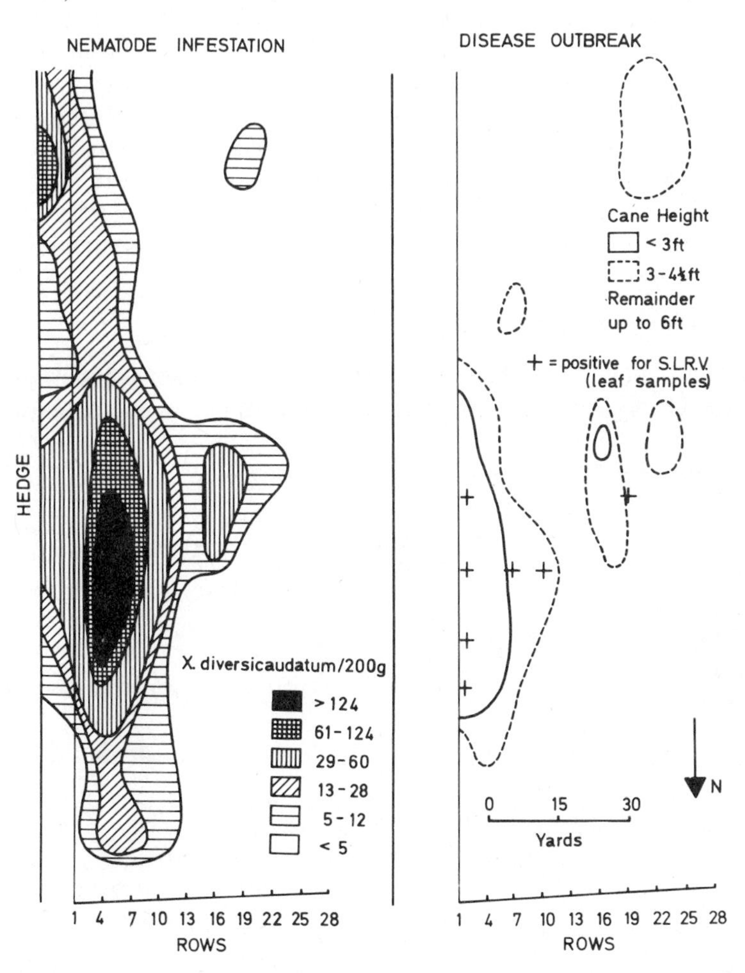

Figure 1 Population contours of *Xiphinema diversicaudatum* in relation to an outbreak of strawberry latent ringspot virus in Malling Jewel raspberry. (From Taylor and Thomas (1968).)

associated with perennial plants, such as blueberry, cherry, peach and strawberry. Nevertheless, *X. diversicaudatum*, *X. americanum* and *L. elongatus* appear to be among the most polyphagous species in their genera and this may explain why they are successfui as virus vectors. The same considerations apply to the trichodorid nematodes, most of which seem to be highly polyphagous and feed on many arable weed species as well as causing root injury and virus infection in a wide range of crops.

The preference of some nematode species for perennial plants may be as much a result of their sensitivity to disturbance of the soil environment as of the unsuitability of annual crops as such. However, individual host–plant preferences undoubtedly exist. In pot experiments (Thomas, 1969; Table 1), many annual weed species were good hosts of *L. elongatus*, whereas among crop plants tested, strawberry, grasses and clovers were the only good hosts. In Scotland, *L. elongatus* is most numerous and outbreaks of disease caused by TBRV and/or RRV are most serious, in crops planted after strawberry or after grass/clover mixtures or where there is a previous history of very weedy crops. Consequently, disease problems can be decreased by employing crop rotations and other practices that diminish nematode and weed populations.

Wild plants as hosts of nematode-borne viruses

Not only do vector nematodes feed on a wide range of plants but most nematode-borne viruses themselves have wide natural and experimental host

Table 1 Crop and wild plants as hosts of *Longidorus elongatus**

Plant species	*L. elongatus* Increase in nematode no.	*L. elongatus* Eggs in uteri
Crop plants		
Barley	0	+
Cocksfoot, ryegrass	+	+
Beet, carrot, lettuce, onion, pea, potato	0	0
Cabbage	+	0
Clovers (red, white)	+	+
Raspberry	0	+
Strawberry	+++	+++
Wild plants		
Agropyron repens	0	0
Brassica sinapis	+	+
Lamium amplexicaule	+	+
Matricaria matricarioides	+	0
Mentha arvensis	+	+
Poa annua	++	++
Senecio vulgaris	+	0
Stellaria media	++	++
Veronica officinalis	++	++

* Summarized from Thomas (1969).
The number of pluses indicates the relative rate of multiplication or number of eggs in uteri.

ranges, and this seems to be an essential feature in their ecology. TRV infected more than 400 plant species in more than 50 dicotyledonous and monocotyledonous families (Schmelzer, 1957); AMV infected 93 out of 136 species in 26 families (Schmelzer, 1963a); TBRV infected 76 out of 107 species in 25 families (Schmelzer, 1963b); CLRV infected 62 out of 104 species in 23 families (Schmelzer, 1966); and SLRV infected 126 out of 167 species in 27 families (Schmelzer, 1969). The natural host ranges of these viruses are also extremely wide and many natural hosts are infected symptomlessly; for example, Murant and Lister (1967) detected TBRV in 11 symptomless weed species (many of which also contained RRV), and Taylor and Thomas (1968) detected AMV and/or SLRV in 11 symptomless weed species. Tuite (1960) detected TobRV in eight weed species, six of which were symptomless. Natural infection with tobraviruses has been reported in more than 100 species (e.g. Noordam, 1956; Bos and van der Want, 1962). Table 2 summarizes some of this information. Therefore, although suitable crop rotations may diminish nematode populations and thus alleviate nematode-borne virus disease, attempts to *eradicate* the viruses from soil by rotational measures are seldom successful. The only exception to this generalization occurred with the hop strain of AMV, which is unusual because it seems to have no other natural hosts (McNamara *et al.*, 1973).

Experimental host ranges are usually determined by inoculating viruses manually to the leaves of test plants, but natural infections occur via the roots and plants infected in this way are not always invaded systemically. Cooper and Harrison (1973) found that TRV was confined to the roots of most weed species tested from one field. It seems possible that nematode-borne viruses

Table 2 Some natural weed hosts of nepoviruses and tobacco rattle virus in Britain*

Weed species	*AMV*	*RRV*	*SLRV*	*TBRV*	*TRV*
Capsella bursa-pastoris	+	+	+	+	+
Lamium amplexicaule	+	+	+	+	•
Mentha arvensis	+	•	+	•	•
Polygonum aviculare	•	•	•	+	+
Senecio vulgaris	+	+	+	+	+
Stellaria media	+	+	+	+	+
Taraxacum officinale	+	•	+	•	•
Tripleurospermum maritimum	•	+	•	+	•
Veronica agrestis	•	+	•	+	•
Veronica officinalis	+	•	+	•	•
Viola arvensis	•	•	•	•	+

* From Murant and Lister (1967), Taylor and Thomas (1968) and Cooper and Harrison (1973).
+ = commonly infected naturally.
• = not, or insufficiently, tested.

are perpetuated in soils by means of localized infections of plant roots, as happens with the fungus-transmitted virus, tobacco necrosis.

Involvement of wild plants in virus spread

Virus-vector nematodes do not form cysts resistant to desiccation and are easily killed when infested soil is air-dried. Therefore, they are not transported efficiently in soil attached to farm implements. Moreover, they migrate only slowly through soils. Harrison and Winslow (1961) calculated that *X. diversicaudatum* had spread from a hedgerow into a woodland at an average rate of only 30 cm year^{-1} over a period of 75 years. Nematode vectors, unlike aerial ones, therefore, are unable to transmit viruses efficiently over a distance and much of the dissemination of nematode-borne viruses must occur within infected plant tissue. Use of infected propagating material is obviously an important way in which the viruses may spread in commerce. Spread of GFLV throughout the world undoubtedly occurred in this way and there is good evidence that in Britain infected planting material was formerly responsible for the widespread dissemination of RRV and TBRV in raspberry and strawberry, and of the hop strain of AMV in hop. The wide host ranges of the viruses and the fact that many plants can be infected without showing symptoms make this form of spread a continuing problem in the commercial exchange of plant material.

However, because the viruses are primarily pathogens of wild plants it is to be expected that they have other, more natural, means of spreading. In fact, nepoviruses characteristically infect the seed (often more than 50%) of many of their host plants, including many wild species (Table 3). Among the tobraviruses, TRV is seed-transmitted in several species, although to only a small proportion of the progeny, and PEBV is seed-transmitted in pea. Seedlings infected via the seed, especially those infected with nepoviruses, commonly show no symptoms.

In raspberry and strawberry, and probably in other species, seeds become infected with RRV and TBRV through either gamete, the proportion of seeds infected being greatest when both parents are infected (Lister and Murant, 1967). Thus, theoretically at least, the viruses can spread in infected pollen and may cause infected seed to be set by healthy plants remote from the original virus source. However, Lister and Murant (1967) found that pollen from infected raspberry plants competed poorly with virus-free pollen for fertilization, so that this method of spread may be unimportant, at least in raspberry. There is no evidence that nepoviruses infect plants pollinated with infected pollen.

Some of the viruses placed in the nepovirus group are not proved to be nematode-borne, but are included because they are serologically related to viruses that have known nematode vectors, or because their biological and

Table 3 Percentage transmission of nepoviruses and tobraviruses through seed of various hosts

	Nepoviruses							Tobraviruses	
Host plant	*AMV**	*CLRV**	*RRV**	*SLRV***	*TobRV†*	*TBRV**	*TomRV‡*	*TRV§*	*PEBV††*
Beta vulgaris	13	–	50	–	–	56	–	–	–
Capsella bursa-pastoris	34	–	3	4	–	83	–	2	–
Chenopodium album	80	–	–	–	–	37	–	0	–
Fragaria x *ananassa*	7	–	49	–	–	34	55	–	–
Glycine max	5	100	4–8	–	54–78	83	76	–	–
Lamium amplexicaule	5–25	–	–	73	–	10–50	–	2	–
Myosotis arvensis	20–99	–	–	–	–	100	–	6	–
Nicotiana tabacum	–	–	–	–	5	–	–	–	–
Pisum sativum	–	–	–	–	–	–	–	–	37
Rubus idaeus	0	–	18	75	–	5–18	–	–	–
Senecio vulgaris	2	–	0	20	–	17	–	1	–
Stellaria media	58	–	30	97	–	66	0	0	–
Viola arvensis	–	1–6	0	–	–	–	–	2–10	–

* Lister and Murant (1967).
** Murant and Goold (1969).
† Valleau (1932); Desjardins *et al.* (1954).
‡ Kahn (1956); Mellor and Stace-Smith (1963); Lister and Murant (1967).
§ Lister and Murant (1967); Cooper and Harrison (1973).
†† Bos and van der Want (1962).
– = not tested.

physical properties are typical of the group. It is not impossible, however, that some may spread in other ways. For example, although CLRV is reported to be transmitted by *X. diversicaudatum* and *X. coxi* in glasshouse experiments, there is no evidence that it is transmitted by nematodes in the field and Cooper (1976, *1981*) suggested that it may be disseminated only in seed and pollen. Indeed, recent field evidence suggests that spread of CLRV in walnut is by pollination of trees by infected pollen (Mircetich *et al.*, 1980).

Dispersal in weed seed and possibly pollen is therefore presumably the main method of long-distance spread of nematode-borne viruses in nature. Direct experimental evidence for this is difficult to obtain but Cooper and Harrison (1973) found that, although trichodorid nematodes appeared in coastal sand dunes at an early stage in their colonization by marram grass (*Ammophila arenaria*), TRV was not detected until the plant community became more diverse and contained several broad-leaved species, including *Viola arvensis*, in which TRV is seed-transmitted.

Wild plants as a means of persistence of nematode-borne viruses in soils

An important factor in the ecology of nematode-borne viruses is the length of time they are retained by the vector nematodes. RRV and TBRV are retained for only 8–9 weeks in *L. elongatus* kept in fallow soil (Murant and Lister, 1967), whereas viruses transmitted by *Xiphinema* spp. are retained for many months: for example, 8–11 months for AMV in *X. diversicaudatum* kept on a virus-immune plant (Harrison and Winslow, 1961), 8 months for GFLV in *X. index* kept in fallow soil (Taylor and Raski, 1964) and over 11 months for TobRV in *X. americanum* kept in fallow soil (Bergeson *et al.*, 1964). TRV can persist in vector nematodes for months or years (van Hoof, 1970). However, all vector nematodes lose the viruses on moulting and there is no evidence that any of the viruses multiplies in its vector.

Viruses that are soon lost by their vectors during periods of fallow must somehow survive in infected plant material. TRV-infected *Stellaria media* and *Viola arvensis* plants may survive throughout the winter in Britain to act as sources of the virus in the spring (Cooper and Harrison, 1973) and this probably happens with other nematode-borne viruses too. The viruses may also survive in perennating structures such as rhizomes or runners. Survival in weed seed is also of obvious importance: when virus-infested soils were air-dried to kill vector nematodes and then remoistened, up to 33% of the *Stellaria media* seeds that germinated and up to 17% of those of all other weed species, were infected with TBRV (Table 4; Murant and Lister, 1967).

Murant and Taylor (1965) found that, after a period of winter fallow between strawberry crops, *L. elongatus* in the soil lost ability to transmit RRV and TBRV. However, they became viruliferous again during the following season, presumably by feeding on weed seedlings germinating from infected seed. In glasshouse experiments (Murant and Lister, 1967) *L. elongatus* carrying RRV and TBRV were rendered viruliferous or non-viruliferous respectively by allowing the weeds to grow or by removing them from the soil as soon as they emerged. The presence of infected weed seeds is probably essential for the survival of RRV and TBRV in soils and this in turn means that the nematodes must feed readily on weed species in which the viruses are seed-borne. Thus the ecology of the viruses is closely bound up with that of their vectors. Survival in weed seed seems less important for TRV and for nepoviruses transmitted by *Xiphinema* spp., which are retained by the nematodes for many months: AMV, SLRV and TobRV were found in a much smaller proportion of arable weed seeds germinating in virus-infested soils than were RRV and TBRV (Murant and Lister, 1967). However, Ramsdell and Myers (1978) considered that infection of weeds and seed transmission were probably important in the ecology of peach rosette mosaic virus, a nepovirus transmitted by *X. americanum*.

Table 4 Percentage of virus-infected seeds in soils from Scottish outbreaks of tomato black ring virus*

	Site number			
Species	*1*	*2*	*3*	*4*
Capsella bursa-pastoris	20–100	56	28	0
Lamium amplexicaule	8–74	8	–	–
Stellaria media	20–55	19	32–49	17
Veronica agrestis	1–10	3	11–49	0

* From Murant and Lister (1967).

One way of avoiding infection of raspberry by RRV is to plant resistant varieties and this is successful even though a resistance-breaking strain of the virus occurs. Murant, Taylor and Chambers (1968) suggested that resistance-breaking isolates are possibly uncommon in the field because they are infrequently seed-transmitted and thus survive poorly in soils. Interestingly, recent studies on the genome properties of RRV (Harrison *et al.*, 1974; Hanada and Harrison, 1977) show that the resistance-breaking and poor seed-transmission characters occur in the same part of the genome, i.e. they are genetically linked. Indeed they may be different expressions of the same gene.

Conclusion

Wild plants play an important part in the ecology of nematode-borne viruses, but in this respect these viruses are not unique, for all the viruses that cause disease in economic plants presumably originated in the wild population and many still depend on wild plants for their perpetuation. The nematode-borne viruses merely provide a particularly elegant example of this relationship.

References

Bergeson, G. B., Athow, K. L., Laviolette, F. A. & Thomasine, M. (1964). Transmission, movement and vector relationships of tobacco ringspot virus in soybean. *Phytopathology* **54,** 723–728.

Bos, L. & van der Want, J. P. H. (1962). Early browning of pea, a disease caused by a soil- and seed-borne virus. *Tijdschrift over Planteziekten* **68,** 368–390.

Cohn, E. (1975). Relations between *Xiphinema* and *Longidorus* and their host plants. In *Nematode Vectors of Plant Viruses*, pp. 365–386. F. Lamberti, C. E. Taylor & J. W. Seinhorst. London and New York: Plenum Press.

Cooper, J. I. (1976). The possible epidemiological significance of pollen and seed transmission in the cherry leaf roll virus/*Betula* spp., complex. *Mitteilungen aus der Biologischen Bundesanstalt für Land- und Forstwirtschaft* (Berlin–Dahlem) **170,** 17–22.

Cooper, J. I. (*1981*). The possible role of amenity trees and shrubs as virus reservoirs in the United Kingdom. In *Pests, Pathogens and Vegetation*, pp. 79–87. J. M. Thresh. London: Pitman.

Cooper, J. I. & Harrison, B. D. (1973). The role of weed hosts and the distribution and activity of vector nematodes in the ecology of tobacco rattle virus. *Annals of Applied Biology* **73,** 53–66.

Desjardins, P. R., Latterell, R. L. & Mitchell, J. E. (1954). Seed transmission of tobacco-ringspot virus in Lincoln variety of soybean. *Phytopathology* **44,** 86.

Hanada, K. & Harrison, B. D. (1977). Effects of virus genotype and temperature on seed transmission of nepoviruses. *Annals of Applied Biology* **85,** 79–92.

Harrison, B. D. (1964). Specific nematode vectors for serologically distinctive forms of raspberry ringspot and tomato black ring viruses. *Virology* **22,** 544–550.

Harrison, B. D. (1977). Ecology and control of viruses with soil-inhabiting vectors. *Annual Review of Phytopathology* **15,** 331–360.

Harrison, B. D. & Murant, A. F. (1977). Nepovirus group. *CMI/AAB Descriptions of Plant Viruses* No. 185.

Harrison, B. D. & Robinson, D. J. (1978). The tobraviruses. *Advances in Virus Research* **23,** 25–77.

Harrison, B. D. & Winslow, R. D. (1961). Laboratory and field studies on the relation of arabis mosaic virus to its nematode vector *Xiphinema diversicaudatum* (Micoletzky). *Annals of Applied Biology* **49,** 621–633.

Harrison, B. D., Murant, A. F., Mayo, M. A. & Roberts, I. M. (1974). Distribution of determinants for symptom production, host range and nematode transmissibility between the two RNA components of raspberry ringspot virus. *Journal of General Virology* **22,** 233–247.

Hewitt, W. B., Raski, D. J. & Goheen, A. C. (1958). Nematode vector of soil-borne fanleaf virus of grapevines. *Phytopathology* **48,** 586–595.

Kahn, R. P. (1956). Seed transmission of the tomato-ringspot virus in the Lincoln variety of soybeans. *Phytopathology* **46,** 295.

Lister, R. M. & Murant, A. F. (1967). Seed-transmission of nematode-borne viruses. *Annals of Applied Biology* **59,** 49–62.

McNamara, D. G., Ormerod, P. J., Pitcher, R. S. & Thresh, J. M. (1973). In *Proceedings of the 7th British Insecticide and Fungicide Conference, Brighton* **2,** pp. 597–602.

McNamara, D. G. & Flegg, J. J. M. (*1981*). The distribution of virus–vector nematodes in Great Britain in relation to past and present vegetation. In *Pests, Pathogens and Vegetation*, pp. 225–235. J. M. Thresh. London: Pitman.

Mellor, F. C. & Stace-Smith, R. (1963). Reaction of strawberry to a ringspot virus from raspberry. *Canadian Journal of Botany* **41,** 865–870.

Mircetich, S. M., Sanborn, R. R. & Ramos, D. E. (1980). Natural spread, graft transmission and possible etiology of walnut blackline disease. *Phytopathology* **70,** 962–968.

Murant, A. F. (1970). The importance of wild plants in the ecology of nematode-transmitted plant viruses. *Outlook on Agriculture* **6,** 114–121.

Murant, A. F. (1981, in press). Nepoviruses. In *Handbook of Plant Virus Infections and Comparative Diagnosis*, pp. 197–238. Ed. E. Kurstak. Amsterdam: North Holland.

Murant, A. F. & Goold, R. A. (1969). Strawberry latent ringspot virus. *Annual Report of the Scottish Horticultural Research Institute for 1968*, p. 48.

Murant, A. F. & Lister, R. M. (1967). Seed-transmission in the ecology of nematode-borne viruses. *Annals of Applied Biology* **59,** 63–76.

Murant, A. F. & Taylor, C. E. (1965). Treatment of soil with chemicals to prevent transmission of tomato black ring and raspberry ringspot viruses by *Longidorus elongatus* (de Man). *Annals of Applied Biology* **55,** 227–237.

Murant, A. F., Taylor, C. E. & Chambers, J. (1968). Properties, relationships and transmission of a strain of raspberry ringspot virus infecting raspberry cultivars immune to the common Scottish strain. *Annals of Applied Biology* **61,** 175–186.

Noordam, D. (1956). Waardplanten en toetsplanten van het ratelvirus van de tabak. *Tijdschrift over Plantenziekten* **62,** 219–225.

Pitcher, R. S. & Jha, A. (1961). On the distribution and infectivity with arabis mosaic virus of a dagger nematode. *Plant Pathology* **10,** 67–71.

Ramsdell, D. C. & Myers, R. L. (1978). Epidemiology of peach rosette mosaic virus in a Concord grape vineyard. *Phytopathology* **68,** 447–450.

Schmelzer, K. (1957). Untersuchungen über den Wirtspflanzenkreis des Tabak-mauche-Virus. *Phytopathologische Zeitschrift* **30,** 281–314.

Schmelzer, K. (1963a). Untersuchungen an Viren der Zier- und Wildgehölze. 2. Mitteilung. Virosen an *Forsythia, Lonicera, Ligustrum* und *Laburnum. Phytopathologische Zeitschrift* **46,** 105–138.

Schmelzer, K. (1963b). Untersuchungen an Viren der Zier- und Wildgehölze. 3. Mitteilung. Virosen an *Robinia, Caryopteris, Ptelea* und anderen Gattungen. *Phytopathologische Zeitschrift* **46,** 235–268.

Schmelzer, K. (1966). Untersuchungen an Viren der Zier- und Wildgehölze. 5. Mitteilung. Virosen an *Populus* und *Sambucus. Phytopathologische Zeitschrift* **55,** 317–351.

Schmelzer, K. (1969). Das latente Erdbeerringflecken-Virus aus *Euonymus, Robinia* und *Aesculus. Phytopathologische Zeitschrift* **66,** 1–24.

Taylor, C. E. & Murant, A. F. (1969). Transmission of strains of raspberry ringspot and tomato black ring viruses by *Longidorus elongatus* (de Man). *Annals of Applied Biology* **64,** 43–48.

Taylor, C. E. & Raski, D. J. (1964). On the transmission of grape fanleaf by *Xiphinema index*. *Nematologica* **10,** 489–495.

Taylor, C. E. & Thomas, P. R. (1968). The association of *Xiphinema diversicaudatum* (Micoletsky) with strawberry latent ringspot and arabis mosaic viruses in a raspberry plantation. *Annals of Applied Biology* **62,** 147–157.

Thomas, P. R. (1969). Crop and weed plants compared as hosts of viruliferous *Longidorus elongatus* (de Man). *Plant Pathology* **18,** 23–28.

Tuite, J. (1960). The natural occurrence of tobacco ringspot virus. *Phytopathology* **50,** 296–298.

Valleau, W. D. (1932). Seed transmission and sterility studies of two strains of tobacco ringspot. *Bulletin of the Kentucky Agricultural Experiment Station* **327,** 43–80.

Van Hoof, H. A. (1970). Some observations on retention of tobacco rattle virus in nematodes. *Netherlands Journal of Plant Pathology* **76,** 329–330.

Section 4 **Arthropod pests**

Wild plants in the ecology of insect pests

H F van Emden
Departments of Agriculture & Horticulture and Zoology, The University, Reading, Berkshire RG6 2AT

This paper discusses the inter-relationships of wild plants and insect pests with the emphasis on weed plants outside the crop. Clearly some of the relationships apply equally for weeds within the crop, but this topic is dealt with in more detail elsewhere (Way and Cammell, *1981*).

The literature on wild plants and insect pests has already been reviewed in relation to hedgerows (van Emden, 1965a; Lewis, 1965a), to weed control in agriculture (van Emden, 1970), to the diversity-stability argument (van Emden and Williams, 1974) and to multiple cropping (van Emden, 1977). It must be emphasized that any judgement on the value of hedges and other areas of uncultivated land in the agricultural landscape cannot be based solely on their role in the ecology of insect pests. Other issues involved include aesthetic ones, the need to control soil erosion or provide windbreaks and the place of areas such as roadside verges in conserving woodland fauna (Pollard, 1968a) and as nesting sites for birds.

In this paper, the relationships between wild plants and insect pests of crops are reviewed in relation to two interacting classifications. Firstly, the topic is subdivided on a seasonal basis. Some of the relationships involve pests early in the season before the crop is available, and these are distinguished from both the actual infestation of the crop and the subsequent development of the infestation as the crop season progresses.

Secondly, a distinction is made between relationships having a biological or physical basis. 'Biological effects' are defined as those where particular species of wild plants are used for feeding by insects. 'Physical effects' are defined as those which are largely microclimatic; the species composition of the wild flora is relatively unimportant and feeding by the insects is not involved.

Before the crop emerges: biological effects

Most farmers are aware that wild plants can act as a reservoir of pests, and there are numerous references to weeds as alternative food for crop pests. Between 1939 and 1963 there were 442 such references (van Emden, 1965a), over half concerning just two groups of pests, the Homoptera (particularly leafhoppers and aphids) and the Coleoptera (particularly leaf beetles, flea beetles and weevils). Muenscher's (1955) textbook on weeds lists 30 pest/weed associations regarded as of major economic importance. Alternative host plants are clearly particularly important where the cropping season is short and followed by a cold or dry season of fallow or total replacement by a different cropping system (e.g. the summer/winter alternation of soybean and wheat across enormous areas of monoculture in southern Brazil).

Of the 442 references tabulated by van Emden (1965a), 100 concern cereals. Related wild plants (i.e. grasses) are particularly abundant and ubiquitous in arable areas, and are becoming increasingly so with use of selective herbicides on cereals (Fryer, *1981*). Similar botanical affinities between crops and weeds influence the abundance of pests in cotton (wild Malvaceae) and potatoes in America and Europe, where wild Solanaceae are common hedgerow and roadside plants. Hoffmann (1949) traced four weevil outbreaks to locally abundant weeds, in each case in the same family as the affected crop plant. Destruction of related weeds is a common pest control recommendation (e.g. Noll, 1942; Pearson, 1958). Such recommendations should not, however, suggest that less attention should be paid to wild plants unrelated to the crop. Kronenberg (1941) advised the destruction of stinging nettles (*Urtica dioica*) harbouring the weevil *Phyllobius pomaceus* near strawberry beds. Certain capsid bugs feed on the young foliage of many weeds, and the cotton aphid (*Aphis gossypii*) feeds on various unrelated weeds in and around cotton fields (Young and Garrison, 1949). Many aphids also show seasonal host alternation, nearly always between unrelated species (Eastop, *1981*).

Weed reservoirs of pests clearly make crop rotation less efficient as a control measure, and weeds are not just a source of pests but also of many pathogens transmitted by insects. The role of wild plants in the epidemiology of plant virus diseases is discussed by Thresh (*1981*).

The use of wild plants by crop pests can occasionally be exploited in a variation of the technique of 'trap cropping' (van Emden, 1977). The most elegant example is the planting of brome grass (*Bromus* spp.) around cereals, trapping a large proportion of incoming wheat stem sawflies (*Cephus cinctus*). The elegant features of using brome are that the sawfly larvae fail to complete their development in the stems of brome and die, and that this happens after any internal parasites have emerged.

Little is known about the importance of wild plants in the quantitative population dynamics of crop pests, but they must represent an arena for intra-specific competition and may also maintain genetic diversity in pest populations, delaying the appearance of insecticide resistance and the breakdown of host plant resistance (Eastop, *1981*). Thus in pest population terms, events on wild plants may not automatically be to the detriment of the farmer (*see also* the example of brome grass cited above). The frit fly (*Oscinella frit*) returns to wild grasses for overwintering each year; widely fluctuating populations are produced in different seasons from the ears of oats. Southwood and Jepson (1962) estimated the average annual total production in England and Wales of *O. frit* from oat panicles as 7.1×10^{12} flies contrasted with 1.0×10^{12} flies from the overwintering generation in grasslands. The shortage of stems at a suitable growth stage in grassland appears to reduce and stabilize the population of flies emerging in the spring irrespective of the number of flies returning each autumn from oat fields (van Emden and Way, 1973).

Before the crop emerges: physical effects

Uncultivated land provides plant debris and other forms of physical shelter for pests hibernating in an inactive state. Tischler (1950) analysed the hibernation habits of crop pests, and concluded that insects which overwintered as adults (especially bugs and beetles) nearly all left the crop to hibernate in the shelter of uncultivated land, and that such insects included the most important pests of beet, crucifers and legumes. Many of the wingless weevils of top and soft fruit hibernate in litter in woodlands and hedgerows (Anon, 1977).

Initial infestation: biological effects

The pest reservoir on uncultivated land is thus often the source of the initial infestation, and the insects move onto the crops for diverse reasons. Wild plants and crops may be part of an obligatory seasonal alternation (e.g. the heteroecious aphids) or the invasion may represent the dispersal of a hibernating population (e.g. flea beetles, *see* below). Insects feeding on weeds may be 'forced' off the latter by the early maturity of the weed flora (e.g. many capsid bugs), by lack of weed food caused by the insects' own attack (Kanervo, 1947) or even by the use of weedkillers on roadsides (Steiner, 1945). Jacobson (1946) observed spring populations of the pentatomid *Chlorochroa sayi* on patches of *Salsola* sp. (Chenopodiaceae) in uncultivated land adjacent to a wheat field in Canada. Samples of threshed grain from the field at the end of the season showed a gradient of shrivelled grain extending over 3 km from the weed patches; indeed seed lots with satisfactory viability could only be obtained more than 0.8 km into the field. This suggests that the failure of many workers to find damage gradients for other pests away from weed sources may be due to the areas studied being too small.

Initial infestation: physical effects

Edge concentrations of pests are often the result of physical characteristics of uncultivated land. Movement of flea beetles from plant debris outside the crop results in spring concentrations of adults and crop damage at the edges of fields (Wolfenbarger, 1940; Moreton, 1945).

It is characteristic of many of the acalyptrate Diptera (e.g. melon fly (*Dacus cucurbitae*), cabbage root fly (*Delia brassicae*), carrot fly (*Psila rosae*)), that larval damage to crops falls off away from the headlands where the adults 'roost' at night or during adverse weather. Wright and Ashby (1946) produced a three-dimensional diagram of decreasing carrot fly infestation away from the field edges with particularly severe attack near the corner where the effects of two headlands overlapped. The opportunity this roosting habit gives of achieving control outside the crop was stressed early in the development of chemical control measures against the pest (Baker *et al.*, 1942). (*See also* Wainhouse and Coaker (*1981*).)

A different pattern of infestation results when dispersing insects are deposited from air currents near taller vegetation such as hedgerows and windbreaks (Lewis, 1965a), where turbulence caused by the irregularity in topography creates downward eddies onto the crop. This phenomenon has been investigated in some detail by Lewis (1965b, 1968), who reproduced the effect with paper discs released into an air-stream and by measuring snow accumulations near barriers. Accumulations of wind-borne organisms and other objects occur close to the barrier on the windward side, and some distance behind it to leeward; the exact zones of maximum deposition vary with the height and permeability of the windbreak. For typical hedgerows, such zones can be expected about 1 × hedgerow height to windward and 3–6 × height to leeward. Such distributions have been strikingly illustrated by the incidence of plants killed by lettuce root aphid (*Pemphigus bursarius*) or infected with an aphid-borne virus around artificial or living windbreaks (Lewis, 1965c; Quiot *et al.*, 1979; Thresh, *1981*).

Development of the infestation: biological effects

Just as uncultivated land maintains a pest reservoir when crop hosts of a pest are absent, so there is also a natural enemy reservoir where predators and parasites are provided with prey when this is absent or scarce on the crop area, including the period following an insecticide application. The prey on uncultivated land can be part of the population of a pest species, and this provides yet another reason for not viewing the pest reservoir in a negative way, since few people would argue that the component of biological control can be ignored in pest control on the crop. Indeed, the aim of preserving a reservoir of beneficial insects by modifying chemical usage led to the original definition of 'integrated control' (Stern *et al.*, 1959).

Some parasites and many predators are highly polyphagous. *Apanteles fulvipes*, an important parasite of the gypsy moth (*Lymantria dispar*) has at least 28 species of host (Györfi, 1951), and 41 species of prey are recorded for *Adalia bipunctata* (Schilder and Schilder, 1928). Nineteen of the secondary hosts of *A. fulvipes* occur on non-economic plants and Hodek (1967) describes how aphids on shrubs and herbaceous weeds maintain ladybirds at various times of the year in different habitats.

At certain times in fallowing regimes or when insecticides are used intensively, prey on uncultivated land may be essential to perpetuate natural enemies in the locality. There are additional instances where use of prey on uncultivated land is part of an obligatory seasonal alternation in the life-cycle of the natural enemy. This occurs with the parasites (*Angitia* spp.) of the diamond-back moth (*Plutella maculipennis*), which is a potential but rarely evident pest of brassicas in Britain (van Emden, 1965a). Since the 1930s it has been realized that *Angitia* spp. are important in keeping diamond-back moth below pest status in Britain, and that the parasites cannot overwinter in *P. maculipennis* (Hardy, 1938). The larvae of the moth overwinter spun up in a cocoon and the adult parasites emerge in early autumn when no young caterpillars are available. As the adults do not survive the winter, it was clear that another lepidopteran species was used as a winter 'bridge', but it was 20 years later that O. W. Richards (unpublished) located *A. fenestralis* over-wintering in *Swammerdamia lutarea* on hawthorn. There has been little systematic study of parasites of insects on uncultivated plants, and such examples are rare; others will perhaps be discovered when the life cycles of parasites are disrupted following changes in land usage.

Wild flowers are extremely important sources of adult food for insects. This is perhaps not generally recognized, although the abundance and diversity of insects visiting patches of wild flowers are often exploited by insect collectors. Umbelliferous flowers seem especially attractive, and Allen (1954) records huge numbers and diversity of species of parasitic Hymenoptera from flowers of this family. The female parasites collected at flowers are mainly immature (van Emden, 1963) and pollen feeding has been shown to be essential for egg maturation in several groups of predators and parasites (Schneider-Orelli, 1945; Voûte, 1946). Early work by Thorpe and Caudle (1938) on parasites of the pine shoot moth (*Rhyacionia buoliana*) may be more generally applicable: newly-emerged female parasites were repelled by the scent of the host plant (pine oil) until their eggs were mature. The need to provide suitable flowers on uncultivated land for adult predators and parasites has been recognized in biological control programmes (Wolcott, 1941a, 1941b, 1942; Leius, 1960). In Russia, high parasitization rates of cabbage cutworm (*Mamestra brassicae*) have been achieved by sowing umbelliferous plants near cabbage fields (Kopvillem, 1960). The routine use of herbicides in modern agriculture successfully keeps crop fields very clean of flowering weeds and wild plants outside the crop are often the only source of flowers for beneficial insects.

Unfortunately, flower feeding is similarly essential for the adults of a number of pest species, e.g. root flies (Petherbridge *et al.*, 1942; Miles, 1951). Eight cow parsley (*Anthriscus sylvestris*) plants per metre of hedgerow produce sufficient nectar to feed at least 2000 cabbage root fly from emergence to oviposition (Coaker and Finch, 1973). However, flower feeding seems much more generally important for beneficial insects than for plant pests.

Development of the infestation: physical effects

The deposition of insects from air currents behind windbreaks has been mentioned in connection with the arrival of pests on the crop. Beneficial insects similarly accumulate behind windbreaks, and Lewis (1965b) includes Syrphidae, Chrysopidae and parasitic Hymenoptera as examples. Nevertheless Lewis (1968) appears convinced that, on balance 'windbreaks may create some new pest problems and certainly accentuate existing ones'.

Other physical effects occur at the edges of crops close to uncultivated land. Here crop plants may suffer strong competition, including shading, from adjacent dense uncultivated vegetation or trees. Aphids on edge plants in a crop appear to suffer both from the bombardment of large drops dripping from tree foliage in wet weather and the weakening of the plant by shade and competition for water (van Emden, 1965b). Shade may also affect the distribution on the crop of both pests (e.g. Knowlton, 1948) and beneficial insects (e.g. Arthur, 1945).

Conclusions

I have attempted to describe and order the more important ways in which weeds and other wild plants influence the ecology and importance of crop pests. The inter-relationships involved are summarized diagrammatically in Fig. 1.

The many ways in which wild plants interact with crop pests and the balance of potentially harmful and beneficial outcomes make it difficult to decide how the farmer should view wild plants in relation to his pest problems. The influence of wild plants on the status of insects as pests of crops will often be on a regional rather than a local scale. Even with perfect weed control within the crop, no part of the field is entirely free from a wider influence. This makes it almost impossible to test the effect of wild plants experimentally, but three general conclusions are possible:

(a) The accumulation of insects around windbreaks is on balance harmful.
(b) The practice of improving road vision for motorists by cutting verges is

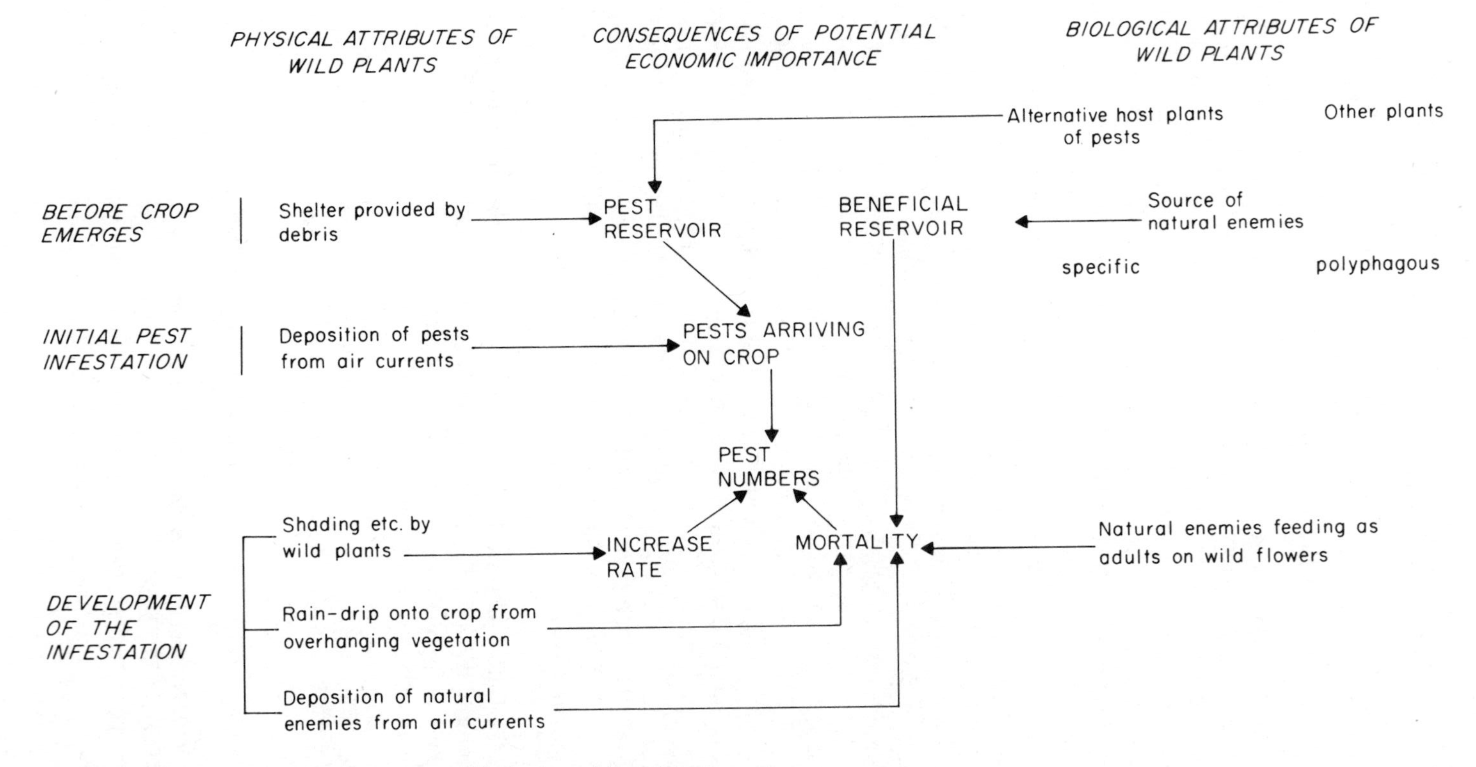

Figure 1 Inter-relationships between wild plants and insect pests.

likely to remove the beneficial element of wild flowers while retaining the pest reservoir of herbivores on the foliage.
(c) A farmer's own wild plants are probably a nuisance to him from the pest control standpoint, yet any biological component of pest control requires that other farmers keep theirs!

Effects of adjacent wild plants on crop pests can be shown by field experiments, but local within-field phenomena (van Emden, 1965b; Pollard, 1968b) or even differences between contrasting areas of farmland 5 km apart (Pollard, 1971) must be on too small a scale to reveal any really important impact of wild plants. Attempts to contrast pest problems in the vegetational mosaic of Great Britain with problems of vast intensive monocultures elsewhere in the world are also not very helpful, since a difference in abundance of wild plants is only one of many differences involved.

The increasing intensification of farming in Britain, involving the removal of hedgerows (Moore *et al.*, 1967) and the extensive use of herbicides (Fryer, *1981*) has resulted in noticeable changes in the abundance of wild plants in agricultural areas. There is no sign that this has markedly changed the type or severity of pest problems experienced by farmers. If wild plants are on balance harmful in pest terms, then it is perhaps surprising that such landscape changes have not been accompanied by a decline in the need to use insecticides. On the other hand, the use of insecticides has not increased to an extent that would lead us to invoke the loss of beneficial effects of wild plants as an explanation! It would be a strange coincidence if beneficial and harmful effects were mutually cancelling over a wide range of land usage; it is much more likely that wild plants are still sufficiently abundant not to be limiting in either direction. Certainly beneficial insects still occur on crops and act as an important buffer reducing the need for pesticide application on cereals, for example, and keeping many herbivores below pest status. Such 'background' biological control is dependent on some wild plants in the agricultural landscape but at present it is impossible to guess at the minimum extent and distribution of wild flora which must be retained to maintain this vital biological component of the agro-ecosystem.

References

Allen, P. (1954). Studies of insects living on plants of the family Umbelliferae with special reference to flies of the family Agromyzidae. *DIC Thesis*, Imperial College, London.

Anon. (1977). Wingless weevils. *Ministry of Agriculture, Fisheries and Food, UK, Advisory Leaflet* No. 57.

Arthur, D. R. (1945). The development of artificially introduced infestations of *Aphidius granarius* Marsh., under field conditions. *Bulletin of Entomological Research* **36,** 291–295.

Baker, F. T., Ketteringham, I. E., Bray, S. P. V. & White, J. H. (1942). Observations on the biology of the carrot fly (*Psila rosae* Fab.): assembling and oviposition. *Annals of Applied Biology* **29,** 115–125.

Coaker, T. H. & Finch, S. (1973). The association of the cabbage rootfly with its food and host plants. In *Insect/Plant Relationships*, pp. 119–128. H. F. van Emden. Oxford: Blackwell.

Eastop, V. F. (*1981*). Wild hosts of aphid pests. In *Pests, Pathogens and Vegetation*, pp. 285–298. J. M. Thresh. London: Pitman.

van Emden, H. F. (1963). Observations on the effect of flowers on the activity of parasitic Hymenoptera. *Entomologist's Monthly Magazine* **98** (1962), 265–270.

van Emden, H. F. (1965a). The role of uncultivated land in the biology of crop pests and beneficial insects. *Scientific Horticulture* **17,** 121–136.

van Emden, H. F. (1965b). The effect of uncultivated land on the distribution of cabbage aphid (*Brevicoryne brassicae*) on an adjacent crop. *Journal of Applied Ecology* **2,** 171–196.

van Emden, H. F. (1970). Insects, weeds and plant health. In *Proceedings of the 10th British Weed Control Conference, Brighton*, pp. 942–952.

van Emden, H. F. (1977). Insect-pest management in multiple cropping systems — a strategy. In *Cropping Systems Research and Development for the Asian Rice Farmer*, pp. 309–323. Philippines: IRRI.

van Emden, H. F. & Way, M. J. (1973). Plants in the population dynamics of insects. In *Insect/Plant Relationships*, pp. 181–199. H. F. van Emden. Oxford: Blackwell.

van Emden, H. F. & Williams, G. (1974). Insect stability and diversity in agro-ecosystems. *Annual Review of Entomology* **19,** 455–475.

Fryer, J. D. (*1981*). Weed control practices and changing weed problems. In *Pests, Pathogens and Vegetation*, pp. 403–414. J. M. Thresh. London: Pitman.

Györfi, J. (1951). Die Schlupfwespen und der Unterwuchs des Waldes. *Zeitschrift für Angewandte Entomologie* **33,** 32–47.

Hardy, J. E. (1938). *Plutella maculipennis* Curt., its natural and biological control in England. *Bulletin of Entomological Research* **29,** 343–372.

Hodek, I. (1957). Bionomics and ecology of predaceous Coccinellidae. *Annual Review of Entomology* **12,** 79–104.

Hoffmann, A. (1949). La flore spontanée et la pullulation des insectes nuisibles aux cultures. *Revue de Pathologie Végétale et d'Entomologie Agricole de France* **28,** 159–162.

Jacobson, L. A. (1946). The effect of say stinkbug on wheat. *Canadian Entomologist* **77,** 200.

Kanervo, V. (1947). Über das Massenauftreten der Gammaeule *Phytometra gamma* L. (Lep. Noctuidae), im Sommer 1946 in Finnland. *Annales Entomologici Fennici* **13,** 89–104.

Knowlton, G. F. (1948). *Empoasca filamenta* damage. *Journal of Economic Entomology* **41,** 115.

Kopvillem, H. G. (1960). [Nectar plants for the attraction of entomophagous insects.] *Zashchita Rastenii ok Vrediteleĭ i Bolezneĭ* **5,** 33–34. (In Russian.)

Kronenberg, H. G. (1941). *Phyllobius urticae* de yeer schadelijk aan aarbeien in Kennemerland. *Tijdschrift over Plantenziekten* **47,** 186–193.

Leius, K. (1960). Attractiveness of different foods and flowers to the adults of some hymenopterous parasites. *Canadian Entomologist* **42,** 369–376.

Lewis, T. (1965a). The effects of shelter on the distribution of insect pests. *Scientific Horticulture* **17,** 74–84.

Lewis, T. (1965b). The effects of an artificial windbreak on the aerial distribution of flying insects. *Annals of Applied Biology* **55,** 503–512.

Lewis, T. (1965c). The effect of an artificial windbreak on the distribution of aphids in a lettuce crop. *Annals of Applied Biology* **55,** 513–518.

Lewis, T. (1968). Windbreaks, shelter and insect distribution. *Span* **11** (reprint), 4 pp.

Miles, M. (1951). Factors affecting the behaviour and activity of the cabbage root fly (*Erioischia brassicae* Bché). *Annals of Applied Biology* **38,** 425–432.

Moore, N. W., Hooper, M. D. & Davis, B. N. K. (1967). Hedges. I. Introduction and reconnaissance studies. *Journal of Applied Ecology* **4,** 201–220.

Moreton, B. D. (1945). On the migration of flea beetles (*Phyllotreta* spp.) (Col., Chrysomelidae) attacking *Brassica* crops. *Entomologist's Monthly Magazine* **81,** 59–60.

Muenscher, W. C. (1955). *Weeds*. New York: Macmillan.

Noll, I. (1942). Erdflohschaden und Erdflohbekämpfung. *Kranke Pflanze* **19,** 67–70.

Pearson, E. O. (1958). *The Insect Pests of Cotton in Tropical Africa*. London: Empire Cotton Growing Corporation and Commonwealth Institute of Entomology.

Petherbridge, F. R., Wright, D. W. & Davies, P. G. (1942). Investigations on the biology and control of the carrot fly (*Psila rosae* F.). *Annals of Applied Biology* **29,** 380–392.

Pollard, E. (1968a). Hedges. IV. A comparison between the Carabidae of a hedge and field site and those of a woodland glade. *Journal of Applied Ecology* **5,** 649–657.

Pollard, E. (1968b). Hedges. II. The effect of removal of the bottom flora of a hawthorn hedgerow on the fauna of the hawthorn. *Journal of Applied Ecology* **5,** 109–123.

Pollard, E. (1971). Hedges. VI. Habitat diversity and crop pests: a study of *Brevicoryne brassicae* and its syrphid predators. *Journal of Applied Ecology* **8,** 751–780.

Quiot, J. B., Verbrugghe, M., Labonne, G., Leclant, F. & Marrou, J. (1979). Ecologie et épidémiologie du virus de la mosaïque du concombre dans le sud-est de la France. IV. Influence des brise-vent sur la répartition des contaminations virales dans une culture protégée. *Annales de Phytopathologie* **11,** 307–324.

Schilder, F. A. & Schilder, M. (1928). Die Nahrung der Coccinelliden und ihre Beziehung zur Verwandtschaft der Arten. *Arbeiten aus der Biologischen Reichanstalt für Land- und Forstwirtschaft, Berlin* **16,** 215–282.

Schneider-Orelli, O. (1945). Bienenweide und Schädlingsbekämpfung. *Schweizerische Bienenzeitung* **9,** 423–429.

Southwood, T. R. E. & Jepson, W. F. (1962). The productivity of grasslands in England for *Oscinella frit* (L.) (Chloropidae) and other stem-boring Diptera. *Bulletin of Entomological Research* **53,** 395–407.

Steiner, H. M. (1945). Ground cover sprays to kill insects and weeds in peach orchards. *Journal of Economic Entomology* **38,** 117–119.

Stern, V. M., Smith, R. F., van den Bosch, R. & Hagen, K. S. (1959). The integrated control concept. *Hilgardia* **29,** 81–101.

Thorpe, W. H. & Caudle, H. B. (1938). A study of the olfactory responses of insect parasites to the food plant of their host. *Parasitology* **30,** 523–528.

Thresh, J. M. (*1981*). The role of weeds and wild plants in the epidemiology of plant virus diseases. In *Pests, Pathogens and Vegetation*, pp. 53–70. J. M. Thresh. London: Pitman.

Tischler, W. (1950). Überwinterungsverhältnisse der landwirtschaftlichen schädlinge. *Zeitschrift für Angewandte Entomologie* **32,** 184–194.

Voûte, A. D. (1946). Regulation of the density of the insect populations in virgin forests and cultivated woods. *Archives Neérlandaises de Zoologie* **7,** 435–470.

Wainhouse, D. & Coaker, T. H. (*1981*). The distribution of carrot fly (*Psila rosae*) in relation to the flora of field boundaries. In *Pests, Pathogens and Vegetation*, pp. 263–272. J. M. Thresh. London: Pitman.

Way, M. J. & Cammell, M. E. (*1981*). Effects of weeds and weed control on invertebrate pest ecology. In *Pests, Pathogens and Vegetation,* pp. 443–458. J. M. Thresh. London: Pitman.

Wolcott, G. N. (1941a). The dispersion of *Larra americana* Saussure in Puerto Rico. *Revista de Agricultura de Puerto Rico* **33,** 607–608.

Wolcott, G. N. (1941b). The establishment in Peurto Rico of *Larra americana* Saussure. *Journal of Economic Entomology* **34,** 53–56.

Wolcott, G. N. (1942). The requirements of parasites for more than hosts. *Science* **96,** 317–318.

Wolfenbarger, D. O. (1940). Relative prevalence of potato flea beetle injuries in fields adjoining uncultivated areas. *Annals of the Entomological Society of America* **33,** 391–394.

Wright, D. W. & Ashby, D. G. (1946). The control of the carrot fly (*Psila rosae* Fab.) (Diptera) with DDT. *Bulletin of Entomological Research* **36,** 253–268.

Young, M. T. & Garrison, G. L. (1949). Aphid collections at Tallulah, Louisiana, from 1941 to 1947. *Journal of Economic Entomology* **42,** 993–994.

The distribution of carrot fly (*Psila rosae*) in relation to the flora of field boundaries

D Wainhouse* and T H Coaker
Department of Applied Biology, University of Cambridge CB2 3DX

Introduction

Earlier studies on carrot fly (*Psila rosae*) showed that abundant flies in non-crop habitats around carrot fields resulted in high levels of larval damage to adjacent carrots (Baker *et al.*, 1942; Barnes, 1942; Petherbridge and Wright, 1943). Cultural control recommendations were made (Petherbridge *et al.*, 1942; Wright and Ashby, 1946; Whitcomb, 1929; Watkins and Miner, 1943; Petherbridge *et al.*, 1945), but as far as is known were never practised on a commercial scale and became unnecessary when organochlorine insecticides were introduced during the 1950s. Following the development of organochlorine resistance (Wright and Coaker, 1968; Gostick and Baker, 1968) there has been concern over the total reliance on insecticides. Moreover, although control is still achieved with organophosphorus insecticides it is inadequate on crops left in the ground over winter (Wright and Coaker, 1968).

The importance of non-crop habitats in carrot fly ecology is now being re-examined to determine how their flora affects the distribution of flies in a commercial carrot growing area.

The study site and description of field boundaries

In the study area at Feltwell Fen, Norfolk (Ordnance Survey Ref TL 650880) carrots (predominantly cv. Autumn King) occupied about 15% of the arable area and were treated with phorate at sowing and chlorfenvinphos as a mid-season spray (Maskell and Gair, 1973). The 6–9 ha fields were divided by ditches and/or windbreaks forming the non-crop habitat (Fig. 1). The ditches (D1–4), which were up to 2 m deep and 5 m wide, were covered by diverse herbs on their sides and verges. The windbreaks, alone or with ditches (W1–3; WD1–3), consisted of trees with an understorey of bushes and their mean height varied from 4 m (W2) to 16 m (WD2). There was an underlying

* Present address: Forestry Commission Research Station, Alice Holt Lodge, Wrecclesham, Farnham, Surrey.

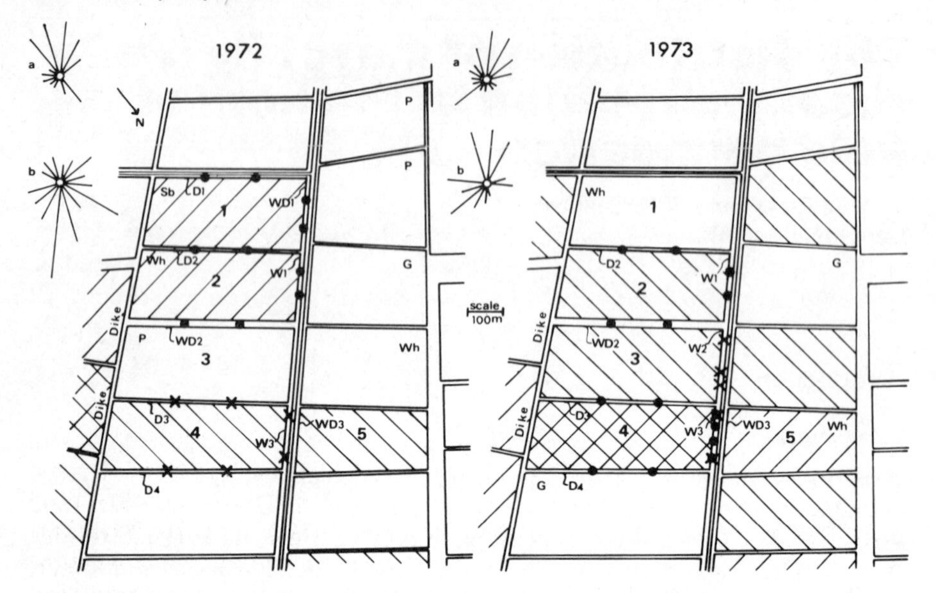

Figure 1 Distribution of crops and trapping points (6 traps/point) for the first (●) and second (X 1972; ● & X 1973) generations of flies in 1972 and 1973.
▨ carrots previous year; ▧ carrots current year; D, Ditch; W, Windbreak; WD, Windbreak and ditch; Sb, Sugar beet; Wh, Wheat; P, Potatoes; G, Grass.

Wind roses

a, First generation	3 May–12 June 1972	3 May–21 June 1973
b, Second generation	7 August–17 October 1972	1 August–13 September 1973

herb layer often several metres wide. The mean number of species in the herb layer (excluding grasses) and in the tree layer at the three types of boundary during 1972/3 is shown in Table 1; a complete list of species is given elsewhere (Wainhouse, 1975). Two herb species were selected for more intensive study, i.e. *Urtica dioica* (common nettle), which was frequently the most abundant component of the herb layer and can provide shelter for the flies (Coppock, 1974), and *Conium maculatum* (hemlock), a host plant (Petherbridge *et al.*, 1942) that could contribute directly to populations of *P. rosae* at boundaries. Both plants were present at all boundaries in the study area.

Table 1 Mean number of species in the herb and tree layers at boundaries studied (Fig. 1).

Boundary	*Herb layer:-*		*Tree layer:-*	
	mean	*range*	*mean*	*range*
D: Ditches	47.0	35–57		
WD: Ditches and windbreaks	32.6	26–41	4.3	4–5
W: Windbreaks	22.0	20–24	5.6	4–7

During the first and second generations of flies in 1972 and 1973 the frequency of *U. dioica* in 30-cm-square random quadrats was determined at boundaries where flies were trapped and the total number of species that occurred in these quadrats was taken as an index of floral diversity at the boundary. The density of *C. maculatum* was assessed by counting all individuals when in flower or by estimating density from sixty 60-cm-square random quadrats. In July after the first generation of *P. rosae* adults, the herb layer was usually cut by the farmer and its regrowth during the second generation provided conditions similar to those during the first.

The main species forming the windbreaks were *Sambucus nigra*, *Alnus glutinosa*, *Salix* spp. and *Populus* spp. At each windbreak the mean number of trees and bushes per 100 m was used as a measure of its density.

Distribution of adult *P. rosae* in relation to field boundaries

In 1972 and 1973, first and second generation flies were caught at boundaries with yellow water-traps (Hawkes, 1969) and a Dietrick-type sampler (Southwood, 1966). There were two to four groups of six water-traps per boundary (Fig. 1), each group comprising two traps 3 m apart on the two edges and in the middle of each boundary. Two-minute suction samples in the herb layer covered about 100 m on each side of the boundary. The trap samples were collected during the morning every 2–7 days.

The boundaries sampled enclosed one or more contiguous fields from which flies emerged from overwintering pupae (first generation) or the current year's crop (second generation). In 1973, first-generation flies were also sampled at boundaries surrounding a non-source from which carrots were absent in 1972.

Wind direction frequency during the trapping periods was estimated to the nearest 22.5° from the mean directions for 0800–2000 hours on each trapping day (Fig. 1).

Captures of flies at boundaries (Figs. 2, 3 and 4) show that overall, water-traps caught about three males to one female in the first generation and nearly equal numbers of the sexes during the second generation. The reasons for this change, which occurred at all types of boundary, are unknown. Suction samples caught about 40% more males than females in both generations, probably because males are about 40% more active (Wainhouse, 1977). Also, suction sampling appeared to be less efficient than water traps at WD2 and W1 in the first generation of 1973, and at all boundaries during the second generation. In the first generation, this probably arose from the difficulty in sampling the herb layer where dense stands of *C. maculatum* were up to 1.5 m tall. During the second generation the increased shelter provided by the carrot crop may have resulted in a more even partitioning of flies between the crop and herb layer in which water traps have a greater 'catching power' (Southwood, 1966).

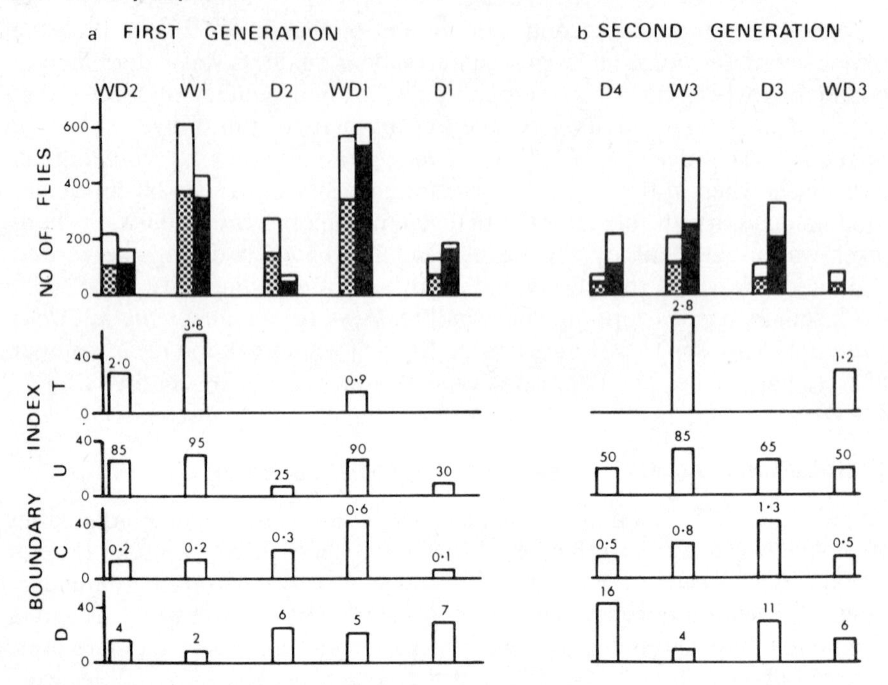

Figure 2 Number of flies trapped and boundary indices in 1972

Suction samples (total catch) ♀□, ♂▩
Water traps (mean number per group of 6 traps) ♀□, ♂■
Figures over histograms represent: T, mean number of trees and bushes per 100 m;
U, frequency of *U. dioica*;
C, density of *C. maculatum* m^{-2}.
D, number of plant species in quadrats.
These values for each boundary are transformed to a percentage of their total over all boundaries (see text) to give the boundary index.

Relationship between abundance of flies and the composition of boundaries

For each trapping period, the frequency of *U. dioica*, diversity index, density of *C. maculatum* and the number of trees and bushes per 100 m measured at each boundary were separately transformed to a percentage of their total over all boundaries around the same carrot field source of flies. This gave four basic indices for each boundary called respectively U, D, C and T (Figs. 2, 3 and 4), which could be taken singly or added together in all permutations to provide 15 indices (Table 2). For each trapping period the captures of flies

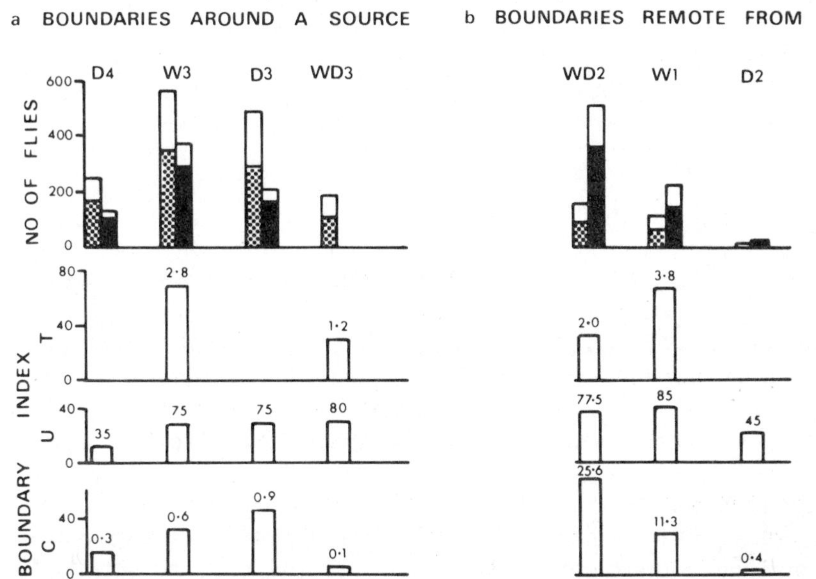

Figure 3 Number of first generation flies trapped and boundary indices in 19?3 (symbols as for Fig. 2).

(males and females) in water-traps or suction samples (y) at each boundary were fitted to the regression equation $\log_{10} y = a + b \log_{10} x + 1$ for each of the 15 indices (x). Thus for each index (Table 2) there were eight regressions corresponding to the four trapping periods and two trapping methods. There were no significant differences in slope between the eight lines in each index (Kozak, 1972), indicating a relationship between catch and index that was similar in both generations in the two years and was unaffected by trapping method. Eight lines with a common slope were fitted to the data for each index and the regression sums of squares expressed as a percentage of the total sum of squares corrected for group means (Table 2). The regressions for index U are shown in Fig. 5.

The analysis indicated that the distribution of *U. dioica* (U) between boundaries had the greatest influence on fly abundance. It appears that the use of other boundary parameters would not improve on a prediction of fly abundance at boundaries based on U alone. Windbreak density (T) also appears to be a significant feature of boundaries, but less important than U and inferior to most compound indices in accounting for variation in the number of flies between boundaries. C and D appear to be unimportant, although catches at WD 2, W1 and D2 during the first generation in 1973 appear to be correlated with abundance of *C. maculatum* (C) (Fig. 3b). This conclusion is supported by estimates of the number of flies emerging from these plants. However, these data were omitted from the analysis because the boundaries were not adjacent to a carrot field source of flies.

Importance of alternative host plants

In a part of the study area (Fields 2, 3 and 4) the numbers of flies emerging from carrot crops and from *C. maculatum* in the boundaries around these three fields were estimated from the number of larval mines on the roots prior to emergence of the first generation in 1973 and 1974. One fly per mine was assumed although this may over-estimate the total, but not the relative population from these two hosts (Wright and Ashby, 1946; Jones and Coaker, 1980).

Emergence from *C. maculatum* contributed about 10% of the estimated total emergence of *P. rosae* in 1973. In that year carrots were grown only in Field 4 and high populations of *C. maculatum* occurred at the boundaries. In contrast, in 1974 carrots were grown in Fields 2, 3 and 4, and at the boundaries there were lower densities of *C. maculatum*, which contributed only about 4% of the total population of flies in the sample area.

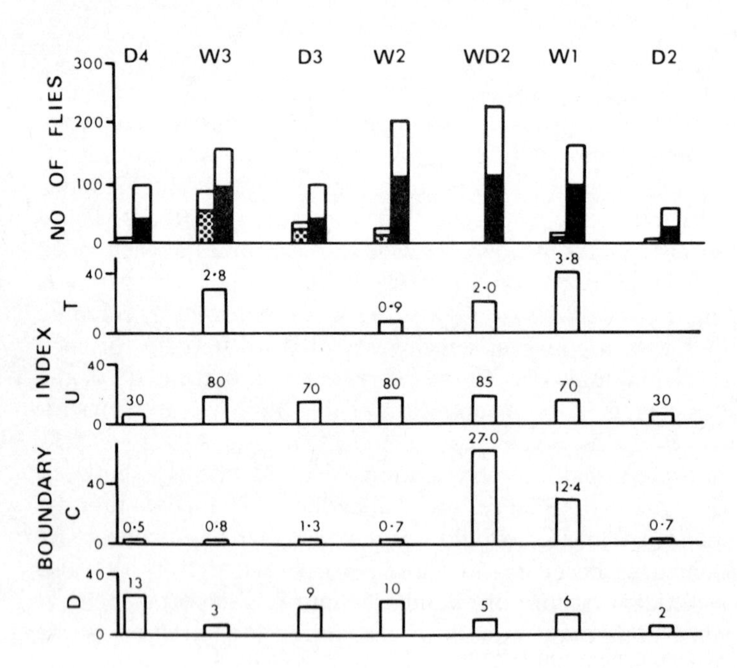

Figure 4 Number of second generation flies trapped and boundary indices in 1973 (symbols as for Fig. 2).

Discussion

The results support those in the 1940s which showed that adult *P. rosae* accumulated at field boundaries adjacent to their emergence sites. Flies were not consistently more abundant at boundaries separating adjacent carrot fields than at boundaries along the edge of one field, suggesting that abundance was not determined by the number of flies emerging in the immediate vicinity of the boundary. Aggregation, therefore, appears to be an active process. Moreover, the aggregation of flies at boundaries was unrelated to the prevailing wind direction and therefore does not support the view of Baker *et al.* (1942) that flies accumulate passively downwind of the source. Flies appear to discriminate between boundaries. Such discrimination may

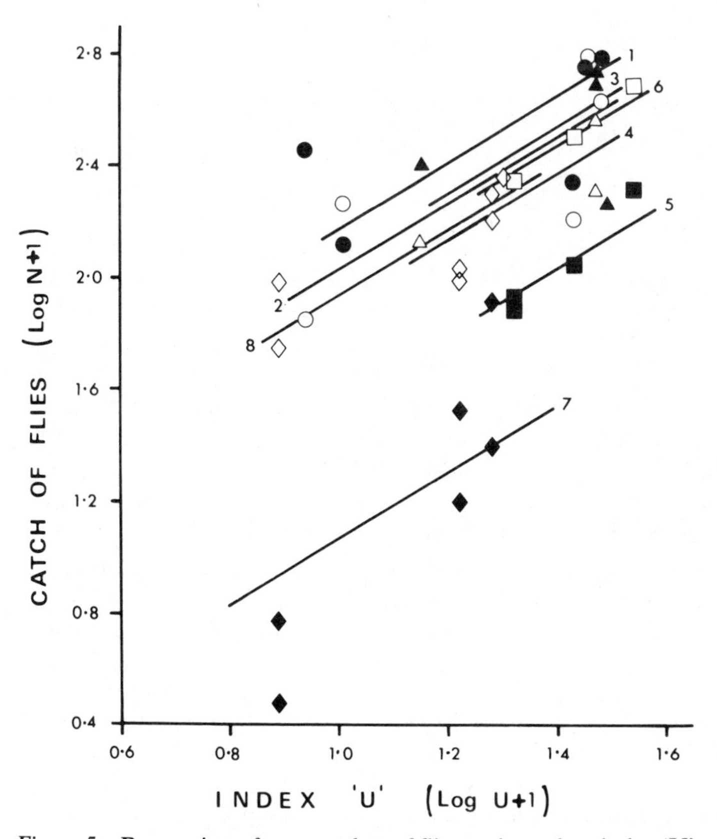

Figure 5 Regression of trap catches of flies on boundary index 'U'.
Closed symbols = suction trap catches; open symbols = water trap catches.
1 (●) and 2 (○) first generation 1972
3 (▲) and 4 (△) second generation 1972
5 (■) and 6 (□) first generation 1973
7 (◆) and 8 (◇) second generation 1973

Table 2 Relationship between trap catches of flies and the various boundary indices

Boundary index	*Common slope*	$\frac{\textit{Regression SS}}{\textit{Total SS}} \times 100$	*P*
U	1.18	54	***
U + D + T + C	1.03	51	***
U + C + T	0.66	50	***
U + T	0.55	47	***
U + D + T	1.02	47	***
T + C	0.43	43	***
C + D + T	0.81	42	***
U + C	0.82	41	***
T + D	0.69	33	***
T	0.20	32	**
U + D + C	0.90	23	**
U + D	0.98	14	*
C	0.29	13	NS
D	−0.29	5	NS
C + D	0.23	3	NS

Indices:
U = Frequency of *U. dioica*
T = Mean number of trees and bushes per 100 m
C = Density of *C. maculatum* m^{-2}
D = Number of plant species in quadrats
*** $P<0.001$ ** $P<0.01$ * $P<0.05$

occur initially by flies responding hypsotactically to trees and bushes in the boundary (Städler, 1972; Brunel, 1977), but there was no evidence of flies aggregating at boundaries with the tallest windbreaks as suggested by Städler (1972). If hypsotactic movement occurs preferentially to tall windbreaks then it must be followed by the redistribution of flies to other boundaries.

It seems from our results that the importance of trees and bushes in exclusively providing shelter for flies (Baker *et al.*, 1942; Brunel, 1977) has been over-emphasized. In the fenland habitat, the herb layer provides important shelter for flies and its species composition influences their abundance.

The presence of *U. dioica* in the herb layer was the most important of the components studied so that, although flies were usually abundant at windbreaks, this was largely due to the composition of the underlying herb layer. As well as providing physical shelter for the flies, *U. dioica* could also provide a food source, especially in the second generation. Many Diptera commonly feed on nectar and pollen, and in *Delia brassicae* this is essential for completion of egg development (Finch and Coaker, 1969). Little is known about the food requirements, feeding behaviour or longevity of *P. rosae* in the field. Floral diversity (D) reflects the abundance of flowering plants at boundaries (Table 1) and is perhaps an indicator of food availability; it appeared to have no influence on fly distribution.

The index analysis suggested that *C. maculatum* was also relatively unimportant although catches appeared to be correlated with its abundance at boundaries remote from a carrot field emergence site (Fig. 3(b)). Although the majority of flies emerged from carrots, up to 10% of first generation flies could derive from *C. maculatum* and this could be important if crop isolation were attempted as part of a management strategy (Wheatley and Percival, 1974).

An alternative strategy may be to simplify the field boundaries since the absence of shelter sites for flies can reduce attack by *P. rosae* (Van't Sant and Brader, 1972; Dabrowski and Legutowska, 1976). Windbreaks could not be removed from fenland farms since they reduce wind erosion of soil and the consequent need to resow crops. Moreover, our results suggest that their removal would not significantly decrease the concentration of flies around carrot fields. The removal of *U. dioica* could have an important effect on adult distribution, but flies might then seek alternative shelter sites within the herb layer. Removal of the total herb layer would drastically affect the preferred microclimate sought by the flies when they aggregate in the herb layer but its effect on the natural enemies of *P. rosae* (Burns, 1979) and also on game birds (e.g., pheasants, *Phasianus colchicus*) may be equally drastic, and further study of these aspects of habitat management is needed.

This work was supported by the ARC and our sincere thanks are due also to Mr R. Gair of ADAS, Mr T. Wooten of Weasenham Farms Ltd, Mr R. Gifford and Mr I. White.

References

Baker, F. T., Ketteringham, I. E., Bray, S. P. V. & White, J. H. (1942). Observations on the biology of the carrot fly (*Psila rosae* F.) assembling and oviposition. *Annals of Applied Biology* **29,** 115–125.

Barnes, H. F. (1942). Studies of fluctuations in insect populations. IX. The carrot fly (*Psila rosae*) in 1936–41. *Journal of Animal Ecology* **2,** 69–81.

Brunel, E. (1977). Etude de l'attraction periodique de femelles de *Psila rosae* Fabr. par la plante-hôte et influence de la végétation environnante. *Comportement des insectes et milieu trophique. Colloques Internationaux du CNRS 13–17 Sept. 1976*, No. 265, pp. 373–389.

Burns, A. J. (1979). The natural mortality of the carrot fly (*Psila rosae* (F)). *PhD Thesis*, University of Cambridge.

Coppock, L. J. (1974). Notes on the biology of carrot fly in eastern England. *Plant Pathology* **23,** 93–100.

Dabrowski, Z. T. & Legutowska, H. (1976). Podstawy integrowanego zwalezania polysnicy marchwianki (*Psila rosae* F.). In *Materialy z XVI Sesji Naukowej Instytutu*

Ochrony Roslin, 12–14 Luty 1976 r. Poznan, pp. 201–219.Poland: Instytut Ochrony Roslin.

Finch, S. & Coaker, T. H. (1969). Comparison of the nutritive values of carbohydrates and related compounds to *Erioischia brassicae. Entomologia experimentalis et applicata* **12,** 441–453.

Gostick, K. G. & Baker, P. M. (1968). Dieldrin resistant carrot fly in England. *Plant Pathology* **17,** 182–183.

Hawkes, C. (1969). The behaviour and ecology of the adult cabbage root fly *Erioischia brassicae* (Bouché). *PhD Thesis*, University of Birmingham.

Jones, O. T. & Coaker, T. H. (1980). Dispersive movements of carrot fly (*Psila rosae*) larvae and factors affecting it. *Annals of Applied Biology* **94,** 143–152.

Kozak, A. (1972). A simple method to test parallelism and coincidence for curvilinear, multilinear and multiple curvilinear regressions. *IUFRO 3rd Conference of the Advisory Group of Forest Statisticians, Jouy-en-Joses France, Sept 7–11, 1970.* Institut National de la Recherches Agronomique, 1972.

Maskell, F. E. & Gair, R. (1973). Experiments on the chemical control of carrot fly in carrots in East Anglia in 1968–72. In *Proceedings of the 7th British Insecticide and Fungicide Conference*, pp. 513–524.

Petherbridge, F. R. & Wright, D. W. (1943). Further investigations on the biology and control of the carrot fly (*Psila rosae* F.). *Annals of Applied Biology* **30,** 348–358.

Petherbridge, F. R., Wright, D. W. & Ashby, D. G. (1945). The biology and control of carrot fly. *Annals of Applied Biology* **32,** 262–264.

Petherbridge, F. R., Wright, D. W. & Davies, P. C. (1942). Investigations on the biology and control of the carrot fly (*Psila rosae* F.). *Annals of Applied Biology* **29,** 380–392.

Sant, L. Van't & Brader, L. (1972). Oekologische waarnemingen als hulpmiddel bij de bescherming van wortelen tegen de aantasting door de wortelvlieg *Psila rosae. Entomologische Berichten* **32,** 187–188.

Southwood, T. R. E. (1966). *Ecological methods*. London: Methuen.

Städler, E. (1972). Über die Orientierung und das Wirtswahlverhalten der Möhrenfliege, *Psila rosae* F. (Diptera Psilidae). II. Imagenes. *Zeitschrift für angewandte Entomologie* **70,** 29–61.

Wainhouse, D. (1975). The ecology and behaviour of the carrot fly *Psila rosae* (F). *PhD Thesis*, University of Cambridge.

Wainhouse, D. (1977). Rhythmic activity of adult carrot fly, *Psila rosae. Physiological Entomology* **2,** 323–329.

Watkins, T. C. & Miner, F. D. (1943). Flight habits of the carrot rust flies suggest possible method of control. *Journal of Economic Entomology* **36,** 586–588.

Wheatley, G. A. & Percival, A. L. (1974). Control of carrot fly: possible organophosphorus resistance in carrot fly populations. *Report of the National Vegetable Research Station for 1973* (1974), pp. 77–78.

Whitcomb, W. D. (1929). Observations on the carrot rust fly (*Psila rosae* Fab.) in Massachusetts. *Journal of Economic Entomology* **22,** 672–675.

Wright, D. W. & Ashby, D. G. (1946). Bionomics of the carrot fly (*Psila rosae* F.). I. The infestation and sampling of carrot crops. *Annals of Applied Biology* **33,** 69–77.

Wright, D. W. & Coaker, T. H. (1968). Development of dieldrin resistance in carrot fly in England. *Plant Pathology* **17,** 178–181.

Windbreaks as a source of orchard pests and predators

M G Solomon
East Malling Research Station, Maidstone, Kent ME19 6BJ

Introduction

Some orchards have windbreaks that are simply old hedgerows allowed to grow tall. Some of these hedgerows are relics of ancient woodland, though most were planted when agricultural land was enclosed at various times during the past 400 years, particularly in the eighteenth century (Pennington, 1969; Trask, *1981*). This kind of hedgerow-windbreak often contains many plant species (Pollard *et al.*, 1974). Single-species windbreaks were also planted around orchards, myrobalan plum (*Prunus cerasifera*) being frequently used. These old windbreaks are rarely pruned, and often have many herbaceous species at the base. During the last 10–15 years, however, rather different kinds of windbreak have been extensively planted in the fruit growing region of south-east England. These too are usually single-species windbreaks, but they are carefully managed, with annual pruning and herbicide applications to prevent ground herbs from competing with the windbreak trees for water. Poplar (*Populus* spp.), willow (mostly *Salix alba* and *S. viminalis* varieties), and cypress (mostly X *Cupressocyparis leylandii*) have commonly been used, but the majority of these modern windbreaks consist of alder (*Alnus* spp.). The native English alder (*Alnus glutinosa*) has been used extensively during the past 15 years, but it grows erratically in dry soils. *Alnus cordata* and especially *Alnus incana* perform better in dry soils and are now preferred (Baxter, 1979).

Such windbreaks have long been used in the Netherlands but their value has only more recently become widely appreciated by English fruit growers. The main reason for planting and maintaining windbreaks is to filter the wind and reduce its speed, thus decreasing wind damage to the fruit trees and crop, and reducing water loss. Pesticide application is also easier in sheltered conditions.

Windbreaks can influence the orchard and hop garden fauna by causing flying insects to be deposited to leeward (e.g. Lewis and Dibley, 1970), and even more importantly by acting as a source or reservoir of pests and beneficial insects. Aspects of this source effect are discussed here.

Most orchards in England are sprayed routinely with broad-spectrum pesticides. Consequently local sources of pests or predators are not very

Figure 1 (Top) old hedgerow-windbreak beside apple orchard. (Bottom) old hedgerow-windbreak beside hop garden, photographed in September shortly after the hop bines had been cut down and removed for harvest.

significant, since sprays are applied whether the pest is present or not. However, for various reasons, English growers are likely to rely increasingly on systems of integrated pest management. Under such systems the effects of natural controlling agents are maximized and the grower applies pesticides only when monitoring of the crop indicates that a particular pest is likely to

Figure 2 (Top) six-year-old alder windbreak beside apple orchard. (Bottom) poplar windbreak beside apple orchard, end view.

cause economic damage. The pesticide then used is not necessarily the most effective in killing the pest, but one that is not too damaging to predators and parasites. For fruit in Britain, this system is most advanced for apple orchards (Carden, 1977; Easterbrook *et al.*, 1979), but is also being developed for pear (Solomon *et al.*, 1979) and hop (Campbell, 1978). Under these new conditions, any local source of potentially colonizing pests or predators is likely to be significant.

Windbreaks as a source of pests

The importance of the various windbreak species as local sources of pests is influenced by several factors, such as the host range and mobility of the pest species, and farm management practices applied to the windbreak.

The host-range of the pest

Some pests are restricted to a single species of fruit-tree host and to non-cultivated members of the same genus. An example is the apple sawfly (*Hoplocampa testudinea*), infestations of which usually spread slowly through orchards from year to year, so that sources in windbreaks can be significant. However, only old hedgerow-windbreaks contain crab apple (*Malus sylvestris*) and the main source of the sawfly is usually unsprayed apple cultivars, often in gardens close to orchards. The cherry blackfly (*Myzus cerasi*) is another pest associated with a single fruit species and its wild relatives. Wild cherries occur in hedgerow-windbreaks, but they are not very important as local sources of blackfly. This aphid moves from cherry to herbaceous plants in mid-summer, returning to cherry in autumn. Wild cherry, therefore, can only be an indirect source of the pest.

A windbreak is likely to be particularly significant for host-alternating pests if it contains the alternate host plant, so that the pests can move easily and directly from windbreak to nearby fruit tree. The apple aphids *Rhopalosiphum insertum* and *Dysaphis plantaginea* move respectively onto grasses and plantains (*Plantago* spp.) during the summer (Fig. 3(a)). Grasses are so abundant that those beneath windbreaks are not specially significant. However, plantains along windbreaks, particularly old hedgerows, may be sufficiently numerous to influence the success of the summer aphid generations.

Some fruit pests occur on diverse host plants and in these cases a windbreak is unlikely to be a major source of reasonably mobile pests. The tortricid moth *Archips podana*, a pest of apple, also feeds on many other deciduous trees and shrubs, so that it moves into orchards from hedgerow-windbreaks and also from woodland and other trees. Windbreaks can also play a small role in the winter dispersal of the pear psylla (*Psylla pyricola*). There is evidence that part of the overwintering generation of this psyllid disperses a short distance

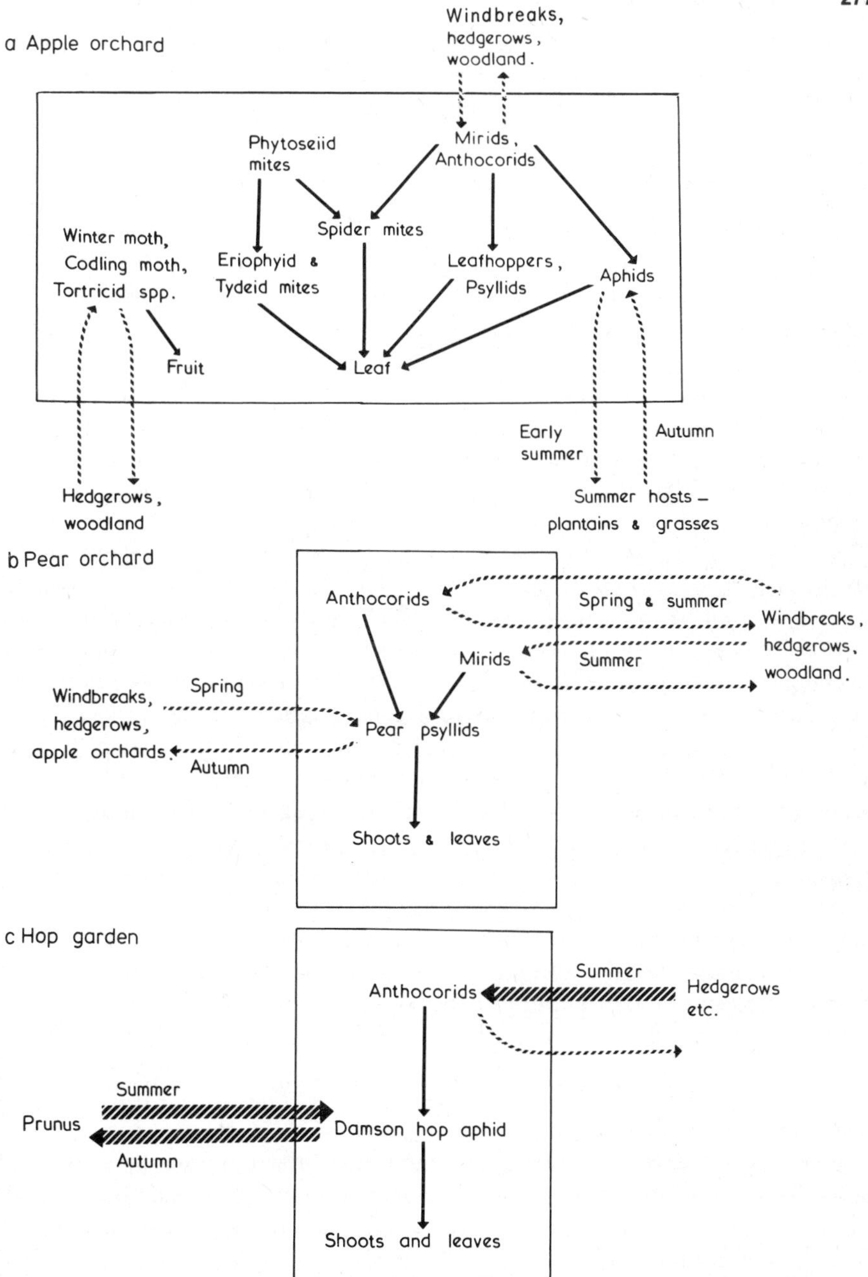

Figure 3 Movements (shown by broken lines) of pests and predators between windbreaks, etc. and (a) apple orchards, (b) pear orchards and (c) hop gardens. Only the more important regular movements are shown. Feeding relationships (solid lines) between phytophagous and predacious arthropods are those that exist in plantings under integrated pest management.

from pear orchards in autumn, returning the following February. The winter may be spent on surrounding windbreaks but equally on almost any other trees, including adjacent apple orchards (Solomon *et al.*, 1979).

The mobility of the pest

The significance of windbreaks is least for pest species (including several aphids) that move large distances. The hop-damson aphid (*Phorodon humuli*) overwinters on various *Prunus* spp., and moves to hop in mid-summer, returning to *Prunus* in autumn (Fig. 3(c)). The distances covered by the aphid are large; Campbell (1977) estimated that most of those arriving in hop gardens had come from sources within an hour's flight, i.e. up to 16–32 km away, and Taylor *et al.* (1979) estimated the median distance travelled by aphids returning to *Prunus* as 15–20 km. Thus although a *Prunus* windbreak might contribute to the regional level of *P. humuli*, it is unlikely to be a specific threat to local hop plantings. Certainly growers cannot expect to avoid *P. humuli* by removing *Prunus* windbreaks on the farm or even over the whole locality.

A windbreak is more likely to be a local source of pests that move only short distances. Phytophagous mites spread rather slowly between trees, although some may be wind-blown. The fruit tree red spider mite (*Panonychus ulmi*) is sometimes abundant on *Prunus* windbreaks, which are then a source of infestation for adjacent orchards. Only slightly more mobile is the winter moth (*Operophtera brumata*), in which the larvae can disperse by drifting on the wind (Holliday, 1977), but the adult females are wingless. This pest of apple and other fruit also feeds on wild *Prunus* spp., beech, oak, and others, so a hedgerow-windbreak containing these species can be a local source.

Windbreaks may also form the base from which birds make short feeding visits into orchards. The redpoll (*Acanthis flaminea*) often feeds on alder cones and occasionally damages nearby plum flowers and fruitlets (Flegg, *1981*).

Management practices

The management of windbreaks can influence their potential as sources of pests. The application of pesticides, by design or as drift from adjacent fruit trees, can induce pest outbreaks on the windbreak by killing natural enemies. *Panonychus ulmi* sometimes becomes abundant on *Prunus* or alder windbreaks in these circumstances (Olroyd, 1971), although the mite is rare on unsprayed alder.

Applying herbicides along the base of a windbreak can also influence its potential as a source of pests. Removing plantains and docks (*Rumex* spp.) eliminates food plants of the aphid *Dysaphis plantaginea* and dock sawfly

(*Ametostegia glabrata*). This is likely to be more important with sawfly, which mainly attacks apples near the docks from which it originates.

Windbreaks as a source of predators

The same general principles outlined for pests apply in lesser degree to predators. In particular, predators tend to feed on several different species of prey, and to distribute themselves on vegetation in response to the availability of prey rather than the plant species.

The fruit and hop pests that are most amenable to regulation by predators in an integrated pest management programme are the spider mite *P. ulmi* and possibly aphids on apple, the pear psylla, and *Phorodon humuli* on hop. Many predators attack *P. ulmi* (Collyer, 1953) but the most important are phytoseiid mites, mirids and anthocorids (Fig. 3(a)). Anthocorids are also important predators of *Psylla pyricola* in USA, Canada and England (Brunner and Burts, 1975; Solomon *et al.*, 1979) and of *P. humuli* (Zohren, 1970; Campbell, 1978); Fig. 3(a) and (b). Phytoseiids are the least mobile of these predators and anthocorids the most.

Of the phytoseiids found on unsprayed apple trees, *Typhlodromus pyri* is the species that most readily colonizes orchards sprayed with selective pesticides (Solomon, 1975a; Easterbrook *et al.*, 1979). An orchard under integrated pest management is likely to be a useful source of wind-blown colonizers for nearby orchards. However, no single windbreak species seems to be an important source of this predator, although it occurs on several species of plant that occur in hedgerow-windbreaks (Chant, 1959). Some other phytoseiid species may be very numerous on other species of tree of potential value as windbreaks. *Amblyseius finlandicus* is often abundant on horse chestnut (*Aesculus hippocastanum*) and *Typhlodromus aberrans* on hornbeam (*Carpinus betulus*). *Typhlodromus aberrans* is often found on unsprayed apple trees, but it has not been found on apple trees sprayed with selective pesticides. It might play a role in such orchards if hornbeam were to be used as a windbreak. Attempts to introduce *A. finlandicus* to integrated pest management trees at East Malling have not been successful, although this species appears to be important in the early years of establishing integrated pest management in apple orchards in the Netherlands (Gruys, 1975).

The species of mirid that have most regularly colonized selectively-sprayed orchards in Kent are *Blepharidopterus angulatus, Phytocoris tiliae, Pilophorus perplexus, Heterotoma planicornis, Atractotomus mali, Malacocoris chlorizans* and *Psallus ambiguus.* All these species are present on a wide range of hedgerow trees such as oak (*Quercus robur*) and hawthorn (*Crataegus monogyna*). Of these mirid species, *B. angulatus* is usually the most abundant in apple orchards sprayed with selective pesticides, and the most important as a regulator of *P. ulmi* (Collyer, 1953; Solomon, 1975b). This predator is particularly abundant on alder, where it feeds principally on the aphid *Pterocallis alni*, but also on leafhoppers. The widespread planting of alder windbreaks in south-east England during the past 15 years has established a considerable reservoir of *B. angulatus* close to many orchards. *B. angulatus* becomes adult around August, when numbers of the aphid *P. alni* are usually declining. Most of the adult *B. angulatus* then leave alder, and may colonize nearby apple orchards. In contrast, *B. angulatus* already established on apple remain and lay eggs during August and September, provided sufficient spider mite is present as a food supply.

In an experiment at East Malling to investigate the relationship between the level of *P. ulmi* on apple trees and the rate of colonization by predators, *B. angulatus* colonized plots during the first year of a selective spray programme, in response to mean *P. ulmi* densities as low as two per leaf (Solomon, 1975b). On these plots the predator settled at about four per tree: on other plots in the same experiment, where *P. ulmi* averaged eight per leaf, *B. angulatus* colonized at the rate of ten per tree. This higher rate of colonization was similar to that recorded by Muir (1966) in response to *P. ulmi* numbers exceeding 100 per leaf. That trial was carried out before alder windbreaks were planted in the area, but the 'colonization' plots were immediately adjacent to plots already containing populations of *B. angulatus*.

Some evidence that colonization by this predator is slower in the absence of nearby alder or other sources is provided by a trial carried out in an orchard with no alder windbreak within 1000 m, and no woodland or mixed hedgerow within 800 m (Solomon, 1975a). During the first year after the cessation of broad-spectrum pesticide application *P. ulmi* numbers exceeded ten per leaf, but the numbers of colonizing *B. angulatus* averaged only about one per tree.

The predacious anthocorids *Anthocoris nemorum* and *A. nemoralis* are also usually present on alder windbreaks, but no similar relationship is apparent between their rate of colonization of apple plots and the numbers on nearby windbreaks. This may be because these anthocorids are present on such a wide range of herbaceous and woody plants in addition to alder.

Early in the season, however, anthocorids concentrate on a particular host plant. During the time that willow (mostly *Salix caprea* and hybrids) is flowering, *A. nemorum* and *A. nemoralis* occur on it in very large numbers (Sands, 1957; Hill, 1957; Anderson, 1962). This flowering period is very early, usually late March or early April, and the anthocorids leave the willow after only a week or so. This is before overwintering eggs of *P. ulmi* on apple have hatched. Apple aphids, and the apple psylla *Psylla mali*, however, are just beginning to hatch at this time, and therefore anthocorids from willow may colonize apple orchards in response to high numbers of these species. The pear psylla hatches earlier than the apple psylla and aphids, so willow in windbreaks is more likely to be significant as a source of anthocorids for pear orchards.

Various selections of *Salix alba* are used as windbreaks, and anthocorids concentrate during the flowering period on these also, though not at such high densities as on more isolated willow trees.

In the case of hop, the other crop in which anthocorids are important predators, the early season migration from willow is not significant since the aerial part of the hop plant has not yet begun to grow.

The other plant species used extensively as windbreaks — poplar, and various conifers, chiefly Leyland cypress — are unlikely to be important sources of predators in orchards. Sampling on these species has yielded few predators of fruit tree pests.

Conclusion

An orchard may be regarded as a simple monoculture, but grass, windbreaks, hedges, and nearby woodland add to the diversity of the ecosystem. A frequently discussed aspect of the relationship between crops and pests is the influence of ecosystem diversity on the overall stability of the system (e.g. Way, 1977). It is generally concluded that there is no simple relationship between diversity and stability, and that a few key relationships may account for stability. Certainly in orchards sprayed with selective pesticides a few key predators restrict some pests to a low level (Solomon, 1975a; Solomon *et al.*, 1979; Campbell, 1978). It seems impossible to generalize about the effects of the increased diversity represented by windbreaks; these effects vary with species of windbreak.

With old hedgerow-windbreaks containing many tree species, this increased diversity can provide a source of several important pests and predators. Adding these predators to the system would be likely to facilitate biological control, although nearby sources of pests would add to the management problems. In apple orchards in particular, a regular influx of such pests as the winter moth *Operophtera brumata* and various tortricid species would necessitate the regular inclusion of pre-blossom and mid-summer insecticides in an integrated pest management programme.

The situation is different with recently planted single-species windbreaks. Alder, poplar, and willow, have very few phytophagous arthropods in common with fruit trees, and they pose no threat to pest management in orchards. Willow is an early-season source of anthocorids, and alder a source of predacious mirids, particularly *B. angulatus*.

References

Anderson, N. H. (1962). Bionomics of six species of *Anthocoris* (Heteroptera: Anthocoridae) in England. *Transactions of the Royal Entomological Society of London* **114,** 67–95.

Baxter, S. M. (Ed.) (1979). *Windbreaks*. Pinner: Ministry of Agriculture, Fisheries and Food.

Brunner, J. F. & Burts, E. C. (1975). Searching behaviour and growth rates of *Anthocoris nemoralis* (Hemiptera: Anthocoridae), a predator of the pear psylla, *Psylla pyricola*. *Annals of the Entomological Society of America* **68,** 311–315.

Campbell, C. A. M. (1977). Distribution of damson-hop aphid (*Phorodon humuli*) migrants on hops in relation to hop variety and wind shelter. *Annals of Applied Biology* **87,** 315–325.

Campbell, C. A. M. (1978). Regulation of the damson-hop aphid (*Phorodon humuli* (Schrank)) on hops (*Humulus lupulus* L.) by predators. *Journal of Horticultural Science* **53,** 235–242.

Carden, P. W. (1977). Supervised control of apple pests in the United Kingdom. In *Proceedings of the 1977 British Crop Protection Conference — Pests and Diseases*, pp. 359–367.

Chant, D. A. (1959). Phytoseiid mites (Acarina: Phytoseiidae). Part I. Bionomics of seven species in Southeastern England. Part II. A taxonomic review of the family Phytoseiidae, with descriptions of thirty-eight new species. *Canadian Entomologist*, Supplement 12.

Collyer, E. (1953). Biology of some predatory insects and mites associated with fruit tree red spider mite (*Metatetranychus ulmi* (Koch))in south-eastern England. IV. The predator–mite relationship. *Journal of Horticultural Science* **28,** 246–259.

Easterbrook, M. A., Souter, E. F., Solomon, M. G. & Cranham, J. E. (1979). Trials on integrated pest management in English apple orchards. In *Proceedings of the 1979 British Crop Protection Conference — Pests and Diseases*, pp. 61–67.

Flegg, J. J. M. (*1981*). Crop damage by birds. In *Pests, Pathogens and Vegetation*, pp. 365–373. J. M. Thresh. London: Pitman.

Gruys, P. (1975). Integrated control in orchards in the Netherlands. In *Proceedings of the 5th Symposium on Integrated Control in Orchards,* pp. 59–68. IOBC/WPRS.

Hill, A. R. (1957). The biology of *Anthocoris nemorum* (L.) in Scotland (Hemiptera: Anthocoridae). *Transactions of the Royal Entomological Society of London* **109,** 379–394.

Holliday, N. J. (1977). Population ecology of winter moth (*Operophtera brumata*) on apple in relation to larval dispersal and time of bud burst. *Journal of Applied Ecology* **14,** 803–813.

Lewis, T. & Dibley, G. C. (1970). Air movement near windbreaks and a hypothesis of the mechanism of the accumulation of airborne insects. *Annals of Applied Biology* **66,** 477–484.

Muir, R. C. (1966). The effects of sprays on the fauna of apple trees. IV. The recolonisation of orchard plots by the predatory mirid *Blepharidopterus angulatus* and its effect on populations of *Panonychus ulmi*. *Journal of Applied Ecology* **3,** 269–276.

Olroyd, K. (1971). Red spider eggs found in alder windbreaks. *The Grower* **76,** 871.

Pennington, W. (1969). *The History of British Vegetation*. London: English Universities Press.

Pollard, E., Hooper, M. D. & Moore, N. W. (1974). *Hedges*. Gasgow: Collins.

Sands, W. A. (1957). The immature stages of some British Anthocoridae (Hemiptera). *Transactions of the Royal Entomological Society of London* **109,** 295–310.

Solomon, M. G. (1975a). Establishment of predators in an apple orchard. *Report of the East Malling Research Station for 1974*, p. 130.

Solomon, M. G. (1975b). The colonisation of an apple orchard by predators of the fruit tree red spider mite. *Annals of Applied Biology* **80,** 119–122.

Solomon, M. G., Cranham, J. E., Easterbrook, M. A. & Souter, E. F. (1979). Pear sucker *Psylla pyricola* Fors., control by pesticides and predators. *Report of the East Malling Research Station for 1978*, pp. 123–125.

Taylor, L. R., Woiwood, I. P. & Taylor, R. A. J. (1979). The migratory ambit of the hop aphid and its significance in aphid population dynamics. *Journal of Animal Ecology* **48,** 955–972.

Trask, A. B. (*1981*). Changing patterns of land use in England and Wales. In *Pests, Pathogens and Vegetation*, pp. 39–49. J. M. Thresh. London: Pitman.

Way, M. J. (1977). Pest and disease status in mixed stands vs. monocultures; the relevance of ecosystem stability. In *Origin of Pest, Parasite, Disease and Weed Problems*, pp. 127–138. J. M. Cherrett & G. R. Sagar. Oxford: Blackwell.

Zohren, E. (1970). Möglichkeiten einer integrierten Bekämpfung von Hopfenschädlingen. *Zeitschrift für Angewandte Entomologie* **65,** 412–419.

The wild hosts of aphid pests

V F Eastop
Entomology Department, British Museum (Natural History), Cromwell Road, London SW7 5BD

Introduction

Many host plant lists of aphids have been published and Börner (1952), Patch (1938) and Richards (1976) are in constant use by students of aphid biology. These three standard works had different aims: Börner critically assessed the European aphid fauna; Patch collated all published host plant records; and Richards listed only the hosts from which the aphids were originally described. Consequently the most widely consulted (Patch, 1938) is not only rather out of date, but contains many records based on wrongly identified aphids, wrongly identified plants, vagrants and other lapses and calamities to which compilations are prone. The confusion of host-specific pests with similar aphids feeding on wild plants, or failure to recognize a pest feeding on wild plants, can render surveys almost meaningless and complicate possible control measures. The main purpose of this paper is to reassess the literature and list the plants usually colonized by the main aphid pests.

Aphid life cycles

Aphids are renowned for their complicated life cycles. About 10% of species alternate between a primary host, on which the sexuales give rise to overwintering eggs and their resultant spring generations, and secondary hosts, on which parthenogenetic generations spend the summer and eventually give rise to autumn migrants which return to the primary host. These heteroecious aphids are thus normally found, even on perennial crops, for only part of the year. The other 90% of aphid species may remain on the same types of plants throughout the year.

Major groups of aphids have characteristic biologies, i.e. taxonomy has predictive value for aphid biology. Host alternation occurs only in some subfamilies, and has apparently evolved separately on several occasions, as it may be achieved in different ways in different subfamilies. For instance, alate males of alternating Aphidinae fly from secondary to primary hosts, while in other groups sexuparae fly from the secondary hosts to produce apterous males on the primary host. Most aphids have annual cycles, but many Fordini and Adelgidae have two-year cycles. *Picea* is the primary host of all alternating Adelgidae: *Pineus* have *Pinus* as secondary hosts, while the other genera of adelgids go to other genera of conifers. No Lachninae, Chaitophorinae, Drepanosiphinae or Greenideinae are known to alternate. Most genera of alternating Aphidinae have characteristic primary hosts: those of *Myzus* and *Brachycaudus* are *Prunus*, those of *Ovatus* and *Dysaphis* are *Malus* and *Crataegus*, etc. The alternate hosts of heteroecious aphids are tabulated and cross-referenced elsewhere (Eastop, 1977).

The general biology of nearly all Western Palaearctic aphids is known: we know which species alternate between primary and secondary hosts, and which do not. However, while most of the population of some alternating aphids returns annually to the primary host, almost regardless of conditions, some individuals may remain on the secondary host in the autumn reproducing parthenogenetically. In mild winters a large proportion of both heteroecious and autoecious species may persist this way. The relative importance of the different methods of overwintering is uncertain for a number of pests, including the cereal aphids *Sitobion avenae* (*see* Hand and Williams (*1981*) and *Metopolophium dirhodum*). More parthenogenetic individuals are likely to survive mild than severe winters, and more in maritime than continental climates, but these are just some of the factors causing aphids to occur sporadically in both time and space (Taylor and Taylor, 1977). There are also qualitative uncertainties. Numerous *Myzus persicae* sometimes overwinter parthenogenetically on *Malva* spp. and *Artemisia vulgaris* in the Thames Valley and on *Uritica urens* in the Midlands and Northern Britain, but it is uncertain whether these populations belong to the same gene pool.

A few species can support many different aphids and have been called 'reserve hosts' (Stroyan, 1957). Examples are *Capsella bursa-pastoris*, *Polygonum* spp. and *Rumex* spp., on which unexpected aphids often occur.

Even when the biology of a pest has been determined in the laboratory, it may not go through exactly the same cycle in the field. For instance, the well-known life cycle of the black bean aphid (*Aphis fabae*), alternating between spindle trees *Euonymus europaeus* and broad beans, seldom occurs except in text books and laboratories! Broad beans are harvested and the plants ploughed under long before the return migration begins in southern Britain. Alatae fly from broad beans to *Chenopodium* spp. and other plants, establishing colonies which produce the return migrants to spindle.

The pests of orchards, long leys and permanent pastures may spend the whole year within or around the crop, but short-term arable crops grown in rotation are colonized from outside sources. These may be adjacent or remote crops as with some biennials or annuals having overlapping generations, or the insects may come from wild plants. The alternative hosts may be closely related crops, as with *Brassica*.

Most aphids have an evidently restricted range of host plants, but eleven species have exceptionally wide yet distinctive host ranges. This is apparent from Table 1, which is a summary of a more extensive data matrix (available

Table 1 The host preferences of twelve polyphagous aphid species*

			Dicotyledons (system of Cronquist, 1968)														
	Lower plants	*Gymnosperms*	*Magnolidae*	*Hamamelidae*	*Caryophyllidae*	*Malvaceae*	*Cucurbitaceae*	*Cruciferae*	*Other Dilleniidae*	*Rosaceae*	*Leguminosae*	*Other Rosidae*	*Solanaceae*	*Compositae*	*Other Asteridae*	*Monocotyledons*	*Total*
Aphis citricola	–	1	7	4	3	1	1	–	5	41	1	32	2	31	24	2	155
Aphis craccivora	–	1	–	–	35	7	2	6	8	6	117	31	3	12	9	10	247
Aphis fabae	–	–	13	9	46	3	1	7	9	4	16	49	12	25	39	18	251
Aphis gossypii	–	–	8	11	21	81	72	5	25	24	34	139	52	47	118	59	696
Aulacorthum circumflexum	6	1	4	1	4	1	–	2	7	2	1	7	4	1	4	11	56
Aulacorthum solani	1	–	9	3	10	2	3	1	7	6	6	35	14	14	21	15	147
Macrosiphum euphorbiae	–	–	8	11	15	8	3	5	5	20	6	27	27	25	31	30	221
Myzus ascalonicus	–	–	2	2	9	–	–	4	4	4	1	7	1	5	12	7	58
Myzus ornatus	–	–	2	2	8	2	–	2	13	5	4	12	1	11	18	3	83
Myzus persicae	–	3	8	7	38	17	2	68	13	23	11	40	79	35	65	21	430
Toxoptera aurantii	–	–	9	–	2	1	–	–	57	6	5	70	1	4	25	2	182
Total	7	6	70	50	191	123	84	100	153	141	202	449	196	210	366	178	2526

* As indicated by the number of host plant family/country or state combinations in British Museum (Natural History) collection.

in the British Museum (Natural History)) for the number of countries or states from which each polyphagous aphid occurs on each plant family. This presentation of the data from the collection has an element of frequency of occurrence without information from wild plants being swamped by data from a few economically important crops such as potatoes and cotton. 'Polyphagous' aphids are rarely found on the small-grain cereals and *Aulacorthum circumflexum* is the only polyphagous aphid commonly occurring on ferns (11% of host family/region records of this aphid). It is also found on monocotyledons (20%), as are *A. solani* (10%), *Macrosiphum euphorbiae* (14%) and *Myzus ascalonicus* (12%). Most of the *A. gossypii* (8%) from monocotyledons feed in the flower heads, whereas many of the other aphids live on the leaves. *A. craccivora* (4%), *Myzus ornatus* (4%), *M. persicae* (8%) and *Toxoptera aurantii* are less often found on monocotyledons. *Aphis craccivora* is found on many plants in many families, but has a strong preference (47% of samples) for Leguminosae. *A. citricola* is mostly found on Rosaceae (26%), Rutaceae (12%) and Compositae (20%). *Toxoptera aurantii* which lives mostly on the young growth of shrubs, is commonly found on Rosidae (45%) and more often (32%) on Dilleniidae than other polyphagous aphids. *Myzus persicae* occurs more often on Asteridae (42%) and less often on Rosidae (17%) than the other polyphagous species. Polyphagous aphids are rarely found on Gymnosperms and are not common on Magnoliidae or Hamamelidae, but unexpected plants may be important, especially under extreme conditions, e.g. *Aphis craccivora* on *Euphorbia hirta* during the dry season in northern Nigeria.

Aphid pests

About 250 species of aphids occasionally occur in numbers sufficient to worry growers or the general public, but more than 100 of these are specific to ornamental plants and of little importance except to a few specialist nurseries (and the local sales of garden pesticides!). About 150 species of aphids may be numerous on more widely cultivated plants and they are listed below with their host ranges. About 40 species of aphids are regularly the target of control measures and are dealt with in more detail. Although only about 10% of all aphids are heteroecious, these host-alternating species comprise 42% (63 out of 151) of the species regarded as pests. Moreover, heteroecious aphids have, on average, more secondary hosts than autoecious aphids. The secondary hosts of aphids are rarely closely related taxonomically to their primary hosts. Different generations of a heteroecious aphid may differ in their host preferences and this may restrict the development of extreme specificity.

Acyrthosiphon *gossypii* is recorded as a pest of *Gossypium*, *Phaseolus*,

Sesbania and *Vicia*. The similar aphids described from *Malva* and *Lepidium* may also be *A. gossypii*, but it is not proven experimentally that the Malvaceae- and Leguminosae-feeders are really the same species. *A. kondoi*, a pest of *Medicago* and *Trifolium*, also lives on *Astragalus*, *Lens* and *Melilotus*. *A. malvae* lives on *Malva*, Geraniaceae and Potentilleae and Sanguisorbeae of the Rosaceae. It is not known whether the strawberry-feeding *A. rogersii* is a distinct (sub-) species specific to *Fragaria* or whether any member of the group found feeding on *Fragaria* is assumed to be *rogersii*. *A. pisum*, the pea aphid, occurs on many herbaceous legumes including *Lathyrus*, *Lotus*, *Medicago*, *Melilotus*, *Pisum*, *Trifolium* and *Vicia*. Similar-looking aphids occur on shrubs and have been called *A. spartii* because they could not be transferred to herbs, but no means of distinguishing them morphologically have been given. Some of the populations from herbs are very different from one another and it is possible that some can also feed on shrubs, when they would be called *spartii*. A distinct species *A. ononis* occurs on *Ononis*, on which *A. pisum* is sometimes also found.

***Amphorophora** agathonica* seems to be specific to *Rubus idaeus* s.-sp. *strigosus*. *R. i. strigosus* var. *peramoneus* is the host of another species, *A. tigwatensa*. *A. idaei* is specific to *R. idaeus s.str.*, so brambles (*R. fruticosus*) are not an alternative host for raspberry aphids, but are the host of a distinct species, *A. rubi*, with which both *A. agathonica* and *A. idaei* have often been confused in the past. *A. rubicumberlandi* lives on *R. leucodermis* and *R. occidentalis*, while *A. rubitoxica* is found mostly on *R. ursinus*, but sometimes on *R. palmatus* and *R. vitifolius*.

Anuraphis species alternate between *Pyrus* and the roots of their secondary hosts. *A. subterranea* occurs on both *Heracleum* and parsnip in the summer. The name *Anuraphis* has been used for species subsequently placed in *Brachycaudus* and *Dysaphis*.

***Aphis** citricola* (=*spiraecola*) is common on Pyroideae, *Citrus*, *Viburnum* and *Eupatorium*, but also occurs on many other plants in many families. *A. craccivora* lives mostly on Leguminosae, but in drier places the dry season is often passed on wild plants, particularly *Euphorbia hirta* in Northern Nigeria. *Aphis fabae* overwinters in the egg stage on *Euonymus*, *Viburnum* and *Philadelphus*. Summer colonies of *A. fabae* occur on many plants: *Chenopodium album* is often infested after the broad bean harvest. *Cirsium vulgare* seems to be a common summer host of *A. fabae* but *C. arvense* and *Arctium lappa* are colonized by different black aphids of uncertain host range and identity. One of the reasons for the lack of pesticide resistance in *A. fabae* may be that so much of the population colonizes wild plants. *Aphis glycines* from soybeans will also colonize *Pueraria javanica* which is planted as a shade tree. *Aphis gossypii* may overwinter on *Rhamnus*, Bignoniaceae and probably also other plants, but many populations are permanently parthenogenetic. It is not known whether *Aphis grossulariae*, which alternates between *Grossularia* and *Epilobium*, is really distinct from *A. epilobii*, which overwinters on *Epilobium*, or whether both overwintering strategies occur within the same gene pool. *A. pomi* lives on wild or hedgerow *Crataegus* at least as readily as on cultivated *Malus* which may delay the advent of pesticide resistance (p. 296).

***Aulacorthum** solani* is perhaps the most polyphagous of all aphids (Table 1). It shows little preference for any particular group of angiosperms and small colonies occur throughout the year on diverse hosts. Some of these populations, however, are probably permanently parthenogenetic populations, representing several distinct lineages. *A.* (*Neomyzus*) *circumflexum*, probably of south east Asian origin, also feeds on many plants including ferns.

Brachycaudus species overwinter as eggs on *Prunus* if they alternate. It is uncertain whether some *B. persicae* (=*persicaecola*) alternate to Boraginaceae, and some *B.* (*Appelia*) *schwarzi* alternate to *Tragopogon*. Similarly there are *B.* (*Thuleaphis*) species on *Prunus* (*Amygdalus*) in the Middle East and on Polygonaceae, but the number of species involved is not certain.

***Brevicoryne** brassicae*, the mealy cabbage aphid, feeds on many genera of the tribe Brassiceae and more rarely on other Cruciferae.

***Callaphis** juglandis* seem to occur only on *Juglans regia*.

***Capitophorus** elaeagni* alternates from Elaeagnaceae to *Carduus*, *Cirsium* and *Cryptostemma* or is anholocyclic on the latter in addition to the globe artichoke.

***Cavariella** aegopodii* alternates from *Salix* to many wild Umbelliferae in addition to fennel and carrots, but is rarely found on *Heracleum*, which is the host of two other species of *Cavariella*.

***Ceratovacuna** lanigera* is a pest of sugar cane in South East Asia. Some *Ceratovacuna* alternate from *Styrax* to Gramineae, and *C. lanigera* may alternate in some areas.

***Cerataphis** palmae* (=*variabilis*) feeds on a number of palms in addition to *Areca*, cocoa, oil palm and raffia palm and *C. orchidearum* occurs on several ornamental orchids in addition to *Vanilla*. However, both have been confused with *C. lataniae* which probably has a restricted host range on palms. There are many taxonomic uncertainties in this group.

***Chaetosiphon** fragaefolii* occurs only on cultivated strawberries but *C. thomasi* occurs on strawberry, *Potentilla* and roses in North America. Somewhat intermediate-looking populations have led to *thomasi* appearing in the synonymy of *fragaefolii* and, consequently, *Potentilla* and roses have been listed as hosts of the latter. Even if it is established that *C. fragaefolii* is only a strawberry-specific biotype of a more polyphagous North American species, roses and *Potentilla* need not be considered as possible hosts of European strawberry aphid unless further genotypes are introduced from North America.

Cryptomyzus *galeopsidis* alternates from currants to Labiatae, particularly *Lamium purpureum* and *Galeopsis*; while *C. ribis* goes from redcurrant to *Stachys*. Populations of both species remain on *Ribes*, and the *C. galeopsidis* which do not alternate have been given subspecific status, as they may be genetically isolated by the early production of sexuales.

Diuraphis *noxia* is a pest of barley and wheat in southern Russia, the Mediterranean and Middle Eastern countries, and was recently introduced to South Africa. Similar-looking species have been described from wild grasses in both Europe and America, but it is not certain that they are really all distinct. The role of wild grasses in the biology of *D. noxia* is thus equally uncertain.

Dysaphis species mostly alternate from *Crataegus*, *Malus* or *Pyrus* to underground parts and leaf axils of Umbelliferae and more rarely, other plants. Permanently anholocyclic pests of fennel and celery occur in the warmer parts of the world, where they also feed on other Umbelliferae and *Rumex*. *Dysaphis pyri* lives on *Galium* in the summer but also wanders to many other species on which it does not reproduce.

Elatobium *abietinum* occurs on *Picea abies* (=*excelsa*) and *P. sitchensis*: records from other coniferous genera may represent exceptional occurrences or other species.

Eriosoma species alternate between *Ulmus* and their secondary hosts, and it is likely that even *E. lanigerum* does this somewhere, perhaps in eastern Asia. The taxonomy of the *E.* (*Schizoneura*) species on the roots of *Pyrus* and close allies is unsatisfactory and thus their host range is uncertain.

Eucallipterus and *Tiliaphis* species live only on *Tilia* and are 'pests' only in dry summers when their honeydew soils parked cars and makes pavements slippery after showers of rain.

Hyadaphis species alternate between *Lonicera* and Umbelliferae but permanently parthenogenetic populations probably predominate in the warmer parts of the world. Taxonomic uncertainty in the *H. foeniculi* group obscures the host range of the individual species.

Hyalopterus species alternate from *Prunus* to *Phragmites*. Populations on almonds and apricots may not belong to the same species that occurs on plums.

Hyperomyzus species alternate from *Ribes* to Compositae and more rarely to members of other families. The European species seem specific to a few secondary hosts, but the biology of several American species has not been elucidated.

Hysteroneura *setariae* alternates from *Prunus* to Gramineae in North America and probably in some of the other temperate regions to which it has been introduced. It is recorded from many Gramineae and also Cyperaceae and young palm fronds.

Macrosiphum *euphorbiae* and *M. pallidum* probably alternate from rose over part of their geographical range and both have many summer hosts.

Melanaphis *pyraria* alternates from *Pyrus* to *Arrhenatherum elaitus* and perhaps other grasses. *M. sacchari* is mostly only found on *Saccharum* and *Sorghum*, and rarely on other Gramineae.

Metopolophium *dirhodum* alternates from *Rosa* to many Gramineae, including wheat and many wild grasses (*see* Hand and Williams (*1981*)).

Myzus *ascalonicus* is one of the most polyphagous aphids (Table 1). *M. cerasi* alternates from *Prunus* (*Cerasus*) to *Galium*, Cruciferae, *Euphrasia* and *Veronica*, and it is not certain that only one species is involved. *M. persicae* alternates from *Prunus* to many plants, but is permanently parthenogenetic on its secondary hosts over much of the world. *M. varians* alternates between *Prunus persica* and *Clematis*.

Nasonovia *ribisnigri* alternates from *Ribes* to the flower heads of Compositae including *Crepis*, *Hieracium*, *Lactuca*, *Lampsana* and also to *Euphrasia* and *Veronica*. In recent years it has also been found feeding in the sticky flower heads of *Nicotiana* and *Petunia*.

Nearctaphis *bakeri* overwinters on *Crataegus* and *Malus* and flies to *Trifolium*, other legumes and other plants including *Castilleia*, *Valeriana*, *Capsella* and *Veronica*. It has only recently been introduced to Europe. Other *Nearctaphis* alternate between Pyroideae and various plants, especially Leguminosae.

Neotoxoptera *oliveri* lives on *Allium*, violets and Caryophyllaceae and some other plants. Other *Neotoxoptera* also occur on some of these hosts. Some of the rather similar-looking *Myzus* (*Nectarosiphon*) species such as *ascalonicus* and *certus* also show a preference for *Allium-Viola* and Caryophyllaceae.

Ovatus species alternate between *Crataegus* and more rarely *Malus* to Labiatae. *O. crataegarius* is the mint aphid and probably also occurs on other Labiatae, but the genus is taxonomically difficult.

Pemphigus *bursarius* alternates from galls on poplar to roots of lettuce, sow-thistle and probably other Compositae. Different *Pemphigus* occur on the roots of carrots and other Umbelliferae and beet and other Chenopodiaceae, but the taxonomy of the group is not sound enough to evaluate properly the role of wild plants in the biology of each pest.

Pentalonia *nigronervosa*, the banana aphid, also lives on Zingiberidaceae, Marantaceae and Araceae. Specimens from Araceae with more slender siphunculi have been described as *P. caladii*, but are apparently a nutritionally-induced form of *P. nigronervosa* (Hardy, 1931).

Phorodon *humuli* migrates from *Prunus* to *Humulus* over a long period in spring and early summer. It is an unusual aphid in that when overcrowded it produces progressively smaller apterae, but few alatae until the autumn. Large populations occur on wild hops in June, but are preyed upon by anthocorids, coccinellids and syrphids and few remain to give rise to return migrants in the autumn. As a result, most of each year's sexual generation probably originates from frequently sprayed cultivated hops.

Pineus. At least one member of the *P. pini/laevis* group has a much wider host range than indicated in the literature and many species of *Pinus* have been infested in recent outbreaks in South America and Central and East Africa.

Rhodobium *porosum* occurs on strawberry and roses in North America, but is only known from roses elsewhere for reasons not yet apparent.

Rhopalosiphoninus *latysiphon* is anholocyclic on the etiolated parts of various plants over much of its range, but may overwinter as eggs on *Deutzia* in the Far East. *R.* (*Arthromyzus*) *staphyleae* alternates from *Staphylea* to the etiolated parts of plants in Central Europe and permanently parthenogenetic populations have been recognized as a distinct subspecies, *tulipaellus* in Western Europe.

Rhopalosiphum *maidis*, a pest of barley in temperate regions and of maize and sorghum in the warmer parts of the world, also occurs on *Eleusine*, *Panicum* and *Triticum*. The spring forms of *R. insertum* cause a characteristic crumpling of the leaves of apple, hawthorn, etc., and resemble the summer form of *R. padi*. In summer *R. insertum* goes to the leaf bases of grasses where it is seldom collected. *R. padi* overwinters in the egg stage on *Prunus padus*, but over most of its range is anholocyclic on Gramineae and more rarely on other monocotyledons, and in the winter on dicotyledons such as *Capsella* and *Stellaria*. It occurs on many Gramineae, and most of the North American records of *R. fitchii* from cereals were based on *R. padi*. *R. rufiabdominalis* overwinters as an egg on *Prunus mume* and is recorded from diverse secondary hosts including rice, wheat and potato. It is probably permanently parthenogenetic on many Gramineae and more rarely other plants over most of its range. Many of the Museum samples come from rice or potato, probably both because of the attention given to these important crops and because they are grown in fertile soils supporting larger aphid colonies that persist longer than on less well-nourished wild plants.

***Schizaphis** graminum* feeds on wild grasses as well as cereals, and populations can differ in host ranges. Until recently the biotypes present in North America lived only on grasses and small grain cereals, but recently a sorghum-feeding biotype, the 'sorghum greenbug' was introduced. *S. graminum* has long been known from sorghum in the Middle East. A number of similar species including *S. agrostis, S. holci* and *S. hypersiphonata* have been confused with *S. graminum*, which renders suspect some host records of the latter.

***Sipha** flava*, a well known pest of sugar cane in the New World, also feeds on *Chaetochloa*, *Digitaria*, *Hordeum*, *Oryza*, *Panicum*, *Pennisetum*, *Setaria* and *Sorghum. Sipha maydis* occurs sporadically on wheat in the warmer parts of the world, although it is only known from *Arrhenatherum elatius* in Britain.

***Sitobion** africanum* occurs on maize and various African grasses and shrubs, but the group is poorly known taxonomically and more than one species may be involved. *S. avenae* overwinters on Gramineae, in the egg stage in very cold climates and parthenogenetically in mild climates. Wild and pasture grasses must often be important sources of aphids infesting spring-sown cereals (*see* Watt (*1981*)). *S. fragariae* alternates from *Rubus fruticosus* to Gramineae including *Avena*, *Hordeum* and *Triticum* and is more rarely found on *Fragaria*, *Rosa*, Juncaceae and Liliaceae. A rosaceous primary host is probably necessary in cold climates. *S. graminis* occurs on rice and several wild grasses. *S. luteum* colonizes a number of genera of orchids only. *S. miscanthi* may alternate from a primary host in eastern Asia, but is permanently parthenogenetic over most of its range on Gramineae, including wheat, barley, oats and also *Capsella* and aquatic *Polygonum. S. nigrinectarium*, a pest of pigeon pea, is known for certain only from *Cajanus*, wild plants of which may act as a reservoir of pests for cultivated *C. cajan.*

***Smynthurodes** betae* occurs on the roots of beans, cotton, potato and many other plants, and some populations can be distinguished from one another by chaetotaxy and other structural characters. It is not certain how many taxa are involved or whether populations from wild plants can infest crops.

Tetraneura species alternate from *Ulmus* to the roots of Gramineae, but permanently parthenogenetic populations of the *T. nigriabdominalis* group, perhaps a distinct (sub-) species *bispina*, occur in Africa and other parts of the tropics. Sexuparae are sometimes produced in the southern United States, but it is not known if they produce functional sexuales on elm.

***Thecabius** (Parathecabius) auriculae* is sporadically abundant on *Auricula* roots. Alternation from *Populus* is likely somewhere in the world, but the galls have not yet been identified.

***Therioaphis** trifolii* is found on a number of herbaceous Papilionaceae, particularly *Medicago* and *Trifolium.* Some populations prefer *Trifolium*, others *Medicago*, and some live readily on both. Where a form with restricted host range such as *T. t. maculata* is introduced to western North America and Australia, it is easier to determine the importance of wild hosts than in the Palaearctic, where there are many genotypes which may be differently selected by different seasons, and a number of genotypes with differing host range on the crop.

Tinocallis species are specific to a few species of either *Ulmus*, *Zelkova* or *Lagerstroemia.* Their honeydew soils parked cars and makes pavements slippery in dry areas, much as *Eucallipterus tiliae* does in dry summers in Western Europe.

Toxoptera *aurantii* occurs on citrus, coffee, tea, cocoa and many other tropical and subtropical trees and shrubs. Small colonies can be found on the young growth of wild shrubs but are not as large or numerous as those on the suckers of neglected cultivated citrus or shoots of cultivated coffee and cocoa. *Toxoptera citricidus* is mostly confined to Rutaceae and large colonies are often found on *Calodendron capense*, which may thus be important in citrus-growing areas.

Trama species feeding on the roots of cultivated Compositae can cause wilting in dry summers. There may be many distinct permanently parthenogenetic populations of *Trama* (Blackman, 1980) but not enough is known about their host ranges to predict the importance of different wild Compositae.

Uroleucon *cichorii* lives on *Crepis* and other Liguliflorae in addition to chicory, and infestations probably originate from wild Compositae. *U.* (*Uromelan*) *compositae* occurs on many Compositae in the warmer parts of the world, including *Dahlia*, *Helianthus*, safflower and many wild Compositae, but it is particularly abundant and persistent on *Vernonia* from which other plants are likely to be infested.

Wahlgreniella *nervata* probably alternates between rose and Ericaceae, but may be abundant on roses in the spring in places where it is seldom found on Ericaceae. As British populations were quite common on *Arbutus unedo* for many years but not found on rose until recently, it seems that populations can differ in host plant preferences.

Implications

From the preceding summary it will be apparent that most aphid pests spend part of the year on wild plants or weeds and that with some heteroecious aphids the alternate host is essential. With conspicuous aphids like *Aphis fabae* (p. 289) it is clear that a considerable proportion of the total population is not on crops, but less conspicuous species such as *Myzus persicae* (p. 292) are more difficult to quantify. This aphid is often difficult to find on wild plants in the summer but whether this is due to the problem of examining the greatly increased amount of vegetation, or whether much of the summer population is really on cultivated plants is not known. A further difficulty is encountered when it becomes necessary to compare the relative importance of a sparsely distributed host supporting many aphids and a common host species with few (Hand and Williams, *1981*).

The more of the population living on wild plants, the less will be the overall effect of selection pressures arising in agriculture from the use of aphid-resistant cultivars or chemical and other control measures. The situation will be affected by the degree to which polyphagous aphids develop genotypes with specific host associations. For example, the relative proportion of each year's sexual generation originating from sprayed and unsprayed parthenogenetic progenitors may be more important than the intensity of the treatment as one effective application can eliminate most susceptible members of a population. Thus susceptible genotypes of polyphagous aphids have more chance of surviving from year to year if substantial populations live on plants not exposed to pesticides.

The absence of reports of insecticide-resistance in polyphagous pests which are often sprayed, such as *Aphis citricola*, *A. craccivora*, *A. fabae* and *Aulacorthum solani* (p. 290), and other pests in which populations may spend the whole year on wild plants, such as *Aphis pomi* (p. 290), *Rhopalosiphum insertum, R. maidis*, and *R. padi* (p. 293), also suggest that polyphagous aphids are less likely than other aphids to develop biotypes resistant to pesticides. It may be significant that resistance to insecticides is recorded for *Dysaphis plantaginea* (p. 291) which, unlike the other apple pests named above, does not also overwinter on hawthorn.

Similarly, populations of *Brevicoryne brassicae* (p. 290) exist with varietal preferences, whereas different populations of the more polyphagous *Myzus persicae* do relatively better or worse on all varieties tested (Blackman, 1976). Polyphagous aphids may be less able to break the immunity of pest-resistant cultivars, but it may of course be correspondingly more difficult to produce one.

Wild plants may be important as hosts of aphids related to pests, acting as reservoirs of natural enemies for the pests. They may also be important as sites for parthenogenetic overwintering and such populations may produce viruliferous winged forms earlier in the season and over a longer period than populations arising from overwintered eggs.

Aphids colonizing Rosaceae often alternate to quite different plants, e.g. Gramineae, Labiatae and Umbelliferae, while most Leguminosae-feeding aphids live only on Leguminosae. Thus immigrants to a leguminous crop are likely to have come from another legume and spread from wild species in this way is a crucial feature in the epidemiology of several virus diseases (e.g.

Stevenson and Hagedorn, 1973). Wild Rosaceae are less likely to be important sources of viruses for crops. This may also explain why a higher proportion of legume viruses than Rosaceae-viruses are transmitted by aphids, despite there being nearly three times as many species of aphids living on Rosaceae as on Leguminosae.

The aphids are obviously a specialized and complex group of pests and their study raises special problems. Some of the taxonomic difficulties that are encountered are apparent from the foregoing sections. Other problems are due to the great mobility of aphids that necessitates a regional approach involving studies over a very wide area for a full understanding of their population dynamics (Taylor, 1977).

Further difficulties are due to the great differences in the behaviour of some species between regions, or between seasons within the same region. For example, *Myzus persicae* may either overwinter in the egg stage on *Prunus*, or entirely parthenogenetically on herbs. It can also overwinter mainly parthenogenetically yet produce a few males, which add the ability to overwinter parthenogenetically to the gene pool overwintering in the egg stage (Blackman, 1976). Similarly, *Hyperomyzus lactucae* regularly alternates between *Ribes* spp. and *Sonchus* spp. in Britain and yet persists anholocyclically on *Sonchus* throughout the year in Australia where it is the main vector of lettuce necrotic yellows virus (Martin and Randles, *1981*). Additional complexities of this type are evident from detailed work on two cereal aphids (*Sitobion avenae* and *Metopolophium dirhodum*) reported elsewhere in this volume (Hand and Williams, *1981*; Watt, *1981*).

References

Blackman, R. L. (1976). Biological approaches to the control of aphids. *Philosophical Transactions of the Royal Society, London* B **274,** 473–488.

Blackman, R. L. (1980). Chromosome numbers in the Aphididae and their taxonomic significance. *Systematic Entomology* **5,** 7–25.

Börner, C. (1952). Europae centralis aphides. *Mitteilungen der Thüringischen Botanischen Gessellschaft* **3,** 1–488.

Cronquist, A. (1968). *The Evolution and Classification of Flowering Plants*. London: Nelson.

Eastop, V. F. (1977). Worldwide importance of aphids as virus vectors. In *Aphids as Virus Vectors*, pp. 3–62. K. F. Harris & K. Maramorosch. New York: Academic Press.

Hand, S. C. & Williams, C. T. (*1981*). The overwintering of the rose-grain aphid (*Metopolophium dirhodum*) on wild roses. In *Pests, Pathogens and Vegetation,* pp. 307–314. J. M. Thresh. London: Pitman.

Hardy, G. H. (1931). Aphididae in Australia. *Proceedings of the Royal Society of Queensland* **43,** 31–36.

Martin, D. K. & Randles, J. W. (*1981*). Interrelationships between wild host plant and aphid vector in the epidemiology of lettuce necrotic yellows. In *Pests, Pathogens and Vegetation*, pp. 479–486. J. M. Thresh. London: Pitman.

Patch, E. M. (1934). Food-plant catalogue of the aphids of the World. *Bulletin of the Maine Agricultural Experiment Station* **393,** 1–431.

Richards, W. R. (1976). A host index for species of Aphidoidea described during 1935 to 1969. *Canadian Entomologist* **108,** 499–550.

Stevenson, W. R. & Hagedorn, D. J. (1973). Overwintering of pea seedborne mosaic virus in hairy vetch, *Vicia villosa. Plant Disease Reporter* **57**, 349–352.

Stroyan, H. L. G. (1957). *A Revision of the British Species of* Sappaphis *Matsumura, Part 1.* London: HMSO.

Taylor, L. R. (1977). Migration and the spatial dynamics of an aphid, *Myzus persicae. Journal of Animal Ecology* **46,** 411–423.

Taylor, L. R. and Taylor, R. A. J. (1977). Aggregation, migration and population mechanics. *Nature* (London) **265,** 415–442.

Watt, A. D. (*1981*). Wild grasses and the grain aphid (*Sitobion avenae*), In *Pests, Pathogens and Vegetation*, pp. 299–305. J. M. Thresh. London: Pitman.

Wild grasses and the grain aphid (*Sitobion avenae*)

A D Watt*
School of Biological Sciences, University of East Anglia, Norwich, NR4 7TJ

Introduction

Sitobion avenae, the English grain aphid, is a pest of cereals, on which it may reach high densities and cause damage in 'outbreak years' such as 1976 and 1977 (Carter *et al.*, 1980; Vickerman and Wratten, 1979). Wild plants are important to cereal aphids as overwintering hosts both to holocyclic populations, e.g. *Metopolophium dirhodum* on rose (Hand and Williams, *1981*) and *S. avenae* on grasses, and to anholocyclic populations, e.g. *S. avenae*, *Rhopalosiphum padi* and *M. dirhodum* on grasses (Dean, 1974a). Such relationships are discussed further by Eastop (*1981*). Grasses also act as hosts for cereal aphids during the summer months, and as such provide an opportunity to study them in an environment other than on their crop hosts. This paper presents data on the abundance of *S. avenae* on grass species and on wheat during the summers of 1976, 1977 and 1978.

The abundance of *S. avenae* on wild grasses

In 1976 a few tillers of four grass species were sampled for *S. avenae*. These grasses were interspersed and were situated between an overgrown hedge and a wheat field at Easton, Norfolk. They were less than 1 m from the wheat crop. In 1977 and 1978, nine and three grass species, respectively, were sampled systematically in two slightly different areas in and around a birch wood in the grounds of the University of East Anglia, Norwich. The grass species in this locality were less interspersed and tended to occur in discrete patches. The nearest wheat crop was over 2 km distant. Up to 100 tillers of each grass were sampled twice weekly. During these three years local wheat fields (cv. Maris Huntsman) were also sampled for cereal aphids at Easton (Carter *et al.*, 1980).

There was considerable variation in aphid density with time (*see* next section) and between different hosts. The peak density of *S. avenae* found on wild grasses ranged from 0.2 tiller^{-1} on *Phleum pratense* to 134.2 tiller^{-1} on *Avena fatua* (Table 1). Aphid densities on grasses were greatest overall in

* Present address: Department of Biology, The University, Southampton SO9 5NH.

1976, the only year in which grasses were sampled alongside a wheat crop. This greater density was probably due to movement of apterous aphids from wheat to the grasses. This suggests that aphid numbers on grasses are usually maintained well below their potential density. In 1977 there were many fewer aphids per tiller and percentage infestation was less on all grasses than on the distant wheat crop. In 1978, although the aphid density on wheat was much less than in the previous two years, the aphid density on grass did not decrease to the same extent; more aphids were found on *Dactylis glomerata* than on wheat, with roughly similar densities on *Poa pratensis*, *Holcus mollis* and wheat.

Factors affecting population development of *S. avenae*

On most grass hosts *S. avenae* was first found shortly before flowering. There was no evidence of aphids overwintering in the localities studied. Winged immigrants of unknown origin founded the aphid colonies on grasses and on cereals in the same area (Carter *et al.*, 1980). Grass heads were rapidly colonized in the same manner as cereal ears (Latteur, 1976). Aphid density rose during and shortly after flowering and then fell sharply as the grass hosts matured. Figure 1 shows aphid population development on *D. glomerata* and

Table 1 Maximum mean densities of *S. avenae* on different grasses.

	Graminaceous hosts	*Aphids per tiller*	*Maximum tillers infested (%)*
1976	Wheat	51.7	—
	Dactylis glomerata (Cocksfoot)	52.8	100
	Avena fatua (Common wild oat)	134.2	100
	Holcus mollis (Creeping soft-grass)	5.8	92
	Agropyron repens (Couch)	9.7	50
1977	Wheat	88.7	100
	Dactylis glomerata	15.2	87
	Holcus mollis	14.3	97
	Arrhenatherum elatius (False oat-grass)	5.9	92
	Poa pratensis	0.5	26
	Trisetum flavescens (Yellow oat-grass)	1.3	50
	Phleum pratense (Timothy)	0.2	17
	Agropyron repens	1.3	38
	Agrostis tenuis (Common bent)	3.1	82
	Festuca rubra (Red fescue)	0.5	18
1978	Wheat	5.0	85
	Dactylis glomerata	10.0	43
	Poa pratensis	4.9	65
	Holcus mollis	4.5	60

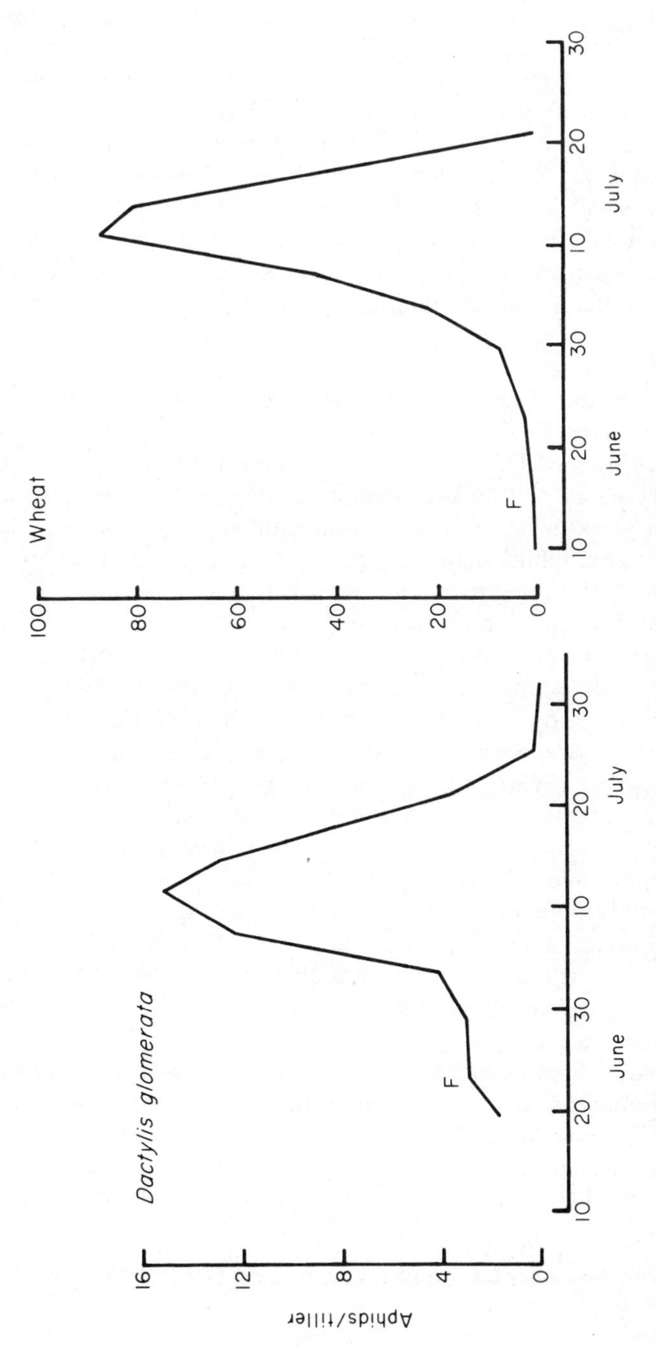

Figure 1 The density of *S. avenae* on *D. glomerata* and wheat in 1977. F denotes onset of flowering.

wheat. On all hosts except *Agropyron repens* peak aphid density occurred during the latter part of flowering or as the grass seeds began to ripen. Since grass species flowered at different times, they were colonized sequentially depending on their development. In 1977 *P. pratensis* and *D. glomerata* were colonized first and *P. pratense* and *A. repens* were colonized last. There was a great deal of overlap in the population development of *S. avenae* on different hosts, so that it was unlikely that progeny born on one grass species colonized the next grass to become suitable.

The time taken for aphid density to increase and then decrease varied between host species. The duration of this period was influenced by the development rate of the individual plants and variation in development within species. In *D. glomerata* the large variation between individual plants led to a long period of infestation. However, several grasses, including *Trisetum flavescens*, developed faster, there was little variation between plants, and aphid numbers rose and fell sharply.

In both 1976 and 1977, *S. avenae* occurred on *A. repens* before flowering and then decreased rapidly in abundance as flowering commenced. *A. repens* was the last grass to flower in both years and consequently was likely to be infested by more aphids than early-flowering grasses due to colonization by aphids born on the other host species. It is tempting to suggest that this has imposed a strong selection pressure which has resulted in *A. repens* becoming unsuitable at a stage which, in other species, is very suitable to *S. avenae*.

It may be assumed from the similarities between aphid population development on grasses and cereals that these hosts play a similar role. On cereals, reproduction in *S. avenae* is greatest on flowering, water-ripe and early milky-ripe plants (Watt, 1979a). Before the eventual reduction in aphid reproduction and longevity has a marked effect on their numbers, alate aphids are produced and leave the cereal crop (Carter *et al.*, 1980). Host plant quality (as indicated by plant growth stage) acts along with aphid density to induce the development of alatae (Watt, 1979b). The production of alatae was also observed on grasses as they matured and as aphid density increased: in 1978 the proportions of fourth instar alatiform nymphs at maximum aphid densities were 0.81 on wheat, 0.93 on *D. glomerata*, 1.00 on *P. pratensis* and 0.89 on *H. mollis*.

These results demonstrate the close inter-relationships between *S. avenae* and its graminaceous hosts. They underline the fact that *S. avenae* gives rise to short and severe infestations at the time of flowering and seed development, posing serious difficulties in pest control (McLean *et al.*, 1977; Taylor, 1977) and they also emphasize the need to understand and predict outbreaks.

Factors affecting abundance of *S. avenae* on different hosts in different years

Most investigations of cereal aphid outbreaks have centred on the role of natural enemies and possible methods of forecasting (Carter and Dewar, in press; Jones, 1972; Vickerman, 1977). The importance of the size and timing of the alate immigration into the crop is unknown (Dean, 1974b).

In 1977 and 1978 different aphid densities were observed on different hosts. Peak aphid density was correlated with the number of alate aphids colonizing each grass species (Fig. 2). In 1978, lower aphid densities on *D. glomerata* and *H. mollis* resulted from the fewer colonizing alatae compared with 1977. Similarly on wheat, the alate immigration was greater in the outbreak years of 1976 and 1977 than in 1978 when few aphids occurred on wheat (Table 2).

These patterns of population development of *S. avenae* outlined above were found to be dependent on the growth stage of their hosts. This suggested that the timing of aphid colonization might be important in determining aphid abundance. This idea was supported by the differences in the abundance of *S. avenae* on *P. pratensis* in 1977 and 1978. *P. pratensis* flowered early in

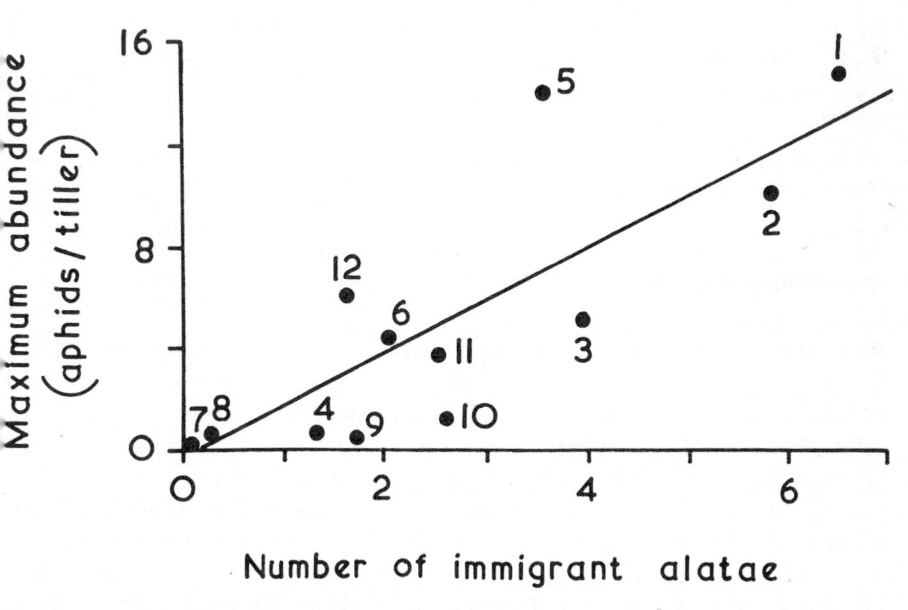

Figure 2 The relationship between maximum density of *S. avenae* on different grasses and the number of immigrant alatae per 100 tillers found on three dates when immigration was at a maximum. $y = 2.1x - 0.5$; $R = 0.82$; $P < 0.001$.

1 = *Dactylis*, 1977
2 = *Dactylis*, 1978
3 = *Poa*, 1978
4 = *Poa*, 1977
5 = *Holcus*, 1977
6 = *Holcus*, 1978
7 = *Phleum*, 1977
8 = *Festuca*, 1977
9 = *Agropyron*, 1977
10 = *Trisetum*, 1977
11 = *Agrostis*, 1977
12 = *Arrhenatherum*, 1977

Note that these results are not comparable with those in Table 2.

1977 but in 1978 it flowered later in relation to other grasses and after *D. glomerata*. More alatae colonized *P. pratensis* in 1978 than in 1977, arriving when it was more suitable to *S. avenae*, and consequently aphid density reached a higher level. Furthermore, it was observed that aphid densities on the three grasses sampled did not decrease to the same extent as those on wheat between 1977 and 1978. This was probably a consequence of the extended period over which grasses began flowering, resulting in them being suitable for colonization for a longer period than cereals. The importance of the timing of immigration could also be seen on cereals. In the two outbreak years of 1976 and 1977 not only were there more immigrants than in 1978, but they arrived in the crop at an earlier stage (Table 2).

Table 2 The number of immigrant *S. avenae* alatae found on wheat during the first three successful sampling occasions and maximum mean densities reached.

	Immigrants			*Subsequent population*	
	Alatae per 100 tillers	*Dates*	*Crop stage (Feekes scale)*	*Maximum mean density (aphids/tiller)*	*Date*
1976	0.89	19 May–2 June	9–10	51.7	30 June
1977	1.13	26 May–9 June	7–9	88.7	11 July
1978	0.50	8 June–22 June	10.1–10.5.1	5.0	3 August

Concluding remarks

This initial study indicates the importance of the size and timing of the immigration of *S. avenae* in determining aphid abundance on grasses and cereals, although their relative importance remains unknown. It would be of interest to see whether severe aphid outbreaks occur in years when immigration is high but late. Certainly high cereal aphid densities on cereals do not always occur in years when aerial aphid density is high (Dean, 1974a). Finally, a great deal is now known about *S. avenae* and its population dynamics on its summer hosts. What is less well understood is the biology and abundance of *S. avenae* and other cereal aphids on grasses at other seasons, when the number of aphids eventually colonizing cereals is largely determined (*see* Hand and Williams (*1981*)).

References

Carter, N. & Dewar, A. (in press). The development of forecasting systems for cereal aphid outbreaks in Europe. In *Proceedings of the IXth International Congress of Plant Protection, 1979.*

Carter, N., McLean, I. F. G., Watt, A. D. & Dixon, A. F. G. (1980). Cereal aphids — a case study and review. *Applied Biology* **5,** 272–348.

Dean, G. J. (1974a). The overwintering and abundance of cereal aphids. *Annals of Applied Biology* **76,** 1–7.

Dean, G. J. (1974b). The four dimensions of cereal aphids. *Annals of Applied Biology* **77,** 74–78.

Eastop, V. F. (*1981*). Wild hosts of aphid pests. In *Pests, Pathogens and Vegetation*, pp. 285–298. J. M. Thresh. London: Pitman.

Hand, S. C. & Williams, C. T. (*1981*). The overwintering of the rose-grain aphid (*Metopolophium dirhodum*) on wild roses. In *Pests, Pathogens and Vegetation*, pp. 307–314. J. M. Thresh. London: Pitman.

Jones, M. G. (1972). Cereal aphids, their parasites and predators caught in cages over oat and wheat crops. *Annals of Applied Biology* **72,** 13–25.

Latteur, G. (1976). Les pucerons des céréales: biologie, nuisance, ennemis. *Centre de Recherches Agronomiques de l'Etat Gembloux*, *Memoire* No. 3.

McLean, I., Carter, N. & Watt, A. (1977). Pests out of control? *New Scientist* **76,** 74–75.

Taylor, L. R. (1977). Aphid forecasting and the Rothamsted Insect Survey. *Journal of the Royal Agricultural Society* **138,** 75–97.

Vickerman, G. P. (1977). Monitoring and forecasting insect pests of cereals. In *Proceedings of the 1977 British Crop Protection Conference, Brighton*, pp. 227–234.

Vickerman, G. P. & Wratten, S. D. (1979). The biology and pest status of cereal aphids (*Hemiptera: Aphididae*) in Europe: a review. *Bulletin of Entomological Research* **69,** 1–32.

Watt, A. D. (1979a). The effect of cereal growth stages on the reproductive activity of *Sitobion avenae* and *Metopolophium dirhodum. Annals of Applied Biology* **91,** 147–157.

Watt, A. D. (1979b). Life history strategies of cereal aphids. *PhD Thesis*, University of East Anglia.

The overwintering of the rose-grain aphid *Metopolophium dirhodum* on wild roses

S C Hand and C T Williams
Department of Biology, Building 44, The University, Southampton SO9 5NH and The Game Conservancy, Fordingbridge, Hampshire SP6 1EF

Introduction

The rose-grain aphid, *Metopolophium dirhodum*, is an important pest of cereals in Europe (Vickerman and Wratten, 1979) and was particularly abundant in Britain in 1979. It reduces cereal yields both directly (Wratten, 1975) and by transmitting barley yellow dwarf virus (e.g. Plumb, 1978). Despite its importance very little is known quantitatively about its biology, particularly during the winter.

Hille Ris Lambers (1947) considered that *M. dirhodum* was able to overwinter only holocyclically on its primary hosts, *Rosa* spp., but recently it has been suggested that it mainly overwinters anholocyclically on Gramineae (Dean, 1978; Elkhider, 1979).

In the present work surveys were carried out both in the autumn and in the spring to obtain estimates of the numbers of *M. dirhodum* on *Rosa* spp. on a farm in Hampshire. Samples were also taken from Gramineae to provide similar estimates of the numbers overwintering on these secondary hosts, but these data will be published elsewhere.

The morph nomenclature is that of Hille Ris Lambers (1966) and Blackman (1974).

Methods

The work was carried out on South Allenford Farm (Grid Ref. SU 090 175) at Damerham, near Fordingbridge, Hampshire. This is a mixed arable–livestock farm of about 750 ha, of which about 12.5 ha is woodland, and about 4 ha is open chalk scrubland, with isolated bushes (mainly *Rosa* spp.) growing in rank grass. An area of *c.* 250 ha of woodland adjoins the farm.

Identification

The *Metopolophium* on rose included *M. dirhodum* and a recently separated species, superficially resembling it, which also alternates between *Rosa* and Gramineae (H. L. G. Stroyan, MAFF Harpenden Laboratory, Herts. personal communication). No attempt was made to separate these species in the field. However, laboratory identification showed that 100% and *c.* 90% of the *Metopolophium* were *M. dirhodum* in spring (1979) and autumn (1978 and 1979), respectively. The second species represented such a small proportion of the *Metopolophium* found that its presence does not affect conclusions on the overwintering of *M. dirhodum* drawn from the results of the surveys.

Rose surveys

The *Rosa* spp. were mainly dog rose (*R. canina* agg.) although field rose (*R. arvensis*) and sweet briar (*R. rubiginosa* agg.) also occurred. Rose bushes were found both in field boundaries (particularly hedgerows), and in woodland and exposed chalk scrubland. Different sampling methods were used for the two types of habitat.

Field boundaries

In autumn 1978, spring 1979 and autumn 1979 samples were taken from field boundaries (hedgerows, fencelines and wood edges). The boundaries on the farm were numbered and those to be examined were selected at random. Records were made of the boundaries' orientation, average height and width of the vegetation, the principal plant species present, appearance in relation to cutting, and the adjoining crops. Along each field boundary one-metre lengths were marked and examined at 15-pace intervals. In each selected metre the height and species of any *Rosa* present were recorded before it was cut and examined for aphids. After examination the number of leaves or buds was counted. The aphids present were identified and the number and morph of *Metopolophium* were recorded per leaf and leaflet. All *Metopolophium* and any unidentified aphids were reared to the adult stage in the laboratory. In the case of woods bordering fields, rose to a depth of 1 m into the wood was examined. If some rose was inaccessible in a hedge the amount and height were estimated. The hedges were examined in approximately the same order in each survey and each time the sample points were moved 5 paces, so that no portion of hedge was examined more than once. In autumn 1979, in three hedgerows where rose was abundant, samples were taken only from every second or third selected metre with rose. Those hedges which had few aphids in autumn 1978 were not resampled the following spring, and the number of aphids was estimated from the survival of the aphids in the 'high density' hedges.

These surveys provide data on the number (and presence) of aphids per metre of field boundary, and hence estimates of the total number of 'field-boundary' *Metopolophium* on the farm.

Woodland and scrubland
Woodland and scrubland rose was sampled in the spring and autumn of 1979. A starting point was selected near the centre of each area and roses were examined in several directions from this. For the one area of scrubland on the farm a very high proportion of the rose was sampled (*c.* 30% and *c.* 70% in spring and autumn, respectively). In spring 1979 only one wood was examined; in autumn 1979 two were sampled and each was divided into areas of 'dense' and 'open' woodland. For each bush sampled the position, morph and age of *Metopolophium* were recorded on up to 250 leaves or open buds. For large bushes the proportion of leaves remaining and thus the total number of aphids on the bush was estimated. The height of the bush and its position in relation to other vegetation were recorded. In one wood, forty 20 × 20 m quadrats were inspected to give the ratio of dense to open wood, and rose bushes were counted in a set of 20 random 10 × 10 m quadrats to provide estimates of their density. Assuming all woods on the farm had the same amount of rose it was possible to estimate the total number of 'woodland' *Metopolophium* in the area.

Results

Distribution of aphids on rose

Only one sample of *R. arvensis* was taken and this had *Metopolophium* present. There was no evidence that *Metopolophium* showed a preference for either of the other two species of *Rosa* found. In dense wood in autumn 1979, seven of 13 bushes of *R. canina* examined had *Metopolophium* compared to three out of six *R. rubiginosa*. The only bush with *M. dirhodum* in scrubland was *R. rubiginosa* (six bushes of *R. canina* and nine of *R. rubiginosa* were sampled). The amount of *R. rubiginosa* in field boundaries was small.

Distribution between habitats
In spring 1979 rose in woodland had a higher proportion of leaves infested with *Metopolophium* than rose elsewhere. In the autumn survey a significantly higher proportion of leaves ($P < 0.001$) was infested in dense woodland than in open woodland or hedges, between which there was no significant difference (Table 1). In spring 1979 no *Metopolophium* were found on scrubland rose, and in autumn the proportion of infested leaves in scrubland was significantly less ($P < 0.05$) than that in the other habitats (Table 1).

In woodland, rose bushes varied in height from about 0.5 m to 6.0 m, but there was no apparent relationship between the proportion of leaves infested with *Metopolophium* and the height of the different bushes.

The hedgerows on the farm were of two main types:

Type 1: Tall (≥ 4 m), thick hedges with diverse woody plant species and not regularly cut back (unless on one side only). This included wood edges bordering fields.

Type II: Low (< 4 m), dense hedges with two or three woody plant species predominating and recently and regularly cut back on the top and at least one side.

In autumn of both 1978 and 1979, the proportion of metre lengths having rose with *Metopolophium* was significantly higher in Type I hedges than in Type II (1978, $\chi^2 = 24.6$, $P < 0.001$; 1979, $\chi^2 = 8.7$, $p < 0.01$). Despite there being significantly fewer metres with rose in the Type I than in the Type II hedges sampled ($\chi^2 = 6.77$, $P < 0.01$), the autumn 1978 results showed that even when those metres with no rose are taken into account, Type I hedges had a significantly higher proportion of metres of field boundary with *Metopolophium* present than Type II ($\chi^2 = 4.1$, $P < 0.05$). This latter difference was not significant in autumn 1979, probably as a result of the lower numbers of *Metopolophium* found.

Type I hedges were of two kinds: tall hedgerows and wood edges. In the autumn of 1978, wood edges had a significantly lower proportion of metres of rose with *Metopolophium* than tall hedgerows ($\chi^2 = 5.9$, $P < 0.05$) but this result is due to a single tall hedgerow which had a large amount of rose and a high proportion of metres with *Metopolophium*. However, wood edges alone still had a significantly higher proportion of metres of rose with *Metopolophium* than Type II hedges ($\chi^2 = 8.0$, $P < 0.01$).

There was no evidence to suggest that the presence or number of aphids in a metre with rose was related to the number of rose leaves.

Table 1 Percentage *Rosa* leaves and open buds infested with *Metopolophium* on the study farm in spring and autumn 1979

Season	*Habitat*	*No. leaves or open buds examined*	*Leaves or buds infested*		*Significance**
			Number	*%*	
Spring	Woodland (dense + open)	2112	94	4.45	A
	Field Boundaries	9082	23	0.25	B
	Scrubland	1321	0	0	B
Autumn	Dense Woodland	2256	54	2.39	a
	Open Woodland	1455	6	0.41	b
	Field Boundaries	21 663	80	0.37	b
	Scrubland	2323	2	0.09	c

* Significance (χ^2): items with the same letter did not differ significantly
Significance: A/B ($P<0.001$)
a/b,c ($P<0.001$)
b/c ($P<0.05$)
No comparison was made between seasons.

Distribution within plants

In spring, *Metopolophium* were usually found between unfolding leaflets and on and between stipules. In autumn, they were found on the abaxial surface of the leaves (including those almost ready to fall) but not on unfolding leaflets. In autumn 1979 the height above ground level of the *Metopolophium* was recorded. In both woodland and field boundaries there was no tendency for the aphids to be found at any particular bush height. *Metopolophium* on the leaves and leaflets of the rose were aggregated. This was because in spring and autumn the majority of the aphids were fundatrigeniae and oviparae respectively, both produced in clusters from a parent. Fifteen of the 27 males found in the autumn surveys were found with oviparae, which is a high number considering that only 1–2% of the thousands of leaves examined had oviparae. Nine males were found with adult oviparae only and three with nymphal oviparae only. A comparison of this ratio with the relative frequencies of leaves with these morphs showed that there was a significant tendency for males to be found with adult rather than nymphal oviparae ($\chi^2 = 10.7$, $P < 0.01$).

Number of aphids on rose on the farm

The total number of 'field boundary' *Metopolophium* on the farm was estimated as *c.* 71 000 in autumn 1978, *c.* 22 000 in spring 1979 and *c.* 26 000 in autumn 1979 (Table 2). Similar estimates of the total numbers of 'woodland' *Metopolophium* were *c.* 210 000 in spring 1979 and *c.* 47 000 in autumn 1979 (Table 2). Estimates of the total number of 'scrubland' *Metopolophium* on the farm, were, by comparison, very small; none in spring 1979 and only 10 in autumn (Table 2). No woodland or scrubland samples were taken in autumn 1978.

Combining all the habitats and allowing for 10% of the autumn 1979 *Metopolophium* not being *M. dirhodum*, gives an overall estimate of the number of *M. dirhodum* on rose on the farm of *c.* 232 000 in spring 1979 and *c.* 66 000 in autumn 1979.

Attempts were made to carry out the surveys at similar stages of development of the aphid populations. However, it should be stressed that the above estimates could be influenced by small differences in sampling dates. A more reliable estimate of seasonal and annual changes in the population was therefore obtained by considering the abundance of aphids, in the metres of hedgerow with rose, in terms of presence or absence. On this basis there was a significant decline in the proportion of metres of rose with *Metopolophium* both between autumn 1978 and spring 1979 ($\chi^2 = 33.9$, $P < 0.001$), and between autumn 1978 and autumn 1979 ($\chi^2 = 7.6$, $P < 0.01$). The latter result shows that despite the very large numbers of *M. dirhodum* present on cereal crops in Britain in mid-summer 1979, the numbers of gynoparae successfully colonizing rose in the autumn of that year had apparently

Table 2 Estimated numbers of *Metopolophium* overwintering on rose on the study farm in autumn 1978, spring 1979 and autumn 1979.

Field boundaries	*Metres sampled:*			*Average no. of aphids/m of field boundary ± SE*	*Total no. of aphids on rose on farm (estimates/habitat)*
	total	*with Rosa*	*with aphids*		
Autumn 1978 (23/10–10/11)	499	104	51	1.38 ± 0.347	71 000
Spring 1979 (27/4–10/5)	299 (499)	48 (104)	6 (9)	(0.433)	22 000
Autumn 1979 (25/10–11/11)	633	103 (142)	32 (45)	(0.504 ± 0.113)	26 000
Woodland	*No. of bushes sampled*		*Average density of aphids/m²:*		
			in dense wood	*in open wood*	
Spring 1979 (1/5–3/5)	12		33.6	0	210 000
Autumn 1979 (2/11–9/11)	51		0.70	0.056	47 000
Scrubland	*No. of bushes sampled*	*% of bushes sampled*	*Total no. of aphids*		
Spring 1979 (11/5)	6	*c.* 30	0		0
Autumn 1979 (5/11)	15	*c.* 70	7		10

Numbers in brackets are estimates from sub-samples.

declined compared to the previous autumn. This suggests that particularly heavy mortality occurred in late summer or autumn.

Discussion

The results of these surveys showed that densities of *Metopolophium* (mainly *M. dirhodum*) were highest on rose in woodland and that rose in 'dense' woodland had more *Metopolophium* than rose in 'open' woodland. A greater proportion of the rose in tall, unmanaged hedgerows was colonized compared with rose in short, managed ones; open scrubland rose was unsuitable. These differences suggested that the height of the surrounding bushes and trees and the degree of shelter they provided strongly influenced colonization, reproductive success and/or survival of *Metopolophium*. Shelter may be particularly important in reducing windspeed, enabling gynoparae to control their flight direction, landing and take-off, and possibly facilitating the detection of the bushes. It may prevent their displacement from suitable leaves before they have reproduced and may also provide more favourable conditions for reproduction and survival by gynoparae and subsequent generations. However, other factors may also influence the colonization of *Rosa* by *M. dirhodum* and the subsequent success of the holocyclic population. For example, there may be differences in the nutritional quality of rose leaves in the different habitats. In the autumn the timing of colonization by the gynoparae appeared to be critical. On some bushes and in some habitats the leaves dropped before the oviparae became mature. Variation in the timing of leaf-fall in different habitats is probably influenced by several factors, such as age of the rose shoots, the degree to which they are sheltered and the extent of shading during the summer.

Gramineae occupied nearly the whole area of the study farm, as cereal and grass crops, weeds and volunteers in other crops, or as wild plants in hedgerows, woodland and scrubland. It was therefore difficult to quantify the relative importance of holocyclic and anholocyclic overwintering as the population density of *M. dirhodum* overwintering on Gramineae required to give a 'farm population' equal in size to that overwintering on *Rosa* spp. was extremely low. For example, in spring 1979 only one *M. dirhodum* per 32 m^2 of Gramineae on the study farm would have sufficed. Nevertheless there was no evidence from vacuum net samples that *M. dirhodum* was present on Gramineae on the study farm, even at this low density during late winter and spring 1979.

The spring migration from *Rosa* spp. on the farm started during the week beginning 13 May in 1979 (Williams, unpublished). This coincided with the first *M. dirhodum* alatae caught that year in the Rothamsted Insect Survey suction traps. The first *M. dirhodum* alatae were found on Gramineae in the study area on 18 May (Vickerman, personal communication).

It appears, therefore, that the more important mode of overwintering was holocyclic (on *Rosa*) as opposed to anholocyclic (on Gramineae) on this study farm, at least in the 1978/79 winter. This was probably partly a result of the weather conditions during the period. Autumn 1978 was relatively mild and dry (as was autumn 1979) which probably favoured the successful migration to *Rosa* by gynoparae and males. Moreover, the 1978/79 winter was particularly cold. On the basis of results of field and laboratory experiments (Williams, unpublished) the minimum temperatures recorded in the field during this winter should have sufficed to kill a high proportion of *M. dirhodum* exules.

Vickerman (personal communication) recorded *M. dirhodum* overwintering successfully in crops of winter oats and barley in West Sussex during the 1972/73 winter, and Prior (1976) mentions 'a number of finds' of *M. dirhodum* on grasses during winters since 1972. Clearly the importance of holocyclic and anholocyclic overwintering in other situations and years warrants investigation.

The work was carried out whilst we held CASE studentships from NERC (SCH) and SRC (CTW). We thank Dr S. D. Wratten (University of Southampton) and Dr G. P. Vickerman (Game Conservancy) for their help and supervision, Mr R. Shepherd for permission to work on South Allenford Farm and Dr H. L. G. Stroyan for his help in identifying the specimens.

References

Blackman, R. (1974). *Invertebrate Types: Aphids*. London: Ginn and Co.

Dean, G. J. (1978). Observations on the morphs of *Macrosiphum avenae* and *Metopolophium dirhodum* on cereals during the summer and autumn. *Annals of Applied Biology* **89,** 1–7.

Elkhider, E. M. (1979). Studies on environmental control of polymorphism in the rose-grain aphid *Metopolophium dirhodum* (Walker). *PhD Thesis*, University of London.

Hille Ris Lambers, D. (1947). Contributions to a monograph on the Aphididae of Europe, III. *Temminckia* **7,** 179–319.

Hille Ris Lambers, D. (1966). Polymorphism in Aphididae. *Annual Review of Entomology* **11,** 47–78.

Plumb, R. T. (1978). Aphids and virus control on cereals. In *Proceedings of the 1977 British Crop Protection Conference — Pests and Diseases* **3,** pp. 903–913.

Prior, R. N. B. (1976). Keys to the British species of *Metopolophium* (Aphididae) with one new species. *Systematic Entomology* **1,** 271–279.

Wratten, S. D. (1975). The nature of the effects of the aphids *Sitobion avenae* and *Metopolophium dirhodum* on the growth of wheat. *Annals of Applied Biology* **79,** 27–34.

Vickerman, G. P. & Wratten, S. D. (1979). The biology and pest status of cereal aphids (Hemiptera: Aphididae) in Europe: a review. *Bulletin of Entomological Research* **69,** 1–32.

The interaction of wild vegetation and crops in leaf-cutting ant attack

J M Cherrett
Department of Applied Zoology, University College of North Wales, Bangor, Gwynedd, LL57 2UW

Introduction

The larger members of the Tribe Attini, the fungus-growing ants, can cut pieces out of living leaves. These leaf-cutting ants are mainly restricted to the genera *Atta* and *Acromyrmex*. Within each genus, some species use mainly dicotyledonous plants and others monocotyledonous ones. However, each species exploits many species of plants, so these ants are truly polyphagous.

Figure 1 *Atta cephalotes* cutting a leaf in primary tropical rain forest, Guyana.

Cherrett (1968a) observed that in eight weeks a single nest of *At. cephalotes* in primary rain forest in Guyana attacked 36 of 72 (50%) woody plant species recorded in the area. Similarly, Rockwood (1976), in a 12-month study in Costa Rica of three *At. cephalotes* colonies in evergreen oak forest, and three *At. colombica* colonies in riparian forests, recorded that up to 77% and 67% respectively of woody plant species in their territories were cut. In all the studies, the extent of exploitation of these plant species was very variable, indicating selectivity.

The ability of leaf-cutting ants to damage many plant species, including crops, makes them important pests. Cramer (1967) cited them as one of only five groups of 'polyphagous pests', estimating that they were responsible for annual losses of one thousand million US dollars.

As leaf-cutting ants appear to be so well pre-adapted to exploit crops, it would be of great interest to know what happens when natural vegetation is replaced by crops. Unfortunately, although forest clearing in South America continues apace (Bunting, *1981*), there has been little detailed ecological work on the implications for ants. It is hoped that this paper will stimulate more systematic studies.

Species favoured by crop conditions

Most of the region where leaf-cutting ants occur was originally forested, and when this is cleared for agriculture the microclimate is changed, the vegetation alters, and cultivation techniques are introduced. These changes seem to favour some leaf-cutting ant species at the expense of others.

Acromyrmex octospinosus in Trinidad

Cherrett (1968b) described some of the consequences for leaf-cutting ants of replacing tropical rain forest in Trinidad by citrus and cocoa plantations. *At. cephalotes*, which builds large nests 100–150 m^2 in surface area, and *Ac. octospinosus* which builds much smaller nests (2 m^2) are the only leaf-cutting ants found on the island. Paired comparisons of tropical rain forest and nearby agricultural plantations showed that *At. cephalotes* dominated in the forest and *Ac. octospinosus* in plantations. Moreover, the *At. cephalotes* colonies in the plantations were all young and small. This distribution resulted from two processes:

(1) The preference of *Ac. octospinosus* queens for clearer areas when seeking a nesting site; and
(2) Man's pest control activities. These are more effective against the large, conspicuous, slowly-reproducing *Atta* colonies than against the smaller, less conspicuous, more rapidly-reproducing ('r-type') *Acromyrmex* colonies.

Atta sexdens in Brazil

Where *At. cephalotes* and *At. sexdens* co-exist, *At. sexdens* has the reputation of being a follower of agriculture (Eidmann, 1935) in much the same way that *Ac. octospinosus* does in Trinidad. Some recent observations in the Rio Jari forestry project in Amazonia support this view. Five 20 minute survey walks in newly-felled tropical rain forest revealed three *At. sexdens*, and three *At. cephalotes* nests, whilst six comparable walks in nearby plantations of *Gmelina arborea*, more than six years old, revealed sixteen *At. sexdens* and no *At. cephalotes* nests ($P = 0.013$ using a Fisher exact test). The reasons for this species shift are unknown, but *At. sexdens* may be better adapted to drier, more open conditions (Weber, 1969).

The increased importance of grass-cutters

In many areas of South America, forest and scrubland are being cleared to create pastures. The increased availability of monocotyledonous plants and the increased productivity when high-yielding, exotic grasses are introduced, permit grass-cutting ants to extend their range and population density at the expense of other species.

At. capiguara, the most recently named of the fourteen *Atta* species, was first described in 1944 from a site within the present city of São Paulo. It makes large distinctive nests and cuts only grasses, so if it had previously been a common species it would almost certainly have been reported much earlier. By 1959, it had been recorded from three municipalities of São Paulo state, and by 1966 from 115 (Mariconi, 1970). By 1967, Amante claimed that in the State it was a prime pest of pastures, reducing their carrying capacity by 800 000 head of cattle. He also recorded the species from the States of Mato Grosso and Minas Gerais, and it has since been found in Paraguay. Amante (1972) thought that in recent years, *At. capiguara* had greatly extended its range and population density, for up to ten nests per hectare occur in certain pasture types (Gonçalves, 1971). The original forest contained *At. sexdens* and *At. laevigata*, but when it was cut down and replaced by pasture, *At. capiguara* replaced *At. sexdens*. *At. laevigata*, which can also utilize monocotyledonous plants, then accounted for about 10% of all nests.

Similarly, with the grass-cutter *Acromyrmex landolti*, Cherrett, Pollard and Turner (1974) observed a mean density of three nests per hectare in original unimproved savanna in Guyana, only one in experimentally-planted *Digitaria setivalva* pastures, but ninety-five in a susceptible *Paspalum notatum* pasture, all sites being within a few hundred metres. This illustrates how an initial, sparse population of ants living in indigenous vegetation can increase thirty-fold when the pasture is improved and planted to susceptible higher-yielding grasses. Other workers have recorded much higher *Ac. landolti* populations in managed grasslands; maximum nest populations per hectare in *Panicum maximum* reached 4400 in Paraguay (Fowler and Robinson, 1977), 5930 in Venezuela (Labrador *et al.*, 1972) and 6000 in Peru (Kelderman, unpublished). However, these densities are probably overestimates as *Ac. landolti* may have more than one entrance per nest (Espina and Timaure, 1977). The recent upsurge of *Ac. landolti* to the point at which grasslands are damaged and susceptible species become uneconomic to grow, presents a serious threat to rangeland management in widely scattered parts of South America.

Figure 2 Living nest of *Atta vollenweideri* in the Paraguay Chaco. Note in the background the clusters of trees ('wood nuclei') which have originated on dead ant nests.

Jonkman (1979a), using aerial photographs, surveyed 80 000 km^2 in Paraguay for nests of the predominantly grass-cutting species *At. vollenweideri*. They were restricted to a few main areas, where their mean density was 0.2 ha^{-1}, with a maximum of c. 4. Intensive studies at one 39 ha site, where a series of aerial photographs was available from 1944 to 1973, showed that the nest density rose during that period from 0.6 to 4.5 ha^{-1}; Jonkman (1979b), attributed this to increased human activity, especially burning and over-grazing with cattle. Vegetation is usually absent from the soil overlying a living nest. After its death and physical collapse, the nest becomes invaded by grasses, cacti, palms and young trees (Jonkman, 1978). These grow on the old nests to form 'wood-nuclei', which probably accelerate the succession from pasture to woodland. However, the increasing woodland is associated with decreasing numbers of *At. vollenweideri* nests, which exist in open woodland at a lower density. If there is a stable climax, it is not clear what it is.

In these examples of ant species favoured by crop conditions, it is important to try and identify the reservoir wild vegetation from which the upsurging ant infestations developed. In some cases, as with *At. sexdens* in *Gmelina* plantations in Rio Jari, or *Ac. landolti* in Guyana, the ants were present in the original vegetation in reduced numbers. In other cases, as for example, *At. capiguara* in São Paulo, the original source is not clear.

Species invading from reservoirs of wild vegetation

Leaf-cutting ants can forage considerable distances from their nests. Cherrett and Lewis (1974) quote typical examples of *Ac. octospinosus* foraging 20–30 m, and *At. cephalotes* 100–150 m. Accordingly, crops can be severely damaged by ants foraging into them from nests in surrounding uncultivated land (Belt, 1874). Lewis and Norton (1973) wrote 'Nests of *At. cephalotes* nearer to crops than 200–250 m, the maximum foraging distance of this species, are a potential threat'. Agricultural systems which involve a small-scale patchwork of crops and primary or secondary wild vegetation are, therefore, particularly at risk from leaf-cutting ant damage, especially where fragmented ownership results in variable standards of control.

Once a leaf-cutting ant colony has grown to maturity, winged male and female sexuals are produced annually, and after their mating flight, the fertilized queen begins to dig her new nest. Moser (1967) found that females of *At. texana* flew the equivalent of 10.4 km on a flight mill, and Cherrett (1969) observed that queens of *At. cephalotes* flew at least 9.6 km in the field. Whenever a species is invading crops in an area it has not previously occupied, or where it cannot develop mature reproducing colonies within the crop, young colonies which develop must be the result of fertile queens flying in from reservoir vegetation.

Species persisting from previous wild vegetation

In the laboratory, leaf-cutting ant queens can live for many years: one queen *At. sexdens* survived for 22 years (Weber, 1972). In the field, Jonkman (1979b) used successive aerial photographs to calculate a mean life expectancy of ten years for a well established nest of *At. vollenweideri*. Mature nests of *Atta* species are formidable structures. One of *At. vollenweideri* excavated by Jonkman (1977) was approximately 35 m^2 in surface area, contained some 3000 chambers, and reached a depth of nearly 5 m.

Crops are planted on a given area soon after wild vegetation has been cleared, where crop areas are expanding or where a system of rotation such as 'slash and burn' is employed which involves a secondary bush fallow. Here, the wild vegetation is outside the crop in a temporal sense.

The ants nest so deep underground that they are little affected by clearing and burning of the wild vegetation cover. After the initial disturbance, the surface of the nest is soon rebuilt and the ants continue foraging in what is now a habitat virtually devoid of green leaves. After several weeks, the

Figure 3 Bulldozer section through a nest of *Atta vollenweideri* in the Paraguay Chaco. Note the pile of excavated soil and fungus garden chambers.

Figure 4 Using swing fog to kill leaf-cutting ant nests in 'slash and burn' agriculture in Guyana.

devastating effect of the ants on newly-transplanted trees, emerging crop seedlings or sprouting tubers can be imagined. As the colonies are so long-lived, they can persist for many years in the new habitat, although it may not be an ideal one, and fertilized queens may be unable to establish further colonies in it.

At. cephalotes can survive forest clearing in this way, as I have noted in Trinidad, Guyana and Rio Jari. Planting a crop such as *Gmelina* or *Pinus* trees, citrus, cocoa or vegetables to replace the original forest will result in a smaller biomass, part of which will consist of grasses that *At. cephalotes* does not cut. A nest density adjusted to forest conditions is therefore likely to do considerable damage in the crops which replace it, especially when the forest was secondary (as it would be in a slash and burn culture). This is because leaf-cutting ant populations are normally higher in secondary than in primary forest (Haines, 1978).

Apparent preferences for introduced vegetation

Leaf-cutting ants can forage so far from their nests that they must often have a choice of crops and/or natural vegetation. Many observers have remarked on their apparent preference for crop species, most of which are introduced. As long ago as 1874, Belt wrote 'None of the indigenous trees appear so suitable for them as the introduced ones. Through long ages the trees and the ants of tropical America have been modified together'. Belt was a pioneer in the study of plant defences against herbivore attack. He thought that leaf-cutting ants had exerted an important selective pressure in tropical America, and wrote 'The leaf-cutting ants are confined to tropical America; and we can easily understand that trees and vegetables introduced from foreign lands where these ants are unknown could not have acquired, excepting accidentally, and without any reference to the ants, any protection against their attacks, and now they are most eagerly sought by them'.

Although we more readily notice attacks on cultivated than on wild plants, it does seem that some crop species are particularly vulnerable. The reasons for this require much more detailed investigation, but some of the defence mechanisms against leaf-cutting ant attack can be outlined.

Chemical deterrents

Cherrett and Seaforth (1968) showed that old leaves of grapefruit (*Citrus paradisi*) were cut less than young ones. Littledyke and Cherrett (1978), surveying responses to old and young leaves of six plant species, found that paper discs dipped in young leaf extract were consistently preferred to old, and that their acceptability depended on the balance between the acceptable and inhibitory chemical components. For *At. cephalotes*, the inhibitors were mainly found in lipid fractions, and dewaxing the leaf cuticle of old leaves usually made them more acceptable.

In addition to this evidence of chemical defences to reduce or prevent leaf-cutting, Littledyke and Cherrett (1976) showed that chemical deterrents even more readily reduce the direct consumption of plant sap by the ants for food. But since the fungus grown on the cut leaves is consumed by the ants, and especially by the larvae as a protein source (Quinlan and Cherrett, 1979), they can to some extent 'sidestep' chemical inhibitors in the leaves. Imported plant species, which had previously relied on chemical inhibitors to deter other insects, may find themselves relatively defenceless against leaf-cutting ants.

Latex

If chemical deterrents within plant material are generally less effective against leaf-cutting ants than they are against most other insect herbivores, and if the ants are a significant selective pressure, then natural vegetation subject to ant attack might have developed other defences. Stradling (1978), re-analysing some data of Cherrett (1968a), concluded that those tree species in tropical rain forest which were little attacked were significantly more likely to contain a latex system than were susceptible species. He also showed experimentally that the sticky latex oozing from cuts in the leaves of *Euphorbia leucocephala* inhibited further cutting by sticking up the ants' mouthparts. It would be interesting to know if laticiferous plants are more common in New World tropical rain forests where leaf-cutting ants abound, than in comparable forests in the Old World where they are absent. Few cultivated crops are laticiferous.

Toughness

Cherrett (1972) showed that the toughness of leaves as measured by dry weight per unit area influenced the speed of an ant's cutting, and that, generally, only the larger individuals cut the toughest material. He also showed that the size of pieces cut from tough leaves was smaller than from softer leaves, and concluded that the tougher a plant's leaves, the less will be its losses. It would be interesting to know if, on average, selective crop breeding reduces leaf toughness. Certainly young Marsh White Seedless grapefruit leaves, which were frequently attacked, were less tough than leaves collected by *At. cephalotes* workers in tropical rain forest (Cherrett, 1972).

Defensive ants

Leston (1973, 1978) has shown that in both the Old and New World tropics there are dominant ant species, which monopolize individual trees, exclude other ants, and form an 'ant mosaic'. He concluded that there was sufficient evidence to warrant a thorough investigation of the role of the ants *Azteca* and *Dolichoderus* spp. in protecting cocoa from leaf-cutting ant attack. Belt (1874) had concluded that the *Pseudomyrma* ants defending bull's horn acacia trees in Central America deter leaf-cutting ant attack, and he drew attention to similar defenders on *Cecropia*, *Melastoma* and *Passiflora*. More recently, Jutsum, Cherrett and Fisher (1981) have demonstrated that *Azteca* sp. prevents *At. cephalotes* from defoliating trees in which it has built its carton nests, and that the ants *Dolichoderus bidens* and *Crematogaster brevispinosa* may similarly restrict attack.

Many crop plants, however, do not possess extra-floral nectaries or special protein-producing organs to attract a defensive ant fauna, and the clean cultivation practised in agricultural monocultures does not offer good sites for the nests of defending ants. Moreover, farmers often kill defending ant colonies because some tend and encourage populations of Homoptera or attack farm workers during harvesting operations.

For all these reasons, introduced vegetation, especially crops bred for high yield or succulence, can be expected to be vulnerable to leaf-cutting ant attack, and experience suggests that they are.

Conclusions

Wild vegetation outside crops provides a reservoir of leaf-cutting ants with a source of workers which can forage into the crop from its margins, even if crop conditions are unfavourable for colony development. If crop conditions are favourable for the ant species, then sexuals can fly in over considerable distances and establish colonies. Similar possibilities exist when wild vegetation precedes a crop on a given site, and the colony survives the process of clearing and planting. Once in a crop, their defoliating activities and the large size of *Atta* nests affect the performance of the crop, and the speed of succession in the case of grasslands.

At the boundaries of introduced crops and wild vegetation, for a variety of reasons, the crops are more vulnerable to leaf-cutting ant attack and consequently bear the brunt of the defoliating activities of those colonies which have a choice of food.

Acknowledgments

These studies have been supported by a grant from the Overseas Development Administration of the Foreign and Commonwealth Office, and I am grateful to past members of the Ant Control Unit in Bangor for valuable discussions and to Mr J. Hobart for his comments on the paper. Ant cultures were maintained under MAFF licence No. HH 10899A/67 issued under the Destructive Pests and Diseases of Plants Order 1965.

References

Amante, E. (1967). A formiga saúva *Atta capiguara*, praga das pastagens. *Biológico* **33,** 113–120.

Amante, E. (1972). Influência de alguns factores microclimáticos sobre a formiga saúva *Atta laevigata* (F. Smith, 1858), *Atta sexdens rubropilosa* Forel, 1908, *Atta bisphaerica* Forel, 1908, e *Atta capiguara* Gonçalves, 1944 (Hymenoptera, Formicidae), em formigueiros localizados no estado de São Paulo. *Tese, Doutor em Agronômia, Escola Superior de Agricultura 'Luiz de Queiroz', Universidade de São Paulo, em Piracicaba.*

Belt, T. (1874). *The Naturalist in Nicaragua*. London: E. Bumpus.

Bunting, A. H. (*1981*). Changing patterns of land use: global trends. In *Pests, Pathogens and Vegetation,* pp. 23–37. J. M. Thresh. London: Pitman.

Cherrett, J. M. (1968a). The foraging behaviour of *Atta cephalotes* L. (Hymenoptera, Formicidae). I. Foraging pattern and plant species attacked in tropical rain forest. *Journal of Animal Ecology* **37,** 387–403.

Cherrett, J. M. (1968b). Some aspects of the distribution of pest species of leaf-cutting ants in the Caribbean. In *Proceedings of the American Society for Horticultural Science, Tropical Region* **12,** 295–310.

Cherrett, J. M. (1969). A flight record for queens of *Atta cephalotes* L. (Hym., Formicidae). *Entomologists' Monthly Magazine* **104,** 255–256.

Cherrett, J. M. (1972). Some factors involved in the selection of vegetable substrate by *Atta cephalotes* (L.) (Hymenoptera: Formicidae) in tropical rain forest. *Journal of Animal Ecology* **41,** 647–660.

Cherrett, J. M. & Lewis, T. (1974). Control of insects by exploiting their behaviour. In *Biology in Pest and Disease Control*, pp. 130–146. D. Price Jones & M. E. Solomon. Oxford: Blackwell.

Cherrett, J. M. & Seaforth, C. E. (1968). Phytochemical arrestants for the leaf-cutting ants, *Atta cephalotes* (L.) and *Acromyrmex octospinosus* (Reich) with some notes on the ants' response. *Bulletin of Entomological Research* **59,** 615–625.

Cherrett, J. M., Pollard, G. V. & Turner, J. A. (1974). Preliminary observations on *Acromyrmex landolti* (For.) and *Atta laevigata* (Fr. Smith) as pasture pests in Guyana. *Tropical Agriculture, Trinidad* **51,** 69–74.

Cramer, H. H. (1967). *Plant Protection and World Crop Production*. Leverkusen: Bayer, Pflanzenschutz.

Eidmann, H. (1935). Zur Kenntnis der Blattschneiderameise *Atta sexdens* L. insbesondere ihrer Ökologie. *Zeitschrift für Angewandte Entomologie* **22,** 185–241.

Espina, E. R. & Timaure, A. (1977). Caracteristicas de los nidos de *Acromyrmex landolti* (Forel) en el oeste de Venezuela. *Revista de la Facultad de Agronomia, Universidad del Zulia* No. 4, 53–62.

Fowler, H. G. & Robinson, S. W. (1977). Foraging and grass selection by the grass-cutting ant *Acromyrmex landolti fracticornis* (Forel) (Hymenoptera: Formicidae) in habitats of introduced forage grasses in Paraguay. *Bulletin of Entomological Research* **67,** 659–666.

Gonçalves, C. R. (1971). As saúvas de Mato Grosso, Brasil (Hymenoptera, Formicidae). *Archivos do Museu Nacional, Rio de Janeiro* **54,** 249–253.

Haines, B. L. (1978). Element and energy flows through colonies of the leaf-cutting ant *Atta colombica*, in Panama. *Biotropica* **10,** 270–277.

Jonkman, J. C. M. (1977). Biology and ecology of the leaf-cutting ant *Atta vollenweideri*, Forel 1893 (Hym: Formicidae) and its impact in Paraguayan pastures. *Thesis* Universiteits bibliotheck, Leiden.

Jonkman, J. C. M. (1978). Nests of the leaf-cutting ant *Atta vollenweideri* as accelerators of succession in pastures. *Zeitschrift für Angewandte Entomologie* **86,** 25–34.

Jonkman, J. C. M. (1979a). Distribution and densities of nests of the leaf-cutting ant *Atta vollenweideri* Forel, 1893 in Paraguay. *Zeitschrift für Angewandte Entomologie* **88,** 27–43.

Jonkman, J. C. M. (1979b). Population dynamics of leaf-cutting ant nests in a Paraguayan pasture. *Zeitschrift für angewandte Entomologie* **87,** 281–293.

Jutsum, A. R., Cherrett, J. M. & Fisher, M. (1981). Interactions between the fauna of citrus trees in Trinidad and the ants *Atta cephalotes* and *Azteca* sp. *Journal of Applied Ecology* **18,** 187–195.

Labrador, J. R., Martinez, I. J. Q. & Mora, A. (1972). *Acromyrmex landolti* Forel, plaga del Pasto Guinea (*Panicum maximum*) en el estado Zulia. *Jornadas Agronomicas, Universidad del Zulia* No. 8, 12 pp.

Leston, D. (1973). The ant mosaic — tropical tree crops and the limiting of pests and diseases. *Pest Articles and News Summaries* **19,** 311–341.

Leston, D. (1978). A Neotropical ant mosaic. *Annals of the Entomological Society of America* **71,** 649–653.

Lewis, T. & Norton, G. A. (1973). Aerial baiting to control leaf-cutting ants (Formicidae: Attini) in Trinidad. III. Economic implications. *Bulletin of Entomological Research* **63,** 289–303.

Littledyke, M. & Cherrett, J. M. (1976). Direct ingestion of plant sap from cut leaves by the leaf-cutting ants *Atta cephalotes* (L.) and *Acromyrmex octospinosus* (Reich) (Formicidae, Attini). *Bulletin of Entomological Research* **66,** 205–217.

Littledyke, M. & Cherrett, J. M. (1978). Defence mechanisms in young and old leaves against cutting by the leaf-cutting ants *Atta cephalotes* (L.) and *Acromyrmex octospinosus* (Reich) (Hymenoptera: Formicidae). *Bulletin of Entomological Research* **64,** 263–273.

Mariconi, F. A. M. (1970). *As Saúvas*. São Paulo: Editôra Agronômica 'Ceres'.

Moser, J. C. (1967). Mating activities of *Atta texana* (Hymenoptera, Formicidae). *Insectes Sociaux* **14,** 295–312.

Quinlan, R. J. & Cherrett, J. M. (1979). The role of fungus in the diet of the leaf-cutting ant *Atta cephalotes* (L.). *Ecological Entomology* **4,** 151–160.

Rockwood, L. L. (1976). Plant selection and foraging patterns in two species of leaf-cutting ants (*Atta*). *Ecology* **57,** 48–61.

Stradling, D. J. (1978). The influence of size on foraging in the ant *Atta cephalotes* and the effect of some plant defence mechanisms. *Journal of Animal Ecology* **47,** 173–188.

Weber, N. A. (1969). Ecological relations of three *Atta* species in Panama. *Ecology* **50,** 141–147.

Weber, N. A. (1972). Gardening ants the Attines. *Memoirs of the American Philosophical Society* **92,** 1–146.

Wild plants in the ecology of the Desert Locust

R C Rainey*
Formerly Senior Entomologist, Desert Locust Survey, East Africa High Commission

Introduction

Wild plants provide a large and probably overwhelming proportion of the food of the Desert Locust (*Schistocerca gregaria*) and this natural vegetation is effectively exploited by the continuously nomadic populations of the pest, in a manner which involves both their flight behaviour and specific features of the environment provided by the winds and weather in which they fly. Vegetation types and structure have also been envisaged as playing a major role in the mobilization of scattered Desert Locust populations into gregarious swarms — the process of 'gregarization' which is so conspicuously manifested by this species. However, earlier ideas on the role of gregarization in the genesis of Desert Locust plagues have had to be radically revised in the light of half a century of carefully documented collective experience (Hemming *et al.*, 1979; Rainey and Betts, 1979).

Wild plants in the locust diet

Most crop damage by Desert Locusts is due to the highly-mobile swarms of long-lived flying adults that are capable of reaching the Persian Gulf (and perhaps India) from the Horn of Africa within a month or even less. They can live for more than six months, in individual swarms sometimes hundreds of square kilometres in extent, with a biomass of about 1 tonne ha^{-1} and a daily feeding rate of the same order (Fig. 1). Damage can be correspondingly devastating but is also characteristically sporadic. This is because the combined areas of the swarms, even in a major invasion, may be only of the order of 1% of the gross extent of the region invaded, and because the scattered crops commonly cover a total area which is small compared with that of the overall matrix of wild vegetation.

No quantitative estimate is possible of the overall proportion of wild plants in the diet of the Desert Locust but it is certainly large. This became clear when estimates were made of the relative vulnerability of different regions

* Present address: Old Risborough Road, Stoke Mandeville, Bucks HP22 5XJ.

Figure 1 A Desert Locust swarm over Hargeisa Airport, Somali Republic, 3 August 1960 (Photograph by A. J. Wood, Desert Locust Control Organization of Eastern Africa). The scale of airborne biomass is illustrated by the kill of some 20 000 tonnes of the locusts which was assessed following the application to this swarm of 38 tonnes of concentrated insecticide from the air, and represented probably the greater part (though certainly not the whole) of the swarm, which covered an area of about 130 km^2 (Joyce, *in* Rainey 1963). Most of this biomass had been produced by feeding on wild plants during May–July 1960 in areas of the Somali Horn too dry for cropping.

and different countries to locust damage. They were made initially to assess appropriate relative contributions to Desert Locust control costs, and involved weighting the probability of locust infestation in each unit area by the percentage area of cultivation (Anon., 1955; Bullen, 1969).

The particular importance of wild plants in the ecology of the Desert Locust arises, however, from the fact that much of the breeding of the species (and hence the feeding of the nymphal stages) is in arid areas too dry for crops. Thus the geographically well-defined major breeding areas are characterized by scanty and erratic rainfall. Typically this averages 80–400 mm year^{-1}, with coefficients of variation of about 70%, and about one year in four giving an annual total of less than 50% of the long-term mean. Such conditions largely preclude rain-grown crops, although some are produced by utilizing run-off water from upland areas with higher rainfall. These cropping systems can be important in human terms, but the areas involved are small compared with those affected by the occasional widespread and heavy rains, covering hundreds of thousands of, usually arid, square kilometres which are so characteristically and successfully exploited by the Desert Locust. Moreover, adult swarms in transit between breeding areas can feed sufficiently even in dry-season *Acacia-Commiphora* thorn-bush to be able to augment their fat reserves in the course of long-distance migration (Waloff, 1960).

Migration, desert rains and wild food-plants

The occasional exceptionally widespread and heavy rains encountered in desert regions are often accompanied by a correspondingly exceptional arrival of Desert Locusts. These subsequently lay their eggs in the soil, which then contains the moisture essential for embryonic development, and the resulting nymphs are similarly provided with appropriate food by the flush of vegetation following the rain. An explanation has been found for this close and apparently purposeful association between migratory flight, rain and the appearance of food-plants. This is associated with the displacement of the flying swarms in the down-wind direction, which is apparent from hourly fixes of swarm position together with instrumental observations of the wind at the

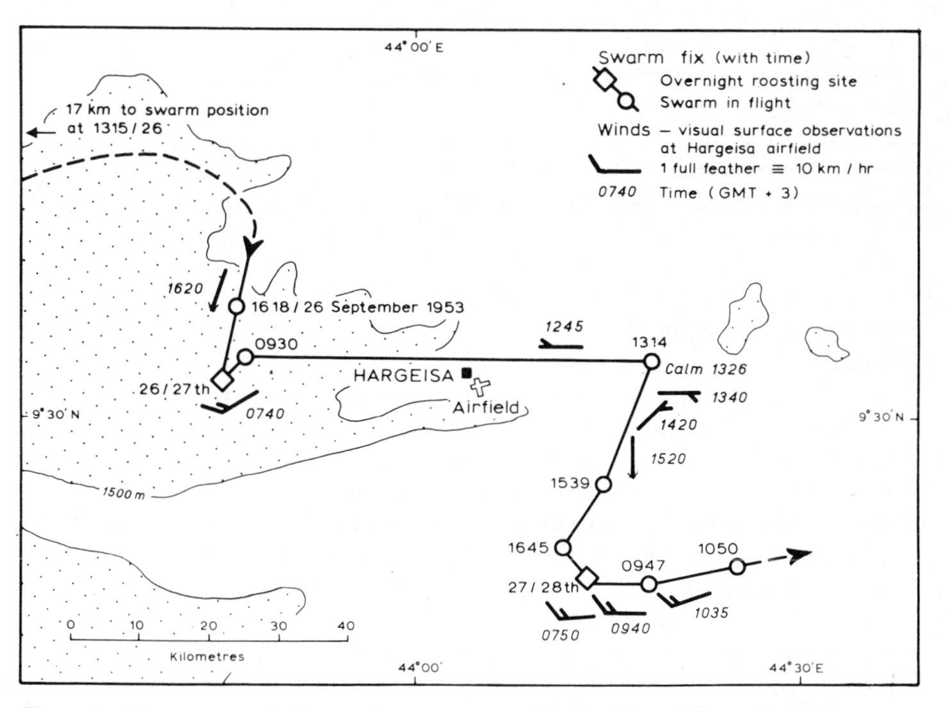

Figure 2 Movements of a Desert Locust swarm with the ebb and flow of opposing winds at the inter-tropical Front over the nothern Somali Republic 26–28 September 1953; aircraft and wind observations by H. J. Sayer and J. E. Allen (Rainey, 1976).

time and altitude of the flying locusts. The root-mean-square difference of only 15° between the direction of swarm-displacement and that of the corresponding wind was no more than was to be expected from the errors of estimation involved in obtaining comparable values of both. Moreover, changes in wind-direction, on time-scales ranging from hours (Fig. 2) to days and seasons, were associated with corresponding changes in the direction of swarm-displacement (Rainey, 1963, 1976).

Such consistently down-wind displacement necessarily means, in meteorological terms, overall movement towards and with zones of low-level wind-convergence (areas around which winds show net inflow). Low-level wind-convergence is necessarily associated with a corresponding ascent of air, and is a necessary, though not sufficient, condition for precipitation. This in essence appears to be the mechanism of the geographical and seasonal patterns of Desert Locust migration (Rainey, 1951). The convergence zones associated with the rains of these regions include, in particular, the semi-permanent inter-tropical convergence zone (ITCZ, Fig. 3) and Red Sea convergence zone (RSCZ), and more mobile and temporary fronts and other zones of convergence of the travelling westerly depressions of higher latitudes. These same wind-systems are involved somewhat differently in the transport of rust spores, as for example between Israel and eastern Africa (Dinoor, *1981*). Southward movement of locust swarms has been followed in May from Israel and neighbouring countries towards the ITCZ across the Red Sea and Egypt into Sudan. There are also particularly well documented locust movements from eastern Ethiopia to Israel and Jordan in January/February, under the influence of the RSCZ and of spells of warm southerly wind associated with the passage of depressions from the west (Rainey, 1963, etc.). These effects of depressions are also paralleled in the long-distance movement of beet leafhoppers in southwestern USA discussed by Thresh (*1981*). Further analogies between the geographical displacement of rust spores and of migrant insects in North America, as well as in Africa, are suggested in Rainey (1977).

Wind systems are thus features of the atmospheric environment which dominate much of the ecology of the Desert Locust. Under their influence, breeding by this outstandingly nomadic species often, and perhaps usually, occurs more than 1000 kilometres from the birthplace of the parent locusts, and by being able to develop in such widely separated areas of good rains, the nymphs of successive generations repeatedly exploit vegetation of a luxuriance in striking contrast with the austerity suggested by annual rainfall figures.

Gregarization and vegetation

By analogy with other species of locust, it was originally considered, and is

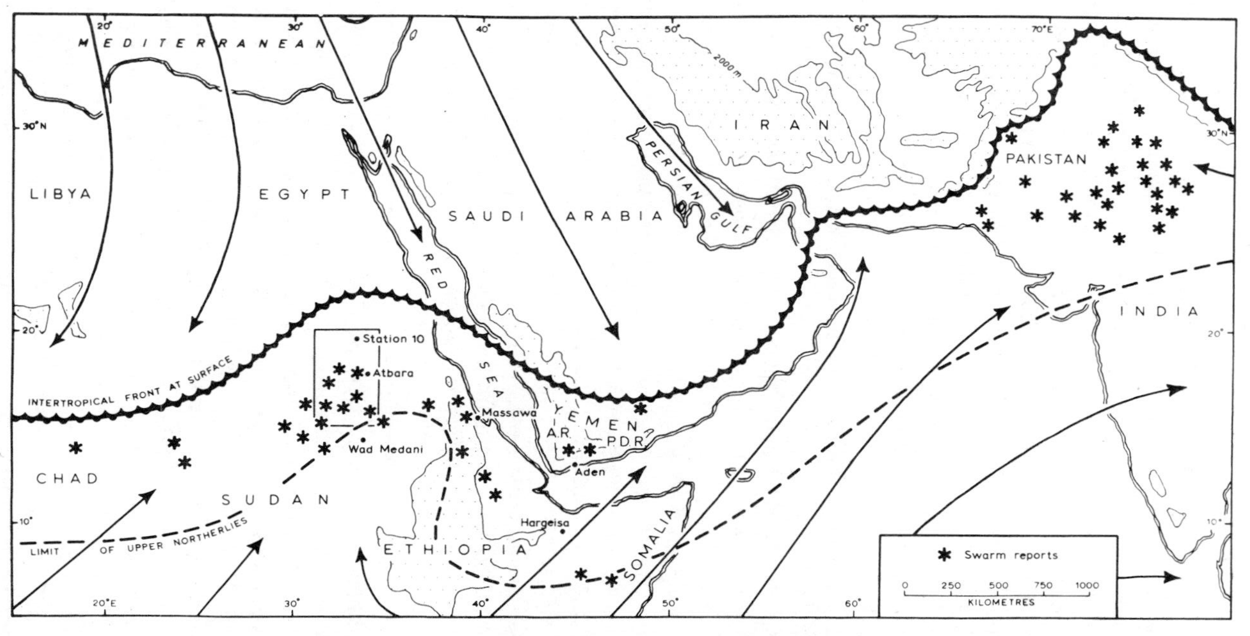

Figure 3 Desert Locust swarms in the inter-tropical convergence zone (ITCZ); swarm distribution and winds 12–31 July 1950, typical of the usual situation at this season with the ITCZ relatively stationary. *See* Fig. 5.8 of Rainey (1976) for sample of further details of the ITCZ in the rectangular area marked in northern Sudan.

still widely believed (Uvarov, 1977; Hemming *et al.*, 1979), that during quiescent periods of plague recession this species persists as small and diffuse populations of solitary-living individuals, and that new plagues of gregarious, swarming locusts arise by the process of gregarization, as a result of exceptional ecological conditions (particularly including vegetation factors). It was envisaged that these brought about a substantial increase in locust numbers, and then caused crowding. This led to the establishment and development of mutual stimulation, with characteristic sequential changes in behaviour, colour and structure, culminating in the typical swarms of gregarious flying locusts. From the early 1920s, field evidence of gregarization has been accumulated for the Desert Locust. Build-up of scattered locusts was observed in the flush of vegetation following exceptional rains, with encounters between previously solitary-living nymphs on sunlit basking sites provided by bare patches amid the vegetation. This started mutual stimulation which increased as the initial area of favourable habitat contracted due to drying out. The locust population became concentrated and typical locust swarms duly appeared.

New beginnings of Desert Locust plagues are rare events, and early on it was suggested that the exceptional ecological conditions originally thought to be required for successful gregarization, and further envisaged as occurring only in special outbreak areas of very limited extent (as indeed remains largely true for certain other species of locust), might be sufficiently exacting to make it practicable to prevent, by human intervention, the development of these special ecological conditions. This was an early concept of ecological control, which was attempted for the Red Locust (*Nomadacris septemfasciata*) in the classical outbreak areas of this species (Gunn, 1958, *et seq.*), and is recently reported to have contributed significantly to the control of the Oriental Migratory Locust (*Locusta migratoria manilensis*) in China (Guyer *et al.*, 1977).

A special long-term international Desert Locust Ecological Survey was accordingly established by FAO, UNESCO and the World Meteorological Organization in 1958. The primary objective was to make it possible for all the more important Desert Locust habitats across the vast territory exploited by the species, from India to the Atlantic, to be 'seen by a single pair of eyes', in the words of B. P. Uvarov who was a prime mover in this as in so many other projects in locust research and control. In this way it was hoped to provide useful conclusions on the ecological characteristics of these habitats. The Ecological Survey operated for six years, with funding from the United Nations Special Fund from 1960, and worked during this period in Senegal, Mauritania, Mali, Niger, Algeria, Nigeria, Chad, Sudan, Ethiopia, Saudi Arabia, Yemen, South Arabian Federation, Iran, Pakistan and India.

Uvarov always hoped for ecological control of the Desert Locust by operating against what he envisaged as initial gregarization, and this was expressed, for example, in his final paper, to the International Congress of

Entomology in Moscow in 1968 (Uvarov, 1977). The findings of the Ecological Survey (published by FAO in an extensive series of regional reports, and summarized by the Team Leader, G. B. Popov, in 1965) have certainly helped materially in the recognition, for example, on the fringes of the Sahara, of particular habitats in which scattered Desert Locusts have been effectively sought, under the guidance both of Popov and of the late Professor Pasquier, doyen of locust research in the Sahara (Baron, 1972; Abdallahi *et al.*, 1979).

The term 'outbreak area' is still applied to these areas, although perhaps somewhat loosely. The conclusion of an FAO expert panel under Uvarov's chairmanship nearly 25 years ago (FAO, 1956) that the Desert Locust has no permanent, static outbreak areas, has in fact been fully confirmed by the findings of the Ecological Survey and other subsequent experience (FAO, 1979). Nevertheless, the value for finding Desert Locusts of recognizing and characterizing these particular Saharan sites has been enhanced by the accumulating evidence of the striking mobility of recession populations of this species, so that these sites are now to be regarded as favoured halting-places (and often breeding areas) for potentially dangerous populations on the move, rather than as independent starting-points for future trouble. Thus '. . . it has been found that nearly all occurrences of Desert Locusts in numbers during periods of recession . . . can be linked together by particular sequences of migration . . . for which good precedents are available from the same areas and seasons in earlier years. This means that every occurrence of Desert Locusts in numbers during a recession can help to provide warning of where and when the next occurrence can be expected; and each such occurrence moreover provides an opportunity for preventive control operations to make a significant contribution towards prolonging the recession' (FAO, 1979).

Furthermore, each of the 27 cases of gregarization which had been observed (or inferred) up to 1967 included circumstantial evidence of a possible connection with known earlier gregarious populations (Waloff, 1966; Rainey and Betts, 1979). None of these cases of gregarization appear to have played any major part in the overall development of the plague, since substantial gregarious populations were already present elsewhere in each case. Gregarization has often occurred without upsurge of the plague, and in most and perhaps all cases appears merely to have followed a temporary dissociation of the locust population concerned. Inhibiting the process of gregarization would accordingly no longer appear to be a useful objective.

Another earlier conclusion, which was supported by the findings of the Ecological Survey, was on particular effects of vegetation structure on locust flight behaviour. The marked reduction of flight activity within grass clumps or stands was initially studied in some detail in the Rukwa valley outbreak area of the Red Locust (Rainey *et al.*, 1957) and noted as relevant also to the retention of stragglers from Desert Locust swarms (Waloff, 1957). Thus

Pennisetum cultivations had long been recognized as a favoured haunt of scattered Desert Locusts on the Red Sea coast (Maxwell-Darling, 1936). Numbers of scattered locusts were found to increase sharply in such cultivations on dates when the area concerned was being traversed by Desert Locust swarms (Morris, 1955). It would appear that in such vegetation gregarious behaviour is markedly inhibited by a restriction of mutual stimulation, and that this may be a factor in the observed retention of locusts within such areas.

Attention should be particularly directed to one of the Ecological Survey's earliest findings. This received little attention at the time, but has recently been given special significance by results obtained elsewhere in the systematic long-range aircraft exploration of wind-systems to locate and assess airborne concentrations of other major insect pests (Rainey, 1976; Greenbank *et al.*, 1980). Thus in July 1958, in the western Sahara, the Ecological Survey party reached the inter-tropical front (ITF), where the westerly monsoon was meeting drier, hotter and more hazy easterly winds. There they found the first swarm to be encountered after two months in the field in Senegal and Mauritania. The Survey's observations, moreover, indicated that the swarm was moving to and fro with the ebb and flow of the opposing winds at the front. This was exactly as John Sayer had shown swarms behaved at the ITF over the Somali Horn five years earlier (Fig. 2). With appropriately equipped aircraft, like the DC-3 used by Greenbank *et al.* (1980), carrying airborne insect-detecting radar as well as Doppler wind-finding equipment, the ITF is indeed likely to be a particularly good place to search for flying locusts, and not only for large swarms of major plague periods. Hind-sight makes clear some relevant features of the two Desert Locust upsurges which have occurred since the early 1960s. Thus important locust populations went missing at crucial stages of these build-ups, between May and November, 1977 (Rainey, 1973; Rainey and Betts, 1979), and between June and August, 1977 (Roffey, 1979). These are likely to have spent many weeks in flight in the ITF over uninhabited areas of the north-eastern Sahara. They seem to have profited yet again from the flush of desert vegetation following unusual rains attributable to surges of this same weather system.

Plant terpenoids and locust maturation

Finally, back again to the Somali Horn, where another aspect of the flush of desert vegetation after the rains has provided yet another tantalizing lead which awaits follow-up. In this area sexual maturation of the winged locusts is often delayed by 3–5 months. It is then followed by rapid and synchronous maturation, with egg-laying beginning within a few days at sites several hundred kilometres apart. This is associated with the onset of the rains, but maturation of the gonads begins before the rains actually break. In well-fed

Desert Locusts in the laboratory, a single contact with any of the terpenoids derived from *Commiphora myrrhae* (α- or β-pinene, limonene, or eugenol) hastens and synchronizes the onset of reproductive activity. This species (which provides myrrh), and other aromatic shrubs (such as *Boswellia*, which provides frankincense) occur in the characteristic thorn-bush vegetation of these important locust breeding areas. Their terpenoids reach maximum concentrations at bud-burst a week or more before the rains break. It has been suggested (Carlisle *et al.*, 1965) that contact with these terpenoids triggers the onset of maturation in the locusts, and that other aromatic shrubs like *Artemisia* and *Peucedanum* may have similar effects on locusts in other major breeding areas. This suggests yet another role of wild plants in the ecology of the Desert Locust.

Acknowledgements

Incompleteness of information on this often elusive migrant still allows for differences in interpretation, illustrated by the range of opinion included in the short bibliography below. I am grateful to my colleagues Elizabeth Betts and George Popov for commenting from differing points of view on the present paper.

Summary

Wild plants provide most of the food of the Desert Locust, the flight behaviour of which enables it to locate and exploit the temporary, but often extensive, ephemeral vegetation which follows the occasional rains in its arid habitats. Vegetation structure in what were originally envisaged as ‘outbreak areas’ contributes to the retention of stragglers from the characteristically mobile adult populations of all phases of this species. Terpenoids from desert shrubs at bud-burst may trigger the sexual maturation of the locusts.

References

Abdallahi, O. M. S., Skaf, R., Castel, J. M. & Ndiaye, A. (1979). OCLALAV and its environment: a regional international organization for the control of migrant pests. *Philosophical Transactions of the Royal Society of London* B **287,** 269–276.

Anon. (1955). In *Report of the Commission on the Desert Locust Control Organisation*. F. Mudie. Nairobi: East Africa High Commission. 87 pp.

Baron, S. (1972). *The Desert Locust*. London: Eyre Methuen.

Bullen, F. T. (1969). The distribution of damage potential of the Desert Locust. London. *Anti-Locust Memoir* 10, 44 pp. + 29 maps.

Carlisle, D. B., Ellis, P. E. & Betts, E. (1965). The influence of aromatic shrubs on sexual maturation in the Desert Locust. *Journal of Insect Physiology* **11,** 1541–1558.

Dinoor, A. (*1981*). Epidemics caused by fungal pathogens in wild and crop plants. In *Pests, Pathogens and Vegetation*, pp. 143–158. J. M. Thresh. London: Pitman.

FAO (1956). Report of the panel of experts on long-term policy of Desert Locust control, London 1955. *Food and Agriculture Organization of the United Nations Meeting Report* 1956/11.

FAO (1979). Possible further improvements in Desert Locust control during plagues and recessions. App. VI in *Report 23rd Session FAO Desert Locust Control Committee*, No. AGP/1979/M/4, pp. 57–59. Food and Agriculture Organization of the United Nations.

Greenbank, D. O., Schaefer, G. W. & Rainey, R. C. (1980). Spruce budworm moth flight and dispersal: new understanding from canopy observations, radar, and aircraft. *Memoirs of the Entomological Society of Canada* **110,** 49 pp.

Gunn, D. L. (1958). Possibilities of prevention of Red Locust plagues in Africa. In *Proceedings of the 10th International Congress Entomology, Montreal*, **3,** pp. 41–47.

Guyer, G. E. *et al.* (1977). Insect control in the People's Republic of China. *Washington National Academy of Sciences, CSCPRC Report,* No. 2, 218 pp.

Hemming, C. F., Popov, G. B., Roffey, J. & Waloff, Z. (1979). Characteristics of Desert Locust plague upsurges. *Philosophical Transactions of the Royal Society of London* B **287,** 375–386.

Maxwell-Darling, R. C. (1936). The outbreak centres of *Schistocerca gregaria* on the Red Sea coast of the Sudan. *Bulletin of Entomological Research* **27,** 37–66.

Morris, H. J. (1955). Unpublished data of Sudan Desert Locust Survey. Quoted in Rainey (1963).

Popov, G. B. (1965). Review of the work of the Desert Locust Ecological Survey. *Rome: FAO*, No. UNSF/DL/ES/8, 80 pp. + 20 photographs & 5 figures.

Rainey, R. C. (1951). Weather and the movements of locust swarms: a new hypothesis. *Nature* (London) **168,** 1057–1060.

Rainey, R. C. (1963). Meteorology and the migration of Desert Locusts. *World Meteorological Organization Technical Notes* No. 54, 117 pp. (also as *Anti-Locust Memoir* 7).

Rainey, R. C. (1973). Airborne pests and the atmospheric environment. *Weather, London* **28,** 224–239.

Rainey, R. C. (1976). Flight behaviour and features of the atmospheric environment. In *Proceedings of the 7th Symposium of the Royal Entomological Society, London*, pp. 75–112.

Rainey, R. C. (1977). Rainfall: scarce resource in 'opportunity country'. *Philosophical Transactions of the Royal Society of London* B **278,** 439–455.

Rainey, R. C. & Betts, E. (1979). Continuity in major populations of migrant pests: the Desert Locust and the African armyworm. *Philosophical Transactions of the Royal Society of London* B **287,** 359–374.

Rainey, R. C., Waloff, Z. & Burnett, G. F. (1957). The behaviour of the Red Locust in relation to the topography, meteorology and vegetation of the Rukwa rift valley, Tanganyika. *Anti-Locust Bulletin* **26,** 96 pp.

Roffey, J. (1979). Developments in the Desert Locust situation during 1976–9. *Philosophical Transactions of the Royal Society of London* B **287,** 479–485.

Thresh, J. M. (*1981*). The role of weeds and wild plants in the epidemiology of plant virus diseases. In *Pests, Pathogens and Vegetation*, pp. 53–70. J. M. Thresh. London: Pitman.

Uvarov, B. (1977). Current and future problems of acridology. In *Grasshoppers and Locusts: a Handbook of General Acridology* **2,** pp. 524–531. London: *Centre for Overseas Pest Research.*

Waloff, Z. (1957). Field observations on Desert Locusts. In Rainey, Waloff & Burnett (1957).

Waloff, Z. (1960). Fat content changes in migrating Desert Locusts. *Verhandlungen XI Internationaler Kongress für Entomologie, Wien* **1,** 735.

Waloff, Z. (1966). The upsurges and recessions of the Desert Locust plague: an historical survey. *Anti-Locust Memoir* **8,** 111 pp. London.

Section 5 **Vertebrate pests**

The role of wild plants in the ecology of mammalian crop pests

L M Gosling
Coypu Research Laboratory, Ministry of Agriculture, Fisheries and Food, Jupiter Road, Norwich, Norfolk NR6 6SP

Introduction

This paper explores the role of wild plants in the ecology of mammalian crop pests and considers how wild plants influence observed patterns of crop damage.

Of the many levels at which these topics can be discussed two have been selected as most interesting and because they pose outstanding interdisciplinary problems. The first adopts an evolutionary approach and the second is concerned with demography and particularly the role of the food supply in population processes.

An outstanding feature of mammalian herbivory, and one that has presumably arisen through the evolutionary interaction between plant and herbivore, is the great diversity of plants eaten by most species. For example, one *Microtus* species investigated ate 35 of the 45 species that are present in its habitat (Bergeron and Juillet, 1979) and a howler monkey (*Alouatta palliata*) ate 61 out of 96 tree species available (Glander, in Rockwood and Glander, 1979). Such species are conventionally termed 'generalist' herbivores. While there are mammals that specialize on only a few plants (termed 'specialists' in this paper) such as the mountain viscacha (*Lagidium peruanum*) which feeds mainly on six plants (Pearson, 1948) or the better known koala (*Phascolarctos cinereus*) and giant panda (*Ailuropoda melanoleuca*), the generalist strategy typifies the mammals to an extent that is not approached by other animal groups.

An important consideration in the evolution of the generalist strategy is that mammals are usually both active (*cf.* insects) and confined to small areas (*cf.* birds) throughout the year and must feed diversely to survive. However, wild plants have also played an active and complex role in the evolution of the generalist strategy and this is the first theme explored in this paper. The second is concerned with the role of wild plants in mammalian population dynamics both in natural and agricultural environments. This theme involves the role of the food supply in natural population processes and the consequences of integrating crops and wild plants into the diet.

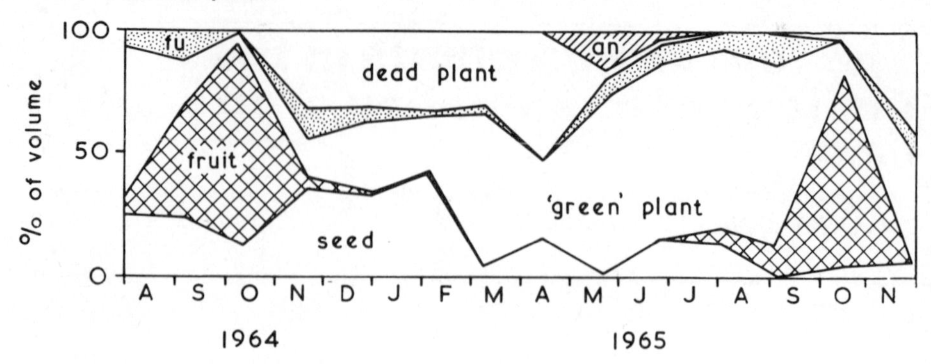

Figure 1 The year round diet of the bank vole, *Clethrionomys glareolus*, a generalist herbivore (from Watts, 1968): fu = fungi, an = animal.

The evolution of generalist mammalian herbivory

General evolutionary considerations

The way in which a herbivore exploits its food supply is shaped, in an evolutionary sense, by its own nutritional requirements and by the constraints imposed by the plants consumed. Most importantly these constraints take the form of an array of defences that limit both the individual herbivore and whole populations. The selection pressures for the defences are the patterns of herbivory at any point in the evolutionary interaction.

However, the pattern of feeding cannot remain stable because the defences also constitute selection pressures for herbivore adaptations that resist or circumvent the plant's defences. This sort of evolutionary interaction has been characterized most recently by Dawkins and J. R. Krebs (1979), as an 'arms-race' with counter-adaptation following adaptation in an escalating sequence.

A prediction from the concept of the arms-race is that a modern herbivore would find it easy to feed amongst the primitive and, presumably, weakly-defended flowering plants of the Cretaceous. While this must remain hypothetical, crop geneticists may have produced an analogous situation in breeding modern crops. These contain higher nutrient concentrations than their wild precursors, reduced defences and are sown for ease of harvest. They thus become useful and available to man, but are vulnerable to other mammalian herbivores with similar requirements (*see also* Cherrett, *1981*).

Plant defences

In the absence of plant defences, herbivores might simply forage in a way that most readily satisfies their nutrient requirements. These requirements for protein, carbohydrate, minerals, etc., have been documented in great detail for domesticated mammals (e.g. P. McDonald *et al.*, 1973). It is assumed that the same nutrients are needed by wild mammals and this work is not reviewed here. Instead, the discussion will concentrate on the constraints, imposed by plants, on the ability of herbivores to realize these requirements.

Plants possess a wide array of defences against mammalian herbivores including anatomical structures such as thorns and silicaceous inclusions, mutualistic associations with ants, indigestible structural components and a large and diverse range of chemical defences. The significance of the sharp and potentially injurious thorns of plants need not be elaborated. However, associations with ants, for example by trees of the genus *Acacia*, merit comment. The ants benefit by obtaining shelter and food (from nectaries and other nutritive structures) while the plants receive protection against insect and mammalian herbivores. Indeed, *Acacia cornigera* cannot survive after experimental removal of its ant colonies (Janzen, 1967). 'Ant-acacias' compose about 10% of the Central American acacias and only the species lacking ants contain highly toxic secondary compounds (Rehr *et al.*, 1973). As suggested by Janzen (1980) for multiple defence systems in general, the absence of these compounds from the foliage of the 'ant-acacias' hints that the energetic cost of maintaining both defence systems might be prohibitive.

The chemical defences of plants against herbivorous mammals are widely documented because of their significance in animal husbandry, pharmacology and toxicology. Many toxins (the so-called 'secondary compounds') exist in nature at concentrations that are potentially lethal or damaging to herbivores. They have been comprehensively reviewed (e.g. Levin, 1976) and only a few examples are mentioned here.

One group of toxic phenolic compounds are the isoflavenes, some of which disrupt the oestrus cycle in sheep that eat subterranean clover, *Trifolium subterraneum* (Braden and I. W. McDonald, 1970). Another simple phenolic compound, coumarin, causes liver damage in several mammals and is lethal at high concentrations. Similarly the phenol gossypol leads to depressed appetite, weakness and lethargy; death is common when the toxin exceeds 0.02–0.03% of the diet (Berardi and Goldblatt, 1969).

Two plant defences that may be interrelated are the use of cellulose as a structural element and the possession of anti-microbial toxins. Cellulose is not the only structural polysaccharide that could be used by plants since many others exist. However, cellulose is highly resistant to digestion by herbivores, which may account for its evolutionary success (Janzen, 1980). One of the mammalian counter-adaptations that will be mentioned later is that some mammals possess diverse gut micro-organisms which can degrade cellulose. These exist in a mutualistic relationship with the herbivore which gains the digestible break-down products. However, plants have responded by evolving specific toxins with an anti-microbial action. Examples are the oxygenated monoterpenes and sesquiterpenes in juniper leaves that inhibit cellulose, starch and dry matter digestion in mule deer (*Odocoileus hemionus*) (Schwartz *et al.*, 1980a).

Finally it has been suggested that plants gain protection against herbivores by association with other species. For example, the palatable grass *Themeda triandra* is less frequently grazed by some herbivores when associated with unpalatable species (McNaughton, 1978). Such dependent, and in other cases, interdependent, associations are called defence guilds (Atsatt and O'Dowd, 1976).

Herbivore adaptations

From the foregoing it might be predicted that mammals must be able to detect, select and use the nutrients they require while detecting and avoiding the effect of plant defences. In practice a compromise between these two activities is likely and this raises the possibility that a feeding strategy could be modelled using an optimization approach. The obstacles to such a procedure will be outlined in a later section, meanwhile, the emphasis is on the mechanisms involved in natural patterns of herbivory.

A primary adaptation is the ability to detect and select particular nutrients in balanced proportions. It is often difficult to determine whether a wild animal is actually selecting nutrients and the most critical information comes from laboratory experiments. For example, laboratory rats can detect subtly different foods when given a choice and then adjust the relative amounts eaten (Barnett and Spencer, 1953), and can detect vitamin and mineral deficiencies (Rogers, 1967; Rozin, 1967). All available food is sampled until the animals find a type that contains the desired constituent (Rozin and Rogers, 1967).

Laboratory rats can also detect and avoid toxins, for example selenium, in choice experiments (Franke and Potter, 1936). In the absence of choice, laboratory rats soon refuse to eat and eventually starve (Glick and Joslyn, 1970). The ability to avoid toxins has been demonstrated in nature, for example by howler monkeys (Rockwood and Glander, 1979), which avoided the mature leaves of most *Gliricidia sepium* trees, but ate from the small minority of the trees of this species that were subsequently shown to lack certain toxic alkaloids. Kendall and Sherwood (1975) similarly demonstrated that the palatability of clones of *Phalaris arundinacea* to meadow voles was inversely related to alkaloid content while Schwartz, Regelin and Nagy (1980b) showed that mule deer avoided juniper species with a high content of toxic terpenes. Swain (1977) reviewed the avoidance responses of mammals to the bitter and astringent tastes that characterize many secondary metabolites.

Learning plays a prominent role in food selection by mammals (Garcia and Hankins, 1977; Rozin, 1977). E. A. Bell and Janzen (1971) showed that representatives of three rodent genera could learn to reject macuna seeds (containing L-dopa) after ingesting less than 1 g. An adverse effect need not be immediate for such learning to occur. Rats learn to reject a specific food if the negative physiological response occurs within 6.5 hours of ingestion (Revusky, 1968). The formation of 'search images' for particular food items has been studied mainly in birds. Tinbergen (1960) suggested this concept after observing a sudden improvement in the ability of titmice (*Parus* spp.) to find a particular prey. When a search image is established the searching route is altered to increase the chance of finding a similar item nearby (Tinbergen *et al.*, 1967; Croze, 1970). Similar behaviour seems likely to occur in mammals particularly where the food items are discrete and cryptic. Finally, mammals can obtain dietary information in social encounters; for example food preferences are readily transmitted from adult to weanling rats (Galef, 1977).

The mammalian feeding apparatus is anatomically very conservative when compared, for example, with the great diversity of highly specialized insect mouth parts. The teeth of rodents, to take a group that includes 344 living genera (Simpson, 1945), are remarkably similar considering the diversity of their diet. However, this apparent uniformity conceals highly specific adaptations. An example is the specialized lower incisors of the kangaroo rat *Dipodomys microps*, which allow the animal to strip hypersaline outer tissues of *Atriplex* leaves and eat the starch-rich interior (Kenagy, 1972). The different muzzle widths of grazing ungulates in East Africa allow them to select differing proportions of leaf blade, sheath and stem from a grass sward (Gwynne and R. H. V. Bell, 1968). Muzzle width also appears to be of critical significance to such occupants of East African scrubland as gerenuk (*Litocranius walleri*) that delicately remove leaves from between needle-sharp *Acacia* thorns. Adaptations of the tongue may serve similar functions, a well known example being that of the giraffe (*Giraffa camelopardalis*) which is long and prehensile and similarly able to remove herbage while avoiding at least the older and tougher thorns that in some areas are a near-universal feature of the large range of trees and shrubs on which it feeds (Leuthold and Leuthold, 1972).

Avoiding the ingestion of toxins is clearly only one part of an adaptive syndrome because many animals eat toxic plants. For example, the mountain viscacha feeds mainly on six plants and three of these are species of *Senecio* containing high concentrations of pyrrolizidine alkaloids (Pearson, 1948); *Peromyscus leucopus* can eat *Prunus* seeds which contain cyanogenic glycosides (Whitaker, 1963) and junipers containing toxic monoterpenes sometimes form an important year-round component of mule deer diet (Anderson *et al.*, 1965).

Herbivores avoid the effects of toxins by detoxification systems based on the microsomal and bacterial enzymes (Freeland and Janzen, 1974). The microsomal enzymes are located on endoplasmic reticula and operate mainly in the liver and kidneys. They act firstly by oxidation, reduction or hydrolysis, and secondly, by conjugation to produce molecules that can be excreted in the bile or urine (Mandel, 1972; Williams, 1959). The enzymes normally exist at low concentrations and cannot function if large amounts of toxin are ingested in the initial feed. However, when the quantity of toxin eaten is gradually increased from small initial doses new enzyme is synthesized with a correspondingly increased detoxification capacity (Conney and Burns, 1972).

Detoxification by bacterial enzymes occurs in the gut and some of the complex gut structures that are generally regarded as adaptations for housing a cellulose-degrading gut flora may have a detoxification function. The guts of rodents, in which detoxification of oxalates is important, have a complex stomach structure while in contrast, the white laboratory rat has a simple gut (Shirley and Schmidt-Nielson, 1967). As in the case of the microsomal enzyme system the ability of the gut flora to degrade a toxin depends on experience. A large initial dose can be fatal while gradually increasing doses allow selection for bacteria that can live with and degrade the toxin. Thus only sheep that have experienced pyrrolizidine alkaloids can degrade them (Langan and Smith, 1970). The adaptation of a mutualistic association with a gut flora, consisting of bacteria and protozoa, that can degrade complex molecules into components which can be easily absorbed, has developed independently in several mammalian groups. The ruminant artiodactyls are particularly well known in this respect, as are their complex gut structures which allow prolonged digestion. These structures show special adaptations to varying diet both in gross and fine structures as demonstrated in the East African ruminants (Hofmann, 1968; Hofmann and Stewart, 1972), although the extent to which these adaptations combat toxins is unknown (*cf.* Shirley and Schmidt-Nielson, 1967).

Lagomorphs, rodents and primates also possess an array of structures that contain specialized floras, an enlarged caecum being common. In some cases the breakdown products of bacterial action are absorbed during a second passage of food through the gut following ingestion of faeces. Such coprophagy is very common in lagomorphs and rodents (Morot, 1882; J. S. Watson and Taylor, 1955; Gosling, 1979; Kenagy and Hoyt, 1980).

Theoretical contexts for mammalian feeding strategies

The generalist solution

Two main theoretical models seek to explain feeding at the strategic level. The first is optimal foraging theory, a branch of optimality theory that owes its early development to MacArthur and Pianka (1966) and Emlen (1966). Subsequent developments are reviewed by Schoener (1971), Pyke, Pulliam and Charnov (1977) and J. R. Krebs (1978) among others. This theory predicts that foraging animals will attempt to maximize the intake of some nutrient (usually total energy) or minimize the time spent in obtaining a particular food requirement. Westoby (1974, 1978) and Pulliam (1975) extended this approach by considering the selection of an optimal mixture of nutrients within the diet. The second model, presented most explicitly by Freeland and Janzen (1974), concentrates on the avoidance, dilution and detoxification of plant secondary compounds as determinants of feeding patterns.

Both optimal foraging theory and toxin avoidance explain why herbivores might tend to become generalists. MacArthur and Pianka (1966) modelled the circumstances in which an animal will tend to diversify its diet. Figure 2 illustrates the optimal behaviour of a foraging animal in a habitat with food items of variable quality. Optimal behaviour is defined simply as that which maximizes nutrient intake during feeding time. As the animal incorporates poor quality food items into its diet it reduces the time spent travelling between items, but average food quality declines. As more items are included an optimal point is reached where nutrient intake is maximal and beyond which it declines even though items are found very quickly. Emlen (1966) considered a related model in which animals were assumed to maximize nutrient intake by deciding whether to eat or leave items according to their quality and abundance. Further developments are described by Schoener (1971), Pyke, Pulliam and Charnov (1977) and others.

In the MacArthur and Pianka model the optimal number of food items varies with habitat quality (=average quality of food items) and generally an

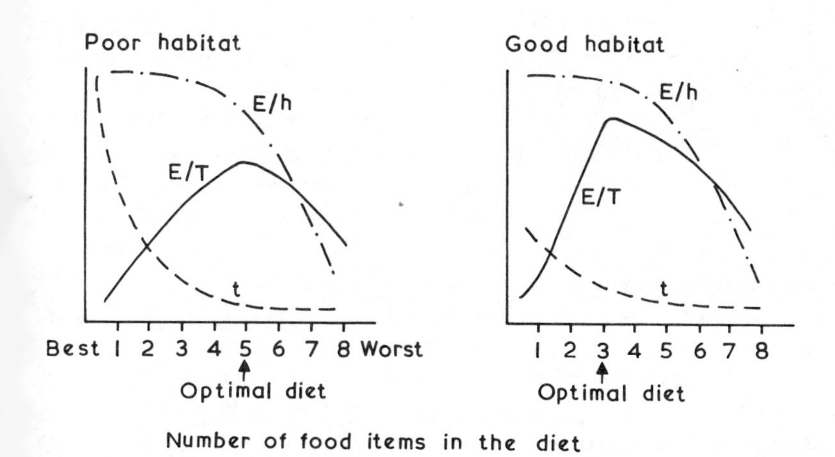

Figure 2 A graphical model of the number of food items in a diet in poor (left) and good (right) quality habitat. The model shows the consequences of progressively incorporating poorer quality food items into the diet where *t* is the time spent travelling between items, E/h the nutrient gain per item and E/T the nutrient gain per time unit expended (modified by J. R. Krebs (1978) from MacArthur and Pianka, 1966). Note that the optimum number of food items required to maximize nutrient intake is greatest in poor habitat, i.e., the animal adopts a more generalist strategy. In crop situations the optimum number of items would normally be one.

animal will increase the number of items as quality declines (Fig. 2(right)). Thus in an environment where quality fluctuates or is unstable there will be selection for generalist foragers. An analagous process occurs during the development of feeding diversity in individuals. For example, deer mice (*Peromyscus*) feed more diversely when experimentally subjected to fluctuations in the type and abundance of available food (Gray and Tardiff, 1979).

Freeland and Janzen (1974) list predictions about feeding behaviour based largely on the strategy to be expected from minimizing the effect of toxins. However, many of the predictions have analagous relevance to other plant defences. They predict that a general herbivore should:

(1) treat new foods cautiously,
(2) be able to learn to eat or reject new foods from very small samples,
(3) be able to find and eat highly specific nutrients (to 'fuel' the detoxification systems),
(4) feed mainly on staple foods while continuously sampling others,
(5) feed on familiar plants for as long as possible,
(6) prefer plants with low toxin content,
(7) have a searching strategy and body size that compromises between maximizing the number of food types available and maximizing food consumption.

Their review of the literature on observed feeding behaviour indicated general agreement with the predictions although inevitably only a few of the observations constituted real tests.

A factor that emerged in the review was that much of the feeding behaviour observed was also clearly related to nutrient selection. Thus the ability of animals to correct highly specific nutritional imbalances (Rogers, 1967; Rozin, 1967) might be adaptive for both general nutritional reasons and to maintain primed detoxification systems. Similarly it seems unnecessarily restrictive to suggest that an animal would select a plant because it has a low toxin content without taking into account its nutrient content, the time involved in finding and handling it, and other factors that are often considered in optimal foraging theory which affect the costs and benefits of feeding on a particular food in a given environment. Very few feeding studies have considered many of these factors or even both the nutrient and toxin content of herbage. An exception is the work by Oates, Waterman and Choo (1980) who found that the leaf-monkey *Presbytis johnii* selected foliage that was characterized by low tannin and fibre content and high digestibility. Belovsky (1978) found that the foraging behaviour of moose (*Alces alces*) was close to that predicted purely on the basis of maximizing nutrient intake (he used a linear programming technique related to that adopted by Westoby, 1974) and concluded that there was no need to take account of the effects of plant toxins. However, he considered very broad diet categories, e.g. 'aquatic plants', and it might be that such effects are demonstrable only at finer resolution.

Both optimal foraging and anti-toxin theory provide good reasons for animals to sample food items. In the first case animals might learn the identity of future profitable items and in the second they would prime their detoxification systems. Paradoxically however, an animal must forage sub-optimally in order to sample in areas which by definition must be relatively unprofitable. This effect has been demonstrated in captive great tits (*Parus major*) which spend more than the optimal time in experimentally created sub-optimal food patches (Smith and Sweatman, 1974; J. R. Krebs and Cowie, 1976).

It is significant that the detailed composition of the diet of generalist herbivores often varies between different locations. Nowhere is this plasticity better shown than in the diets of introduced animals. In many cases the feeding regime adopted bears many similarities to that which might be expected in a plant community that had co-evolved with the herbivores. Examples occur in the feeding behaviour of rabbits (*Oryctolagus cuniculus*) throughout their extensive introduced range. A well documented case is that of rabbits living in the herbaceous flora of a sand dune system in England (Bhadresa, 1977). These animals ate at least 20 plant species and showed distinct dietary preferences, particularly for the grass *Festuca rubra*, while other species such as *Ammophila arenaria* were avoided to a relatively large extent (Bhadresa, 1977). Coypus (*Myocastor coypus*) similarly adapt to local variations in their extensive introduced range (Hillbricht and Ryszkowski, 1961; Gosling, 1974; Willner *et al.*, 1979) but in this case they are assisted by the rather conservative composition of the wetland plant communities they inhabit. These examples, of course, are biased because they refer only to those species that survive in novel circumstances. However, they demonstrate that there is sometimes considerable flexibility in feeding behaviour and particularly in the ability to utilize nutrients and avoid toxins in a novel array of food plants.

Why should a generalist feeding pattern occur more universally in mammals than in other animal groups? The answer may be quite simply a consequence of the evolution of homeothermy and the retention of terrestrial locomotion. Mammals usually remain active throughout the year (hibernation is a rare phenomenon) and are sedentary. In contrast, many invertebrates and lower vertebrates spend periods of food shortage in dormant or quiescent states, while the second major group of homeotherms, the birds, have greatly enhanced mobility through the evolution of flight. An extreme, but common, use of flight is intercontinental migration which allows bird species to maintain similar specialist diets throughout the year. Mammals, however, must survive on the resources of relatively restricted areas and, consequently, there has been selection for the ability to feed on the plants, or the parts of plants, that provide optimum food in the different seasons. The selection pressures of radical seasonal change in the site of essential nutrients might thus be an important determinant of the generalist feeding habit and one that could be assessed by quantitative measurement of seasonal changes in optimal foraging and in toxin avoidance.

It is clear, intuitively, how the nutrient foraging and defence avoidance requirements could interact but less easy to see how these two aspects can be modelled. Some aspects are compatible, for example, the inhibition of deer rumen micro-organisms by various plant secondary compounds (Nagy *et al.*, 1964; Nagy and Tengerdy, 1968; Oh, *et al.*, 1967; Schwartz *et al.*, 1980a) and could be expressed in the same units as in conventional optimal foraging theory. Thus a deer would start to lose nutrient units, for example calories, as its consumption of a particular secondary compound exceeded a critical value and digestion was inhibited. In other words, the gains from efficient foraging and the losses from ingestion of toxins can be expressed in the same 'currency' (to use the terminology of optimality theory). There are obvious difficulties in obtaining some measurements but no fundamental modelling problems.

This is not the case when risks are involved that cannot be expressed in the same currency. For example, there are strong reasons for a herbivore to avoid eating gossypol, which can cause death (Levin, 1971), and for guinea pigs to avoid indospicine, which causes them to abort (Hutton *et al.*, 1958). Both sorts of feeding clearly reduce the inclusive fitness of the herbivores concerned but the loss involved cannot be expressed in the sort of currency that could be simply used in an optimal foraging model. In the terms used previously, risk and nutrient acquisition or loss involve incompatible currencies except when the currency is fitness; but in this case a model would cease to refer to the area currently covered by optimal foraging theory.

In spite of this fundamental obstacle to a comprehensive model it is still possible to predict many patterns of feeding behaviour and also to identify situations where the predictions of optimal foraging theory are more or less applicable. For example, it would be surprising if studies of species that risk death or abortion when feeding on particular plants, should yield results consistent with predictions based on nutrient content. However, when herbivores feed on crops in which anti-herbivore defences have been systematically removed by selective breeding, then animals should approach the simple predictions of optimal foraging theory more closely.

The specialist solution

One response to the selection pressure of escalating plant defences is the development of specific counter adaptations to a few species, so that the herbivore wins the 'arms race' by attacking a limited array of defences. This concept is instructive in energetic as well as in evolutionary terms: for example, there is a high metabolic cost in maintaining 'primed' detoxification systems for a wide range of potential toxins, many of which are seldom encountered (Janzen, 1980). Presumably specialist herbivores have specialized behavioural and physiological adaptations to the defences of the plants in their diet. The few plants that form the diet of the mountain viscacha contain high concentrations of pyrrolizidene alkaloids, which would be lethal except for the existence of appropriate detoxification processes. The koala feeds exclusively on the foliage of trees of the genus *Eucalyptus* and, through the ability to detoxify the highly toxic volatile oils that they contain, gains access to a large and relatively unexploited food supply (Degabriele, 1980).

Specialist herbivory is rare amongst the mammals despite its apparent advantages. The usual explanation for this is that single food sources are unreliable as demonstrated with pandas that feed largely from the young leaves of a few bamboo species. These flower synchronously at very long intervals, often exceeding 100 years, and after flowering exist mainly as seeds until new bamboo groves develop. Under these conditions of acute food shortage many pandas starve, as reported recently in China (Anon, 1980). Pandas must have survived such vicissitudes in the past but they are clearly vulnerable to extinction, particularly were this factor to occur in chance combination with another, such as disease.

Patterns of crop damage

The diversity of plants eaten is an obvious corollary of a generalist strategy. The adaptations of mammals to a generalized plant diet should theoretically permit the utilization of crops when these fall within the range of food types in the wild plant diet. Closer parallels might also be expected; for example, a crop is likely to be eaten at the same season as a similar wild plant. These predictions have been considered in relation to the patterns of crop feeding by coypu, which have been intensively studied in England (Gosling, 1974). Although an alien, its wetland habitat and feeding behaviour are similar to those of its original range (Murua *et al.*, 1981).

Coypus in East Anglia feed on almost the entire array of arable crops available (Norris, 1967; Gosling 1974) and, predictably, individual crops are eaten at the same growth stages as are their wild relatives. Thus cereals and wild grasses are grazed in the spring and, in the late summer, coypu stomach contents contain both the ripening seeds of wheat and barley, and the seeds of many wild grasses. Taylor and M. G. Green (1976) found similar relationships in their study of East African rodents.

However, simple relationships between crops and wild plants can break down because of the different growth cycles of wild plants and crops and because of land-use practices such as harvesting. Thus sugar beet is harvested in the autumn and early winter in East Anglia and so is unavailable during most winter months when coypus eat the roots of wild plants. Instead they eat the swollen storage stems of brassicas, particularly kale in late winter, even though a food supply of this type does not form part of the wild plant diet either in England or South America (Ellis, 1963; Gosling, 1974; Murua *et al.*, 1981).

The utilization of crops at times of year that are apparently inappropriate, is possible because of the flexibility and opportunism of generalist feeding behaviour. However, preference for an inappropriate food requires further explanation and two main factors seem responsible. The first is that crop defences have been effectively removed so that herbivores are freed from any nutritional constraint, such as reduced digestibility, and any other risk, such as death. Secondly, the concentration of nutrients in the crop is higher than in the wild plant and, from the maximizing assumption of nutrient gain per unit of time, exclusive feeding on the crop would be predicted.

The role of availability in determining patterns of crop damage is reversed in the winter when many crops are no longer available. Thus the diet of red deer (*Cervus elaphus*) in Denmark contained a significant proportion of arable crops only in the summer (Jensen, 1968). This constituent was replaced by grasses in an area where crops were relatively unavailable. Broader patterns of habitat manipulation also influence the impact of mammals, as with red deer in Scotland. The widespread removal of natural woodland and its replacement with conifer plantations has, not unexpectedly, resulted in substantial damage by browsing and bark-stripping (Mitchell *et al.*, 1977).

The diversity of crop plants available and their growth cycles also affect the impact of mammalian pests. A series of plants that grow and mature sequentially provides ideal conditions for a herbivore that can utilize one or more staple foods while sampling those that follow. Feeding on a succession of crops has been observed in the coypu and also the cotton rat (*Sigmodon hispidus*). During an outbreak of cotton rats in Honduras severe damage to rice, cotton, sugar-cane, maize, sorghum, sesame, melons and to the bark of cashew nut trees, occurred in succession from August to December (Espinoza and Rowe, 1979).

In general the literature on crop damage suggests that mammalian crop pests feed on crops as though attempting to maximize nutrient intake. One of the few exceptions to this generalization is that coypus, even when feeding on a crop that to the individual is super-abundant, still include a proportion of non-crop items in their diet. Two interpretations are possible: either these items provide essential nutrients or they are needed to prime detoxification systems. A second exception is that some animals show consistent differences in preference between different varieties of the same crops. Hares (*Lepus capensis*) for example preferred particular varieties of turnip (Hewson, 1977) and this was explained in terms of palatability. Finally, particular crop areas may be preferred because the feeding animal is at least risk to predators. Hewson (1977) gives this plausible explanation for the observation that hares always fed at the edges of trial plots, near the narrow strip of fallow that separated the plots, where approaching predators would be easily detected.

The local high intensity of crop damage seems explicable in terms of the concept of a specific 'search image' (Tinbergen, 1960) combined with the optimal foraging assumption of maximizing nutrient intake. Examples are the complete removal of cereals by rabbits, East African rodents (Taylor and Green, 1976) and coypus (Gosling, unpublished) from particular patches of fields, and similar behaviour towards sugar beet seeds which are excavated by wood mice, *Apodemus sylvaticus* (R. Green, 1978; Pelz, 1979). In general such damage occurs in runs of consecutive plants or seeds along rows. Hypothetically the herbivore locates one plant by chance while sampling novel food items, receives relevant physiological feedback (nutrient high, toxins low), establishes a search image for the item and then concentrates its search for others nearby. Such a sequence seem particularly likely when the food item is not immediately obvious as with ripe cereals and sugar beet seeds. In these cases coypus and wood mice must respectively bite through or pull down the cereal stalk, and excavate the seed before they can feed. Uniformly spaced crop stands clearly provide an ideal arrangement of food items when feeding mechanisms of this type are employed.

Role of wild plants in the population dynamics of mammalian crop pests

General considerations: the food supply

As in the previous case it is only possible to predict events in agricultural situations by first understanding the interrelationships between the herbivore and its food supply in natural situations. At a fundamental level there has been extensive exploration of the factors that limit populations of mammalian herbivores. In general terms it is almost a truism that the food supply is ultimately involved; for example, the variable mammalian biomass of the East African savannah is highly correlated with the primary production available as food (Coe *et al.*, 1976, and others). Primary production, in turn, is dependent on rainfall and soil nutrients in these areas and, in less hospitable climates, on temperature.

The main impetus of research on population limitation has been directed towards the factors that limit density in particular areas and especially towards understanding the mechanisms that prevent unlimited population increase. Such limitation appears as a series of checks or declines in number that occur at variable intervals (although they are sometimes very regular in particular populations). These restricted fluctuations are commonly explained by density-dependent regulation (Nicholson, 1954). However, there is no consensus about the relative importance of food, social behaviour and predation in these events. Lack (1954) argued that populations were normally limited by the food supply and there are a number of cases which support this conclusion; for example of buffalo (*Syncerus caffer*) in the Serengeti region of Tanzania (Sinclair, 1977). Pearson (1966) and Gibb (1977), working on mice and rabbits respectively, are among those who emphasize the importance of predation in determining the amplitude of population fluctuations or in prolonging a period of decline. However, in recent years most attention has been given to the idea of self-regulation through qualitative changes within the population (Chitty, 1952). C. J. Krebs (1978) has reviewed the three main hypotheses. Two of these, the 'behaviour hypothesis' and the 'polymorphic behaviour hypothesis', suggest that spacing-out behaviour affects key population processes. The first hypothesis places greater weight on the environmental influence on spacing behaviour. A. Watson and Moss (1970) have also pointed out that extrinsic factors, particularly the food supply, predators and weather may interact with spacing behaviour to limit population increase.

The population changes of the wood mouse, *Apodemus sylvaticus*, are probably the best understood in a mammal that can sometimes be an important crop pest. They illustrate how the food supply and behavioural factors interact to produce well defined annual fluctuations. Overwinter survival is positively correlated with the food supply and in particular the acorn crop (Fig. 3(left)). However, although breeding starts in the spring numbers do not increase until after a variable, and density-dependent delay that may be influenced by territorial behaviour (Watts, 1969; Flowerdew, 1974). As a result the autumn population density is similar in successive years (Fig. 3(right)). The most important ultimate factor that determines this density might be the long-term food supply.

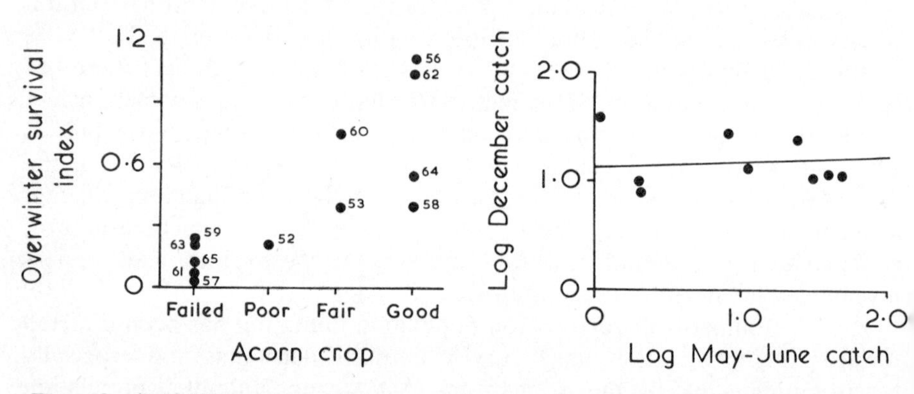

Figure 3 (left) The relationship between the overwintering survival of wood mice *Apodemus sylvaticus* in relation to the acorn crop of the previous autumn (right) the relationship between numbers of mice in May–June and December; the similar numbers caught annually in December suggest the operation of an efficient density-dependent regulating process (from Watts, 1969).

Experimentally increased food supplies have been shown to affect populations of several rodents. Unseasonal breeding has been induced in *Apodemus sylvaticus* (Watts, 1970; Smyth, 1966; Hansson, 1971), *Clethrionomys glareolus* (Andrzejewski, 1975) and two *Peromyscus* species (Jameson, 1953).

Flowerdew (1972) showed improved survival of wood mice when given supplementary food. However, conflicting experimental results are available for other species. For example, C. J. Krebs and DeLong (1965) found supplementary feeding had little effect on a population of *Microtus californicus*.

An important area for future investigation is the relationship between social behaviour and the individual's food supply. For example, how does the quality of the defended food supply affect the intensity of agonistic behaviour of male wood mice?

A general difficulty in many investigations of population limitation is the assumption that the mechanisms involved, for example spacing behaviour, are functionally concerned with population processes. In fact consideration of the level at which selection operates indicates that this is unlikely. A more profitable approach might be to consider the adaptive significance of spacing behaviour for individuals, or kin, and to test whether population changes can be predicted from these adaptive traits.

Wild plants as shelter

There is little doubt that wild plants are important in providing shelter from inclement weather and predators. Red deer in Scotland shelter in long vegetation, particularly during cold weather (Staines, 1976) when the heat loss from individual deer in exposed situations is twice that of deer in woodland (Grace and Easterbee, 1979).

Birney, Grant and Baird (1976) showed that normal *Microtus ochrogastor* and *M. pennsylvanicus* cycles only occur when grassland cover exceeds a critical threshold. Population density remains low and relatively stable when cover is low. They list several features that might contribute to this effect including protection from predators (when at low densities), the food supply and a benign micro-environment. Many mammalian herbivores employ crypsis as an anti-predator strategy and often such behaviour is dependent on particular properties of wild plants. M. G. Green and Taylor (1975) showed experimentally that removing vegetation cover led to increased predation, mainly by birds, on several species of rodents in East African grassland.

Population dynamics in agricultural situations

Most agricultural systems are characterized by extreme instability. Periods of extreme abundance are followed, after harvest, by periods of food scarcity. Wild plants in limited natural or semi-natural communities within the crop area or as weeds in fields generally differ both quantitatively and qualitatively from extensive natural plant communities whose composition is a consequence of the co-evolution of its components and their herbivores.

In such situations mammalian generalist herbivores face the general problems outlined previously, that is, how to gain access to an array of food plants that is sufficiently diverse to maintain an adequate year-round nutrient intake and minimize the impact of toxins and other defences. Unfortunately there are very few relevant field studies. Descriptions of crop damage and investigations of natural diets are common but those of the way in which mammals integrate the wild plant and crop components of their diet are rare. Exceptions are studies of *Arvicanthis niloticus* and other rodent pests in Kenya (Taylor, 1968; Taylor and M. G. Green, 1976; M. G. Green and Taylor, 1975), of the wood mouse in England (R. Green, 1979) and of red deer in Denmark (Jensen, 1968).

The study by Taylor and Green is a particularly clear example of the dietary interaction of wild plants and crops and of the response by the pest population. *Arvicanthis* began breeding 2 months after the start of the rains (in a water-limited environment) at the time when seeds of wild plants started to appear in the diet (Fig. 4). This event, rather than the consumption of cereals which started one or more months later, appeared to trigger the onset of reproduction, and its cessation appeared linked with a dietary switch to vegetative plant tissue.

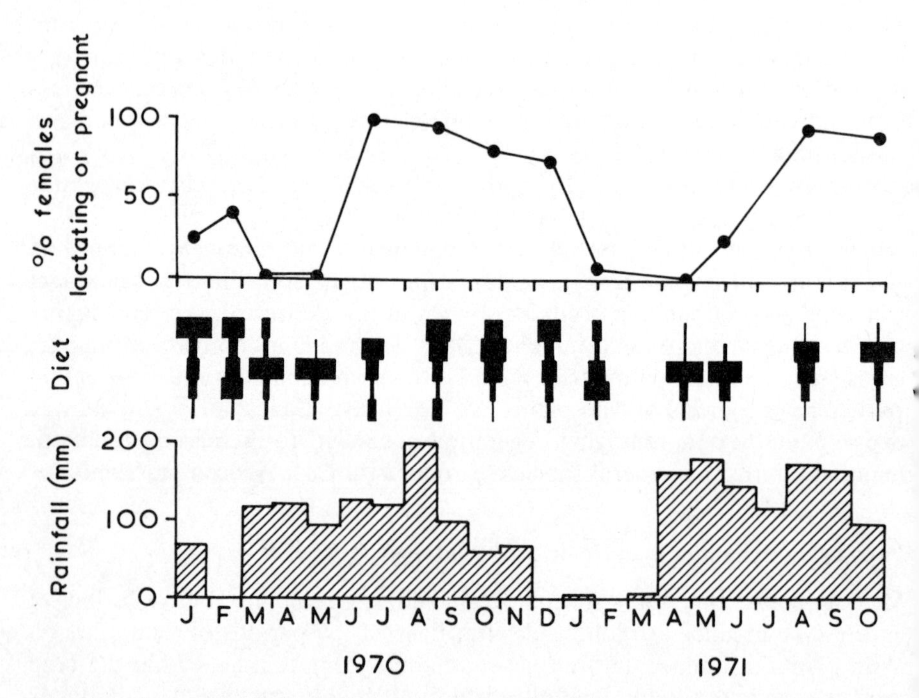

Figure 4 Seasonal relationships between rainfall and the reproduction and diet of the mouse *Arvicanthis niloticus* in arable farmland in Kenya. The diet index was based on an estimate of the volume of each constituent: C = cereals; S = seeds; G = grass; W = weeds; I = insects (adapted from Taylor and Green, 1975).

The availability of the wild plants that effect *Arvicanthis* reproduction, and crop production depends on the quantity of rain. In 1961 a year of particularly heavy rain, there was massive production of both weeds and crops and this was followed by a widespread eruption in the number of field rodents (Taylor, 1968). Similar events have been recorded in California (Brooks and Barnes, 1972) and Bulgaria (Straka and Gerasimov, 1971), suggesting that rodent populations are often food-limited in arable areas.

Arvicanthis niloticus in Kenya is an important crop pest that is at least partially dependent on wild plants. However, the study of wood-mice (R. Green, 1978 and 1979), which sometimes cause significant damage to sugar beet crops, yielded rather different results. Wood-mice had previously been shown to make short excursions from hedgerows into fields where they sometimes bred before returning to adjacent woodland after harvest (Kikkawa, 1964; Bergstedt, 1966; Flowerdew, 1976). However the mice studied by Green (1979) and Pelz (1979), in England and Germany respectively, lived in open fields throughout the year. The wild plants that formed an important part of their diet consisted largely of plants growing within crops, particularly annual meadow grass (*Poa annua*) and chickweed (*Stellaria media*). Surprisingly, the annual cycle of numbers in this open-field situation was rather similar to that described in woodland (Tanton, 1965; Bergstedt, 1965; Watts, 1969; Crawley, 1970). Some population processes might thus remain similar even when the species composition of the food supply is drastically altered, although Pelz (1979) has suggested the alternative that low summer densities are a result of a poor food supply and not of density-dependent regulation.

Green's wood mouse study raises a number of intriguing points about the role of wild plants which deserve further study. It has generally been assumed that hedgerows provide refuges for crop pests such as wood-mice, but this is clearly not always so. The shelter belt in Green's study appeared to be generally avoided and its removal, in the sort of field enlargement programme that is currently pursued (*see* Trask, *1981*) would have affected mainly the non-pest voles, *Clethrionomys giareolus* and *Microtus agrestis*. It has also been shown that such areas provide suitable habitat for mouse predators such as the weasel (*Mustela nivalis*) (Moors, 1974). This suggests that areas of wild vegetation might sometimes play a positive role in limiting populations of mammalian pests, as van Emden (1974, *1981*) has discussed for insect pests.

Conclusions

The role of wild plants in the ecology of mammalian crop pests has been explored using two main themes. The first was the role of wild plants, and in particular their defence systems, in shaping the generalist strategy that typifies the mammalian herbivore. Secondly, the role of the food supply in mammalian population limitation was briefly reviewed. Both areas of investigation have revealed complex adaptive syndromes that are imperfectly expressed in current theory, but which have clear relevance for any attempt to predict both patterns of crop damage and the population dynamics of crop pests.

It might be expected that the generalist mammalian herbivore would conform with the predictions of optimal foraging theory; i.e. maximizing nutrient intake, or minimizing time spent feeding, except when constrained by plant defences. However, plant defences are ubiquitous and such opportunities must be extremely rare in natural systems so that observed feeding behaviour should normally be a compromise between obtaining nutrients and avoiding defences. There is some evidence for such a compromise in the literature, but few studies constitute real tests of this hypothesis and, in general, the literature on these two areas of theory has remained disappointingly separate.

A central point that emerges in pursuing both evolutionary and demographic themes is the extreme flexibility of mammalian feeding patterns. Mammals show few of the constraints exhibited by groups with more specialized feeding behaviour and crops can thus be incorporated into a basically opportunistic strategy. This ability assumes special significance in the dynamics of crop pests which, in the few cases that have been investigated, show a remarkable ability to integrate wild plants and crops in a year-round food supply.

References

Anderson, A. E., Snyder, W. A. & Brown, G. W. (1965). Stomach content analysis related to condition in mule deer, Guadalupe Mountains, New Mexico. *Journal of Wildlife Management* **29,** 352–365

Andrzejewski, R. (1975). Supplementary food and winter dynamics of bank vole populations. *Acta Theriologica* **20,** 23–40.

Anon. (1980). A blooming disaster for starving pandas. *New Scientist* **86,** 133.

Atsatt, P. R. & O'Dowd, D. J. (1976). Plant defence guilds. *Science* **193,** 24–29.

Barnett, S. A. & Spencer, M. M. (1953). Experiments on the food preferences of wild rats (*Rattus norvegicus* Berkenhout). *Journal of Hygiene* **51,** 16–34.

Bell, E. A. & Janzen, D. H. (1971). Medical and ecological considerations of L-dopa and 5-HTP in seeds. *Nature* (London) **229,** 136–137.

Belovsky, G. E. (1978). Diet optimization in a generalist herbivore: the moose. *Theoretical Population Biology* **14,** 105–134.

Berardi, L. C. & Goldblatt, L. A. (1969). Gossypol. In *Toxic Constituents of Plant Foodstuffs*, pp. 212–266. I. E. Liener. New York: Academic Press.

Bergeron, J. M. & Juillet, J. (1979) L'alimentation estivale du campagnol des champs, *Microtus pennsylvanicus*, Ord. *Canadian Journal of Zoology* **57,** 2028–2032.

Bergstedt, B. (1965). Distribution, reproduction, growth and dynamics of the rodent species *Clethrionomys glareolus* (Schreber), *Apodemus flavicollis* (Melchior) and *Apodemus sylvaticus* (Linne) in southern Sweden. *Oikos* **16,** 132–160.

Bergstedt, B. (1966). Home ranges and movements of *Clethrionomys glareolus, Apodemus flavicollis* and *Apodemus sylvaticus* in Sweden. *Oikos* **17,** 150–157.

Bhadresa, R. (1977). Food preferences of rabbits *Oryctolagus cuniculus* L. at Holkham sand dunes, Norfolk. *Journal of Applied Ecology* **14,** 287–291.

Birney, E. C., Grant, W. E. & Baird, D. D. (1976). Importance of vegetative cover to cycles of *Microtus* populations. *Ecology* **57,** 1043–1051.

Braden, A. W. H. & McDonald, I. W. (1970). Disorders of grazing animals due to plant constituents. In *Australian Grasslands*, pp. 381–392. R. M. Moore. Canberra: Australian National University Press.

Brooks, J. E. & Barnes, A. M. (1972). An outbreak and decline of Norway rat populations in California rice fields. *California Vector Views* **19,** 6–14.

Cherrett, J. M. (*1981*). The interaction of wild vegetation and crops in leaf-cutting ant attack. In *Pests, Pathogens and Vegetation*, pp. 315–325. J. M. Thresh. London: Pitman.

Chitty, D. (1952). Mortality among voles (*Microtus agrestis*) at Lake Vyrnwy, Montgomeryshire in 1936–39. *Philosophical Transactions of the Royal Society, London, Series B* **236,** 505–552.

Coe, M. J., Cumming, D. H. & Phillipson, J. (1976). Biomass and production of large African herbivores in relation to rainfall and primary production. *Oecologia* **22,** 341–354.

Conney, A. H. & Burns, J. J. (1972). Metabolic interactions among environmental chemicals and drugs. *Science* **179,** 576–586.

Crawley, M. C. (1970). Some population dynamics of the bank vole *Clethrionomys glareolus* and the wood mouse *Apodermus sylvaticus* in mixed woodland. *Journal of Zoology* **160,** 71–89.

Croze, H. (1970). Searching image in carrion crows. *Zeitschrift für Tierpsychologie* **5,** 1–86.

Dawkins R, & Krebs, J. R. (1979). Arms races between and within species. *Proceedings of the Royal Society of London, Series B* **205,** 489–511.

Degabriele, R. (1980). The physiology of the koala. *Scientific American* **243,** 94–99.

Ellis, E. A. (1963). Some effects of selective feeding by the coypu (*Myocastor coypus*) on the vegetation of Broadland. *Transactions of the Norfolk and Norwich Naturalists Society* **20,** 32–35.

van Emden, H. F. (1974). *Pest Control and its Ecology*. Studies in Biology, No. 50. London: Edward Arnold.

van Emden, H. F. (*1981*). Wild plants in the ecology of insect pests. In *Pests, Pathogens and Vegetation*, pp. 251–261 J. M. Thresh. London: Pitman.

Emlen, J. M. (1966). The role of time and energy in food preference. *American Naturalist* **100,** 611–617.

Espinoza, H. R. & Rowe, F. P. (1979). Biology and control of the cotton rat, *Sigmodon hispidus*. *PANS* **25,** 251–256.

Flowerdew, J. R. (1972). The effect of supplementary food on a population of wood mice (*Apodemus sylvaticus*). *Journal of Animal Ecology* **41,** 553–566.

Flowerdew, J. R. (1974). Field and laboratory experiments on the social behaviour and population dynamics of the wood mouse (*Apodemus sylvaticus*). *Journal of Animal Ecology* **43,** 499–511.

Flowerdew, J. R. (1976). The effect of a local increase in food supply on the distribution of woodland mice and voles. *Journal of Zoology* **180,** 509–513.

Franke, K. W. & Potter, V. R. (1936). The ability to discriminate between diets of varying degrees of toxicity. *Science* **83,** 330–332.

Freeland, W. J. & Janzen, D. H. (1974). Strategies in herbivory by mammals: the role of plant secondary compounds. *The American Naturalist* **108,** 269–289.

Galef, B. G. (1977). Mechanisms for the social transmission of acquired food preferences from adult to weanling rats. In *Learning Mechanisms in Food Selection*, pp. 123–148. L. M. Barker, M. R. Best & M. Domjan. Baylor Scientific Press.

Garcia, J. & Hankins, W. G. (1977). On the origin of food aversion paradigms. In *Learning Mechanisms in Food Selection*, pp. 3–19. L. M. Barker, M. R. Best & M. Domjan. Baylor Scientific Press.

Gibb, J. A. (1977). Factors affecting population density in the wild rabbit, *Oryctolagus cuniculus* (L) and their relevance to small mammals. In *Evolutionary Ecology*, pp. 33–46. Stonehouse & C. M. Perrins. London: Macmillan.

Glick, Z. & Joslyn, M. A. (1970). Food intake, depression and other metabolic effects of tannic acid in the rat. *Journal of Nutrition* **100,** 509–515.

Gosling, L. M. (1974). The coypu in East Anglia. *Transactions of the Norfolk and Norwich Naturalists Society* **23,** 49–59.

Gosling, L. M. (1979). The twenty-four hour activity cycle of captive coypus (*Myocastor coypus*). *Journal of Zoology* **187,** 341–367.

Grace, J. & Easterbee, N. (1979). The natural shelter for red deer (*Cervus elaphus*) in a Scottish glen. *Journal of Applied Ecology* **16,** 37–48.

Gray, L. & Tardif, R. R. (1979). Development of feeding diversity in deer mice. *Journal of Comparative and Physiological Psychology* **93,** 1127–1135.

Green, M. G. & Taylor, K. D. (1975). Preliminary experiments in habitat alteration as a means of controlling field rodents in Kenya. *Ecological Bulletin* **19,** 175–186.

Green, R. (1978). Wood mice taking (sugar beet) seed. *Rothamsted Experimental Station Report for 1978*, Part 1, p. 62.

Green, R. (1979). The ecology of wood mice (*Apodemus sylvaticus*) on arable farmland. *Journal of Zoology* **188,** 357–377.

Gwynne, M. D. & Bell, R. H. V. (1968). Selection of vegetation components by grazing ungulates in the Serengeti National Park. *Nature* (London) **220,** 390–393.

Hansson, L. (1971). Small rodent food, feeding and population dynamics. A comparison between granivorous and herbivorous species in Scandinavia. *Oikos* **22,** 183–198.

Hewson, R. (1977). Food selection by brown hares (*Lepus capensis*) on cereal and turnip crops in north-east Scotland. *Journal of Applied Ecology* **14,** 779–785.

Hillbricht, A. & Ryszkowski, L. (1961). Investigation of the utilization and destruction of its habitat by a population of coypu — *Myocastor coypus* Molina bred in semi-captivity. *Ekologia Polska, Series A* **9,** 506–524.

Hofmann, R. R. (1968). Comparisons of rumen and omasum structure in East African game ruminants in relation to their feeding habits. In *Comparative Nutrition of Wild Animals*, pp. 179–194. M. A. Crawford. London: Academic Press.

Hofmann, R. R. & Stewart, D. R. M. (1972). Grazer or browser: a classification based on the stomach structure and feeding habits of East African ruminants. *Mammalia* **36,** 226–240.

Hutton, E. M., Windrum, G. M. & Kratzing, C. C. (1958). Studies on the toxicity of *Indigofera endecaphylla*. *Journal of Nutrition* **64,** 321–338.

Jameson, E. W. (1953). Reproduction in deer mice in the Sierra Nevada, California. *Journal of Mammalogy* **34,** 44–58.

Janzen, D. H. (1967). Interaction of the bull's horn acacia (*Acacia cornigera* L.) with an ant inhabitant (*Pseudomyrmex ferruginea,* F. Smith) in eastern Mexico. *Kansas University Science Bulletin* **67,** 315–358.

Janzen, D. H. (1980). Evolutionary ecology of personal defence. In *Physiological Ecology: an Evolutionary Approach to Resource Use.* C. R. Townsend and P. Calow, Oxford: Blackwell.

Jensen, P. V. (1968). Food selection of the Danish red deer (*Cervus elaphus* L.) as determined by examination of the rumen content. *Danish Review of Game Biology* **5,** 1–44.

Kenagy, G. J. (1972). Saltbush leaves: excision of the hypersaline tissue by a kangaroo rat. *Science* **178,** 1094–1096.

Kenagy, G. J. & Hoyt, D. F. (1980). Reingestion of faeces in rodents and its daily rhythmicity. *Oecologia* (Berlin) **44,** 403–409.

Kendall, W. A. & Sherwood, R. T. (1975). Palatability of leaves of tall fescue and reed canary grass and some of their alkaloids to meadow voles. *Journal of Agronomy* **67,** 667–671.

Kikkawa, J. (1964). Movement, activity and distribution of the small rodents *Clethrionomys glareolus* and *Apodemus sylvaticus* in woodland. *Journal of Animal Ecology* **33,** 259–299.

King, C. M. (1977). Weasel, *Mustela nivalis.* In *The Handbook of British Mammals*, pp. 338–345. G. B. Corbet & H. N. Southern. Oxford: Blackwell.

Krebs, C. J. (1978). A review of the Chitty hypothesis of Population regulation. *Canadian Journal of Zoology* **56,** 2463–2480.

Krebs, C. J. & Delong, K. T. (1965). A *Microtus* population with supplemental food. *Journal of Mammalogy* **46,** 566–573.

Krebs, J. R. (1978). Optimal foraging: decision rules for predators. In *Behavioural Ecology, an Evolutionary Approach*, pp. 23–63. J. R. Krebs & N. B. Davies. Oxford: Blackwell.

Krebs, J. R. & Cowie, R. J. (1976). Foraging strategies in birds. *Ardea* **64,** 98–116.

Lack, D. (1954). *The Natural Regulation of Animal Numbers.* Oxford: Clarendon Press.

Langan, G. W. & Smith, L. W. (1970). Metabolism of pyrrolizidine alkaloids in ovine rumen. I. Formation of 7 alpha-hydroxyl-alpha-methyl-8 alpha-pyrrolizidene from heliotrine and basiocarpine. *Australian Journal of Agricultural Research* **21,** 493–500.

Leuthold, B. M. & Leuthold, W. (1972). Food habits of giraffe in Tsavo National Park, Kenya. *East African Wildlife Journal* **10,** 129–141.

Levin, D. A. (1971). Plant phenolics: an ecological perspective. *American Naturalist* **105,** 157–181.

Levin, D. A. (1976). The chemical defences of plants to pathogens and herbivores. *Annual Review of Ecological Systems* **7,** 121–159.

MacArthur, R. H. & Pianka, E. R. (1966). On optimal use of a patchy environment. *American Naturalist* **100,** 603–609.

McDonald, P. Edwards, R. A. & Greenhalgh, J. F. D. (1973). *Animal Nutrition.* London: Longman.

McNaughton, S. J. (1978). Serengeti ungulates: feeding selectivity influences the effectiveness of plant defence guilds. *Science* **199,** 806–807.

Mandel, H. G. (1972). Pathways of drug biotransformation: biochemical conjugations.

In *Fundamentals of Drug Metabolism and Drug Distribution* B. N. La Du, H. G. Mandel & E. J. Way. Baltimore: Williams and Wilkins.

Mitchell, B., Staines, B. W. & Welch, D. (1977). Ecology of red deer — A research review relevant to their management in Scotland. Cambridge: *Institute of Terrestrial Ecology*, pp. 1–74.

Moors, P. J. (1974). The annual energy budget of a weasel (*Mustela nivalis* L.) population in farmland. *PhD Thesis*, University of Aberdeen. (Cited in King (1977)).

Morot, C. (1882). Des pelotes stomacales des léporides. Mémoire de la Société Centrale de Médicine Vétérinaire (Paris) **12,** Series 1.

Murua, R., Neuman, O. & Dropelmann, Y. J. (1981). Food habits of *Myocastor coypus* (Molina) in Chile. In *Proceedings of International Furbearer Conference, Frostburg, Maryland* (in press).

Nagy, J. G. & Tengerdy, R. P. (1968). Antibacterial action of essential oils of *Artemisia* as an ecological factor. II. Antibacterial action of the volatile oils of *Artemisia tridentata* (big sagebrush) on bacteria from the rumen of mule deer. *Applied Microbiology* **16,** 441–444.

Nagy, J. G., Steinhoff, H. W. & Ward, G. M. (1964). Effects of essential oils of sagebrush on deer rumen microbial function. *Journal of Wildlife Management* **28,** 785–790.

Nicholson, A. J. (1954). An outline of the dynamics of animal populations. *Australian Journal of Zoology* **2,** 9–65.

Norris, J. D. (1967). A campaign against feral coypus (*Myocastor coypus* Molina) in Great Britain. *Journal of Applied Ecology* **4,** 191–199.

Oates, J. F., Waterman, P. G. & Choo, G. M. (1980). Food selection by the south Indian leaf-monkey, *Presbytis johnii*, in relation to leaf chemistry. *Oecologia* **45,** 45–56.

Oh, H. K., Sakai, T., Jones, M. B. & Longhurst, W. M. (1967). Effects of various essential oils isolated from Douglas fir needles upon sheep and deer rumen microbial activity. *Applied Microbiology* **15,** 777–784.

Pearson, O. P. (1948). Life history of mountain viscachas in Peru. *Journal of Mammalogy* **29,** 345–374.

Pearson, O. P. (1966). The prey of carnivores during one cycle of mouse abundance. *Journal of Animal Ecology* **35,** 217–233.

Pelz, H.-J. (1979). Die waldmaus, *Apodemus sylvaticus* L., auf ackerflächen: populationsdynamik, saatschäden und abwehrmöglichkeitein. *Zeitschrift für Angewandte Zoologie* **66,** 261–280.

Pulliam, H. R. (1975). Diet optimization with nutrient constraints. *American Naturalist* **109,** 765–768.

Pyke, G. H., Pulliam, H. R. & Charnov, E. L. (1977). Optimal foraging: a selective review of theory and tests. *Quarterly Review of Biology* **52,** 137–154.

Rehr, S. S., Feeny, P. P. & Janzen, D. H. (1973). Chemical defence in Central American non-ant-acacias. *The Journal of Animal Ecology* **42,** 405–416.

Revusky, S. H. (1968). Aversion to sucrose produced by contingent X-irradiation. *Journal of Comparative and Physiological Psychology* **65,** 17–22.

Rockwood, L. L. & Glander, K. E. (1979). Howling monkeys and leaf-cutting ants: comparative foraging in a tropical deciduous forest. *Biotropica* **11,** 1–10.

Rogers, W. L. (1967). Specificity of specific hungers. *Journal of Comparative and Physiological Psychology* **64,** 49–58.

Rozin, P. (1967). Specific aversions as a component of specific hungers. *Journal of Comparative and Physiological Psychology* **64,** 237–242.

Rozin, P. (1977). The significance of learning mechanisms in food selection: some biology, psychology and sociology of science. In *Learning Mechanisms in Food Selection*, pp. 557–589. L. M. Barker, M. R. Best & M. Domjan. Baylor Scientific Press.

Rozin, P. & Rogers, W. (1967) Novel diet preferences in vitamin deficient rats and rats recovering from vitamin deficiency. *Journal of Comparative and Physiological Psychology* **63,** 421–428.

Schoener, T. W. (1971). Theory of feeding strategies. *Annual Review of Ecological Systems* **11,** 369–404.

Schwartz, C. C., Nagy, J. G. & Regelin, W. L. (1980a). Juniper oil yield, terpenoid concentration, and antimicrobial effects on deer. *Journal of Wildlife Management* **44,** 107–113.

Schwartz, C. C., Regelin, W. L. & Nagy, J. G. (1980b). Deer preference for juniper forage and volatile oil treated foods. *Journal of Wildlife Management* **44,** 114–120.

Shirley, E. K. & Schmidt-Nielson (1967). Oxalate metabolism in the pack rat, sand rat, hamster and white rat. *Journal of Nutrition* **91,** 496–502.

Simpson, G. G. (1945). The principles of classification and a classification of mammals. *Bulletin of the American Museum of Natural History* **85,** 1–350.

Sinclair, A. R. E. (1977). *The African Buffalo.* Chicago: University of Chicago Press.

Smith, J. N. M. & Sweatman, H. P. A. (1974). Food searching behaviour of titmice in patchy environments. *Ecology* **55,** 1216–1232.

Smyth, M. (1966). Winter breeding in woodland mice, *Apodemus sylvaticus*, and voles, *Clethrionomys glareolus* and *Microtus agrestis* near Oxford. *Journal of Animal Ecology* **35,** 471–485.

Staines, B. W. (1976). The use of natural shelter by red deer (*Cervus elaphus*) in relation to weather in north-east Scotland. *Journal of Zoology* **180,** 1–8.

Straka, F. & Gerasimov, S. (1971). Correlations between some climatic factors and the abundance of *Microtus arvalis* in Bulgaria. *Annales Zoologici Fennici* **8,** 113–116.

Swain, T. (1977). Secondary compounds as protective agents. *Annual Review of Plant Physiology* **28,** 479–501.

Tanton, M. T. (1965). Problems of live trapping and population estimation for the wood mouse, *Apodemus sylvaticus* (L). *Journal of Animal Ecology* **34,** 1–22.

Taylor, K. D. (1968). An outbreak of rats in agricultural areas of Kenya in 1962. *East African Agricultural and Forestry Journal* **34,** 66–77.

Taylor, K. D. & Green, M. G. (1976). The influence of rainfall on diet and reproduction in four African rodent species. *Journal of Zoology* **180,** 367–389.

Tinbergen, L. (1960). The natural control of insects in pinewoods. I. Factors influencing the intensity of predation by song birds. *Archives néerlandaises de Zoologie* **13,** 265–343.

Tinbergen, N., Impekoven, M. & Franck, D. (1967). An experiment on spacing out as defence against predation. *Behaviour* **28,** 307–321.

Trask, A. B. (*1981*). Changing patterns of land use in England and Wales. In *Pests, Pathogens and Vegetation*, pp. 39–49. J. M. Thresh. London: Pitman.

Watson, A. & Moss, R. (1970). Dominance, spacing behaviour and aggression in relation to population limitation in vertebrates. In *Animal Populations in Relation to Their Food Resources*, pp. 167–220. A. Watson. Oxford: Blackwell.

Watson, J. S. & Taylor, R. H. (1955). Reingestion in the hare *Lepus europaeus* Pal. *Science* **121,** 314.

Watts, C. H. S. (1968). The food eaten by wood mice, *Apodemus sylvaticus* and bank voles, *Clethrionomys glareolus*, in Wytham Woods, Berkshire. *Journal of Animal Ecology* **37,** 25–41.

Watts, C. H. S. (1969). The regulation of wood mouse (*Apodemus sylvaticus*) numbers in Wytham Woods, Berkshire. *Journal of Animal Ecology* **38,** 285–304.

Watts, C. H. S. (1970). Effect of supplementary food on breeding in woodland rodents. *Journal of Mammalogy* **51,** 169–171.

Westoby, M. (1974). An analysis of diet selection by large generalist herbivores. *American Naturalist* **108,** 290–304.

Westoby, M. (1978). What are the biological bases of varied diets? *American Naturalist* **112,** 627–631.

Whitaker, J. O. (1963). Food of 120 *Peromyscus leucopus* from Ithaca, New York. *Journal of Mammalogy* **44,** 418–419.

Williams, R. T. (1959). *Detoxification Mechanisms.* New York: Wiley.

Willner, G. R., Chapman, J. A. and Pursley, D. (1979). Reproduction, physiological responses, food habits, and abundance of nutria on Maryland marshes. *Wildlife Monographs* **65,** 1–43.

Crop damage by birds

J J M Flegg
East Malling Research Station, Maidstone, Kent ME19 6BJ

Introduction

Changes in the fortunes of bird populations often result from complex causes. Obviously, natural factors such as climatic fluctuations may play a part, but in many cases, especially during the last millennium, the important factors inducing changes of status arise from man's modifications to the environment. Man is now the major user of the landscape in many areas, so it is not surprising that the crops with which he has displaced the natural vegetation are exploited as a food source by birds (and other animals).

In this survey, the patterns of vegetational change are examined first, as their impact is of basic importance. Brevity precludes an exhaustive examination of all bird species causing crop damage: examples of various types of attack are considered, often, but not always, from temperate regions. This does not infer that damage is less extensive, or less costly, in the tropics (a brief glance at *Quelea* damage statistics emphasizes that the opposite may be the case!), but that work on temperate bird problems is generally more extensive. Certainly the general principles derived from a study of the British situation are widely applicable.

Vegetation changes

An estimated 60% of the land surface of Britain was covered by forest 3500 years ago (Stamp, 1969). Much of the lowland forest was oak (*Quercus* spp.) with beech (*Fagus sylvatica*) and ash (*Fraxinus excelsior*) on chalk soils, elm (*Ulmus* spp.) on some clays, and alder (*Alnus glutinosa*) in wet areas. Scots pine (*Pinus sylvestris*) predominated above an altitude of about 300 m, and in the north.

Since that time, there is ample evidence of the huge changes produced by man. By the sixteenth century, a steadily increasing population was housed in timber-framed buildings and defended by a large navy of wooden vessels. Wood was the major domestic and industrial fuel. Further increases in the population, and the onslaught of the industrial revolution, helped reduce forest cover to about 5% during the nineteenth century. Despite extensive reafforestation programmes, there has since been only a marginal increase.

Forest clearances were not solely to provide timber. They allowed, in particular, the development of agriculture (to feed the growing population and for export) and in so doing produced, almost as an incidental consequence, a much more varied landscape than the original forests. Initially, clearances would have been small and widely scattered, following the pattern of the 'dens' of the Kentish Weald, now reflected in many village names. Since medieval times the clearings have extended and coalesced, to produce today, over much of Britain, a network of copses and woodlands, linked by a still extensive (though diminishing) hedgerow network (Trask, *1981*). Within this matrix crops are grown, and 'improved' agricultural land covered 68% of the English landscape in the 1960s (Williamson, 1967). This woodland/hedgerow matrix forms a reservoir and a communication system for wildlife — plants and other animals besides birds — some beneficial to man, some harmful, and others without influence in the context of crop damage.

In recent decades, but on a relatively limited scale, the trend has progressed further, and some expanses of uniformity have reappeared, but as the 'prairie farmlands' of extensive cereal monoculture. These monocultures are often additionally characterized by an impoverished fauna and flora, and occasionally by greater levels of crop damage by pests (Murton, 1971; Potts and Vickerman, 1974).

Broadening the scene, in the tropics, major inroads into the forests due to agriculture or forestry have occurred mainly over the last century and still continue (Bunting, *1981*). Only rarely is agriculture in the tropics at the intensity normal in temperate regions, and forest often regenerates rapidly after 'slash-and-burn' cultivation. Nevertheless, in that (smaller) proportion of the tropical land surface under man's influence, the picture developed above for temperate regions also holds true, although the duration of the agricultural intrusion may be much briefer as small-scale farmers pass through the landscape.

The overall impact of vegetation changes

What has been the impact of these forest clearances on the British avifauna? Flegg (1975) suggested that the nett effect was likely to have been beneficial, although there might have been some losses, notably of larger species like the goshawk (*Accipiter gentilis*) and the honey buzzard (*Pernis apivorus*), which demand extensive forest areas for breeding. Flegg and Bennett (1974) documented the consequences of opening-up the continuous canopy of an oak woodland to provide glade conditions. This almost doubled the small-bird breeding population and increased species diversity by one-third. Such figures cannot necessarily be applied to mediaeval or even earlier forest clearance, as any assessment of the composition and size of the forest avifauna at these times must be purely speculative. However, it seems most likely that many of the farmland species of today were originally birds of forest margins. Clearance would have enhanced the status of many of these 'edge effects' providing more space for nesting territories and more extensive and diverse, feeding habitats (Murton, 1971; Williamson, 1967).

The Common Birds Census report for 1976–77 (Marchant, 1978) detailed farmland and woodland population changes indicating that farmland plots were generally richer, and that more species occurred exclusively on farmland than did so on woodland (Table 1).

Table 1 Bird species of regular occurrence on Common Birds Census plots in Britain and Ireland (derived from Marchant, 1978)

Habitat	*Occurrence*	*Total spp.*	*Exclusive spp.*
Wood	> 36/91 plots	32	7
Farm	> 30/89 plots	39	14

Most farmland birds are characterized by a catholic diet, and seem able to adapt quickly to new foods. Such adaptability is apparent from the rapid spread through populations of blue and great tits (*Parus caeruleus* and *P. major*) of the habit of opening milk bottles to drink the cream (Fisher and Hinde, 1949). There is also recent evidence that magpies (*Pica pica*) now recognize egg cartons and attack 'doorstep deliveries' to eat the eggs within. In comparison, the evolutionary pressures imposed by changing agricultural practices have acted relatively slowly.

Types of bird damage

Table 2 lists bird families associated with the bulk of crop damage reports worldwide, and the part of the crop attacked.

It is possible to partition and assess the various types of damage under headings related broadly to the importance of the crop as an element in the birds' diet. Obviously, the boundaries between these categories of attack are indistinct: many birds causing damage can, and do, feature in more than one of them.

Crop simply augmenting natural resources

Such attacks reflect the natural dietary flexibility of many species, and also the benefits of widespread 'sampling' of potential food sources (Gosling, *1981*). Their occurrence, though widespread, is sporadic, and the resultant damage rarely severe. The list of birds damaging crops occasionally in this way is lengthy, but good examples are damage to ripe fruits by bulbuls (Pycnonotidae) in the tropics, and by blue tits and recently-established feral populations of the ring-necked parakeet (*Psittacula krameri*) in temperate apple orchards.

Crops providing 'better' alternatives to natural resources

One important characteristic of cultivars is that the parts harvested (whether leaves, fruit or seeds) are usually larger and more palatable than those of the wild parental stock from which they were selected. For example, wild cherries (*Prunus avium*) would not have been prominent in the diet of starlings (*Sturnus vulgaris*) and various thrushes (Turdidae) in the distant past, whereas plantations of heavier-cropping, larger-fruited commercial cherry cultivars are vulnerable to damage by starlings in voracious and highly mobile feeding flocks. There are also smaller-scale but persistent attacks from blackbirds (*Turdus merula*) and song thrushes (*T. philomelos*) resident nearby. Similarly, extensive plantings of brassicas or peas are vulnerable to woodpigeon (*Columba palumbus*) attack. Murton (1965) commented that it was not 'bad luck' that the woodpigeon was a pest, but that this was 'an inevitable result of the pre-adaptation of the bird to many of our crops'.

Bullfinches (*Pyrrhula pyrrhula*) are considered in detail elsewhere (Matthews and Flegg, *1981*; Summers, *1981*). They are relatively specialist feeders, with a beak structure adapted to husking buds to reach the flower initials within. During the winter, buds of various wild trees in the family Rosaceae form an important element in their diet. The vast majority of bullfinches survive adequately on these, but in the relatively small centres of the fruit growing industry in south-east and south-west England, heavy damage to the larger buds of fruit cultivars often ensues, particularly when plantations of susceptible varieties (currants, gooseberries, plums, pears and some apples) are grown adjacent to the bullfinches, preferred woodland

Table 2 Some bird families and the damage they cause. (x) indicates damage of secondary importance

Family	*Buds*	*Flowers*	*Leaves*	*Fruit*	*Seeds*
Bulbuls				X	
Crows				X	X
Finches	X	X			
Icterids				X	X
Larks			X		
Parrots		(x)		X	X
Pigeons		(x)	X	X	X
Starlings				X	
Thrushes				X	
Tits				X	
Weavers	(x)	(x)			X
Wildfowl			X		X

habitat. Damage can usually be alleviated, if not eliminated, by planting fruit at a suitable distance from the nearest cover. As yet, our inadequate knowledge of the complex interrelationships between rates of eating, weights of buds and their nutritional content, availability of other foods and 'ripeness' between wild plants and cultivars precludes any simple explanation of why certain cultivars may be attacked (Matthews and Flegg, *1981*).

Crops 'bridging the gap' when natural resources are scarce

The most damaging avian agricultural pest on a world scale is the black-faced dioch (*Quelea quelea*). It occurs in enormous flocks (Fig. 1), feeding, breeding and roosting gregariously, and ranging over much of south and central Africa. *Quelea* damage various cereal crops, particularly guinea corn, both by eating the grains and in causing 'physical' damage by the sheer weight of numbers in the flocks. Cultivated cereals seem not to be preferred foods; smaller seeds of various wild grasses form the bulk of the diet. Even in years of severe damage, only 20% of *Quelea* food intake may be of cereal cultivars (Ward, 1965).

The widespread and vast areas producing wild seed in the rainy seasons are exploited by many animals, but it is only towards the end of dry seasons that food shortage forces *Quelea* flocks into the river valleys and inundation zones

Figure 1 A large flock of Quelea photographed by Dr. Peter Ward at Ngorongoro crater, Tanzania, in July, 1971, (Photograph by courtesy of Mrs. Waina Ward and Centre for Overseas Pest Research).

favoured for agricultural crops. The 'last straw' comes at the onset of the rains, when any remaining seeds promptly germinate and the *Quelea* flocks face starvation. This provokes a merging of already-large flocks, and causes nomadic migrations in search of areas where earlier rains have replenished seed stocks, both wild and cultivated. On these, the vast and ravenously destructive hordes descend: to such effect that in South Africa, damage was devastating in one year despite the poisoning of an estimated 100 million *Quelea* by aerial sprays (Crook and Ward, 1968).

In temperate regions, there may be similar but less spectacular problems. Feare, Dunnet and Patterson (1974) and Feare (1978) attributed damage to late-sown cereals in Scotland to rooks (*Corvus frugilegus*). They suggested that early-sown crops in southern England grew away earlier than in the north and before rooks reached the period in the breeding season when high energy and protein demands could be satisfied by taking late-sown cereal grains.

Skylarks (*Alauda arvensis*) occasionally damage sugarbeet seedlings by eating the cotyledons. Green (1978, 1979) was able to show that grazing beet seedlings was one of the least profitable feeding methods, weed seeds and various beetles being preferred. Damage could be high on fields sown to stand and maintained weed-free by herbicide treatment, so that preferred foods were not available. Where weed seeds and seedlings were abundant, so too were skylarks, although damage to the sugar beet was much less than would have been expected from skylark density.

Crop management diminishing natural resources

The mid-winter shortage of natural and long-lasting wild seed stocks, such as dock (*Rumex* spp.) due to efficient orchard mowing and the maintenance (by herbicide treatment) of cleared strips at the bases of the trees, might influence bullfinch damage levels (Flegg and Matthews, 1979, 1980). The skylark problem outlined above also fits this category.

A further example is linnet (*Acanthis cannabina*) damage to strawberry. Regular herbicide treatments maintain extensive strawberry fruiting plantations free of weeds. In some areas in southern England, fruiting is advanced by covering the rows of plants with polythene tunnel-cloches. After several years of herbicide treatment, soil weed-seed levels are low, and the seeds embedded in the surface of ripening strawberries provide a tempting alternative food source for linnets. In spring, seed numbers are at their lowest, and seeds from the fresh season are available only in minimal quantity, creating a time of hardship for graminivorous birds. Provision of strawberry seeds progressively earlier in the summer must help remedy this situation for those birds able to exploit them, and reports of damage increase.

The redpoll (*A. flaminea*), a species related to the linnet, was introduced into New Zealand by early settlers. Its success was such that Stenhouse (1962) was able to document damage to the flowers and fruitlets of fruit cultivars. In

England, the redpoll feeds on small seeds on the soil surface for much of the winter, and as these become scarcer tends to move to the late-ripening cones of waterside alders (*Alnus* spp.). In recent years, weed-seed numbers in orchards have been much reduced by regular herbicide applications and mowing (Atkinson and White, *1981*), whereas alder windbreaks have been widely planted (Solomon, *1981*). In some areas of Kent, damage to plum flowers and fruitlets has been reported already, and the extensive use of alder windbreaks, although desirable in some respects, may bring with it new hazards in terms of redpoll damage.

Land management providing new resources

Perhaps the best recent examples of the creation by man of bird problems have occurred following land drainage. A natural wetland flora supports, amongst others, a diversity of wildfowl and passerines including, in North America, an Icterid (the red-winged blackbird *Agelaius phoeniceus*), which readily turns its seed-eating capacities to attack cereals. Wildfowl, too, once the wetland area has been reduced to permanent water in wide ditches, surrounded by reclaimed arable fields, will feed with enthusiasm on grain grown nearby. This problem is particularly serious in the North American wildfowl autumn migration 'flyways' (Boyd, 1980).

In Britain, similar problems are more recent in origin, but reports of grain damage increase. In the last few years, the fresh marshes of the Thames estuary, formerly used for rough grazing, have largely been drained. The improved acreage is now devoted to cereal and potato production and Wright and Isaacson (1978) have already reported cereal damage due to Brent geese (*Branta bernicla*). Brent geese were previously considered to be selective feeders, concentrating heavily on the marine grass *Zostera*, but the provision of an alternative food source rapidly demonstrated otherwise. Coupled with increasing goose numbers resulting from a series of successful breeding seasons in the Arctic, this provision, by man, of a readily accessible and acceptable food supply has turned a rigorously protected species into a pest in the eyes of affected cereal growers.

Conclusions

Many birds are well-suited in their numbers, powers of recruitment, mobility, and generally catholic diets, to damage agricultural crops. Fortunately, their full damage potential is rarely realized, and crops normally feature only sporadically in the diet of avian pests, which largely depend, for much or even all of the year on wild plants. Unlike the other groups considered in this volume, there are no 'specialist pests', dependent for all or much of the year on crops. Consequently, the occurrence and level of bird damage is influenced by several more or less natural factors, including climate, amongst which the availability (or otherwise) of natural foods probably predominates. Unnatural factors also play a major role, particularly those related to the location and management of the crop and its surroundings, for example, by extensive herbicide usage. These considerations support the view (Flegg, 1980) that ecologically-based considerations applied *before* new cropping systems are introduced may be more cost-effective in avoiding or minimizing bird damage than physical or chemical attempts at protection once problems have occurred.

References

Atkinson, D. & White, G. C. (*1981*). The effects of weeds and weed control on temperate fruit orchards and their environment. In *Pests, Pathogens and Vegetation*, pp. 415–428. J. M. Thresh. London: Pitman.

Boyd, H. (1980). Relative merits of damage prevention and compensation programmes in the Canadian Prairie Provinces. In *Understanding Agricultural Bird Problems*, pp. 7–19. E. N. Wright. London: British Crop Protection Council.

Bunting, A. H. (*1981*). Changing patterns of land use: global trends. In *Pests, Pathogens and Vegetation*, pp. 23–37. J. M. Thresh. London: Pitman.

Crook, J. H. & Ward, P. (1968). The *Quelea* problem in Africa. In *The Problems of Birds as Pests*, pp. 211–229. R. K. Murton & E. N. Wright. London: Academic Press.

Feare, C. J. (1978). The ecology of damage by rooks (*Corvus frugilegus*). *Annals of Applied Biology* **88,** 329–334.

Feare, C. J., Dunnett, G. M. & Patterson, I. J. (1974). Ecological studies of the rook (*Corvus frugilegus* L.) in north-west Scotland. Food intake and feeding behaviour. *Journal of Applied Ecology* **11,** 867–896.

Fisher, J. & Hinde, R. A. (1949). The opening of milk bottles by birds. *British Birds* **42,** 347–357.

Flegg, J. J. M. (1975). Bird population and distribution changes and the impact of man. *Bird Study* **22,** 191–202.

Flegg, J. J. M. (1980). Biological factors affecting control strategy. In *Understanding Agricultural Bird Problems*, pp. 20–27. E. N. Wright. London: British Crop Protection Council.

Flegg, J. J. M. & Bennett, T. J. (1974). The birds of oak woodland. In *The British Oak*, pp. 324–340. M. G. Morris & F. H. Peering. London: Botanical Society of the British Isles.

Flegg, J. J. M. & Matthews, N. J. (1979). Bullfinch damage. *Report of East Malling Research Station for 1978*, pp. 128–129.

Flegg, J. J. M. & Matthews, N. J. (1980). Bullfinch damage. *Report of East Malling Research Station for 1979*, pp. 129–130.

Green, R. (1978). Birds grazing seedlings. *Rothamsted Experimental Station Report for 1977*, p. 64.

Green, R. (1979). Birds grazing seedlings. *Rothamsted Experimental Station Report for 1978*, pp. 62–63.

Gosling, L. M. (*1981*). The role of wild plants in the ecology of mammalian crop pests. In *Pests, Pathogens and Vegetation*, pp. 341–364. J. M. Thresh. London: Pitman.

Marchant, J. H. (1978). Bird population changes for the years 1976–77. *Bird Study* **25,** 245–252.
Matthews, N. J. & Flegg, J. J. M. (*1981*). Seeds, buds and bullfinches. In *Pests, Pathogens and Vegetation*, pp. 375–383. J. M. Thresh. London: Pitman.
Murton, R. K. (1965). *The Woodpigeon*. London: Collins.
Murton, R. K. (1971). *Man and Birds*. London: Collins.
Potts, G. R. & Vickerman, G. P. (1974). Studies on the cereal ecosystem. *Advances in Ecological Research* **8,** 109–197.
Solomon, M. G. (*1981*). Windbreaks as a source of orchard pests and predators. In *Pests, Pathogens and Vegetation*, pp. 273–283. J. M. Thresh. London: Pitman.
Stamp, L. D. (1969). *Man and the Land*. London: Collins.
Stenhouse, D. (1962). A new habit of the redpoll *Carduelis flammea* in New Zealand. *Ibis* **104,** 250–252.
Summers, D. D. B. (*1981*). Bullfinch (*Pyrrhula pyrrhula*) damage in relation to woodland bud and seed feeding. In *Pests, Pathogens and Vegetation*, pp. 385–391. J. M. Thresh. London: Pitman.
Trask, A. B. (*1981*). Changing patterns of land use in England and Wales. In *Pests, Pathogens and Vegetation*, pp. 39–49. J. M. Thresh. London: Pitman.
Ward, P. (1965). Feeding ecology of the black-faced dioch *Quelea quelea* in Nigeria. *Ibis* **107,** 173–214.
Williamson, K. (1967). The bird community of farmland. *Bird Study* **14,** 210–226.
Wright, E. N. & Isaacson, A. J. (1978). Goose damage to agricultural crops in England. *Annals of Applied Biology* **88,** 334–338.

Seeds, buds and bullfinches

N J Matthews and J J M Flegg
East Malling Research Station, Maidstone, Kent ME19 6BJ

The problem

Damage done by bullfinches (*Pyrrhula pyrrhula*) in winter is widely known but little understood. It is thought that when autumn seed stocks are exhausted or have fallen to the ground, swelling flower-buds are sought as alternative food. These then form the major part of the diet until fresh seeds appear in April and May. Over much of Britain bullfinches feed on buds of hawthorn (*Crataegus monogyna*) and blackthorn (*Prunus spinosa*), but those of commercial fruit trees and bushes are taken where available and considerable damage can be apparent in spring when affected areas show a conspicuous lack of blossom.

Despite the attempts of several research workers and numerous fruit-growers there is, currently, no simple, quick, cost-effective method of preventing damage. Much information on bullfinches is available but little is known about the factors which control the onset and duration of the damage period, the nutrients gained from the buds (only the flower initials in the core of the bud are eaten), and the proportion of the bird's diet which is derived from commercial fruit-buds. In short, it is not known why damage occurs and it is hardly surprising that no 'cure' is yet available.

Autumn feeding

In summer and early autumn the bullfinch will eat the seeds of many plants. Over forty common and widely distributed plant and tree species supply food during the year (Newton, 1967). However, as autumn progresses bullfinches concentrate on certain species, probably because of their availability. The seeds of most species fall during the autumn but some, including nettle (*Urtica dioica*), dock (*Rumex* spp.), bramble (*Rubus* spp.) and ash (*Fraxinus excelsior*), usually retain seed in large quantities. The bullfinch can take seeds from the ground but seems to prefer those remaining on plants. This is hardly surprising since most, if not all such seeds will be easily recognized and clean, and will not have been eaten previously by other birds. Given that seeds are easy to reach, and the bullfinch is very agile, time spent searching for food will be minimal. It is difficult to determine the relative importance of each of the seed types eaten during winter because of the problems of following and recognizing free-living birds individually for more than a few minutes at a time, and because the complexities and logistic problems of aviary experiments make them an unreliable guide.

Field observations and analyses of bullfinch crop contents suggest that the autumn diet consists mainly of nettle, dock, bramble, and ash seeds (Summers, *1981*). These provide for most if not all the needs of the bullfinch at this time. The numbers of such seeds available, however, vary according to several factors including weather, the previous year's seed crop, pollination and the number of plants in the area. Each species, therefore, may or may not set a good seed crop, and may or may not retain these seeds once set. The nettle seed crop is usually exhausted by late December, and ash, although many trees retain their keys throughout January and February, has a particularly variable set of seed between years (Summers, 1979). In winter the number of bramble seeds left for bullfinches by other 'predators' can be influenced by weather. Dock can be scarce or abundant depending on agricultural practices.

It might appear easy to follow the decline in these seed stocks and relate this to severity of bullfinch damage each year. Newton (1964) examined this in the early 1960s, when the ash trees in southern England were producing synchronous biennial seed crops. Thus the autumns of 1961 and 1963 saw virtually no ash keys compared with a super-abundance in the winter of 1962/63. Damage to fruit trees occurred early in 1962 and 1964, but not in 1963. Newton concluded that the ash seed crop was the major factor determining commercial fruit-bud damage. He found a correlation between ash pollen counts and the autumn seed crop and suggested that this could be used as an 'early warning' system (Newton, 1967):

low pollen counts → small ash crop → severe bud damage.

Subsequent fluctuations in ash pollen counts and seed crops have been less extreme, although Summers (1979) again demonstrated a correlation between pollen counts and damage. The pollen counts came from few locations, all distant from fruit plantations, and were prone to considerable biasses caused by such factors as the position of the nearest ash tree and the prevailing wind direction. Moreover, severe damage can undoubtedly occur when ash seeds are available. Flegg and Matthews (1979, 1980) recorded substantial damage in three consecutive winters when ash seeds were not scarce.

Field and aviary problems

Newton showed annual differences in the decline of other seed species in the early 1960s. Bramble and dock, for example, declined more rapidly during the winter of 1963/64 than 1962/63 and thus the abundance of their seeds could have influenced bud damage. There are, however, considerable problems in assessing the availability of seeds to birds. For example, in the case of ash, given enough trees, only a small percentage of their keys might suffice to feed the local bullfinches for weeks or even months. The seeds are so large (30–60 mg dry weight) that the limiting factors involved in their consumption are the breaking up of the kernel in the bill (which can take 30–50 seconds) and the digestion rate. Cropping trees are obvious at a distance and would be well known to most resident birds, and plucking the keys from their branches followed by 'husking' and discarding the wing takes only about 10 seconds. In contrast, bramble and dock seeds weigh only 1–2 mg dry weight but are husked and swallowed very rapidly. With these, searching time is important. They tend to be widely but patchily distributed in winter, both in terms of plants and remaining seeds. Small changes in seed density might be very important in determining the usefulness of these seeds to feeding birds. Declines of seed numbers in sample plots, therefore, should be regarded with caution since they may or may not indicate that the seed supply is of decreasing value as a food source.

Bullfinches can be kept in aviaries and feeding experiments have been conducted at East Malling Research Station and elsewhere. The scope for such experiments is limited by the logistic problems of feeding bullfinches on any of their natural foods in cages that are small enough to allow monitoring of bird weights, food eaten, and collection of faeces, yet large enough to allow a realistic choice of experiments and to hold sufficient food. Collecting fruit-buds for only a few birds is a major task since they are eaten very rapidly, each bird being capable of eating the buds from two to three full-size pear trees day^{-1}. Wild birds with slowly-changing diets are probably at all times 'acclimatized' to eating the available foods, but aviary birds fed on a commercial cage-bird seed mixture need several days to accustom themselves to an imposed diet, however 'natural'.

Feeding experiments, field data and damage

Table 1 lists diets that have maintained one or more bullfinches for a period of time during the winter.

In the case of bud diets, only birds fed on hawthorn in January lost weight rapidly, and longer periods of acclimatization might have allowed most, or even all, to survive without weight loss. It seems unlikely that wild birds would restrict themselves to diets of only one or two food types, so the experiments are probably severe tests in this respect. In contrast, energy requirements of captive birds sheltered from wind and prevented from flying far would be less than those of wild birds.

How do these feeding data relate to the damage in orchards? It seems likely that wild bullfinches could, if necessary, survive at least for a few days on any one of their natural food types. At the outset it should be stressed that only bullfinches in southern England come into regular contact with orchards, and thus the vast majority live without access to commercial fruit-buds. Northern bullfinches eat buds of hawthorn and blackthorn as do southern birds, but very little is known of the importance of these buds to bullfinches, and still less of the extent to which individuals might concentrate or 'specialize' on one particular food source. A study of damage to a particularly vulnerable 2 ha pear orchard in Kent (Flegg and Matthews, 1979) showed a loss of 1.2 million fruit-buds in 1978, mostly during January and February. Although in later months about 100 birds were shown to be visiting the area over a period of days, no more than 20 birds were suspected of taking pear buds on any one day. An estimate of 1000 buds bird^{-1} day^{-1} (if 20 birds had taken all the buds) seems large until it is realized that at a comfortable feeding rate of 12 buds minute^{-1} (well within observed feeding rates) only $1\frac{1}{2}$ hours need be spent in the orchard: only a small portion of an 8–9 hour day. Longer periods were spent in the neighbouring woodland, feeding on other food sources — ash was present throughout January and February — than in the orchard. In this area, pear buds may still have been a 'minor food', though taken in large numbers.

Buds: the part eaten

The feeding data presented above indicate that the nature of the seed crop

Table 1 Bullfinch feeding experiments (Newton, 1964; Matthews and Flegg, unpublished)

Diet	*Month of experiment*	*Weight loss*	*Days on diet*
Seeds			
Dock + ash	December	0	16
Ash	December	0	6
Dock	November + December	++	4
Bramble	February	0	14
Buds			
Hawthorn	January	++	3
Comice	January	+	3
Conference	January	+	3
Hawthorn	February	+	4
Comice	February	0	4
Conference	February	0	4
Hawthorn	March	0	4
Comice	March	0	4
Conference	March	0	4

Weight loss: + variable (mostly gradual); + + rapid; 0 none.

probably influences the onset of damage. However, the Kent study orchard suffered damage for two months in 1979, commencing six weeks later than in 1978 (Flegg and Matthews, 1980). Only minor differences in available seeds were apparent, though bramble was certainly more abundant in 1979. Another previously unmentioned factor may have been important. During the winter, buds swell at a rate partly controlled by temperature, and measurements of fruit-bud size, i.e. the whole bud, the flower initials eaten, and the initial:bud ratio, all showed a three-week lag during the cold winter of 1978/79 compared with 1977/78. Thus, contrary to popular belief, cold weather in late autumn and early winter might delay rather than promote damage. In 1979 Conference pear flower initials swelled from 6 mg dry weight in January to 9 mg in late February, and to 20 mg by the end of March. This increase allows a greater weight of initial to be eaten in a given time and development of the bud changes the nutrient composition of the initial.

The size of initial, ease of husking, and developmental stage are probably all important in determining the preferences shown by bullfinches for buds of particular types of commercial and wild trees as the season progresses (Newton, 1967). Comice pear, for example, is not a preferred variety and examination showed that its flower initials are about 3 mg lighter than Conference and that bud expansion occurs later. Blackthorn and hawthorn flower initials are very small (0.5–2 mg dry weight in January and February) but are still eaten, particularly late in winter. Faster feeding rates help compensate for their small size. One concomitant of bud swell is that nitrogen concentration in the flower initial rises from January onwards. The nitrogen concentration in the bud scales and leaf initials that are discarded so characteristically by bullfinches is much lower and changes only slowly. The birds are not necessarily choosing their buds or the portion of the bud eaten by the nitrogen (or protein) content. However, changes in the proportions of, for example, sugars, might determine the time of year at which each variety is nutritionally most effective. Chemical analyses of buds and seeds are best coupled with faecal analyses during feeding tests, though the results obtained will have to be interpreted carefully in relation to the orchard situation.

As spring approaches, the weather affects both the length of time for which the fruit-buds remain a profitable food — damage usually stops at bud-burst — and the appearance of new food sources. A cold spring might delay both bud-burst and the development of other plant species, i.e. their buds, flowers and ultimately seeds, and so prolong the period of damaging attacks. However, species do not necessarily vary in unison, each reacting independently to the prevailing weather.

Bird 'pressure'

Bird pressure is an important consideration in the defence of orchards. It includes the number of birds involved, their mobility and resilience to population reductions, and the vulnerability of the orchard concerned. The protective measures now used are unsophisticated, and their effectiveness largely depends on these factors.

Any attempt at eradicating the bullfinch from southern England is beset with problems. In a recent survey organized by the British Trust for Ornithology, Sharrock (1976) estimated a total British and Irish population of 600 000 pairs but densities of up to 50 pairs km^{-2} have been recorded in favoured areas (Newton, 1972). Bullfinches have a high reproductive rate and their mobility is sufficient to fill, in time, any small 'gaps' created by trapping or shooting. Even if every fruit grower in Kent killed hundreds of bullfinches each year the total Kent population is unlikely to be materially affected. Additionally, there would be an unfavourable public reaction to this course of action.

Bullfinches are essentially woodland and hedgerow birds, nesting, feeding and normally moving within the cover provided by trees and bushes. Although some birds will fly over wide, open spaces, an orchard which is some hundreds of metres away from good bullfinch habitat is unlikely to be heavily damaged, even if planted with 'preferred' varieties. In contrast, if a portion of woodland is felled and planted with fruit trees providing cover and an even more attractive food source than the trees that were removed, then it is hardly surprising that bullfinches take advantage of the new habitat.

The tendency of bullfinches to take the most easily available seeds and buds first is of great importance to present and future orchard protection techniques. It determines which orchards and which trees within them are most vulnerable. An orchard with one edge adjoining woodland and with open fields on the other sides will show a gradient of vulnerability and consequent degree of attack by feeding birds (Fig. 1). Orchard trees nearest the woodland may be left with very few buds while those on other edges are untouched. Some fruit tree varieties can tolerate the removal of many buds without any effect on crop and in most cases loss of crop could be avoided by a more even distribution of damage across the entire orchard.

More 'natural' food — less damage?

Presently available deterrents such as gas-fired bangers, hanging ribbon, black cotton, and chemical sprays, are not very effective. Many bullfinches are trapped or shot each winter and Newton (1968) suggested that this should be done in autumn, before damage starts, so preserving natural food stocks for those that remain and delaying damage. However, although bullfinches are basically sedentary, the experience of many fruit growers has shown that movement is sufficient to at least partly fill the gaps created. Unless removal of birds is continued through the winter such infilling could ensure that there

Figure 1 A pear orchard in Kent at blossom time when bullfinch damage is most obvious. Damage to Conference (right) is greater than to Comice (left) although both cultivars have been stripped at the far ends of the rows adjoining a belt of woodland. Note the effective weed control in a modern orchard of this type with grassed alleys and bare ground along the tree rows treated annually with residual herbicides (Atkinson and White, *1981*). Note also the accumulation of leaf litter to leeward of the grass strips as described by Harris (*1981*).

are still enough bullfinches to cause extensive damage. In winter there are probably hundreds of bullfinches within reach of any orchard wherever suitable habitat occurs. Ringing studies indicate that they are mobile (Flegg and Matthews, 1979) and probably visit several different orchards over a period of time, although movement for distances greater than 5 km is unusual (Summers, 1979). The damage in orchards, therefore, is often done by many birds and not by 'just a few locals'. Trapping and shooting are very time-consuming, their cost-effectiveness depending largely on the rate and degree of in-filling.

Bullfinch damage would undoubtedly be reduced by removing woodland and hedgerow cover, but this course of action has limitations, not least on environmental grounds. At the other extreme is the contrasting idea, as yet untested, of providing extra food for bullfinches in an attempt to keep them off fruit-buds. Bullfinches are rarely seen at bird tables, so food is perhaps best provided by planting their natural winter food species. Nettle and dock are common orchard 'weeds' and could be easily encouraged, although the compatibility of this suggestion with currend weed control practices would need close examination. A few large ash trees nearby could provide masses of seed, as even in 'poor' years, some ash trees crop heavily. Their presence could reduce bud-eating until most seeds had been eaten, thus perhaps preventing damage from reaching critical levels.

There are two main objections to this suggestion. First, bullfinches from surrounding areas might be attracted to the food, while it lasted, and then, rather than returning to their 'home ranges', remain to feed on nearby fruit-buds. Second, the abundance of food could allow more birds to survive the winter, thereby increasing the potential for damage in subsequent years.

There is little information on the factors limiting bullfinch populations. Ringing recoveries indicate peaks of mortality in mid-winter and in midsummer, but these might not represent the true pattern. Many of these recoveries are biassed purely because man has found the dead bird, and in the case of the bullfinch many of those recovered are shot, trapped, or caught by cats. Such deaths may reflect a different mortality pattern from the bulk of the population, but perhaps one still relevant to fruit-growing areas. It is likely that food shortage before buds are sufficiently large, causes much of the 'natural' winter mortality (Newton, 1964). It is also likely that even in the absence of seeds, most birds surviving to January can last the whole winter — as time passes buds become larger and days longer. Perhaps the best food provision policy might be to encourage plants, such as dock and nettle, that could be covered with netting until natural mortality has reduced the population. When orchard damage starts, they could be uncovered as a new food source, reducing the need to eat fruit-buds. Bullfinches have been seen feeding on dock beneath Conference pear trees in early February before any buds were taken (this study).

Current standards of orchard cleanliness amount to near-freedom from 'weeds' (Atkinson and White, *1981*). This may be forcing bullfinches, deprived of their natural foods, to exploit fruit-buds to an increasing degree. The simplest solution to the problem is to plant varieties vulnerable to bullfinch damage well away from suitable woodland habitat. Financial losses might often be reduced by more effective use of existing deterrents but a reduction in the use of herbicides, so that natural seeds are available at a time when they might be preferred, may also have the desired effect.

References

Atkinson, D. & White, G. C. (*1981*). The effects of weeds and weed control on temperate fruit orchards and their environment. In *Pests, Pathogens and Vegetation*, pp. 415–428. J. M. Thresh. London: Pitman.

Flegg, J. J. M. & Matthews, N. J. (1979). Bullfinch damage. *Report of East Malling Research Station for 1978* (1979), pp. 128–129.

Flegg, J. J. M. & Matthews, N. J. (1980). Bullfinch damage. *Report of East Malling Research Station for 1979* (1980), pp. 129–130.

Harris, D. C. (*1981*). Herbicide management in apple orchards and the fruit rot caused by *Phytophthora syringae*. In *Pests, Pathogens and Vegetation*, pp. 429–436 J. M. Thresh. London: Pitman.

Newton, I. (1964). Bud-eating by bullfinches in relation to the natural food-supply. *Journal of Applied Ecology* **1,** 265–279.

Newton, I. (1967). The feeding ecology of the bullfinch (*Pyrrhula pyrrhula* L.) in southern England. *Journal of Animal Ecology* **36,** 721–744.

Newton, I. (1968). Bullfinches and fruit buds. In *The Problems of Birds as Pests*, pp. 199–209. R. K. Murton & E. N. Wright. London: Academic Press.

Newton, I. (1972). *Finches*. London: Collins.

Sharrock, J. T. R. (1976). *The Atlas of Breeding Birds in Britain and Ireland.* Berkhamsted: T. and A. D. Poyser.

Summers, D. D. B. (1979). Bullfinch dispersal and migration in relation to fruit bud damage. *British Birds* **72,** 249–263.

Summers, D. D. B. (*1981*). Bullfinch (*Pyrrhula pyrrhula*) damage in orchards in relation to woodland bud and seed feeding. In *Pests, Pathogens and Vegetation*, pp. 385–391. J. M. Thresh. London: Pitman.

Bullfinch (*Pyrrhula pyrrhula*) damage in orchards in relation to woodland bud and seed feeding

D D B Summers
ADAS Agricultural Science Service, Worplesdon Laboratory, Tangley Place, Worplesdon, Guildford, Surrey GU3 3LQ

Introduction

The severity of bullfinch (*Pyrrhula pyrrhula*) damage to fruit buds in orchards fluctuates considerably between years. Following the winter of 1961/62, when widespread and relatively distant movements by bullfinches were recorded for the first time in 50 years of bird ringing, and when damage was particularly severe, it was suggested (Summers, 1962) that there could be links between bud damage in orchards, levels of migratory activity and variations in natural food supplies. The findings of Newton (1964) also suggested a relationship between damage and the availability of natural foods and he also related bullfinch numbers to food stocks in deciduous woodland over two winters. Summers (1979) further demonstrated these relationships with 14 years data from ash (*Fraxinus excelsior*) pollen counts, damage assessments and ringing recoveries. This paper relates bullfinch damage in orchards to the birds' diet, assessed by examining gut contents, over a six year period from autumn 1969 to spring 1975.

Methods

Study area

The study area consisted of two adjacent farms at Kirdford, in West Sussex, totalling about 60 ha comprising 61% orchards, 16% deciduous woodland and thorn scrub and 23% mixed agricultural land.

Assessment of bullfinch damage

Each year bud damage was assessed at blossom time, when it was most apparent. About 400 trees were inspected and classified as damaged or undamaged. This simple procedure is adequate to compare differences in damage between years (Summers, unpublished). The behaviour of bullfinches is such that trees at the edge of an orchard, especially near to cover, are the first to be attacked (Wright and Summers, 1960) and, therefore, ten trees at the end of each row, in a pear orchard adjoining a wood, were inspected. This sample gave an exaggerated but sensitive indication of the incidence of damage each year. The orchard contained cultivars of Doyénné du Comice, Packham's Triumph, Conference and William's Bon Chrétien. Conference was used to measure annual damage levels because it is widely grown commercially and often suffers bullfinch damage.

Food eaten by bullfinches

Bullfinches were shot for gut analysis each winter (November to March) from 1969 to 1975. Two hundred and forty-three birds were collected (mean 8.4 $month^{-1}$); it was not possible to collect a larger number without risk of seriously depleting the study population. The aim was to collect 10 bullfinches per month and this was usually achieved in the first fortnight. When they proved difficult to get, sampling continued throughout the month. The contents of proventriculus and gizzard were removed from each bird and stored in 70% alcohol, the food items being identified later and counted under a low power binocular microscope.

Results

Gut contents

Seeds of ash, blackberry, dock and nettle were recorded in all six years of the study. Blackberry was, in fact, recorded in every month of every year but its dietary importance is probably exaggerated by the persistence of the hard seed coat which tends to remain in the gizzard longer than other seeds and may act like grit. The buds of hawthorn and blackthorn were the only species recorded in all years. All species found to be fed on by bullfinches are listed in Table 1 together with their latin names.

Bullfinch damage

The relationship between the consumption of wild buds and seeds and orchard damage is shown in Fig. 1. Damage and the consumption of herbaceous seeds were positively correlated ($r = +0.84$, $P < 0.05$). There was a significant negative correlation between damage and consumption of ash seed, but only when January was considered alone (*see* below).

Annual differences

Patterns of bud feeding varied greatly between years (Table 2).

Table 1 Percentage incidence of food items in bullfinch gut contents (November–March)

	1969–70	*1970–71*	*1971–72*	*1972–73*	*1973–74*	*1974–75*
Seeds						
Dock (*Rumex* spp.)	34	21	33	15	15	20
Nettle (*Urtica dioica*)	20	14	30	13	5	10
Compositae	7		13			
Redshanks (*Polygonum persicaria*)	2		2			30
Plantain (*Plantago* spp.)	18	5	4			
Chickweed (*Stellaria media*)	4	2	2			5
Fathen (*Chenopodium album*)			2		3	25
All seed (*Chenopodium polyspermum*)	4	7	4			
Shepherd's Purse (*Capsella bursa-pastoris*)		2	2			
Blackberry (*Rubus* spp.)	86	62	80	69	78	75
Ash (*Fraxinus excelsior*)	32	33	13	69	90	5
Sycamore (*Acer pseudo-platanus*)			2			
Unidentified	5	7	41		8	20
Buds						
Plum (*Prunus domestica*)	14	2	4		5	5
Pear (*Pyrus communis*)	7	2	2			
Crab Apple (*Malus sylvestris*)	9		2	8		
Hawthorn (*Crataegus monogyna*)	11	17	26	10	13	20
Blackthorn (*Prunus spinosa*)	4	28	15	13	8	5
Unidentified	4	17	17	10		25
Others						
Willow flowers (*Salix* spp.)	2					
Moss (*setae*) (Bryophyta)	7	10	4		5	
Spiders (Araneida)			2	3		
Dipterous larva? (Diptera)	2					

1969–1970

Damage was slightly above the average for the six year period, whereas the total consumption of ash seeds was below the average. Spring was later than in other years of the study and at Efford Experimental Station (70 km westwards) apple blossom (cv. Egremont Russet) was three weeks later than the average. Very few buds were eaten before February and wild buds were eaten by fewer birds than usual, presumably because they were too small (average dry weight of blackthorn buds in February 1970 was 0.54 mg compared with 0.77–0.85 mg at the same time in other years).

Table 2 Seasonal changes in the percentage of bullfinches' guts containing ash seeds and buds

	Ash seed	*All buds**	*Wild buds*	*Cultivated fruit buds*
1969–1970				
November	75	0	0	0
December	39	8	0	8
January	50	0	0	0
February	40	50	20	30
March	0	100	58	42
1970–1971				
November	0	0	0	0
December	40	20	0	0
January	88	25	25	0
February	27	82	45	9
March	0	100	100	10
1971–1972				
November	21	50	0	50
December	22	44	33	22
January	7	43	14	14
February	0	75	75	25
March	0	100	100	11
1972–1973				
November	57	0	0	0
December	100	27	18	0
January	86	14	14	0
February	50	80	60	0
March	25	100	75	0
1973–1974				
November	100	0	0	0
December	100	0	0	0
January	68	11	0	0
February	100	29	24	6
March	75	100	100	0
1974–1975				
November	0	50	0	40
December	0	50	25	50
January	33	67	33	33
February	0	100	100	0

* Some birds contained buds at such an advanced stage of digestion that they could not be identified.

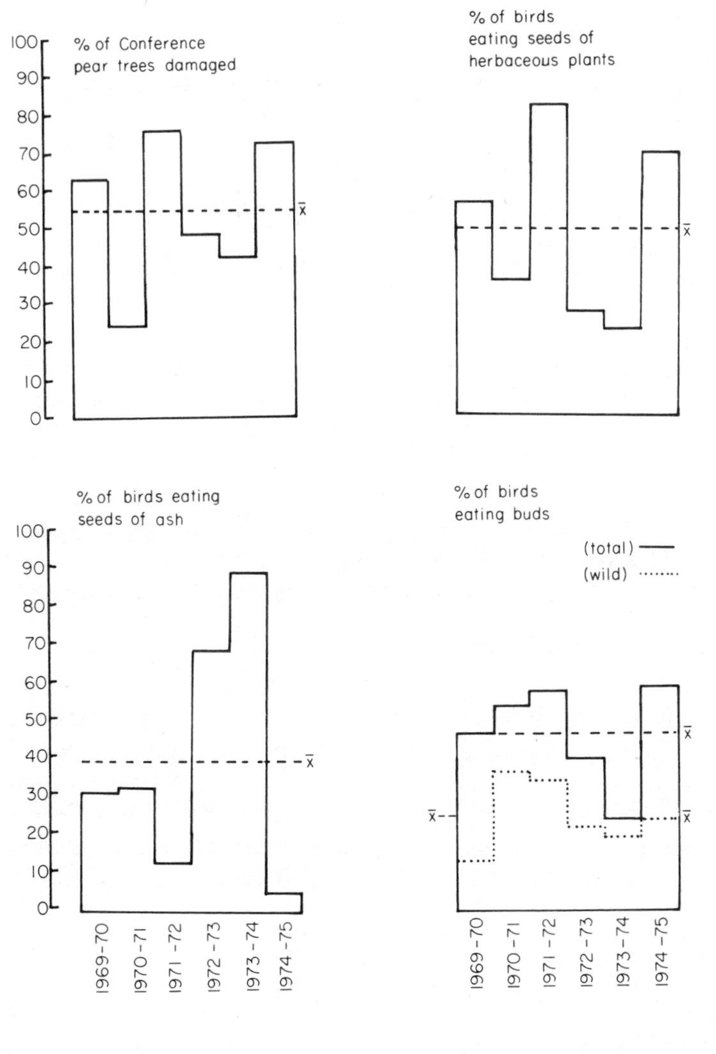

Figure 1 Percentage of pear trees damaged by bullfinches and the main foods eaten each winter from 1969–1970 to 1974–1975.

1970–1971

Damage was the least recorded in the study. Ash seed was abundant but the total number of bullfinches that consumed ash seeds was slightly below the average for the whole study period.

1971–1972

Damage was the highest recorded during the study. It started early with reports during November, and hawthorn and crab apple appeared in the December diet. Ash seed was scarce in the study area and few birds ate it.

1972–1973

Bud damage was below average whereas the consumption of ash seed was high. Few birds fed on the seeds of herbaceous plants and none after December.

1973–1974

Relatively little damage occurred and the overall level of ash seed consumption was the highest recorded during the study. Few birds ate herbaceous plant seeds and there was no significant bud feeding until March, although bud burst was slightly early.

1974–1975

Damage began early and was severe. Ash consumption was the lowest during the study, a small amount being eaten in January only. Bud feeding began in November and was restricted to cultivated trees on which damage would probably have been even worse had bud burst not been very early and most buds eaten in February were wild ones. Blackthorn buds were not weighed in February but mean dry weight in March 1975 was 2.33 mg compared with 0.88 mg in March 1970.

Discussion and conclusions

The availability of ash seed appears to be the most important single factor determining the extent to which bullfinches cause damage in commercial orchards. Between 1959 and 1964 ash trees showed a distinct biennial bearing of seeds and bullfinch damage was equally well defined in its fluctuations during that time (Newton, 1964; Summers, 1979). Since 1964, the biennial regime in ash seed production has broken down and annual fluctuations have been less marked. Consequently, subsidiary factors probably had a relatively greater influence on levels of bud damage in some recent years. Such a factor is the timing of bud swell; for example severe damage occurred in the spring of 1970 when the season was late, whereas in 1974–75 damage, which started very early in the year possibly because ash seed was scarce, did not continue as long as it might have done because bud swell was early and wild buds became big enough to make foraging profitable.

The modest amount of ash seed in the bullfinches' diet during 1970–71, when damage was light, does not appear to support the hypothesis linking ash consumption with damage. However, aviary experiments in which bullfinches were fed pear buds showed that in January they could not maintain themselves on buds alone (Summers, in preparation). Thus ash is a particularly important food item at that time when seeds of herbaceous plants have declined in abundance (Newton, 1964). If January is taken to be the most critical month, and the incidence of ash seed in bullfinches' guts is considered for this month alone, it can be seen (Table 2) that the highest occurrence was recorded in 1971 and there is a significant negative correlation between consumption of ash seed and damage ($r = -0.89$, $P < 0.02$) over the six years.

Blackberry seeds occurred in a high proportion of bullfinch guts in all months every year. Although obviously an important food in the bullfinches' diet, no direct link was apparent between the incidence of feeding on blackberry seeds and damage in orchards. Blackberry seed does not appear to be a favoured food but a standby. This is suggested by the fact that there was no peak time for feeding on blackberry seeds. In some years the incidence of feeding gradually increased during the winter whilst in others it decreased; sometimes the incidence fell then rose again. Nevertheless Matthews and Flegg (*1981*) were able to maintain bullfinches in an aviary for 14 days on blackberry seed.

Acknowledgment

I wish to thank Messrs. J. B. Bradley and E. B. Cruse for allowing me to carry out this project on their farms.

References

Matthews, N. J. & Flegg, J. J. M. (1981). Seeds, birds and bullfinches. In *Pests, Pathogens and Vegetation*, pp. 375–383. J. M. Thresh. London: Pitman.

Newton, I. (1964). Bud-eating by Bullfinches in relation to the natural food supply. *Journal of Applied Ecology* **1,** 265–279.

Summers, D. D. B. (1962). Biology of bullfinches. *Proceedings of the Conference on Damage to Fruit Crops*, pp. 1–13. London: MAFF/NFU.

Summers, D. D. B. (1979). Bullfinch dispersal and migration in relation to fruit bud damage. *British Birds* **72,** 249–263.

Wright, E. N. & Summers, D. D. B. (1960). The biology and economic importance of the bullfinch. *Annals of Applied Biology* **48,** 415–418.

The relevance of 'natural' habitats to starling damage

C J Feare
MAFF, Worplesdon Laboratory, Tangley Place, Worplesdon, Guildford, Surrey GU3 3LQ

Introduction

The feeding activities of starlings (*Sturnus vulgaris*) lead them to cause economic losses in various agricultural situations, (Feare, 1980). In Britain, the loss of cattle food to starlings has increased during the last decade as a result of intensive rearing (Feare and Swannack, 1978; Feare and Wadsworth, 1981) and this has led to severe problems on some farms. Many studies (e.g. Dunnet, 1956; Bailey, 1966; Williamson and Gray, 1975) indicate that starlings feed primarily on invertebrates of grassland. However, they utilize other resources when the food in grassland becomes less readily available, usually as a result of cold weather or snow cover. In Britain, and presumably elsewhere, the occurrence of extensive damage even in mild winters suggests that this is an incomplete picture, and this paper considers the relationship between starlings feeding in grassland (the 'natural' habitat) and on cattle food at a farm in southern England between 1974 and 1978. The feedlots

Figure 1 Cattle and starlings competing for barley at a feeding trough (photograph by J. T. Wadsworth).

where this cattle food was available to starlings consisted of five open-fronted calf yards, each housing twelve calves, where food was presented in open troughs. In each winter the calves in at least two of the yards were given food *ad libitum* and the food eaten by the starlings consisted mainly of crushed barley. The aim of this paper has been to examine the relative use of grassland and feedlots in relation to population size and the time budgets of the birds, to discuss the role of intra-specific competition and to propose an hypothesis involving nutrient requirement in the selection of feeding sites.

Methods

The study area was centred on Bridgets Experimental Husbandry Farm (Bridgets EHF), Martyr Worthy, Hampshire and comprised the farm buildings and adjacent agricultural land extending to 3 km from the buildings. This area was selected initially because over 90% of the land was visible from roads but subsequent work showed that the daytime movements of the birds from the farm were largely restricted to this study area.

The number of birds in each flock was counted, or estimated with large flocks, and the situation of these flocks (type of field, trees, buildings, etc.) was recorded on a fixed transect which involved driving and walking. Where flocks contained both feeding and non-feeding birds the number in each category was recorded to estimate the proportion of all the birds on the farm that was feeding on each transect. The farm was visited at approximately weekly intervals throughout the winter, and on each day at least one transect was undertaken during each of the following time intervals: 0600–0900 hours, 0900–1200 hours, 1200–1500 hours and, if the birds had not left for their roost, after 1500 hours. Each transect involved over 11 km of travel and took approximately one hour. It is assumed in the analysis that the number of flocks missed balanced the number counted twice.

Meteorological data were obtained from daily recordings at Bridgets EHF.

Results

This study confirmed that at Bridgets EHF, as in previous studies, most starlings fed in grassland. Furthermore, Fig. 2 shows that the proportion feeding in the feedlots varied inversely with the proportion feeding in grass fields, since other potential feeding sites, e.g. germinating cereals, spilled grain and stubble, were available only for short periods or were used infrequently. In 1977–1978, the calf feedlots were utilized much less than in previous years, even though the proportion of birds feeding in grassland remained approximately the same. This reduced use of the feedlots may have been due to excessive disturbance on the farm during a building programme, which led more birds than usual to feed on stubble early in the winter and on slurry in the farmyard later.

Figure 3 shows that the number of starlings feeding in grassland was positively correlated with the total number on the farm. This indicates that

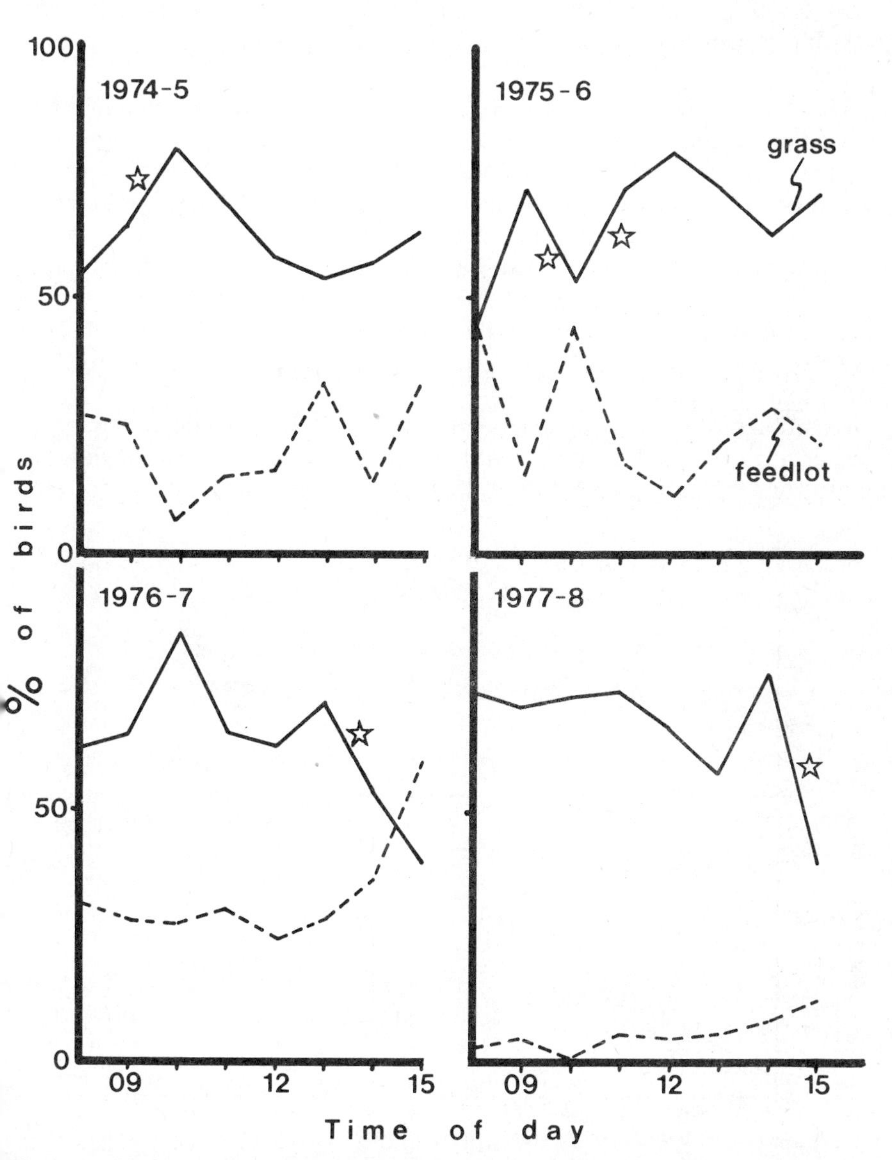

Figure 2 The proportion of starlings at Bridgets EHF that were feeding in grassland and in feedlots throughout the day. Note that in each year the graphs are roughly complementary and that few of the variations between different times are statistically significant ($P < 0.05$, indicated by☆).

the food available in grassland did not impose a density dependent limitation on the number feeding there. By contrast, there was no relation between the total number of birds on the farm and the number utilizing the feedlots ($r_{1974-75} = 0.12$, $n = 36$; $r_{1975-76} = 0.09$, $n = 54$; $r_{1976-77} = 0.12$, $n = 25$; $r_{1977-78} = 0.12$, $n = 40$), although the number recorded there never exceeded 400 birds, which presumably represents a physical limitation imposed by space.

On most days starlings assembled in trees around the farm buildings both after their arrival from the roost in the morning and again before their evening departure. Nevertheless, no statistically significant diurnal variation was found in the proportion of the population that was feeding, although the early morning and late afternoon estimates were the most variable. Overall, the proportion of the birds that were feeding increased as the proportion utilizing grassland increased and these correlations were highly significant in two of the three years considered in this analysis (Fig. 4). Data for 1977–78 were omitted as the method of estimating the proportion feeding differed from that used previously.

Recordings of environmental temperatures were not taken on each transect and, since air temperatures often varied widely between transects made even on the same day, relations between the maximum/minimum air temperatures

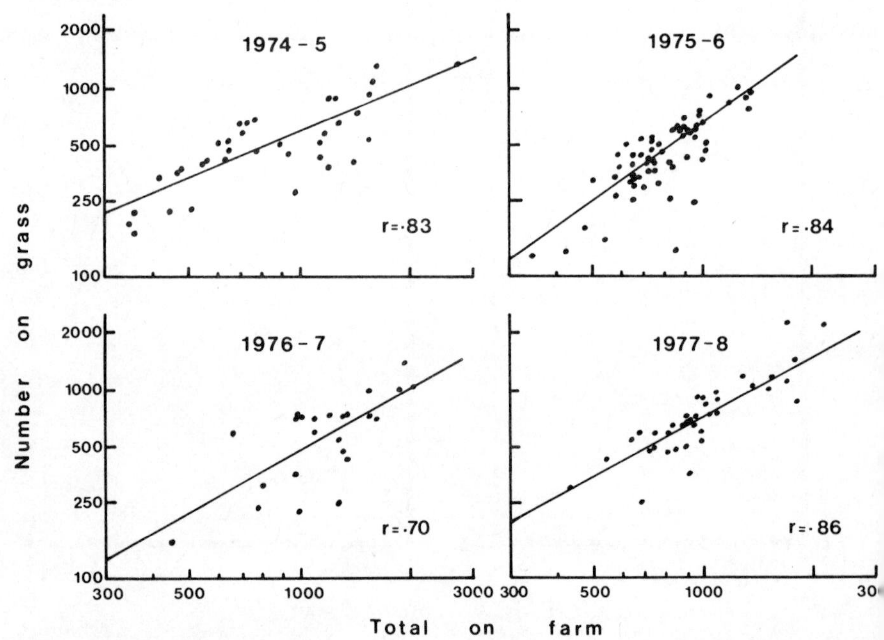

Figure 3 The relation between the number of starlings feeding in grass fields and the total number of starlings on the farm. Both axes are plotted on a logarithmic scale. Even at the highest population levels recorded there was no reduction in the birds' utilization of grassland.

and starling behaviour would be difficult to interpret. Soil temperatures show more diurnal stability than air temperatures and the behaviour of many of the soil invertebrates on which starlings feed may be more related to soil than to air temperature. There were only two years in which the proportion of starlings on grassland was related to soil temperatures at 5 cm ($r_{1975-76}$ = 0.30, n = 62, P <0.05; $r_{1977-78}$ = 0.50, n = 53, P <0.01; in the other years $r_{1974-75}$ = 0.16, n 46; $r_{1976-77}$ = 0.07, n = 26). The only year in which soil temperatures dropped below 0°C on the days when the farm was visited was 1977–78. Sub-zero temperatures then resulted in an almost

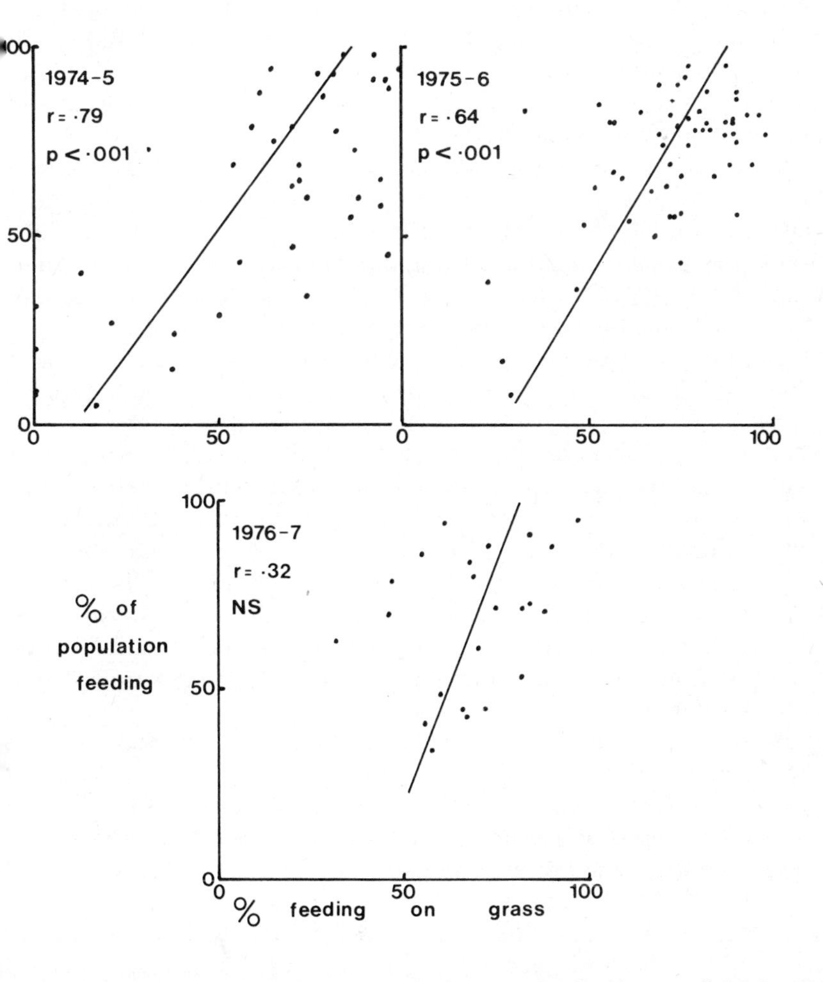

Figure 4 The proportion of the starlings that were feeding in grass fields in relation to the proportion of the total farm population that was feeding. The proportion of the population that was feeding on each transect is assumed to indicate the proportion of the birds' time that is devoted to feeding and the graph shows that birds feeding in grassland spend a high proportion of their time feeding.

complete cessation of grassland feeding (Feare, 1980), whereas the number of birds feeding on spilled grain, manure heaps and in the feedlots increased.

Discussion

At Bridgets EHF, the majority of starlings fed in grass fields but, despite this, those that fed in the feedlots consumed large quantities of cattle food, especially barley (Feare and Swannack, 1978). The data above show that the number of starlings in the 'natural' grassland habitat was not subject to density dependent limitation. Presumably, therefore, the food available in the surface layers of grass fields was not limiting. In this case, why should starlings feed on cattle food at all?

Bailey (1966) recorded lower utilization of grass fields and increasing use of feedlots with declining air temperatures and increased use of 'unnatural' foods during cold weather has been noted by other workers. In this study such a relationship with soil temperature was apparent in only two of the years but in all years starlings fed in the feedlots and caused economic losses, suggesting that factors other than low soil temperatures led the birds to take cattle food.

The high proportions of birds feeding when most of the population was in grass fields (Fig. 4) indicates that time budgets may be involved, with starlings taking longer to obtain their daily food requirement when eating grassland invertebrates than when eating barley. Taitt (1973) showed that a starling's daily requirement of vegetable food (in her case poultry pellets) was equivalent to around 29 g of barley. At an ingestion rate of 0.54 g min^{-1} (Feare and Swannack, 1978) a starling feeding at the feedlots could obtain its daily requirement in about an hour, leaving the remainder of the day for other activities, e.g. preening and resting, which could be done in situations where the birds could be more vigilant, or less vulnerable, to predators. Presumably, starlings searching for less readily available soil or surface-living invertebrates would take longer to obtain their food, as indicated by Fig. 4, although this has yet to be demonstrated for wild starlings by measurements of winter intake rates.

If starlings find cattle food so readily available that such a short feeding time is needed, the original question should perhaps be re-stated: why should starlings feed on grassland invertebrates when cattle food is available?

Feare and Inglis (1979) have shown that there is an optimum density of birds on a cattle trough: below this density feeding rates are decreased by the wariness of the birds, while above the optimum density feeding rates are depressed as a result of conflicts between individuals over feeding space. In this competitive situation males dominate females, which may explain why samples of starlings caught in the Bridgets feedlot were predominantly males. This predominance was greatest in 1976–77 when the farm had a particularly high winter population of starlings (Feare, in press). Furthermore, observations of individually marked birds showed that during the 1976–77 winter, females tended to feed further away from the farm buildings than in 1975–76, when fewer birds were present, and this may have resulted from competitive exclusion by the presumed larger total number of males present in 1976–77 (Feare, in press). Thus competition for feeding space could be a potent factor mitigating against the use of feedlots by all but the most dominant birds. However, the failure of many birds to utilize the feedlots even when the farm population was only around 400 birds (the maximum that could feed in the feedlots — *see* above), and the increased use of feedlots at low soil temperatures in two years, suggests that many starlings prefer invertebrates from grass fields to cereal-based foods. That the grassland-feeding segment of the population contains most of the females has yet to be verified by collecting specimens but this seems probable and prompts a further question: why does this segment of the starling population apparently prefer to feed on the resources available in grass fields, irrespective of the effects of competition for space in the feedlot?

Berthold (1968) has proposed that the recent increase in numbers and extension of the range of starlings in Europe has been facilitated by breeding earlier in the year, ultimately as a result of climatic amelioration, since early breeding results in higher breeding success and may raise the chances of laying a second clutch (Ricklefs and Peters, 1979). Perrins (1970, 1979) argued that the principal determinant of the date of initiation of egg laying is the food available for the female to produce eggs, while Jones and Ward (1976) and Ward (1977), for quelea (*Quelea quelea*) and starlings respectively, were more specific in suggesting that the protein reserves in the female's pectoral muscles determine her egg-laying ability. Therefore, it may be advantageous for female starlings to attempt to maintain high levels of pectoral muscle protein during the winter in order that they may breed earlier in the spring. These high protein levels are more likely to be attained on a diet of invertebrates which are rich in protein (earthworm dry weight is 53–71% protein; Laverack, 1967) and are readily assimilated (Taitt, 1973), rather than on a carbohydrate-rich and less digestible diet consisting primarily of cereals (barley dry weight is 9% protein; Spector, 1956). Furthermore, animal protein probably represents a higher quality protein for starlings since the amino acid constituents will be in proportions more akin to the birds' requirements than are those found in vegetable protein.

Thus while competition for space in the feedlots may exclude some birds from feeding there, a positive selection of grassland feeding sites by other members of the population may be imposed by requirements for specific nutrients. Age-dependent selection of nutrients has been demonstrated in domestic chicks (Kaufman *et al.*, 1978) and Tinbergen and Drent (1980) have suggested that female starlings select food items to obtain different nutrients for their chicks. Laboratory experiments and field observations are now required to test the hypothesis that nutrient selection plays a role in the selection of feeding sites by female starlings in winter.

Acknowledgments

I am grateful to Messrs P. J. Jones and E. J. Mundy, successive Directors of Bridgets EHF, for permission to work on the farm, to Messrs P. A. Scolari and R. Robertson for assistance with the field work and to Miss J. Hogan for helping with the analyses.

References

Bailey, E. P. (1966). Abundance and activity of starlings in winter in northern Utah. *Condor* **68,** 152–162.

Berthold, P. (1968). Die Massenvermehrung des Stars *Sturnus vulgaris* in fortpflanzungs — physiologischer Sicht. *Journal für Ornithologie* **109,** 11–16.

Dunnet, G. M. (1956). The autumn and winter mortality of starlings *Sturnus vulgaris*, in relation to their food supply. *Ibis* **98,** 220–230.

Feare, C. J. (1980). The economics of starling damage. In *Bird Problems in Agriculture*, pp. 39–55. E. N. Wright, I. R. Inglis & C. J. Feare. Croydon: British Crop Protection Council.

Feare, C. J. (in press). Local movements of starlings in winter. In *Proceedings of the XVIIth International Ornithological Congress, Berlin* (in press).

Feare, C. J. & Inglis, I. R. (1979). The effects of reduction of feeding space on the behaviour of captive starlings *Sturnus vulgaris. Ornis Scandinavica* **10,** 42–47.

Feare, C. J. & Swannack, K. P. (1978). Starling damage and its prevention at an open-fronted calf yard. *Animal Production* **26,** 259–265.

Feare, C. J. & Wadsworth, J. T. (1981). Food loss to starlings at complete diet dairy feeding areas. *Animal Production* **32,** 179–183.

Jones, P. J. & Ward, P. (1976). The level of reserve protein as the proximate factor controlling the timing of breeding and clutch-size in the red-billed quelea *Quelea quelea. Ibis* **118,** 546–574.

Kaufman, L. W., Collier, G. & Squibb, R. L. (1978). Selection of an adequate protein-carbohydrate ratio by the domestic chick. *Physiology and Behaviour* **20,** 339–344.

Laverack, M. S. (1967). *The Physiology of Earthwoms.* Oxford: Pergamon.

Perrins, C. M. (1970). The timing of birds' breeding seasons. *Ibis* **112,** 242–255.

Perrins, C. M. (1979). *British Tits.* London: Collins.

Ricklefs, R. E. & Peters, S. (1979). Intraspecific variation in the growth rate of nestling European starlings. *Bird Banding* **50,** 338–348.

Spector, W. S. (1956). *Handbook of Biological Data.* Philadelphia: Saunders.

Taitt, M. J. (1973). Winter food and feeding requirements of the starling. *Bird Study* **20,** 226–236.

Tinbergen, J. & Drent, R. (1980). The starling as a successful forager. In *Bird Problems in Agriculture*, pp. 83–97. E. N. Wright, I. R. Inglis & C. J. Feare. Croydon: British Crop Protection Council.

Ward, P. (1977). Fat and protein reserves of starlings. *Institute of Terrestrial Ecology Annual Report*, pp. 54–56.

Williamson, P. & Gray, L. (1975). Foraging behavior of the starling (*Sturnus vulgaris*) in Maryland. *Condor* **77,** 84–89.

Section 6 **The impact of weeds and weed control practices on crop pests and diseases**

Weed control practices and changing weed problems

J D Fryer
Agricultural Research Council Weed Research Organization, Begbroke Hill, Yarnton, Oxford OX5 1PF

Introduction

A plant becomes a weed when it is unwanted or a nuisance. The same species may be a crop in one situation, a weed in another, or simply a component of a natural plant community. In nature there are no weeds.

Weeds and their control must, therefore, always be related to the activities or desires of man, whether these lie in the achievement of specific cropping systems, his health and safety problems, or the pursuit of recreation and enjoyment. In the context of this volume weeds are considered as part of the overall ecosystem of cropped land and related areas: a factor which, by modifying the environment and serving as alternate hosts or in other ways, may influence the incidence of pests and diseases. Their role in this capacity is often little understood and seldom studied. The direct losses that weeds cause to crop production are of such immediate practical importance that weed scientists and agronomists naturally concentrate their efforts on how to overcome or minimize them. Moreover, the understanding and quantification of the, often more subtle, indirect losses due to the association of weeds, pests and diseases cannot be accomplished by the weed scientist alone. It needs co-operation with plant pathologists, entomologists, nematologists and other specialists, often in complex field investigations — not an impossible task but one that is seldom attempted. Several of the papers in this volume explore this association and I hope that they will lead to the identification of useful work to be done and of research groups with the interest and enthusiasm to do something constructive.

This paper reviews some of the developments that have taken place in weed control technology in recent years and indicates the influence that these have had on the way crops are grown and on weed populations; both of which factors may interact with the incidence of pests and diseases.

The situation as it was

Before the advent of herbicides weed control in crops had to be achieved by hoeing, tillage, hand-pulling, cutting and burning, supported by a host of agronomic practices such as crop rotation, seed rate and grazing management, in other words by all the tenets of so-called 'good husbandry'. The limited effectiveness of these techniques, exemplified by the picturesque landscapes of pre-war Britain, dominated in summer by colourful weeds, often led to serious losses in yield and quality of crops, and indeed dictated what crops could be grown on particular pieces of land. The constraints imposed by weeds, however, had to be accepted and the rotational husbandry and associated cultural practices employed during the eighteenth, nineteenth and first part of the twentieth centuries led, throughout much of northern Europe, to the classical weed flora of temperate arable land, namely aggressive annual dicotyledons such as *Sinapis arvensis*, *Matricaria* spp., *Stellaria media*, *Papaver rhoeas* and perennial grasses of which *Agropyron repens* was probably the most serious. Since effective hoeing could not be achieved on a large scale in wheat, barley and oats, these crops encouraged such weeds and could only be grown successively for two or three years before easily-hoed row crops had to be planted or the land put down to grass. Since Roman times ploughing had been universal as the only way to bury weeds and trash and prepare the ground for the next crop. This was the scene in Britain and northern Europe at the end of World War II. In sub-tropical and tropical agriculture similar constraints were imposed by weeds but the means available for tackling them mechanically were generally much more limited and hand hoeing was, and in many places still is, the rule.

The influence of herbicides

Whilst chemical weed control had been practised in Britain on a small scale in wheat and barley and in some plantation crops since the end of the last century (Strawson, 1903), the efficacy and selectivity of the materials available (e.g. copper salts, inorganic arsenicals, chlorates, sulphuric acid, borax and mineral oils) were so limited that they did little more than alleviate some weed problems in crops grown in the traditional way. It was not until the discovery of the so-called 'phenoxy' or hormone weedkillers, MCPA and 2,4-D, in 1942 and their marketing in 1946, that herbicides began to play a major role in agriculture. Cheap, safe, easy to use, effective against a wide range of dicotyledonous species and highly selective in favour of cereals and other graminaceous crops, they were welcomed by post-war farmers and encouraged the chemical industries of Europe and North America to direct massive resources into extensive synthesis and screening programmes in the hope of finding new active compounds. One success led to another, spurred on by demands by governments and international organizations for increased food production to meet the needs of a growing world population. The achievements of the agrochemical industry, as it came to be known, were such that to-day some 200 synthetic compounds are under practical development or being used as active ingredients in many thousands of herbicide products for a vast range of crops and weeds. It was estimated that world sales of herbicides in 1980 would amount to more than $4000 million (Anon., 1979). In the UK, herbicide sales for agriculture and horticulture amounted to £94.2 million in 1979 compared with £15.5 million and £19.3 million respectively for fungicides and insecticides (Anon., 1979a). The major arable and vegetable crops are virtually all sprayed, many fields more than once (Table 1).

Table 1 Herbicide use in Great Britain

	Percentage of total national area treated with herbicide (1978)
Cereals	103*
Sugar beet	186*
Potatoes	68*
Field beans	98†
Oil seed rape	64†
Fodder roots	55†
Grassland	6*
Vegetable crops and bulbs	103†‡
Dessert, culinary and cider apples	217†‡
Bush fruit	247†
Cane fruit	153†
Strawberries	215†
Grapevines	284†

Figures greater than 100 indicate that some plantings were treated more than once

* British Agrochemicals Association Annual Report 1978/79.
† Ministry of Agriculture, Fisheries and Food Surveys 1972–75
‡ England and Wales statistics only

The impact of herbicides has been much more than to substitute the sprayer for the hoe. Their remarkable ability to kill weeds selectively in crops, and during the fallow period between successive crops, without disturbing the soil has allowed a transformation of crop production technology and a complete re-appraisal of the role of tillage (e.g. Atkinson and White, *1981*). Monoculture cropping and the adoption of husbandry methods based on minimum cultivation are good examples of how herbicides have led to greater specialization and intensification of crop production than was feasible before their introduction. Thus in the UK many arable farms are now based on highly productive near-continuous cereal farming and, overseas, long runs of such crops as maize, cotton and soybeans have become commonplace. The effect on weed populations has been profound. The new husbandry systems made possible by chemical weed control techniques have themselves had an enormous influence on the success or otherwise of individual weed species.

The following sections consider how weeds have responded to the chemical onslaught and to the new husbandry techniques.

Response of weeds to herbicides and intensified crop production

When the same crop and the methods used to grow it and to prepare the land are repeated year after year, the selective pressures on the weed flora tend to be less varied than when crop rotation is practised. Moreover the annual use of a limited range of herbicides appropriate to a given crop favours the selection from the many species in the soil seed bank of those which can tolerate the control measures employed. It might be expected that for both reasons the species diversity of the weed flora in areas of intensive arable agriculture would decline and it is often maintained that this has occurred. However, Fryer and Chancellor (1970a) in reviewing the published literature found little evidence to support this view, although major changes in the species composition of weed floras have of course occurred. Many of the more tolerant and adaptable dicotyledonous species such as *Polygonum* spp., *Stellaria media* and *Veronica* spp. are still abundant and remain very important weeds in Britain whereas many others such as *Sinapis arvensis* and *Papaver rhoeas* have declined greatly in abundance and importance. Taking their place are annual grass weeds notably *Avena* spp. and *Alopecurus myosuroides*, which have increased enormously as predicted by Fryer and Chancellor (1970b) and illustrated in Fig. 1.

That changes of this kind have been common is clear from the literature and the following examples are representative. In France a recent survey (Aymonin, 1976) showed that many plant species of cultivated land are declining following the widespread use of herbicides. In Germany a total of 3200 recordings taken in arable land during 1948–55 and 1958–65 revealed a reduction in easily controlled annual weeds over the 17-year period accompanied by increases in *Agropyron repens*, *Poa annua*, *P. trivialis* and *Avena fatua* (Bachthaler, 1967). In sugar beet fields in Israel many herbicide-susceptible species which were common 30 years ago are no longer troublesome weeds, whereas formerly unimportant species which are tolerant to herbicides have become a major problem (Cohen, 1970). In Czechoslovakia continuous cultivation of winter wheat between 1971 and 1975 increased populations of *Apera spica-venti*, *Tripleurospermum maritimum* ssp. *inodorum* and *Veronica hederifolia*, whereas continuous cropping of spring barley increased *Sinapis arvensis* and *Avena fatua* (Zemanek, 1976). In Hungary a survey of the weed flora of maize and winter cereals was made in 1969–70 and the results compared with those of a national weed survey in 1947. Many typical arable weeds had declined (notably *Convolvulus arvensis*, *Polygonum aviculare* and *Cirsium arvense*) and this was attributed to the use of MCPA and 2,4-D in cereals. Species which had increased included *Ambrosia elatior*, *Anagallis arvensis*, *Polygonum convolvulus*, *Apera spica-venti* and *Matricaria* spp. (Hunyadi, 1973). Also from Hungary comes the information that the weed cover in maize crops decreased significantly over 15 years as a result of introducing large scale mechanized farming including the use of herbicides. The continued application of simazine and atrazine over several years caused undesirable shifts in the weed flora of maize monocultures and increased the population of perennial weeds by 250% (Fekete, 1974).

The use of simazine and atrazine has allowed intensification of maize production in many countries but their repeated use and the lack of crop rotation have led to an upsurge of resistant species. The most important of these are grasses such as *Digitaria sanguinalis*, *Rottboellia exaltata* or *Sorghum halepense*, for example in the Philippines (Anon., 1980), Spain (Villarias, 1976) and the United States (Peters, 1967; Dale and Chandler, 1979).

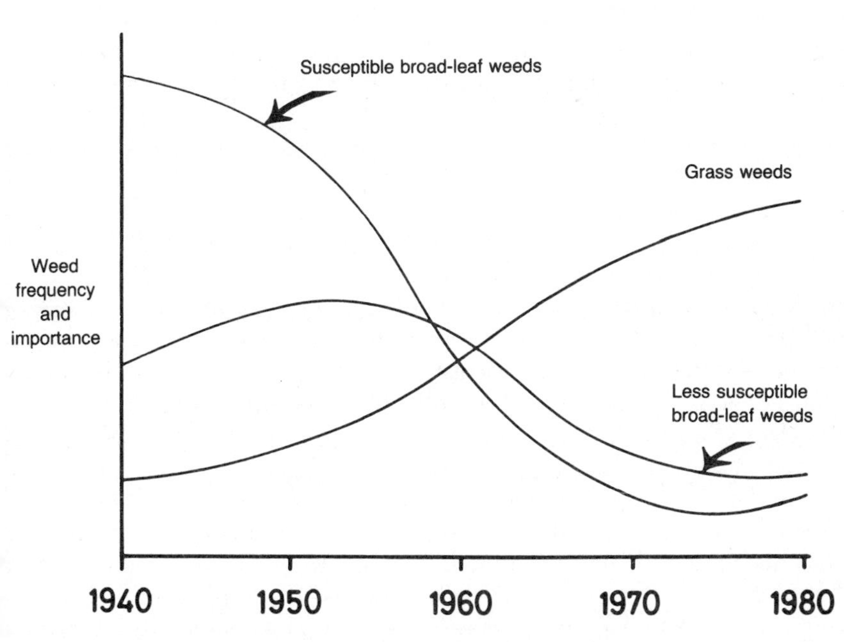

Figure 1 Conjectured changes in the British arable weed flora over four decades (Fryer and Chancellor 1970b).

Response of weeds to minimum cultivation and direct-drilling

The repeated use of herbicides and departure from crop rotations have greatly affected the weed flora as indicated by these few examples. A further consequence of herbicide use affecting weed populations is the reduced level of soil tillage now practised in many crops. In the 1950s the technical feasibility of sowing crop seed directly into untilled soil after the living vegetation had been killed by herbicides became evident. The first experiments in the UK to compare ploughing, reduced cultivation and direct-drilling methods of cereal growing were begun in 1961 (Davies and Cannell, 1975). With the advent of paraquat and the pioneering work of ICI Ltd, direct-drilling of many crops became a practical proposition. This and the widespread adoption of reduced tillage techniques involving disturbance of the soil to a depth only of 3–5 cm is now commonplace in British agriculture and sometimes overseas. It is being encouraged by the increasing cost of energy and several important technical advantages of which timeliness in sowing is perhaps the most important.

Such a major change in soil tillage after centuries of ploughing clearly has important implications for the weed flora (Cussans, 1975). One example is that reduced cultivation and direct-drilling techniques in the UK have allowed a far larger area of land to be planted to winter cereals and for the crops to be sown much earlier in the autumn than is possible with ploughing. Thus weeds now have the opportunity to establish before onset of winter and have a 10-month growing season during which the effectiveness of control measures may break down. Examples of species which are capable of germinating in early autumn are *Stellaria media*, *Matricaria* spp., *Veronica* spp., *Galium aparine*, *Alopecurus myosuroides* and *Poa* spp. (Cussans, 1980). Many weeds can also benefit from the short period between one autumn-sown cereal crop and the next, which gives little opportunity for farmers to cultivate their fields to encourage annual weeds to germinate and to destroy perennials (Cussans, 1976). Moreover, in the absence of ploughing, weed seeds and perennating organs are no longer buried deeply, favouring some species and discouraging others. The consequences for British agriculture have been reviewed by Cussans (1976) and by Cussans *et al.* (1979).

It is clear from the work of these authors that the relationship between tillage and individual weed species is highly complex in cereal growing in Britain and that generalizations are seldom possible. Whilst *Avena fatua* is variable in its response to tillage *Alopecurus myosuroides* and *Poa annua* are consistently favoured by reduced cultivation regimes for cereal production (Cussans *et al.*, 1979). Some dicotyledonous weeds, e.g. *Polygonum* spp., *Sinapis arvensis* and *Papaver rhoeas* appear to be favoured by ploughing (Pollard and Cussans, 1977). Others, e.g. *Matricaria* spp., *Aphanes arvensis*, *Sonchus arvensis*, have been found to be more numerous at low levels of cultivation. For a few species notably *Stellaria media* there seems no clear trend.

In the United States and elsewhere, many severe weed problems have developed when maize has been grown continuously under systems of minimum or zero tillage. In a seven-year experiment in Ohio in which maize growing with and without tillage was investigated, Triplett and Lytle (1972) found that *Apocyanum cannabinum* and several other perennial species, which did not feature where tillage was practised, became a serious problem in untilled plots. From a survey of changes in the weed flora associated with maize production by minimum tillage in the southern United States, Heron *et al.* (1971) reported that the annual grasses *Digitaria* spp. and *Panicum dichotomiflorum* frequently increased and that perennial weeds tended to become serious problems including *Solanum carolinense*, *Physalis* sp., *Ampelamus albidus*, *Campsis radicans* and *Cyperus* spp. There are numerous other examples.

Cyperus spp. are of course amongst the worst weeds in the tropics (Holm *et al.*, 1977) and are frequently associated with crops in which only shallow tillage is practised. The whole range of crops is affected from arable food or fibre crops to tropical plantation crops. The same can be said for many other perennial weeds, particularly in humid climates. In temperate fruit orchards where no tillage is practised, severe infestations of weeds such as *Agropyron repens*, *Convolvulus arvensis, Cirsium arvense, Equisetum arvense* can quickly colonize strips treated with soil-applied herbicides (Davison, 1974). In rubber and other tropical plantation crops the continued use of the same herbicide regime can lead to resistant species becoming dominant. In Malaya, for example, when paraquat was used to control *Ottochloa nodosa*, Pushparajah and Woo (1971) reported the marked development of *Borreria latifolia* and *Cleome ciliata.* When an almost pure stand of *Melastoma malabathricum* was sprayed with amitrol, the treated area was invaded by *Mikania cordata* and *Imperata cylindrica.*

The dynamic nature of weed problems — further examples

In this short review it is only possible to mention a few of the many crops for which the introduction of herbicides and new agronomic practices has led to changes in weed populations. Rather than extend the list by superficial cataloguing of other crops I have selected the following examples to illustrate in rather more detail the dynamism of current weed problems in intensive arable agriculture.

If the forecast trend towards intensification of cereal production in Britain is realized (Anon., 1979b) it seems inevitable that grass weeds of one species or another will remain a dominant problem. A recent survey (Elliott *et al.*, 1979) has shown that *Avena fatua* and to a lesser extent *A. ludoviciana* now occur throughout cereal growing areas in the UK. Population densities are being kept in check by lavish use of herbicides but resurgence would be rapid if the level of profitability of cereal production were to decline to the point that spray programmes had to be cut back. Similarly with *Alopecurus myosuroides* which appears still to be extending its range in both spring and winter cereals. Other grass weeds are making their presence felt increasingly with the widespread growing of winter cereals and the earlier drilling made possible by reduced soil tillage. The traditionally ruderal grass *Bromus sterilis* has, since the dry summer of 1976, become of major concern to cereal growers, spreading from field margins and not being well controlled by herbicides. Where cereal/ley rotations are practised, *Lolium* spp., *Poa* spp. and other pasture grasses have become troublesome weeds in the cereal crop. Another serious problem for the intensive cereal grower is volunteer wheat growing in the barley crop and *vice versa*.

An analogous grass weed problem to that in the UK has recently arisen in India, where wheat production has almost trebled since 1964, when the All-India Wheat Improvement Programme began with the first imports of seed of the new semi-dwarf hybrid wheats bred at CYMMYT in Mexico. Over large areas of North India irrigated wheat is now grown in the cool season followed immediately by rice during the summer season, the sequence being repeated in successive years. This wheat–rice 'rotation' which is vigorously supported by the Government is highly productive and very profitable for the growers, so there is an understandable reluctance to adopt alternative, less attractive crop sequences. The system, however, also suits an annual grass *Phalaris minor* which, prior to intensification of wheat production, was unimportant as a weed. Indeed it remains so in rain-fed areas. Vast areas of irrigated wheat are now heavily infested. In spite of the abundance of cheap labour, the density of *Phalaris minor* can be such that crop failures may occur after even three hand-weedings. In some areas *Avena* spp. and the poisonous *Lolium temulentum* are also serious problems. There is now no alternative but to turn to herbicides, or abandon this highly efficient and productive crop rotation. This example shows that a grass weed problem in cereals is not necessarily the outcome of herbicide use, as is commonly believed on account of the explosion in grass weeds which followed the widespread adoption of herbicides in temperate cereals.

As already noted grass weeds in sub-tropical and tropical graminaceous crops have also become major problems where herbicides have facilitated intensive production systems. Repeated use of triazine herbicides, notably atrazine, has led to massive infestations in maize and sorghum of tolerant species such as *Rottboellia exaltata*, *Sorghum halepense*, *S. bicolor* and *Panicum dichotomifolium*. In rice, grasses such as *Echinochloa crus-galli* that are not always well controlled by available herbicides can become major problems. Species of wild rice which are well suited to intensive cropping systems and tolerant to selective herbicides used in the rice crop can quickly build up into massive infestations. *Oryza punctata* is an example from Swaziland and *O. rufipogon* from the Americas. In Korea repeated use of a number of different herbicides in rice (butachlor, thiobencarb + simetryne, nitrofen and 2,4-D) quickly led to a predominance of *Eleocharis kuroguwai*, *Cyperus serotinus* and *Scirpus juncoides* whilst broadleaved weeds declined from 45% to 10% during the four year period (Ahn *et al.*, 1976).

One may conclude that the intensification of cereal growing which is often, but not always, a product of herbicide use, is liable to result in selective pressures that favour weeds which 'mimic' the crop in their ability to thrive in the agricultural ecosystem provided by the farmer and to tolerate the control measures employed. The ultimate example is when the crop itself becomes a weed. It is of interest to note that Holzner (1978) has drawn attention to the vulnerability of such crop mimics through their close adaptation to the ecological requirements of the crop and their dependence on the crop continuing to be grown.

Concluding remarks

Most of this paper deals with the role of herbicides in the intensification of field crop production and the impact that both they and new crop production methods have had on the weed flora. There are of course many other situations in which herbicides are used: in grassland for weeds which are poisonous, unpalatable or otherwise a constraint on livestock production; in vegetables in which achievement of near 100% weed control is often an essential requirement; in fruit and other plantation crops much of which is now grown in intensive, non-tillage systems in which the aim is total weed control; in forest seedbeds and young plantations; in hardy ornamental stock; in drainage channels and freshwater lakes where traditional methods of weed control are seldom feasible; and in industrial, urban and amenity land of all kinds. The plants growing in such situations form part of the agricultural ecosystem from the point of view of disseminating weeds, pests and pathogens and all are subject both to neglect and to positive management to varying degrees by herbicides or other means.

If the subject of weeds in relation to crop pests and diseases were extended beyond agriculture into other organisms affecting the well-being of man and his animals then weed problems such as water hyacinth (*Eichhornia crassipes*) which provides breeding sites for malaria-carrying mosquitoes, scrub which harbours tsetse flies responsible for trypanosomiasis and plants (e.g. *Ambrosia artemisiifolia*) providing sources of pollen allergens would need to be included in a comprehensive review.

It seems inevitable from the evidence briefly reviewed here that weed populations will continue to respond to changing agronomic practices and control measures, as intensification of crop production progresses to meet the needs of a rapidly growing world population. Up to the present there has been a steady flow of novel herbicides to tackle weed problems as they develop, but the constraints on the agrochemical industry are such that the commercial development of new compounds has slowed down and seems unlikely to be restored to its former level. How then are we going to deal with the weed problems of the future?

The extensive range of existing herbicides will remain the keystone of future control techniques, supplemented by a limited flow of new compounds from industry. However, research must find new rational ways of exploiting the inherent physiological properties of these chemical tools, the current uses of which are largely empirical. This will be achieved in part through a fuller understanding of their action in plants and soil, and in part by harnessing ecological and agronomic approaches to weed control; in other words the realization of integrated control systems based on a detailed knowledge of weed population dynamics in the context of specific agricultural ecosystems and economic objectives. In the UK inter-disciplinary research by the Weed Research Organization has made considerable progress in establishing such systems for two intractable weeds of intensive cereal growing (*Avena fatua* (Cussans, 1976a) and *Agropyron repens* (Cussans and Ayres, 1977)). Marked progress is currently being made with *Alopecurus myosuroides* and mixed dicotyledonous and grass weed communities in winter cereals.

In the future, as our pursuit for knowledge of the behaviour and role of weeds delves deeper into specific weeds and weed communities, we shall undoubtedly learn more about constraints on their growth and development caused by pests and pathogens. The extent to which such knowledge and the organisms themselves can be harnessed to aid weed control in cropped land may be in doubt but two things are certain. First, weed scientists cannot afford to overlook any potentially helpful aids if our weed control technology is to provide a continuing flow of remedies for the changing weed problems of the future. Second, the knowledge gained about the inter-relationships between pests, diseases and weeds should surely be of value to entomologists and plant pathologists in advancing the practice as well as the theory of integrated crop protection.

References

Ahn, S. B., Kim, S. Y. & Kim, K. U. (1976). Effect of repeated annual application of pre-emergence herbicides on paddy field weed populations. In *Proceedings 5th Asian–Pacific Weed Science Society Conference, Tokyo, Japan 1975*, pp. 287–292.

Anon. (1979). A look at world pesticide markets. *Farm Chemicals* **142,** (9), 64–68.

Anon. (1979a). *Annual Report British Agrochemicals Association 1978/79*, pp. 6–8.

Anon. (1979b). *UK Royal Commission on Environmental Pollution. 7th Report on Agriculture and Pollution*, pp. 280. London: HMSO.

Anon. (1980). Weed poses a serious problem. *Infoletter*, No. 44.

Atkinson, D. & White, G. C. (*1981*). The effects of weeds and weed control on temperate fruit orchards and their environment. In *Pests, Pathogens and Vegetation*, pp. 415–428. J. M. Thresh. London: Pitman.

Aymonin, G. G. (1976). La Baisse de la diversité spécifique dans la flore des terres cultivées. In *Comptes Rendus du Ve Colloque International sur l'Ecologie et la Biologie des Mauvaises Herbes, Dijon*, pp. 195–202.

Bachthaler, G. (1967). Changes in arable weed infestation with modern crop husbandry techniques. *Abstract 6th International Congress Plant Protection, Vienna*, pp. 167–168.

Cohen, A. (1970). The influence of chemical weed control on the weed flora in beet fields in Israel. In *Proceedings 2nd International Meeting on Selective Weed Control in Beetcrops, Rotterdam*, pp. 213–214.

Cussans, G. W. (1975). Weed control in reduced cultivation and direct-drilling systems. *Outlook on Agriculture* **8,** 240–242.

Cussans, G. W. (1976a). The influence of changing husbandry on weeds and weed control in arable crops. In *Proceedings 1976 British Crop Protection Conference — Weeds*, pp. 1001–1008.

Cussans, G. W. (1976b). Population dynamics of wild oats in relation to systematic control. *Report of the ARC Weed Research Organization for 1974–75*, No. 6, pp. 47–56.

Cussans, G. W. (1980). Weeds in cereals and their control. *Span* **23,** (1), 30–32.

Cussans, G. W. and Ayres, P. (1977). An experiment in cultural and chemical control of *Agropyron repens* in five successive years of spring barley. In *Proceedings EWRS Symposium on Different Methods of Weed Control and their Integration, Uppsala*, pp. 171–179.

Cussans, G. W., Moss, S. R., Pollard, F. & Wilson, B. J. (1979). Studies of the effects of tillage on annual weed populations. In *Proceedings EWRS Symposium on the Influence of Different Factors in the Development and Control of Weeds, Mainz*, pp. 115–122.

Dale, J. E. & Chandler, J. M. (1979). Herbicide–crop rotation for Johnsongrass (*Sorghum halepense* (L) Pers.) control. *Abstracts Weed Science Society of America*, pp. 29–30.

Davies, D. B. & Cannell, R. Q. (1975). Review of experiments on reduced cultivation and direct-drilling in the United Kingdom, 1957–1974. *Outlook on Agriculture* **8,** 216–220.

Davison, J. G. (1974). The control of bindweed in fruit crops. *Report ARC Weed Research Organization 1972–1973*, No. 5, pp. 56–62.

Elliott, J. G., Church, B. M., Harvey, J. J., Holroyd, J., Hulls, R. H. & Waterson, H. A. (1979). Survey of the presence and methods of control of wild-oats, blackgrass and couch grass in cereal crops in the United Kingdom during 1977. *Journal of Agricultural Science, Cambridge* **92,** 617–634.

Fekete, R. (1974). [Comparative weed investigations in wheat and maize crops cultivated traditionally and treated with weedicides. 2. Changes in the weed vegetation of maize crops.] *Acta Biologica, Szeged* **20,** 37–46.

Fryer, J. D. & Chancellor, R. J. (1970a). Herbicides and our changing weeds. In *Botanical Society of the British Isles Conference Report 1970, 11, The Flora of a Changing Britain*, pp. 105–118.

Fryer, J. D. & Chancellor, R. J. (1970b). Evidence of changing weed populations in arable land. In *Proceedings British Weed Control Conference*, pp. 958–964.

Heron, J. W., Thompson, L. & Slack, C. H. (1971). Weed problems in no-till corn. In *Proceedings 24th Annual Meeting Southern Weed Science Society*, p. 170.

Holm, L. G., Plucknett, D. L., Pancho, J. V. & Herberger, J. P. (1977). *The World's Worst Weeds*. Honolulu: University of Hawaii (for East–West Center).

Holzner, W. (1978). Weed species and weed communities. *Vegetation* **38,** 13–20.

Hunyadi, K. (1973). Weed problems of the south-western arable lands of Hungary. In *Jugoslovenski Simpozium o Borbi protiv Korova u Brdsko-Planinskim, Producjima, Sarajeva*, pp. 61–66.

Peters, R. A. (1976). Changes in the weed population when using a single herbicide in maize monoculture. *Abstract 6th International Congress Plant Protection, Vienna*, pp. 382–383.

Pollard, F. & Cussans, G. W. (1977). The influence of tillage on the weed flora of four sites sown to successive crops of spring barley. In *Proceedings British Crop Protection Conference — Weeds 1976*, pp. 1019–1028.

Pushparajah, E. & Woo, Y. K. (1971). Weed control in rubber plantation. In *Proceedings 3rd Asian–Pacific Weed Science Society Conference, Kuala Lumpur*, pp. 9.

Strawson, G. P. (1903). *Standard Fungicides and Insecticides in Agriculture*. London: Spottiswoode & Co. Ltd.

Triplett, G. B. & Lytle, G. D. (1972). Control and ecology of weeds in continuous corn grown without tillage. *Weed Science* **20,** 453–457.

Villarias, J. L. (1976). Evolution de la flora adventice soumise à la monoculture traitée avec des herbicides. *Comptes Rendus du Ve Colloque International sur l' Ecologie et la Biologie des Mauvaises Herbes, Dijon*, pp. 187–193.

Zemanek, J. (1976). [The influence of annual application of herbicides on the changes of weed communities on ploughland.] *Agrochemia* **16,** (3), 73–76.

The effects of weeds and weed control on temperate fruit orchards and their environment

D Atkinson and G C White
Pomology Department, East Malling Research Station, Maidstone, Kent ME19 6BJ

In reviewing the importance of weeds to world food production Holm (1976) showed that in some rural communities the proportion of working time spent weeding crops could be as high as 74% and losses due to weeds can be almost total. In efforts to avoid the effects of weed competition many of the world's rural populations are condemned to a life of weeding. It is against this background that possible beneficial effects of weeds as hosts for beneficial organisms, and the arguments for their maintenance to retain species diversity in crops or hedges, must be assessed. This paper discusses the effects of weed competition on temperate fruit production and profitability, giving some perspective to other papers dealing with alternative aspects of weeds.

The effects of removing weeds, mechanically, manually, or with herbicides are often ignored or considered neutral to the environment. However, without vegetation, soil conditions are greatly modified and this may influence other parts of the ecosystem. Effects on the environment resulting from the absence of weeds are discussed.

The impact of weed competition

Some form of weed control has been required for as long as crops have been grown. Mechanical cultivation or crop rotations were used until chemical herbicides were introduced (Fryer, *1981*). Chemicals remove vegetation without physically modifying the environment or causing mechanical damage to crop plant root systems. In any situation the need for weed control will be related to the impact of the weeds. Table 1 summarizes some of the effects of weed competition on the growth and yield of temperate tree crops. Effects can vary from none to crop death. Results for diverse tree species show effects ranging from a 15% reduction in growth of willows (Stott, 1980) to a 96% decrease in *Acer platanoides* (Davison and Bailey, 1980).

Effects on fruit yield of tree crops are also substantial and because of additional effects on fruit quality and size, the reduction in marketable yield and financial return can be greater than that simply on crop weight. In a trial started in 1972 with established apple trees (cv. Cox on M.26 rootstock)

Table 1 Effects of weed competition on the growth and yield of tree crops

Crop species	Weed species	Percentage decrease	Reference
Effects on growth			
Apple	Annual weeds	40–67	Atkinson and Holloway (1976)
Acer platanoides	*Matricaria recutita* and other species	30–96	Davison and Bailey (1980)
Philadelphus sp.		68	Davison and Bailey (1980)
Potentilla fruticosa		59–86	Davison and Bailey (1980)
Prunus laurocerasus		47–64	Davison and Bailey (1980)
Chamaecyparis lawsoniana		27–62	Davison and Bailey (1980)
Sitka spruce	*Calluna vulgaris*	50	Read and Jalal (1980)
Willow	*Stellaria media*, *Polygonum persicaria*, *Chenopodium album*	15–88	Stott (1980)
Sitka spruce, *Picea abies*	Various	26	McIntosh (1980)
Picea sitchensis, *Pinus contorta*	Grass weeds	26	Jones and Jones (1980)
Effects on yield			
Apple	Grass	17–66	Atkinson and White (1976)
Apple	Grass	34	O'Kennedy (1974)
Apple	Grass	41	Stott (1976)
Apple	Annual weeds	21–40	Atkinson and Holloway (1976)

competition from either grass weeds or annual weeds (*Senecio vulgaris, Capsella bursa pastoris* and *Poa annua*) growing in the tree row area (herbicide strip) which is normally kept weed-free, reduced crop weight by 16% and 21%, respectively in 1975 (Table 2). In 1976 reductions were 35% and 49%, respectively. The market value of fruit increases with size. Thus, in 1979 mean Cox prices varied from 8.3 pence per kilogram for fruit of 55–60 mm diameter to 19.1 pence per kilogram for fruit of >70 mm diameter. Due to competition from grass or annual weeds financial losses were 25% and 34% in 1975 and 45% and 55% in 1976, respectively. For the two years weeds reduced mean returns by £834 sterling ha^{-1}.

Table 2 Effects of weed competition in the herbicide-treated strip on apple yield and fruit size in 1975 and 1976

	1975			*1976*		
	Herbicide control	*Grass weeds*	*Annual weeds*	*Herbicide control*	*Grass weeds*	*Annual weeds*
Yield (kg $tree^{-1}$)	15.3	12.9	12.1	14.1	9.2	7.2
Fruit size*						
<60 mm	15	30	36	51	79	68
60–65 mm	23	29	30	22	8	20
65–70 mm	40	31	29	17	7	10
>70 mm	22	11	5	10	6	3

* Percentage of fruit in each diameter class.

Vegetation in the inter-row area (the grass alley) also competes with the tree (Atkinson and White, 1980). In a young apple orchard the initial five years' growth and yield were increased 44% and 37%, respectively, by avoiding inter-row grass competition. Other crops are also sensitive to weed competition. Baldwin (1979), reviewing numerous ADAS trials on removing either *Avena fatua* or *Alopecurus myosuroides* noted an average increase in yield of about 20% for a range of cereal varieties. Thus the cost of competing vegetation in an orchard or other crop is likely to be high and any arguments for its retention must consider these adverse effects (Altieri, *1981*).

The effects of weeds on the environment

During cultivation soil conditions are altered by both physical disturbance and because vegetation is absent. After herbicide treatment the soil remains undisturbed, as with grass or weeds, although compared with grass or weed cover the environment will be changed due to:

(1) the presence of herbicide residues or degradation products;
(2) a greatly reduced input of organic matter to the soil;

(3) a reduced 'transport system' through the soil due to the absence of either weed or grass root systems; and
(4) reduced physical protection of the soil from damage caused by rain or machinery.

The magnitude and significance of these factors vary between different crops. With cereals or vegetables, crop cover and organic matter input to the soil are likely to reduce the importance of factors (2) to (4). In top fruit orchards where the organic matter input to soil and root densities are both low and the protection of the soil surface is limited for much of the year all four factors are likely to be important. Fruit trees are most commonly grown with the row area (strip) treated with herbicide and the inter-row area under grass. However, a few orchards remain under overall grass whilst overall herbicide management is increasing. Thus environments with contrasting vegetative cover can be compared under conditions where the absence of vegetation is likely to have maximal impact. These effects are discussed in the remainder of this paper.

Soil physical condition

Effects of long-term herbicide treatments on soil conditions in fruit orchards have been reviewed by Atkinson and Herbert (1979). Without cultivation or vegetative cover soil bulk density increases (Table 3). For diverse soil types and crop species, densities were approximately 9% higher under herbicide than for either cultivation or grass. In cultivated plantations soil densities similar to those with herbicide treatment can rapidly develop as a result of wheel compaction or soil settlement. A difference in bulk density of 0.38 g ml^{-1} between cultivated and herbicide-treated soils at the beginning of a season decreased to 0.16 g ml^{-1} within 1 month and to 0.07 g ml^{-1}

Table 3 Effects of soil management on the bulk density (g ml^{-1}) of surface soil of apple (A) or raspberry (R) plantations*

Position	*Treatment* *Cultivation*	*Grass*	*Herbicide*	*Reference*
Row (A)	–	1.43	1.48	Atkinson and White (1976)
Inter-row (A)	–	1.34	1.54	
Row (A)	1.19	1.22	1.37	Atkinson and Herbert (1978)
Inter-row (A)	1.50	1.43	1.59	
Unspecified (A)	1.45	1.32	1.45	De Kock *et al.* (1977)
Inter-row (R)	1.21	1.06	1.24	Bulfin and Gleeson (1967)

* Derived from Atkinson and Herbert (1979).

in four months, (Soane *et al.*, 1975). An increase in bulk density signifies a loss of total pore space which may have significant biological effects.

For a root to grow, soil must either contain pores of a size greater than the root diameter or be sufficiently malleable to produce adequate pores. Similar considerations apply to the movement of most animals in soil. The suction needed to remove water from soil pores increases with decreasing pore diameter so small pores will be filled with water for longer during the year than large pores. Potential problems, as a result of the loss of pore space (increased bulk density), become important only if the volume of pores in a critical diameter range (hence water holding capacity) becomes reduced at particular times of the year.

Table 4 Effects of orchard soil management on the porosity of surface soil

	Position	*Porosity (total)*	*Percentage of soil volume as pores*		
			*>115 μ**	*30–115 μ**	*<30 μ**
(a) Sandy loam soil					
Herbicide	Row	43.7	14.6	3.8	25.3
	Inter-row	41.2	10.3	2.7	28.2
Grass	Inter-row	48.7	9.7	3.7	35.3
(b) Sandy soil					
Herbicide	Row	48.1	11.3	13.1	23.7
	Inter-row	40.1	9.8	4.9	25.4
Grass	Inter-row	46.1	7.1	6.0	33.0

* Water released at potentials of >2.5, 2.5–10.0 and <10.0 −KPa, respectively

A reduction in total pore volume could occur by the loss of space over either most or only a limited part of the range of pore sizes found in soil. It has usually been suggested that increased bulk density is associated with the loss of large pores, although part of the relevant evidence is from laboratory experiments with artificially compacted soils (Schuurman, 1965). Table 4 presents data for two orchards treated with herbicides for several years. Total porosity was reduced in herbicide-treated inter-row areas compared to grassed inter-rows but this was not always so for herbicide-treated rows where the volume of large pores was high. There was little effect of treatment on the volume of pores of 30–115 μ, while with grass there were many more pores <30 μ diameter. A higher proportion of the water retained by them appeared unavailable to plants, e.g. grassed soil retained 14% of its moisture but herbicide-treated soil only 12% at a water potential of −1.5 MPa, usually taken as the lower limit of water extraction by plant roots.

The large pores found in soil under both herbicide-treated and grassed conditions are formed by animals or roots. Such channels have survived for over 20 years adjacent to the windows of a root observation laboratory at East

Malling (Fig. 1). When an unprotected soil surface is exposed to rain, the soil aggregrates degrade and block the surface pores (Bulfin and Gleeson, 1967). Crusting is alleviated by surface cracking which appears to occur in most uncultivated soils. Grass or weeds protect the soil surface physically from rainfall, while the higher input of organic matter to the soil increases the stability of aggregates. In a five-year-old orchard the percentage of aggregates (0.5–2 mm diameter) from surface soil which resisted five immersions in water

Figure 1 The soil profile visible through an observation window (60 × 50 cm) of a root laboratory in 1961 (left), and the same window in 1972 (right). A number of soil channels created by worms or roots persisted throughout the 11 years.

fell from 56% for soil from a grassed inter-row to 13% for a herbicide-treated row and 6% for a treated inter-row. The degradation of aggregates is important in relation to the ability of soil to re-wet during rainfall. Recent experiments (Gurung, 1979; Farré, 1979) have shown that the permeability of soil to rainwater can be affected by vegetation (Fig. 2) and the way in which herbicide management is initiated. During the growing season soil under grass developed a larger moisture deficit than soil treated overall with herbicide following initial cultivation, but the grassed soil was re-wetted by rainfall to a greater extent (higher water content). As a result the residual soil moisture deficit in the month after heavy rain was highest with overall herbicide management of a previously cultivated soil. Where the herbicide treatment was introduced on a grassed soil or where a cultivated surface was protected with a thin layer of straw, permeability was increased relative to overall herbicide applied following cultivation (Fig. 2).

Light interception by fruit trees in modern orchards is poor and much of the incident radiation falls on the inter-row area which, if grassed, shows increased water loss. Consequently soil under grass or weeds usually develops a higher soil water deficit, (lower soil water potentials), than that under herbicide. These lower soil water potentials under grass have important effects both on tree crops and on other organisms within the ecosystem. For example, nematodes can deform soil only to a limited extent during movement (Wallace, 1968) so their diameter will influence the proportion of

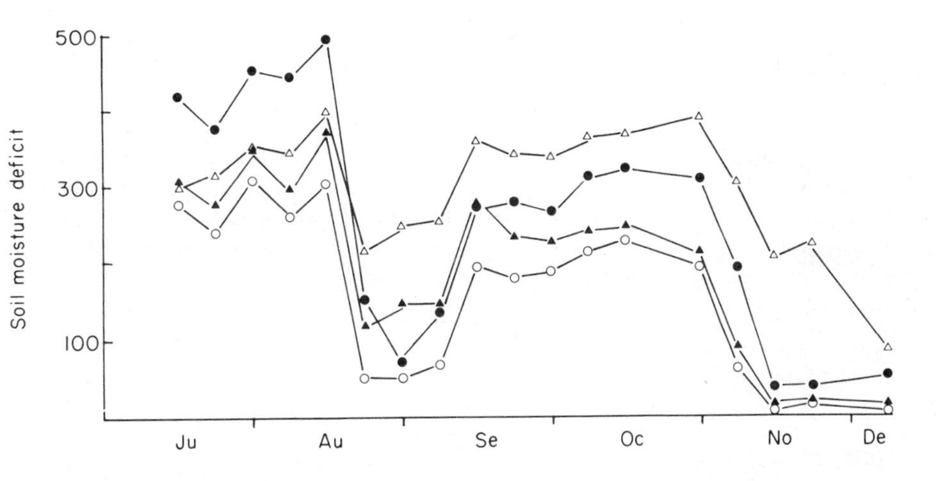

Figure 2 The soil moisture deficit (l tree^{-1}) as affected by soil management,
△ overall herbicide initiated from cultivated soil,
○ overall herbicide initiated from killed grass,
● wide herbicide strip (initiated from killed grass) with grassed alley,
▲ overall herbicide from cultivation but with straw cover.

available soil pore space utilized. Nematodes 10 μ in diameter used 25% of the soil volume but those of 17 μ only 20% (F. G. W. Jones *et al.*, 1969). Plant nematodes move most rapidly through soil pores of 30–100 μ diameter, only the large pores being suited to the longer nematodes. In addition to physical restrictions the presence of water in pores can be important with respect to movement and other activities. Pores of 30–100 μ diameter drain at potentials of −3 to −10 kPa. Potentials lower than these will develop much earlier and persist longer under vegetation than where herbicides are used (Table 5). Thus the effects of changes in soil physical properties on the suitability of the soil environment for root growth and the activities of other soil organisms are complex with important soil factors being affected differently by weeds.

Table 5 Effects of soil management on the soil pores and soil water potential (−MPa) in the top 25 cm of soil during 1975

Treatment	*Position*	*Maximum pore diam. water-filled* (μ)	*Soil water potential*		
		18 June	*18 June*	*23 July*	*12 Sept.*
Grass	Row	(0.23)	1.32	>1.5	>1.5
	Inter-row	(0.24)	1.24	>1.5	>1.5
Herbicide	Row	(3.8)	0.08	0.12	0.20
	Inter-row	(4.3)	0.07	0.10	0.11

Soil temperature

Vegetation affects the temperature of the soil by acting as an insulator so that soil temperatures tend to be higher when the soil is cooling and lower when it is warming. Differences can be as high as 5–6°C at 5 cm depth in the English summer (Atkinson and Wilson, 1980). During the spring frost period bare soil absorbs more heat in the course of the day than that covered with vegetation and re-radiates more to the atmosphere at night, so that air temperature is subsequently greater than above vegetation. During the growing period soil temperatures are generally higher under relatively bare soil than under an extensive vegetation cover.

Nutrient availability

The management of soil with herbicides rather than under grass affects soil chemical properties and nutrient availability; pH is usually decreased and available phosphorus and nitrate increased (Table 6). The difference in pH between a herbicide-treated row and a grassed inter-row can exceed 1 unit. This could be due to:

(1) direct effects of the chemical;
(2) changes in organic matter and associated changes in microbial activity;
(3) changes in exchange capacity.

Table 6 Effects of soil management on pH, 'available' P, K, NO_3 (mg l^{-1}) and organic matter (%) in the top 7 cm of soil

Treatment	*Position**	*pH*	*P*	*K*	NO_3	*Organic matter*
Herbicide	Row[1]	5.1	49	267	7.8	2.14
	Inter-row[2]	6.2	30	255	6.0	3.05
Grass	Inter-row[1]	6.2	24	242	13.8	3.03

* Herbicide treatment applied for (1) 9 years, (2) 3 years.

The latter effect results from alteration of physical structure and/or the mineralization of organic matter. The quantity of herbicide added to the soil is small so that the most probable cause of the effect is an increase in nitrogen transformations as a result of changed organic matter inputs and more available water.

Increased phosphorus availability is likely to result from reduced removal by plants and by direct effects upon availability. Initially grass withdraws much phosphorus from the soil reserves but eventually some of this is recycled and the difference in phosphorus withdrawal between grassed and herbicide-treated soils should fall.

Soil phosphorus occurs in several chemical forms of differing availability.

$$P_{\substack{\text{soil}\\\text{soln}}} \underset{\text{'rapid'}}{\leftrightharpoons} P_{\text{labile}} \underset{\text{'slow'}}{\rightleftharpoons} P_{\text{non-labile}}$$

An increase in the rate of P desorption from the labile pool would increase the amount measured as 'available'. This could result from a change in physical structure or in the dominant chemical form of phosphate $(PO_4)^{3-}$, $(HPO_4)^{2-}$, $(H_2PO_4)^-$ or the associated cation (calcium, iron or aluminium). Phosphate adsorption capacity is reduced in acid sandy soils (Sutton and Gunary 1969) so phosphorus availability may be influenced by pH. Increased movement of phosphorus from the labile pool may lead to loss by leaching from the ecosystem and depletion of total P reserves. Changes in factors such as pH and phosphate, in addition to their effects on crops, may influence the composition of the weed flora which can establish.

Micro-organisms

Effects of herbicide-based soil management on soil micro-organisms again could result from:

(1) direct effects of the herbicide;
(2) reduced organic matter input; or
(3) effects of the herbicide on exudation and other processes occurring at the root-soil interface.

In a recent review Greaves (1979) concluded that effects due to reduced organic matter accounted for most of the reported effects of herbicides on soil micro-organisms in field experiments. Effects of herbicides on microbial activity were few where organic matter was high, either because of crop residues or, in plantation crops, through adding organic matter, e.g. a straw mulch. In assessing effects of herbicides on microbial activity natural variation in both population numbers and their responses poses a major problem; e.g. with apparently similar conditions the herbicide Alloxydim-sodium either reduced, increased or had no effect on the microbially produced nitrogenase activity in pea root nodules suggesting that other factors were even more important (Greaves *et al.*, 1978). Long-term effects of herbicides on soil processes are likely to be complex.

Associations between plants and micro-organisms are of great importance in relation to mineral nutrition and may be affected by soil management. Growing fruit trees in both nurseries and orchards after soil fumigation and with herbicide management can result in almost sterile root systems. The phosphorus content of fruit tree leaves was relatively unaffected by fertilizers but increased from 0.19% under overall herbicide to over 0.26% when grass and abundant irrigation were combined. Under field conditions mycorrhizal infection appears often to occur by root contact (Read *et al.*, 1976). It has been suggested that the above effect on leaf phosphorus might be due to an effect on mycorrhizal establishment and performance (Atkinson and White, 1980).

Effects of herbicides on the morphology and physiology of root systems can influence their susceptibility to diseases such as 'take all' in cereals (*Gaeumannomyces graminis*) (Tottman and Thompson, 1978).

Macro-organisms

Where surface soil permeability is reduced by some types of soil management using herbicides, any process increasing permeability will be important. Earthworms, particularly the large burrowing species, may be crucial. Reviewing literature on herbicide effects on earthworms, Edwards and Stafford (1979) concluded that herbicides were seldom toxic but acted mainly by decreasing plant cover and hence the availability of decaying soil organic matter and roots. In a study of the effects of orchard soil management on soil invertebrates, worm populations under grass, initially 2 224 000 ha^{-1}, fell after herbicide treatment by 31%, 67% and 72% after one, two and three years, respectively (Stringer *et al.*, 1971). Populations of most other soil invertebrate groups were also reduced. Recent studies have shown that the presence of worms and other invertebrates can increase crop growth (Edwards and Lofty, 1978). Thus where the creation of new pores connecting with the soil surface is important, the absence of vegetation, via its effects on soil fauna, seem to be harmful.

In addition to creating new pore space worms are important in recycling leaf material which may influence some diseases, e.g. *Phytophthora syringae*, a major cause of fruit loss in apple during storage in some years, can survive in dead leaves and reduced leaf burial by worms may well facilitate infection (Harris, 1979, *1981*).

Weed populations

The herbicide most appropriate for a particular crop situation depends upon various factors, but toxicity to the weed species and tolerance by the crop are crucial. Weed species resembling the crop are favoured and in orchards woody tree and shrub weeds are increasing in importance. Herbicides will act selectively on the potential range of weed species present and this may have wide-ranging implications.

In addition to selective effects on weed species the use of herbicides has increased the area of orchard land now covered by acrocarpous bryophytes, mainly *Bryum argenteum*. This species, protects the soil surface from some of the adverse effects of rain and may also hinder weed seed establishment so that its effects are generally beneficial.

Crop root distribution

The density and distribution of fruit tree roots are influenced by orchard soil management. Gurung (1979) studied the effects of 13 years of herbicide or grass management on apple root growth and found a 33% reduction in root number under grass but little effect on root distribution. In a 10-year-old orchard, the proportion of roots near the soil surface decreased under grass with an increase in the absolute number at depth (Farré, 1979). These effects are likely to influence the uptake of nutrients and water by the crop as well as the distribution of soil organic matter. A cover crop has previously been shown to influence the movement of phosphorus and calcium through the soil profile (Deist *et al.*, 1973) and so effects on the root distribution of both crop and other species are likely to have wide-ranging effects.

Conclusions

The use of herbicides in removing or reducing competition from weeds or other vegetation can greatly influence tree growth and fruit productivity. However, in the absence of weeds or grass a completely different orchard environment is created. Most of these effects seem due to the lack of soil protection from the adverse effects of rainfall and from changes in the input of soil organic matter. Such differences may well be less extreme in arable and other crops where these factors may have a lesser impact than in fruit orchards. Nevertheless the beneficial effects of increased water and nutrient availability and the absence of root disturbance usually more than offset any adverse effect. In assessing the value of herbicide applications and their environmental impact in any particular situation both types of factors must be taken into consideration and also their effects on pests and diseases. These are likely to be greater than present evidence suggests and justify the detailed studies completed or in progress on the impact of weed control practices on soil-borne diseases of apple and hop (Harris, *1981*; Sewell *1981*).

References

Altieri, M. (*1981*). Crop–weed–insect interactions and the development of pest-stable cropping systems. In *Pests, Pathogens and Vegetation*, pp. 459–466. J. M. Thresh. London: Pitman.

Atkinson, D. & Herbert, R. F. (1978). Long term comparison of the effects of soil management. *Report of East Malling Research Station for 1977*, 53–54.

Atkinson, D. & Herbert, R. F. (1979). A review of long term effects of herbicides: effects on the soil with particular reference to orchard crops. *Annals of Applied Biology* **91,** 125–129.

Atkinson, D. & Holloway, R. I. C. (1976). Weed competition and the performance of established apple trees. In *Proceedings 1976 British Crop Protection Conference — Weeds* **1,** pp. 299–304.

Atkinson, D. & White, G. C. (1976). Soil management with herbicides — the response of soils and plants. In *Proceedings 1976 British Crop Protection Conference — Weeds* **3,** pp. 873–884.

Atkinson, D. and White, G. C. (1980). Some effects of orchard soil management on the mineral nutrition of apple trees. In *The Mineral Nutrition of Fruit Trees*, pp. 241–254. D. Atkinson, J. E. Jackson, R. O. Sharples & W. M. Waller. London: Butterworth.

Atkinson, D. & Wilson, S. A. (1980). The growth and distribution of fruit tree roots: Some consequences for nutrient uptake. In *The Mineral Nutrition of Fruit Trees*, pp. 137–150. D. Atkinson, J. E. Jackson, R. O. Sharples & W. M. Waller. London: Butterworth.

Baldwin, J. H. (1979). The chemical control of wild oats and black grass. A review based on results of ADAS agronomy trials. *ADAS Quarterly Review* **33,** 69–101.

Bulfin, M. & Gleeson, T. (1967). A study of surface soil conditions under a non-cultivation management system. 1. Physical and Chemical Properties. *Irish Journal of Agricultural Research* **6,** 177–188.

Davison, J. G. & Bailey, J. A. (1980). The effects of weeds on the growth of a range of nursery stock species planted as liners and grown for two seasons. In *Proceedings of Weed Control in Forestry Conference*, pp. 13–20. Eds. M. G. Allen, D. Atkinson, R. J. Makepeace, W. J. McCavish, M. G. O'Keefe & D. H. Spencer-Jones. Wellesbourne, Warwick, UK: Association of Applied Biologists.

Diest, J., Kotze, W. A. G. & Joubert, M. (1973). The role of cover crops in the movement of phosphate and calcium in soils. *Deciduous Fruit Grower* **23,** 138–141.

De Kock, J. S., Van Zyl, H. J., Jolly, P. R. & Terblanche, J. H. (1977). The effect of soil preparation practices on the growth and production of apples and the physical and chemical soil properties. *Deciduous Fruit Grower* **27,** 158–164.

Edwards, C. A. & Lofty, J. R. (1978). The influence of arthropods and earthworms upon root growth of direct drilled cereals. *Journal of Applied Ecology* **15,** 789–796.

Edwards, C. A. and Stafford, C. J. (1979). Interactions between herbicides and the soil fauna. *Annals of Applied Biology* **91,** 132–137.

Farré, J. M. (1979). Water use and productivity of fruit trees: effects of soil management and irrigation. *PhD Thesis*, University of London.

Fryer, J. D. (*1981*). Weed control practices and changing weed problems. In *Pests, Pathogens and Vegetation*, pp. 403–414. J. M. Thresh. London: Pitman.

Greaves, M. P. (1979). Long term effects of herbicides on soil micro-organisms. *Annals of Applied Biology* **91,** 129–132.

Greaves, M. P., Lockhart, L. A. & Richardson, W. G. (1978). Measurements of herbicide effects on nitrogen fixation by legumes. In *Proceedings 1978 British Crop Protection Conference — Weeds* **2,** pp. 581–586.

Gurung, H. P. (1979). The influence of soil management on root growth and activity in apple trees. *M.Phil. Thesis*, University of London.

Harris, D. C. (1979). The occurrence of *Phytophthora syringae* in fallen apple leaves. *Annals of Applied Biology* **91,** 309–312.

Harris, D. C. (*1981*). Herbicide management in apple orchards and the fruit rot caused by *Phytophthora syringae.* In *Pests, Pathogens and Vegetation*, pp. 429–436. J. M. Thresh. London: Pitman.

Holm, L. (1976). The importance of weeds in world food production. In *Proceedings 1976 British Crop Protection Conference — Weeds* **3,** pp. 753–771.

Jones, F. G. W., Larbey, D. W. & Parrott, D. M. (1969). The influence of soil structure and moisture on nematodes, especially *Xiphinema, Longidorus, Trichodorus* and *Heterodera* sp. *Soil Biology and Biochemistry* **1,** 153–165.

Jones, R. G. & Jones, R. G. (1980). Weed competition, changing weed composition and tree growth following herbicide application. In *Proceedings of Weed Control in Forestry Conference*, pp. 68–72. Eds. M. G. Allen, D. Atkinson, R. J. Makepeace, W. J. McCavish, M. G. O'Keefe & D. H. Spencer-Jones. Wellesbourne, Warwick, UK: Association of Applied Biologists.

McIntosh, R. (1980). The effect of weed control and fertilization on the growth and nutrient status of Sitka spruce on some upland soils. In *Proceedings of Weed Control in Forestry Conference*, pp. 55–63. Eds. M. G. Allen, D. Atkinson, R. J. Makepeace, W. J., McCavish, M. G. O'Keefe & D. H. Spencer-Jones. Wellesbourne, Warwick, UK: Association of Applied Biology.

O'Kennedy, N. D. (1974). Methods of soil management. An Foras Taluntais, Dublin. *Horticultural Research Report for 1973*, pp. 54–55.

Read, D. J. & Jalal, M. A. F. (1980). The physiological basis of interaction between *Caluna vulgaris*, forest trees and other plant species. In *Proceedings of Weed Control in Forestry Conference*, pp. 21–32. Eds. M. G. Allen, D. Atkinson, R. J. Makepeace, W. J. McCavish, M. G. O'Keefe & D. H. Spencer-Jones. Wellesbourne, Warwick, UK: Association of Applied Biologists.

Read, D. J., Koucheki, H. K. & Hodgson, J. (1976). Vesicular arbuscula mycorrhiza in natural vegetation systems. 1. The occurrence of infection. *New Phytologist* **77,** 641–653.

Schuurman, J. J. (1965). Influence of soil density on root development and growth of oats. *Plant and Soil* **22,** 352–374.

Sewell, G. W. F. (*1981*). Soil-borne fungal pathogens in natural vegetation and weeds of cultivation. In *Pests, Pathogens and Vegetation*, pp. 175–190. J. M. Thresh. London: Pitman.

Soane, B. D., Butson, M. J. & Pigeon, J. D. (1975). Soil machine interactions in zero tillage for cereals and raspberries in Scotland. *Outlook on Agriculture* **8,** 221–226.

Stott, K. G. (1976). The effects of competition from ground covers on apple vigour and yield. *Annals of Applied Biology* **83,** 327–330.

Stott, K. G. (1980). The control of weeds in short rotation coppice willow. In *Proceedings of Weed Control in Forestry Conference*, pp. 33–44. Eds. M. G. Allen,

D. Atkinson, R. J. Makepeace, W. J. McCavish, M. G. O'Keefe & D. H. Spencer-Jones. Wellesbourne, Warwick, UK: Association of Applied Biologists.

Stringer, A., Lyons, C. H. & Millsom, N. (1971). Comparison of soil fauna under sward and bare ground: effect of treatments with simazine. *Report of Long Ashton Research Station for 1970*, pp. 106–108.

Sutton, C. D. & Gunary, D. (1969). Phosphate equilibria in soil. In *Ecological Aspects of the Mineral Nutrition of Plants*, pp. 127–135. I. H. Rorison. Oxford: Blackwell.

Tottman, D. R. & Thompson, W. (1978). The influence of herbicides on the incidence of take-all diseases (*Gaeumannomyces graminis*) on the roots of winter wheat. In *Proceedings 1978 British Crop Protection Conference — Weeds* **2,** pp. 609–616.

Wallace, H. R. (1968). The dynamics of nematode movement. *Annual Reviews of Phytopathology* **6,** 91–114.

Herbicide management in apple orchards and the fruit rot caused by *Phytophthora syringae*

D C Harris
East Malling Research Station, Maidstone, Kent ME19 6BJ

Introduction

The herbicide revolution in orchard management has brought considerable benefits to top fruit production through reduced costs and improved crops (Atkinson and White, *1981*). However, some problems have also followed the new management, of which probably the most striking is the emergence of *Phytophthora syringae* as a major cause of storage rot (Edney, 1978). Virtually unknown in Britain as a storage pathogen since the late 1930s, this fungus was responsible for substantial losses in stored apples and pears in the 1973/1974 season (Fig. 1). Serious damage was done in subsequent harvests

Figure 1 A box of apples (cv. Cox's Orange Pippin) rotted by *Phytophthora syringae* after storage at low temperature.

and this fungus is now considered one of the most important pathological hazards to top fruit storage. There is mounting evidence that the transition from grass to herbicide management is a major factor in the resurgence of this disease.

The pathogen

Phytophthora syringae, a homothallic, soil-borne species was first fully described as a pathogen of lilac (Klebahn, 1909). It causes diseases of citrus, almond, apricot and fennel, but in temperate countries this fungus has attracted most attention as a pathogen of apple and pear. *P. syringae* occurs commonly in orchard soils in England (Sewell and Wilson, 1964) and Holland (Roosje, 1958) and its ability to infect and rot undamaged apple fruits has long been known (Lafferty and Pethybridge, 1922). It also causes a collar rot of apple scion cultivars (Sewell, 1963) and pears (Roosje, 1962). Sewell, Wilson and Dakwa (1974) monitored the activity of Phytophthoras in soils at East Malling Research Station and showed a markedly seasonal pattern. *P. syringae* was quiescent from June to August and was active from September to May, with a maximum from September to November coincident with the harvest of apples and pears. Harris (1979a) showed that fallen apple leaves on bare soil were colonized by *P. syringae* with the production of numerous oospores. The presence of these resistant structures in large numbers explained the burst of activity in late summer and autumn after a period of quiescence, which is characteristic of the fungus. Fruits rotted by *P. syringae* yielded very few oospores, and it was postulated that fallen leaves are the principal substrate for generating inoculum in orchards.

Post-harvest losses of apples and pears due to *P. syringae* occurred in western England and in Wales in 1929 and 1930 (Ogilvie, 1931), in Kent and Sussex in 1936 (Salmon and Ware, 1937), and at about the same time in Northern Ireland (Colhoun, 1938), but following this spate there were no published reports of this disease in Britain until 1973. Several references to fruit rot from Europe suggest, however, that *P. syringae* fruit rot continued there as a noticeable if minor problem (Buddenhagen, 1955; Roosje, 1957; Braun and Schwinn, 1963). In Ireland an outbreak occurred in 1966 (J. A. Kavanagh *et al.*, 1969).

Phytophthora syringae awakes

In the 1973/1974 season fruit rot due to *P. syringae* appeared suddenly and dramatically in east Kent. Long-term stored fruit was the most seriously affected, and losses of up to 40% occurred in some batches (Upstone, 1978). It was soon established that most of this wastage occurred in fruit picked after two unusually wet and windy days in September 1973. The disease reappeared over a wider area in the 1974, 1975 and 1976 crops at levels related to the wetness of the harvest period (Upstone and Gunn, 1978). Soil was often visible on rotted fruit and on containers, and infection apparently resulted from splashing with contaminated soil or soil water before or during harvest, a feature of the disease noted in several earlier reports. However, the most serious aspect of the problem was the development of massive secondary rotting in bulk containers during storage from the primary infections occurring in the field. *P. syringae* can grow, albeit slowly, at the temperatures of refrigerated stores (3–4°C) and can spread extensively as mycelium through adjacent fruits in the humid conditions of storage.

The possibility that a new, more virulent strain of the fungus had evolved was discounted on the grounds that the disease appeared simultaneously over a wide area, and the fear that post-harvest dips used for *Gloeosporium* and brown rot control were spreading the disease was dismissed by Upstone (1978), who showed that the levels of *Phytophthora* fruit rot depended on the origin of the fruit and not on the post-harvest treatment received.

The influence of herbicide management

The importance of soil contamination in the disease and the greater opportunities for mud to be splashed on to low-hanging fruit over bare soil inevitably suggested that herbicide practices had contributed substantially to this 'new' problem (Upstone, 1978). However, not all orchards managed in this way had suffered outbreaks, and it was difficult to understand the abrupt appearance of the disease on such a scale. Herbicides were introduced to British orchards in the early 1960s and were soon widely adopted, but fruit rot did not appear, or was not noticeably apparent, for a decade. The need for wet conditions at harvest alone could not explain this lag since September 1968 was also very wet, and there was no indication then of the losses to come.

The discovery of leaf colonization suggested that there might be ways in which the introduction of herbicides had contributed to the fruit rot problem, other than by encouraging soil mobility. In this, as in any disease, the incidence of infection depends on the concentration of inoculum. *Phytophthora* species can undergo explosive increases of biomass under wet conditions by rapidly exploiting suitable substrates through an asexual cycle of sporangia production and spread by mobile zoospores. With the onset of adverse conditions, especially drying, ephemeral stages of the life cycle such as mycelium, sporangia, zoospores and zoospore cysts are destroyed and the only lasting accumulation of fungus is in the form of resistant, perennating structures. In homothallic species like *P. syringae* the principal survival structure is the oospore, and long-term changes in soil inoculum depend on fluctuations in oospore content. *P. syringae*, as mentioned earlier, is pathogenic to various unrelated plants, and oospores are produced in the tissues of several of them.

The related species *P. cactorum* is pathogenic to a wide range of plants (Nienhaus, 1960) and, under experimental conditions, colonizes and produces oospores in several species common in orchards (Dakwa, 1970) including two grasses, although the extent to which this occurs under natural conditions is not known. *P. syringae* may thus be able to generate oospores by colonizing grass and weeds in orchards, but whether such sources of perennating inoculum can rank in importance with fallen leaves is uncertain. If leaf colonization is the main mechanism of inoculum accumulation in orchards, any change in factors which influence this process will ultimately affect the amount of fungus in the soil.

In a grassed orchard fallen leaves are held temporarily away from the soil surface by herbage which may be several centimetres thick. They eventually sink to soil level where most of them are consumed by earthworms (Raw, 1962). Where there are alternating bands of grass and bare soil many fallen leaves come immediately into direct contact with soil and accumulations occur along the edges of the herbicide-treated areas (*see* photograph on p. 381). It is noticeable that, unlike in the grass sward, much of the leaf debris on bare soil remains at the surface, presumably the consequence of reduced earthworm activity (Stringer, 1969; Hogben, personal communication). In orchards where there is a total herbicide regime leaves can be blown off the orchard to be trapped against the nearest windbreak, but some are deposited in hollows or become entangled in persistent weeds.

Experiments on the colonization of apple leaves after abscission show that they become a decreasingly favourable substrate for *P. syringae* as they senesce. Leaves ageing under orchard conditions and exposed to the fungus at various intervals after abscission produced about half the number of oospores after one week and very few after three weeks. *P. syringae* has evidently little or no competitive saprophytic ability in dead leaf tissues. The grass barrier in an orchard would delay the exposure of fallen leaves to soil-bound *P. syringae* and the opportunities for leaf colonization would be greatly curtailed; conversely, leaf colonization would be greatly enhanced in the absence of this barrier. Moreover, the inoculum-rich leaf litter tends to build up on the soil surface whence infective stages of the pathogen can be splashed on to low-hanging fruit in wet weather. Corke (1965) considered the repercussions of altered leaf disposal resulting from herbicide use on the air-borne disease American gooseberry mildew, although in this instance the effects were on the survival of a pathogen already present in the fallen leaf.

The increase of inoculum levels in orchards through leaf colonization would be a progressive process since most leaf-fall periods in Britain have some wet weather favourable to *P. syringae*. Thus inoculum levels in 1968 may not have increased sufficiently in orchards under herbicide management to have produced noticeable amounts of fruit rot in that year.

A trial has been established at East Malling Research Station to test the long-term effects on *P. syringae* of the change from grass to herbicide management. An orchard grassed throughout, until the winter of 1977/1978 when a herbicide strip regime was introduced, has been assayed twice. In November 1978 the fungus was present, but at very low levels. Within a year there was no change in the fungus content of the herbicide strip samples or in the centre of the grass alley, but there was a marked increase in the samples from the edge of the alley. These preliminary results are consistent with the effects predicted but suggest that inoculum may accumulate in a more complex way than was envisaged.

Discussion

Altman and Campbell (1977) listed four possible ways in which herbicides might increase disease:

(1) by reducing host defences;
(2) by stimulating exudation from host tissues;
(3) by stimulating pathogen growth; and
(4) by inhibiting microorganisms antagonistic to the pathogen.

The nature of fruit rot suggests that the first two mechanisms can be discounted and there is no information on the other possibilities. However, the effects of herbicide are undoubtedly more complex in perennial systems where herbage is permited as a deliberate policy in contrast to the more usual total suppression of non-crop plants as in the systems considered by Altman and Campbell. The view propounded here is that the recrudescence of *P. syringae* results more from indirect effects of herbicides, by increasing the mobility of soil in wet weather and affording ready access for the pathogen to fallen leaves, and, by reducing the disposal of infective leaf litter by earthworms.

If levels of *P. syringae* increase in orchard soils in the way proposed, then those orchards managed with herbicides for the longest period should have experienced most fruit rot. There are insufficient data available to test such a relationship but it is apparent that growers in the van of the herbicide revolution have suffered most from this problem. The apparent spread of the disease in south-east England during the wet harvest years from 1973 to 1976 is also correlated with the widening adoption of herbicides during the 1960s. Importantly, it would follow that orchards which have so far escaped this problem, such as those in the western counties, may have to reckon with the new hazard in the not too distant future.

The history of individual orchards is probably crucial in determining their susceptibility to outbreaks of this disease. Fruit in an orchard which has been managed as strips and then converted to total herbicide would be extremely vulnerable since soil mobility and soil inoculum concentration are high, whilst fruit in an orchard which has been in total herbicide since planting may be less at risk if the levels of *P. syringae* in the soil have been limited through leaf removal. This could explain the lack of a close relationship between types of management and fruit rot incidence (Upstone, personal communication), although in other *Phytophthora* diseases soils are known to vary widely in their conduciveness to the pathogen, and such factors may also operate in fruit rot.

One aspect of the problem which cannot easily be explained is the fluctuation of disease from one wet season to another at the same site (Upstone, 1978), since rapid changes of soil inoculum appear unlikely. However, local weather conditions, such as temperature, may be critical in determining fungus activity, even when water is available (Harris, 1979b) and there may be other influential factors as yet unknown.

The case against herbicides although impressive is still largely circumstantial. The last 70 years have seen many developments in orchard management and some of these, particularly the new generations of chemicals for disease and pest control, may have contributed in some degree to the new orchard environment which has allowed *P. syringae* to become a serious problem. Grass management was itself a development, and followed clean cultivation in the 1930s and cultivation with late summer crops or weeds in the 1950s (White, 1980). Clean cultivation would have provided some of the conditions for maintaining high levels of *P. syringae* as do herbicides, and it is significant that, even though tree size and pruning systems were generally such as to keep fruit away from soil, the last reported outbreaks of *Phytophthora* fruit rot in Britain before the recent epidemics were during the 1930s. In fact, the management vogues of the 1940s and 1950s may have caused a decline in *P. syringae* and it is only in response to a disease threat that we are beginning to appreciate how closely linked can be the fortunes of this soil-borne pathogen with the methods of soil management.

T. Kavanagh (1969) predicted a decline of collar rot caused by *P. cactorum* and *P. syringae* under herbicide due to adverse changes of microclimate at the tree base and the absence of the mechanical damage inevitable with mowing. He could not have foreseen the emergence of *P. syringae* as a major fruit rot pathogen, nor the possibility of increased inoculum of this fungus in the soil since the leaf colonizing phase was not known. Collar rot seems to have declined, although this may be as much due to other changes in the orchard such as high working of scions and the widespread replacement of older and, therefore, more susceptible trees in new, intensive plantings.

References

Altman, J. & Campbell, C. L. (1977). Effect of herbicides on plant diseases. *Annual Review of Phytopathology* **15**, 361–385.

Atkinson, D. & White, G. C. (*1981*). The effects of weeds and weed control on temperate fruit orchards and their environment. In *Pests, Pathogens and Vegetation*, pp. 415–428. J. M. Thresh. London: Pitman.

Braun, H. & Schwinn, F. J. (1963). Fortgeführte Untersuchungen über den Erreger der Kragenfaule des Apfelbaumes (*Phytophthora cactorum*). II. *Phytopathologische Zeitschrift* **47,** 327–370.

Buddenhagen, I. W. (1955). Various aspects of *Phytophthora cactorum* collar-rot of apple trees in the Netherlands. *Tijdschrift over Plantenziekten* **61,** 122–129.

Colhoun, J. (1938). Fungi causing rots of apple fruits in storage in Northern Ireland. *Annals of Applied Biology* **25,** 88–99.

Corke, A. T. K. (1965). American gooseberry mildew on blackcurrants. In *Proceedings of the 3rd British Insecticide and Fungicide Conference*, pp. 336–339.

Dakwa, J. T. (1970). Ecological studies of *Phytophthora cactorum* (Leb. & Cohn) Schroet., with reference to collar rot disease of apple. *PhD Thesis,* University of London, 204 pp.

Edney, K. L. (1978). The infection of apples by *Phytophthora syringae. Annals of Applied Biology* **88,** 31–36.

Harris, D. C. (1979a). The occurrence of *Phytophthora syringae* in fallen apple leaves. *Annals of Applied Biology* **91,** 309–312.

Harris, D. C. (1979b). The suppression of *Phytophthora syringae* in orchard soil by furalaxyl as a means of controlling fruit rot of apple and pear. *Annals of Applied Biology* **91,** 331–336.

Kavanagh, J. A., O'Malley, M. & Simmonds, A. (1969). Some observations on the rotting of apple fruits by *Phytophthora syringae* Kleb. *Irish Journal of Agricultural Research* **8,** 439–441.

Kavanagh, T. (1969). The influence of herbicides on plant disease. I. — Temperate fruit and hops. *Scientific Proceedings, Royal Dublin Society, Series B* **2,** 179–190.

Klebahn, H. (1909). Die neue Zweig- und Knospenkrankheit. In *Krankheit des Flieders*, pp. 18–75. Berlin: Gebrüder Bornträger.

Lafferty, H. A. & Pethybridge, G. H. (1922). On a *Phytophthora* parasitic on apples which has both amphigynous and paragynous antheridia; and on allied species which show the same phenomenon. *Scientific Proceedings of the Royal Dublin Society* **17,** 29–43.

Nienhaus, F. (1960). Das Wirtsspektrum von *Phytophthora cactorum* (Leb. et Cohn) Schroet. *Phytopathologische Zeitschrift* **38,** 33–68.

Ogilvie, L. (1931). A fruit rot of apples and pears due to a variety of *Phytophthora syringae. Report of the Agricultural and Horticultural Research Station*, *University of Bristol for 1930*, pp. 147–148.

Raw, F. (1962). Studies of earthworm populations in orchards. I. Leaf burial in apple orchards. *Annals of Applied Biology* **50,** 389–404.

Roosje, G. S. (1957). Het mycologisch onderzoek. Stambasisrot. *Jaarverslag 1956 Proefstation voor de Fruitteelt in de volle grond*, pp. 46–47.

Roosje, G. S. (1958). Het mycologisch onderzoek. Stambasisrot. *Jaarverslag 1957 Proefstation voor de Fruitteelt in de volle grond*, pp. 40–41.

Roosje, G. S. (1962). *Phytophthora syringae* geisoleerd uit bast van appel en peer. *Tijdschrift over Plantenziekten* **68,** 246–247.

Salmon, E. S. & Ware, W. M. (1937). Department of Mycology. *Journal of the South-Eastern Agricultural College, Wye, Kent* **39,** 17–18.

Sewell, G. W. F. (1963). Branch, stem and collar rot of apple caused by *Phytophthora* species of the '*cactorum* group'. *Nature* (London) **200,** 1229.

Sewell, G. W. F. & Wilson, J. F. (1964). Death of maiden apple trees caused by *Phytophthora syringae* Kleb. and a comparison of the pathogen with *P. cactorum* (L. & C.) Schroet. *Annals of Applied Biology* **53,** 275–280.

Sewell, G. W. F., Wilson, J. F. & Dakwa, J. T. (1974). Seasonal variations in the activity in soil of *Phytophthora cactorum*, *P. syringae* and *P. citricola* in relation to collar rot disease of apple. *Annals of Applied Biology* **76,** 179–186.

Stringer, A. (1969). Comparison of soil fauna under sward and bare ground. *Report of Long Ashton Research Station for 1969*, pp. 96–97.

Upstone, M. E. (1978). *Phytophthora syringae* fruit rot of apples. *Plant Pathology* **27,** 24–30.

Upstone, M. E. & Gunn, E. (1978). Rainfall and the occurrence of *Phytophthora syringae* fruit rot of apples in Kent 1973–75. *Plant Pathology* **27,** 30–35.

White, G. C. (1980). Orchard soil management. *Annual Report of East Malling Research Station for 1979*, pp. 223–225.

Effects of weed grasses on the ecology and distribution of ergot (*Claviceps purpurea*)

P G Mantle
Biochemistry Department, Imperial College, London SW7 2AY

Introduction

Ergot (*Claviceps purpurea*) is a common fungus in Britain, occurring as a range of biological forms ecologically more or less restricted to certain grasses. Some forms readily infect cereals, particularly rye and wheat, reducing seed yield and contaminating the grain with toxic sclerotia. Outbreaks of disease in cereals are often promoted by weed grasses, within or around the crop. Control of cereal weed grasses is, therefore, particularly important.

The ergot life cycle

In Britain the grass and cereal hosts of ergot flower over an approximately six-month period from May to November according to species, circumstances and climate. The fungus naturally infects only the unfertilized ovary (Fig. 1) and by further development of the asexually-sporulating sphacelial stage it prevents the formation of seed by infected florets. As the white parasitic mycelium continues to grow in place of the ovary the unique sclerotial growth-form of the fungus becomes differentiated and further proliferation of this tissue alone forms the characteristic ergot sclerotium. This is a banana-shaped purple/black mass of fungus tissue which provides a means of overwintering during the half year in which no suitable host is available. The sexual stage arises from the sclerotia in the following spring, releasing ascospores which are disseminated in the air and cause primary infection of appropriate host plants, thus completing the cycle.

In uncultivated land where the composition of mixed grass communities is determined by various ecological constraints there are opportunities for strains of the ergot fungus to evolve a wide host range. However, there is a tendency for this to be restricted in agriculture, where pastures are often monocultures of ryegrass and considerable effort is expended on keeping cereals free from arable weed grasses. Ergots in arable land germinate in early June and may continue to eject ascospores for 4–6 weeks. This sequence

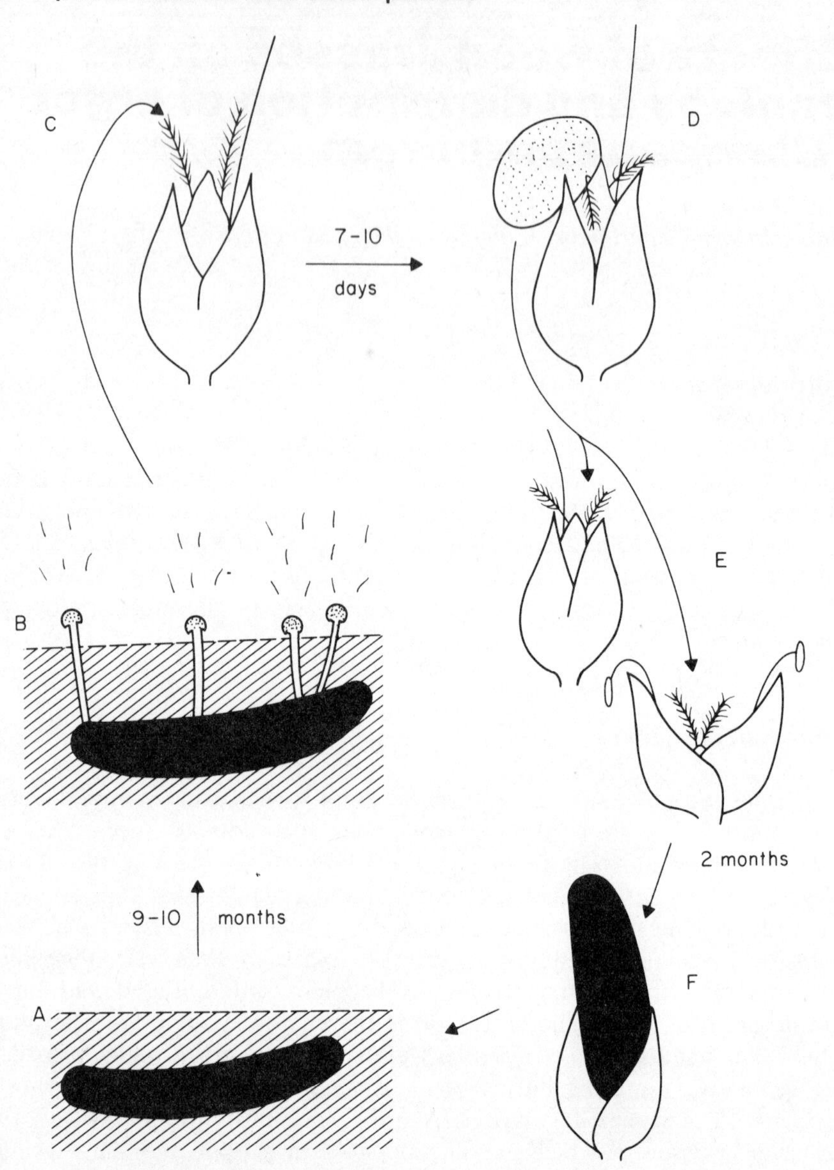

Figure 1 Life cycle of *Claviceps purpurea* in wheat, interpolating a multiplication step on an early-flowering arable weed grass (e.g. black grass).
A. Dormant sclerotium buried superficially in soil. B. Production of ascogenous (sexual spore) stage and discharge of ascospores into the air. C. Ascospores trapped on the stigmas of grass hosts. D. Ovary colonized by the ergot pathogen causing the release by the plant of a sugary exudate containing conidiospores. E. Transmission of spores to other grass or cereal florets facilitated by physical contact, rain splash and insects. F. Differentiation of ergot sclerotium.

s well suited to primary infection of wheat flowering in mid-June, but the equence also permits prior multiplication on an intermediate grass host. The ollowing characteristics of blackgrass (*Alopecurus myosuroides*) make it a lamaging weed of wheat and other arable crops and an important host of ergot, particularly in the southern part of Britain:

1) morphological similarity to wheat
2) flowering height similar to wheat
3) early flowering habit
4) high reproductive capacity in arable crops
5) susceptibility to wheat-infective strains of *C. purpurea*.

Control of blackgrass was necessary for agronomic reasons long before a close association with wheat ergot was demonstrated.

Role of blackgrass in the epidemiology of ergot disease in wheat

Several hundred isolates of ergot fungi from a wide range of hosts and locations in Britain were surveyed 7–8 years ago (Mantle *et al.*, 1977). None exhibited any distinctive morphology, so a diagnostic biochemical parameter was sought. Fortunately, several related alkaloids occur in sclerotia, with various permutations of relative abundance. The biosynthesis of the various ergot alkaloids involves several enzymes, thereby conferring a measure of genetic stability on the metabolic products. It is also well known that each strain of the fungus produces its characteristic spectrum of alkaloids on all host species parasitized and the alkaloid spectrum of a few small ergots is fairly easy to assess by appropriate chromatographic techniques.

Ergots of most common grasses of pasture and uncultivated land were found to contain predominantly the alkaloid ergotamine, whereas wheat and blackgrass ergots were relatively rich in ergotoxine. Thus the crop and its weed appeared to be connected circumstantially with the same strain of ergot. The flowering habit of blackgrass makes it generally unlikely to be infected from wheat but it is ideally suited to become infected itself first and then, a week or so later, produce numerous spores that can infect wheat. These are released as asexually produced conidia suspended in a sweet exudate ('honeydew') from infected florets. Honeydew exudes at about the height of the wheat inflorescence so that the potential for epidemic development is greatly enhanced. Interpolation of an intermediate parasitic stage on black-grass transforms the weak inoculum potential of a few airborne ascospores into the epidemic potential of many thousands, or millions, of conidia.

Surveys of wheat crops in southern England showed a close correlation between the extent of ergot and the degree of crop infestation by blackgrass. Cross-inoculation experiments using the conidia present in honeydew collected from ergotized blackgrass (the use of inoculum from laboratory cultures for assessing host range is open to criticism) invariably gave hig yields of ergots on wheat, barley, rye and triticale, so establishing a very clos connection between wheat, blackgrass and ergot disease.

Host restriction of ergot in other grasses

When extended to other grasses on a countrywide scale this investigatio revealed various relationships between ergots of grasses (weed and culti vated) and wheat (Table 1). On a local scale sharper distinctions between th

Table 1 Infection of wheat by ergot strains from cereals and grasses in England i 1972/1973*

Source of inoculum	*Percentage infection**
Alopecurus myosuroides (Blackgrass)	80
Alopecurus pratensis (Meadow foxtail)	60
Poa annua (Annual poa)	80
Secale cereale (Rye)	70
Hordeum vulgare (Barley)	80
Triticale	85
Dactylis glomerata (Cock's-foot)	0–20
Lolium perenne (Rye-grass)	2–30
Lolium multiflorum (Italian rye-grass)	10
Lolium temulentum (Darnel)	8
Festuca arundinacea (Tall fescue)	20
Arrhenatherum elatius (Oat-grass)	3–10
Poa annua (Annual poa)	3
Anthoxanthum odoratum (Sweet vernal-grass)	0–1
Holcus mollis (Creeping soft-grass)	1
Holcus lanatus (Yorkshire fog)	1
Phleum pratense (Timothy)	1
Spartina townsendii (Cord-grass)	0
Phragmites communis (Reed)	0

* The percentage of inoculated wheat florets producing ergot sclerotia.

susceptibility and resistance of host grasses to the indigenous strain(s) of ergo were evident. By virtue of their separate habitats and times of flowering certain host plants such as *Phragmites*, *Spartina* and *Molinia*, which grow i wet environments of one sort or another, seem to have their own particula strain of ergot which, under natural conditions, probably both cannot an does not infect cereals. The restriction of such strains is, however, no absolute since it has been possible experimentally to persuade isolates from all three hosts to infect cereals though after considerable unnatural effort (Mantle 1969a, 1969b and unpublished). For some reason *Dactylis*, *Lolium* and, to a lesser extent, *Arrhenatherum*, seem to occupy an intermediate position, on a nationwide scale being a source of a cosmopolitan array of strains. *Dactylis* and *Lolium* produce inflorescences throughout summer and autumn and are thus often available to receive inoculum over a prolonged period and from diverse hosts.

Recent trends and future prospects

In Britain, ergot was quite common in the 1950s, 1960s and early 1970s, especially in late-sown wheat. Indeed, the disease presented one of the principal constraints on the commercial introduction of F1 hybrid wheat in this country. In 1975 and 1976 droughts virtually eliminated the pathogen, and possibly created an 'ecological vacuum' facilitating the evolution of pathogenically different forms. However, such is ergot's capacity for epidemic production that in the last few years it has again become common in pasture and waste-land grasses.

It is unwise to conclude that the range of pathological forms of the fungus remains as it was in the early 1970s. Grasses such as ryegrass, cocksfoot, Yorkshire fog and timothy are probably still not a particularly serious threat to cereals. However, early-flowering arable weed grasses such as blackgrass must be controlled, otherwise, given two successive years with a damp June, ergot could quickly recur to cause local epidemics in cereals. These could occur in 1981 following the generally damp summer of 1980 which is likely to have led to a build-up of ergot sclerotia in cereals and grasses. The paucity of information on the epidemiology of this ancient disease would then be further exposed (Watson, 1979).

Control of the ergot disease in cereals may occur automatically in a dry climate but can be achieved elsewhere by good agricultural practice. This should include the control of blackgrass or other arable weed grasses where appropriate, and the general control of seeding grasses around the crop perimeter.

Ergot has been studied for many years and yet problems remain. Has the pattern of ergot strains that are potentially damaging to cereals changed over the past 7–8 years? What, if any, weed grass reservoir of infection was responsible for a recent epidemic in late-sown wheat in the north of Scotland (*The Aberdeen Press and Journal*, 28 September, 1978), where blackgrass is absent? What is the source of inoculum for *Phragmites*, which flowers around September and can frequently be found bearing many sclerotia in December? Is the germination of its ergots delayed in the field? What is the biochemical basis for host restriction in *C. purpurea*? Scientists only need a little knowledge before asking awkward questions and ergot still offers interesting areas for study.

References

Berde, B. & Schild, H. O. (1978). *Ergot Alkaloids and Related Compounds*. Berlin: Springer-Verlag.

Mantle, P. G. (1969a). Studies on *Claviceps purpurea* parasitic on *Phragmites communis*. *Annals of Applied Biology* **63,** 425–434.

Mantle, P. G. (1969b). Development of alkaloid production *in vitro* by a strain of *Claviceps purpurea* from *Spartina townsendii*. *Transactions of the British Mycological Society* **52,** 381–392.

Mantle, P. G., Shaw, S. & Doling, D. A. (1977). Role of weed grasses in the aetiology of ergot disease in wheat. *Annals of Applied Biology* **86,** 339–351.

Watson, R. D. (1979). Ergot (*Claviceps purpurea*) on winter wheat in Scotland. *Federation of British Plant Pathologists, Newsletter* 2, 48–50.

Effects of weeds and weed control on invertebrate pest ecology

M J Way and M E Cammell
Imperial College, Silwood Park, Ascot, Berks SL5 7PY

Introduction

In this paper we refer first to ways in which weeds may influence invertebrate pests of crops:

(a) by acting as alternate or alternative hosts of the pests
(b) by affecting the visual responses of flying (insect) colonists
(c) by affecting the behaviour of pests once they have contacted the weed
(d) by influencing pest mortality, notably that due to natural enemies which themselves may be affected directly by weeds, as in (b) and (c) above, as well as indirectly by hosts or prey feeding on the weeds.

Secondly, we note effects on pests and beneficial species of mechanical and chemical weed control practices, other than through obvious consequences of weed removal associated with (a) to (d) above.

Thirdly, we illustrate how weeds might be influencing pest population dynamics, and by implication how pest abundance in time and space and also pest 'quality' might be affected by weed control, especially by conventional use of chemical herbicides.

Finally, we briefly discuss weeds in the context of integrated pest, disease and weed control.

Weeds as alternate or alternative hosts

Van Emden (1965) gives 442 examples of weeds or wild plants as hosts of crop pest or disease organisms. This evidence is supplemented by further reviews on insects (van Emden, 1970), nematodes (Franklin, 1970) plant viruses (Heathcote, 1970; Murant, *1981*) and on crop pests and diseases of some weeds (Thurston, 1970). Table 1 indicates the serious potential of three common weeds as source hosts.

Conclusions from such general evidence on weeds as hosts of pests are mostly phrased unequivocally; for example, 'any small beneficial contribution weeds (in the crop) may make to pest control is far outweighed by their harmful effect' (van Emden, 1970), and 'when nematodes that damage crops are present in local weeds, it is obvious that weed control is necessary for

Table 1 Numbers of invertebrate pest species and invertebrate-transmitted plant viruses known to occur in three common weed species (from Thurston, 1970)

Weed species	*Number of pest species and viruses*			*Total number of crop species attacked*
	Insects	Nematodes	Viruses	
Stellaria media (chickweed)	4	8	4	>30
Senecio vulgaris (groundsel)	6	5	3	
Chenopodium album (fat hen)	4	8	6	

efficient nematode control' (Franklin, 1970). Yet it is dangerous to generalize because, with a few important exceptions, notably in relation to plant viruses (Heathcote, 1970; Murant, *1981*), there is little information on the frequency of occurrence of pests or pathogens on crop weeds; and there are examples of weeds acting beneficially as alternative hosts which divert pests away from less attractive crop hosts (Heathcote, 1970; Dunning, 1971). We also need to be more aware of subtle ecological effects of weeds before accepting generally that there is no case for modifying the current trend towards weed-free crops (Atkinson and White, *1981*; Fryer, *1981*).

Effects of weeds on the visual attractiveness of a crop to colonizing animals

Most of the major exogenous insect pests of arable crops seem to be species originally adapted to find and exploit the natural equivalent of an arable crop, namely an early or 'immature' stage in plant succession. The crucial visual cues are often yellow-green colours associated with optomotor stimuli as insects fly across spaced-out plants silhouetted against bare soil. That weeds act as camouflage and lessen the contrast between plant and soil, so decreasing colonization, was demonstrated by Smith (1969, 1976a) and Cromartie (1975) for several brassica pests, notably Homoptera. Dempster (1969) similarly recorded decreased oviposition by the small cabbage white butterfly, *Pieris rapae*, on brassicas growing among weeds; the white flowers of one weed species seemingly acting as an added visual deterrent. Therefore 'weed-free' crops created by herbicides have undoubtedly benefited certain pests, especially at early stages of crop growth before there is total ground cover. So, present-day techniques have made many crops visually much more attractive to some insects.

On the other hand, herbicides have created new opportunities of decreasing the visual attractiveness of crops. For example, the wide row spacing originally needed for mechanical weeding is now unnecessary for some crops. Plants can therefore be sown more closely or evenly, thereby creating earlier complete crop cover and hence shortening the period of maximum visual attractiveness. In this context, the traditional mixed cropping practices of many small farmers in the tropics no doubt confer some degree of pest control by maintaining a complete canopy. However, mixed cropping is incompatible with the use of many herbicides and it is being replaced by pesticide-protected monocultures.

Effects of contact with weeds on behaviour of pests

Whether an animal encountering a particular plant stops to feed and/or reproduce may depend on various olfactory/gustatory chemical cues (Gosling, *1981*). As already implied, these may sometimes make weed hosts more attractive than the crop, so diverting a pest away from the latter or at least diluting the pest population. The most significant effects are, however, with non-host weeds that either lack attractive chemical cues or contain compounds which are repellent or toxic. For example, Tahvanainen and Root (1972) attributed decreased numbers of a brassica flea beetle, *Phyllotreta cruciferae*, to the presence of certain non-host crops and weeds containing chemicals that interfered with host-finding and feeding. Grass weeds similarly greatly decreased colonization of a whitefly species on beans even when the grass was merely in the form of a 1 m border around 16 m^2 monocrop plots (Altieri *et al.*, 1977). *Tagetes* spp. have been shown to decrease numbers of root lesion nematodes, *Pratylenchus* spp., in the soil (Franklin, 1970).

Fundamental studies on effects of olfactory and gustatory cues indicate that aphids depart more vigorously from a non-host than from a host (Kennedy, 1966). Plants with highly repellent chemicals might be expected to cause relatively intense 'rebound' and so induce a relatively longer period of subsequent flight before the insect again responds to landing stimuli. Such findings have obvious implications for pest control as does the evidence that some aphids are hyperselective, even of apparently suitable host plants (Kennedy *et al.*, 1959a, 1959b; Muller, 1958). Such evidence points to opportunities for using repellent plant chemicals artificially and for crop protection by selective retention of certain weeds, or through conscious manipulation of mixed cropping using 'repellent' species to protect others. Such systems have been developed and used empirically by some small farmers in the tropics (Altieri, *1981*).

Effects of weeds on natural enemies of pests

Van Emden (1965, 1970) and Zandstra and Motooka (1978) give many examples of natural enemies with free-living stages that feed on the pollen and nectar of wild plants and with predatory/parasitic stages that feed on prey or hosts on wild plants. No doubt the flowers provide essential food for some natural enemies, for example, the adults of some Syrphidae and parasitic Hymenoptera. But these plants occur commonly in hedgerows and wild places where they are not always classed as weeds. It is significant that some adult Syrphidae can fly long distances between feeding sites in wild habitats and oviposition sites in crops (Schneider, 1969), and there may be no differences in syrphid predatory activity between areas where adult food is scarce and where it is abundant (Pollard, 1971). Perhaps the value of flowering weeds in, or close to, crops has been over-estimated.

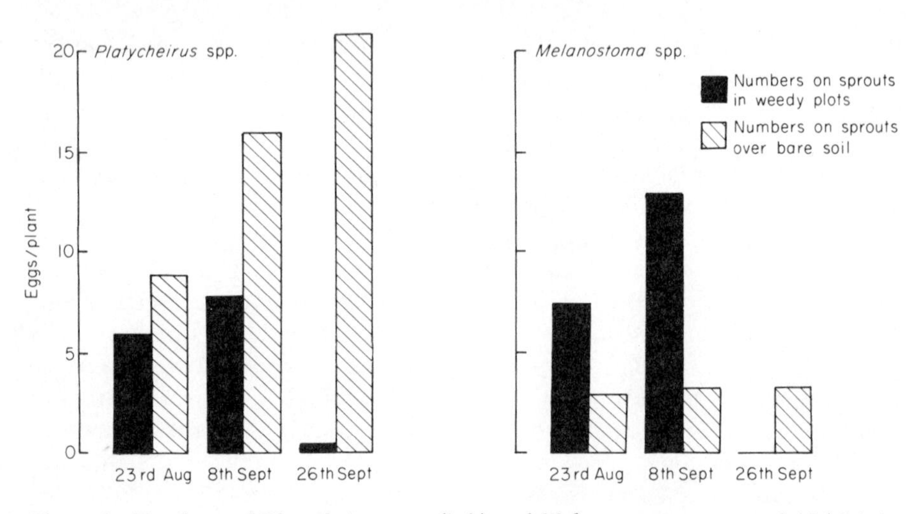

Figure 1 Numbers of *Platycheirus* spp. (left) and *Melanostoma* spp. eggs laid (right) on brussels sprouts in weedy and weed-free plots (after Smith, 1976b).

There is much circumstantial and some more definitive evidence that weeds, or wild plants, can be important food sources for alternate hosts of natural enemies of pests. Thus giant ragweed, *Ambrosia trifida*, is recognized as supporting a host of the first generation of the parasite, *Lydella grisescens*, which then caused notably increased parasitism of European corn borer, *Ostrinia nubilalis*, which appears too late for the first generation of the parasite (Hsiao and Holdaway, 1966). *A. trifida* also supports an important host of parasites of boll weevil, *Anthonomus grandis*, and oriental fruit moth, *Cydia molesta*, (Zandstra and Motooka, 1978). From such evidence of the importance of weeds, van Emden (1970) and Stary (1964) have concluded that, in our present state of limited knowledge, weedy areas should be consciously retained outside crops. Studies by Pierce (1912) and Perrin (1975) have indicated how *A. trifida* and stinging nettle, *Urtica dioica*, might be cut at appropriate times to force their natural enemy populations onto crops, but we must acknowledge totally inadequate knowledge of the potential and practicability of such treatments in agriculture. As indicated later, this stems from inability to analyse critically or undertake relevant experiments to explain the outcome of complex interactions between 'wild' and agricultural systems. Even within one part of one system, for example, a field of brassicas, it has proved impossible to unravel the full implications of weeds for biological control of one aphid species by one group of seemingly important predators. For example, Smith (1969, 1976b) showed that some syrphids (*Melanostoma* spp.) laid many more eggs on brassica plants in weeds than on weed-free brassicas whilst others (*Platycheirus* spp.) laid many less in the presence of weeds (Fig. 1). Furthermore, *Melanostoma* spp. laid many more eggs on weeds in the crop than on the crop plants themselves (Fig. 2). Yet there was no evidence that aphid numbers on the crop were affected or affected differently by syrphid predation in the weed-free and weedy systems.

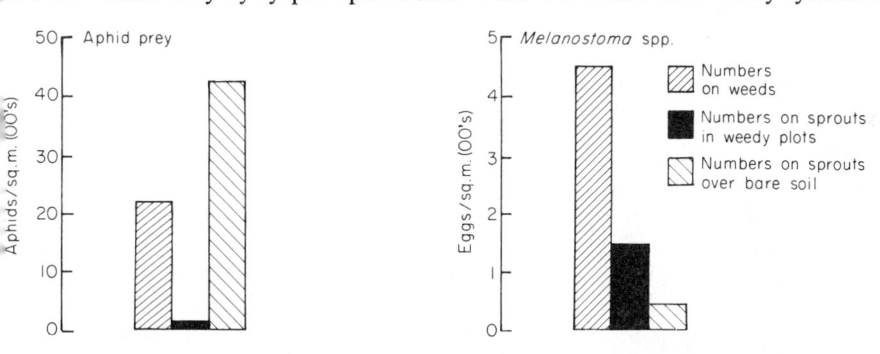

Figure 2 Number of aphid prey (left) and of *Melanostoma* spp. eggs laid (right) on weeds and brussels sprouts in weedy plots, and on sprouts in weed-free plots (after Smith, 1976b).

Effects of weed control practices

The above evidence indicates how weeds may affect pests and their natural enemies and hence, by implication, indicates some of the possible consequences of weed control. Weed control may also have other effects.

Implications of herbicide use

Extensive studies on soil fauna have shown relatively minor direct effects of some herbicides on beneficial or apparently beneficial soil animals (Brown, 1978; Davis, 1965; Edwards, 1970; Edwards and Stafford, 1979). In practice, it may be difficult to distinguish the direct effects of herbicides from the more striking indirect effects associated with the destruction or alteration of the weed flora, changes which mostly decrease populations of the soil animals dependent on the weeds (C. J. S. Fox, 1964). There is no evidence that the direct action of herbicides on soil organisms affects pest incidence or that, in general, herbicides cause changes in soil fauna as striking as those caused by mechanical weed control. Indeed, soil organisms, apart from some Coleoptera and Diptera, were more abundant in herbicide-treated direct-drilled plots than in conventionally cultivated soil (Edwards, 1975).

Direct effects of herbicides on beneficial species such as natural enemies, and on the crop plants' vulnerability to pests, are particularly relevant to pest incidence. Thus some herbicides, notably 2,4-D, can be toxic to predatory mites, Coccinellidae and to certain general predators. The direct toxicity of 2,4-D to Coccinellidae in cereal crops is considered responsible for outbreaks of aphids on cereals in New Brunswick (Adams and Drew, 1965). Some ground predators seem likely to be harmed indirectly by removal of weed cover (Speight and Lawton, 1976) since they are less active or abundant in monocrops than in undersown ones (Dempster and Coaker, 1974; O'Donnell and Coaker, 1975).

Many herbicides influence the growth and physiology of treated crops such that they become more susceptible to pests and/or diseases. 2,4-D, for example, increases susceptibility to wireworm, *Agriotes* spp., by delaying the germination and early growth of wheat (W. B. Fox, 1948). Moreover, this and some other herbicides can improve nutritiousness or otherwise increase susceptibility of the crop, as shown by increases in size, fecundity and abundance on treated crops of pests such as pea aphid, *Acyrthosiphon pisum*, on beans (Maxwell and Harwood, 1960), stem borer, *Chilo suppressalis*, on rice (Ishii and Hirano, 1963) and corn leaf aphid, *Rhopalosiphum maidis*, and European corn borer, *Ostrinia nubilalis*, on maize (Oka and Pimentel, 1976). Herbicides also increase susceptibility of oats to a nematode, *Ditylenchus dipsaci*, (Webster, 1967) and can influence the symptoms due to arthropod-transmitted plant viruses (Heathcote, 1970).

Implications of conventional cultivations

Possible harmful secondary effects of weed control by herbicides and to a lesser extent the direct effects of some herbicides on beneficial species and on the crops' resistance to pests, have recently caused concern. In contrast, mechanical weed controls have long been accepted as the norm and not subjected to much critical analysis, except by implication. For example, many weeds are less well controlled mechanically than by herbicides, so there is greater danger from source host weeds. Moreover, conventional cultivations create oviposition sites for some pests such as wheat bulb fly, *Delia coarctata*, (Bardner *et al.*, 1971) and can greatly decrease survival of some overwintering natural enemies (Way, Murdie and Galley, 1969) (Table 2). This has obvious

Table 2 Numbers of an overwintering hymenopterous parasite and of Syrphidae emerging from cultivated and uncultivated plots on which a brassica crop was grown the previous year (from Way, Murdie and Galley, 1969).

	Numbers emerging m^{-2}	
Species	*Cultivated soil*	*Uncultivated soil*
Diaretiella rapae	12	156
Adult Syrphidae	1	16

implications for non-tillage systems made possible by use of chemical herbicides. Indeed, more careful analyses are needed of the contrasting benefits and disadvantages of direct-drilled and conventional systems. Direct drilling plus chemical herbicides no doubt benefits some pests such as slugs (Edwards, 1975), and, by making simplified crop rotation possible, has exacerbated some nematode and insect problems, e.g. *Diabotrica* spp., which are only pests in maize after maize.

Pest dynamics in relation to weeds and weed control in annual crops

So far, we have discussed weed species as hosts for pests as well as localized effects of weeds and weed control on pests. Such evidence is unlikely to alter

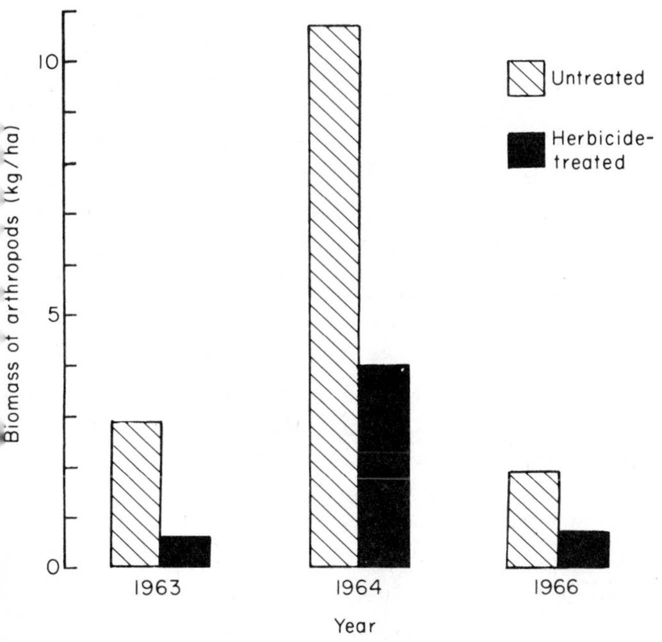

Figure 3 Biomass of arthropods in herbicide-treated and untreated portions of barley fields (after Southwood and Cross, 1969).

the present trend towards increasing dependence on herbicides in view of their outstanding effectiveness and convenience. Yet modern weed control has probably caused the greatest ecological change in British agriculture since the initiation of enclosures in the early 1700s as indicated by data on arthropods such as those of Southwood and Cross (1969) (Fig. 3). Overall it is likely that insect communities in and around agroecosystems have been affected much more by herbicides than by insecticides, as predicted by Way, Glynne-Jones and Johnson (1956). As applied ecologists, we should be especially concerned with the impact of such practices on the population dynamics of pests in space and time. Attempts should be made to assess whether pest status is changing, why it is changing and whether it might justify attempts to modify practices such as weed control. Unfortunately, as already mentioned, we are mostly faced with complex systems where many management practices besides weed control are also continually changing and where the dynamics of any one pest species usually depends on interactions occurring regionally between different crop and 'wild' environments. Despite these obstacles to experimental proof, we should make some studies aimed not only at recording, but also at explaining, possible long-term changes in pest abundance. Some crops and crop pests are already providing 'baseline' evidence.

Cereal aphids

Perhaps the most comprehensive attempt to analyse the impact of modern crop production technology, especially weed control by herbicides, is in southern England at North Farm, Sussex, where large decreases in the diversity and biomass of insects and other organisms have been recorded in herbicide-treated cereal fields (Potts, 1970; Potts and Vickerman, 1974; Vickerman, 1974). There were fewer non-specific predators, and some evidence that cereal aphids are commoner in relatively weed-free conditions (Table 3). It is not known whether the apparent upsurge of cereal aphids in the 1970s is due to a greater realization of their presence, or to 'natural' fluctuation, or to a real and persistent increase in their abundance associated with changing crop production technology (changed varieties, more fertilizers and herbicides). The dominance of cereals in arable farming, the routine use

Table 3 Mean numbers of aphids and staphylinids in plots of winter barley with or without grass weeds (Vickerman, 1974)

	Mean number m^{-2} in plots		
Insect	*Grass weeds*	*Weed-free*	
Aphids	347	442	$P<0.05$
Staphylinids	103	10	$P<0.001$

of herbicides on virtually all cereal land in Britain and many other countries, and the indications that they increase aphid infestations, undoubtedly justify the present attention being given by ecologists to cereal aphids. Clearly, we cannot forego the advantages of herbicides for cereal production (Fryer, *1981*). However, proof and cause of their responsibility for increased aphid attacks could, if substantiated, make it possible to modify control practices, perhaps through adjustments in the kinds of herbicides used and the times, amounts and methods of application.

Sugar beet yellowing viruses

In the last 30 years, yellowing diseases of sugar beet due to beet yellows and/or beet mild yellowing viruses have been combated by:

(1) a combination of measures aimed at eliminating or isolating overwintering seed crop hosts,
(2) controlling aphids, especially *Myzus persicae*, on overwintering crops,
(3) aphicides applied from the time of arrival on root crops in early summer.

The widespread use of pre-emergence and other herbicides on virtually all arable crops has no doubt decreased the abundance of certain widely distributed crop weeds, perhaps the main remaining sources of virus,

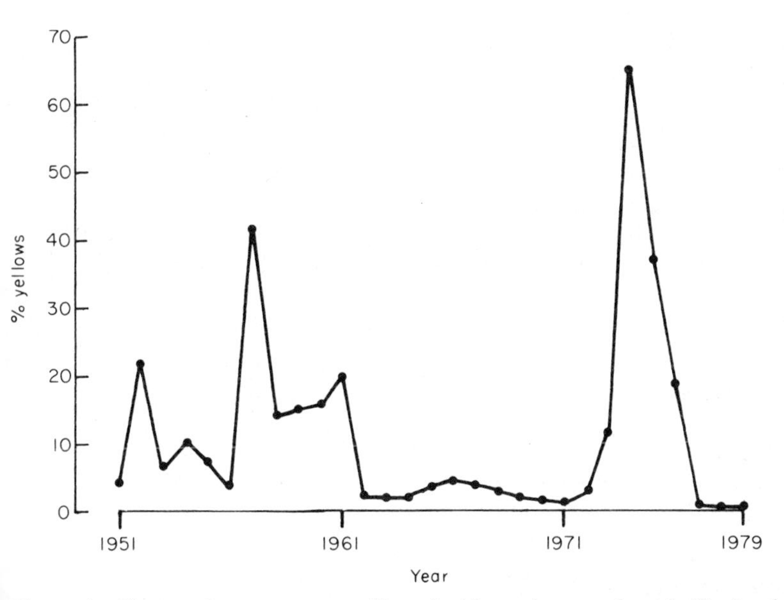

Figure 4 Changes in percentage yellows incidence in sugar beet in England, 1951–1979 (Watson, *et al.*, 1975 and Heathcote, personal communication).

especially at the critical time before spread to crops in early summer. Unfortunately there are few data on possible fluctuations or long-term changes in the relative importance of the two viruses but Fig. 4 shows that there is no evidence for a decrease in overall incidence of yellowing. Perhaps we are overestimating the damaging role of weeds, since specific control practices combined with chemical weed control in arable crops should otherwise have caused an obvious decrease. Assuming that spread within the crop has remained relatively unchanged, it seems that as much yellows as was once transmitted from all crop and weed sources is still being transmitted from fewer weeds. This could indicate some detrimental changes in the ecology of the vector and/or changes in the viruses, presumably associated with changing crop production technology. What these changes might be is largely speculation but those associated with chemical control of weeds could include:

(a) increased selection pressure for production of insecticide-resistant *M. persicae*, now widely prevalent in the UK and elsewhere. The removal of weed hosts from crops not normally treated with aphicides has concentrated selection pressure on the increased proportion of the total *M. persicae* population concentrated on aphicide-treated crops, notably in midsummer. This has similar implications for induced resistance of most pests that have weed hosts (Eastop, *1981*).
(b) scarcity of alternative weed hosts that previously 'diverted' viruliferous aphids from sugar beet crops.
(c) increased selection for strains of vector that preferentially colonize sugar beet, because previously preferred weed hosts are scarce or absent. A very small increase in attractiveness could cause a large change towards choosing to colonize rather than reject sugar beet. Without doubt, many pests have become increasingly adapted to crops, for example, the tomato moth, *Diataraxia oleracea*, once unable to breed on tomato, became established as late instar larvae after early development on weeds, but soon bred solely on tomato (Speyer and Parr, 1948).
(d) sub-lethal action of herbicides which makes the beet crop more attractive to aphids and/or more readily infected with viruses.
(e) direct and/or indirect effects of some herbicides or other agricultural practices on natural enemies, especially those attacking overwintering and spring populations of pests on weeds and overwintered crops.

Perhaps the most serious general implications of the above speculations is the selection pressure created by weed scarcity which favours increasing adaptation of pests to crops and to insecticides.

The black bean aphid — *Aphis fabae*

This pest colonizes a sequence of wild, crop and weed hosts (Eastop, *1981*).

Crops may be heavily colonized during June and July but the aphid survives mainly on weeds during the critical August–September period. Fat hen, *Chenopodium album*, seems to be especially important as a weed host which continues to grow and remains nutritionally adequate at this time even in severe drought conditions as occurred in Britain in 1976. Work during the 1940s and 1950s (Jones and Dunning, 1972; Way, 1967), mostly in eastern England, elucidated a basic two-year cycle of abundance of *A. fabae* on crops. Observations since 1969 (Way and Cammell, 1973; Way *et al.*, 1977 and unpublished) indicate that the cycle has not continued. It is tempting to suggest that this might be associated with improved control of the weed hosts, notably *C. album*, which is upsetting an already vulnerable phase in the life cycle of the pest.

Role of weeds in perennial crops

So far, we have concentrated on weed problems in arable crops. Perennial systems are relatively much more stable, and it is recognized that pests associated with them tend to be much less dispersive and also more influenced by local interactions with natural enemies and competitors (Way, 1978). It is to be expected, therefore, that so-called weeds in a perennial cropping system will play a more specialized role than in arable systems. This is borne out by many examples. In parts of the USA, the weeds in apple orchards provide essential prey for an important predacious mite when its spider mite prey on apples is unavailable early in the season (Croft, 1975). Ground cover plants, normally classed as weeds, have also been shown to be crucially important in the dynamics and biological control of certain insect pests of coconut and oil palms. Thus, in East Africa and in the Solomon Islands, the coreids *Pseudotheraptus wayi* and *Amblypelta cocophaga* can cause serious losses of young nuts except where predatory ants, *Oecophylla* spp., are present in the palms. In East Africa the occurrence, and hence success, of this ant as a predator depends partly on ground-nesting competitor ants, *Anoplolepis* spp., which suppress the tree-nesting *O. longinoda*. Where *Anoplolepis* spp. are present as competitors, their survival depends on a sparse ground cover which seemingly allows insolation of the soil necessary for breeding success (Way, 1953). Damage by the coreid is, therefore, associated with areas where the ground cover is kept sparse by grazing or by weed control. Indeed, there would appear to have been a cycle in some East African plantations associated with changes in the amount of attention given to the crop. Yields began to fail in well-kept plantations because *O. longinoda* is destroyed, so the farmer neglects the plantation, ground vegetation develops, *Anoplolepis* spp. die out, *O. longinoda* returns, the coreid ceases to cause damage, the crop begins to yield well, so the farmer again tends his plantation and the cycle is repeated. In a comparable situation in the Solomon Islands, another ant, *Pheidole* sp., also inimical to *O. smaragdina*, nests at the base of palm trees and similarly can deplete its populations and hence increase damage by the coreid. A simple control (Stapley, 1973) involves treating a circle around the base of each palm tree with paraquat to eliminate weeds from which *Pheidole* spp. obtain honeydew produced by root-feeding Homoptera. This, in an entirely different way, alters the balance in favour of *O. smaragdina* and of the biological control it exerts. Finally, weed cover or artificially established ground cover has provided the basis for control of severely damaging rhinoceros beetles, *Oryctes* spp. of palm plantations in Malaysia (Wood, 1971). Here, the vegetation cover appears to restrict or otherwise affect the ability of beetles to colonize rotting vegetation and tree trunks in which their larvae breed.

Such evidence highlights the potentialities of so-called weeds in perennial systems as a vital component in the dynamics and control of certain pests. Nevertheless, it must be recognized that, in some circumstances, remarkable biological control is obtained in perennial weed-free crop monocultures and might even be benefited by absence of weeds that could divert natural enemies away from their pest hosts on the crop (Way, 1977).

Implications of weeds and weed control for concepts of integrated pest, disease and weed control

Despite abundant evidence that weeds harbour pests, it is remarkable that, with few exceptions, the large decreases or changes in composition of weed populations, and the corresponding changes in associated animal communities and populations caused by herbicides in the last 20 years seem to have had little demonstrable impact on pest incidence in annual crops. This is not to say that wild plants are unimportant; some hedgerow, scrubland or woodland overwintering host species must be the cause of certain pest and disease problems as discussed by many other contributors to this volume. However, there is little evidence of local effects on pests that can be attributed to the dramatic decrease in crop weeds through herbicide use, and we can still only speculate on long-term effects. In this context, it is significant that, ecologically, the early stage in natural succession (initial colonization of soil exposed, for example, by flooding), which corresponds to an arable crop, is recognized as 'unstable' but nevertheless 'robust' (Margalef, 1968). The range of artificialities imposed on arable systems is probably less variable than the diverse mix of biotic and abiotic components that can exist emphemerally in natural homologues. So it might be argued that changed diversity due to artificialities, such as intensive weed control, is much less likely to increase pest abundance in a field than, for example, fertilizer treatment and emphasis on cultivars for high yield which would be strongly selected against in a natural system.

However, there is no cause for longer-term complacency. So far, little attempt has been made to integrate pest, disease and weed control practices even when it is known that there are simple incompatibilities. In important systems, notably cereals, we should continue to follow up experimentally the preliminary evidence that herbicides may benefit pests such as aphids. Moreover, as cropping systems become more artificial, there will be an increasing need for long-term studies such as the North Farm project, for long-term analysis of causes of pest population fluctuations, and for monitoring systems such as the Rothamsted trapping programme (Taylor, 1977, 1979; Way, *et al.*, 1981) which provide standardized indicators of long-term changes in crop populations and communities. It is only through such work that we can gain insights on the causes of changes in overall levels of pest abundance including the role of weeds and of weed control practices.

References

Adams, J. B. & Drew, M. E. (1965). Grain aphids in New Brunswick III. Aphid populations in herbicide-treated oat fields. *Canadian Journal of Zoology* **43,** 789–794.

Altieri, M. A. (*1981*). Crop–weed–insect interactions and the development of pest-stable cropping systems. In *Pests, Pathogens and Vegetation*, pp. 459–466. J. M. Thresh. London: Pitman.

Altieri, M. A., van Schoonhoven, A. & Doll, J. (1977). The ecological role of weeds in insect pest management systems: A review illustrated by bean (*Phaseolus vulgaris*) cropping systems. *PANS* **23**(2), 195–205.

Atkinson, D. & White, G. C. (*1981*). The effects of weeds and weed control on temperate fruit orchards and their environment. In *Pests, Pathogens and Vegetation*, pp. 415–428. J. M. Thresh. London: Pitman.

Bardner, R., Calam, D. H., Greenway, A. R., Griffiths, D. C., Jones, M. G., Lofty, J. R., Scott, G. C. & Wilding, N. (1971). The Wheat Bulb Fly. *Rothamsted Experimental Station Report for 1971, Part 2,* pp. 165–176.

Brown, A. W. A. (1978). *Ecology of Pesticides*. New York: Wiley.

Croft, B. A. (1975). Integrated control of orchard pests in the U.S.A. In *Proceedings of the 5th Symposium on Integrated Control in Orchards, Bolzano OILB/SROP, 1974*, pp. 109–124.

Cromartie, W. J. (1975). The effect of stand size and vegetational background on the colonization of cruciferous plants by herbivorous insects. *Journal of Applied Ecology* **12,** 517–533.

Davis, B. N. K. (1965). The immediate and long-term effects of the herbicide MCPA on soil arthropods. *Bulletin of Entomological Research* **56,** 357–366.

Dempster, J. P. (1969). Some effects of weed control on the numbers of the small cabbage white (*Pieris rapae* L.) on brussels sprouts. *Journal of Applied Ecology* **6,** 339–345.

Dempster, J. P. & Coaker, T. H. (1974). Diversification of crop ecosystems as a means of controlling pests. In *Biology in Pest and Disease Control*, pp. 59–72. D. Price Jones and M. E. Solomon. Oxford: Blackwell.

Dunning, R. A. (1971). Changes in sugar beet husbandry and some effects on pests and their damage. In *Proceedings of the 6th British Insecticide and Fungicide Conference* **1,** pp. 1–8.

Eastop, V. F. (*1981*). Wild hosts of aphid pests. In *Pests, Pathogens and Vegetation*, pp. 285–298. J. M. Thresh. London: Pitman.

Edwards, C. A. (1970). Effects of herbicides on the soil fauna. In *Proceedings of the 10th British Weed Control Conference* **3,** 1052–1062.

Edwards, C. A. (1975). Effects of direct drilling on the soil fauna. *Outlook on Agriculture* **8,** 243–244.

Edwards, C. A. & Stafford, C. J. (1979). Interactions between herbicides and the soil fauna. *Annals of Applied Biology* **91,** 132–137.

van Emden, H. F. (1965). The role of uncultivated land in the biology of crop pests and beneficial insects. *Scientific Horticulture* **17,** 121–136.

van Emden, H. F. (1970). Insects, weeds and plant health. In *Proceedings of the 10th British Weed Control Conference* **3,** 953–957.

Fox, C. J. S. (1964). The effects of five herbicides on the numbers of certain invertebrate animals in grassland soil. *Canadian Journal of Plant Science* **44,** 405–409.

Fox, W. B. (1948). 2,4-D as a factor in increasing wireworm damage of wheat. *Scientific Agriculture* **28,** 423–424.

Franklin, M. T. (1970). Interrelations of nematodes, weeds, herbicides and crops. In *Proceedings of the 10th British Weed Control Conference* **3,** 927–933.

Fryer, J. D. (*1981*). Weed control practices and changing weed problems. In *Pests, Pathogens and Vegetation*, pp. 403–414. J. M. Thresh. London: Pitman.

Gosling, L. M. (*1981*). The role of wild plants in the ecology of mammalian pests. In *Pests, Pathogens and Vegetation*, pp. 341–364. J. M. Thresh. London: Pitman.

Heathcote, G. D. (1970). Weeds, herbicides and plant virus diseases. In *Proceedings of the 10th British Weed Control Conference* **3,** 934–941.

Hsiao, T. H. & Holdaway, F. G. (1966). Seasonal history and host synchronization of *Lydella grisescens* (Diptera: Tachinidae) in Minnesota. *Annals of the Entomological Society of America* **59,** 125–133.

Ishii, S. & Hirano, C. (1963). Growth responses of larvae of the rice stem borer to rice plants treated with 2,4-D. *Entomologia Experimentalis et Applicata* **6,** 257–262.

Jones, F. G. W. & Dunning, R. A. (1972). *Sugar Beet Pests.* 3rd edn. London: *Ministry of Agriculture, Fisheries and Food, Bulletin* No. 162.

Kennedy, J. S. (1966). The balance between antagonistic induction and depression of flight activity in *Aphis fabae* Scopoli. *Journal of Experimental Biology* **45,** 215–228.

Kennedy, J. S., Booth, C. O. & Kershaw, W. J. S. (1959a). Host finding by aphids in the field. I. Gynoparae of *Myzus persicae* (Sulzer). *Annals of Applied Biology* **47,** 410–423.

Kennedy, J. S., Booth, C. O. & Kershaw, W. J. S. (1959b). Host finding by aphids in the field. II. *Aphis fabae* Scop (Gynoparae) and *Brevicoryne brassicae* L.; with a re-appraisal of the role of host finding behaviour in virus spread. *Annals of Applied Biology* **47,** 424–444.

Margalef, R. (1968). *Perspectives in Ecological Theory.* Chicago: University of Chicago Press.

Maxwell, R. C. & Harwood, R. F. (1960). Increased reproduction of pea aphids on broad beans treated with 2,4-D. *Annals of the Entomological Society of America* **53,** 199–205.

Muller, H. J. (1958). The behaviour of *Aphis fabae* in selecting its host plants, especially different varieties of *Vicia faba*. *Entomologia Experimentalis et Applicata* **1,** 66–72.

Murant, A. T. (*1981*). The role of wild plants in the ecology of nematode-borne viruses. In *Pests, Pathogens and Vegetation*, pp. 237–248. J. M. Thresh. London: Pitman.

O'Donnell, M. S. & Coaker, T. H. (1975). Potential of intra-crop diversity for the control of brassica pests. In *Proceedings of the 8th British Insecticide & Fungicide Conference* **1,** 101–107.

Oka, I. N. & Pimentel, D. (1976). Herbicide (2,4-D) increases insect and pathogen pests on corn. *Science* **193,** 239–240.

Perrin, R. M. (1975). The role of the perennial stinging nettle, *Urtica dioica* L. as a reservoir of beneficial natural enemies. *Annals of Applied Biology* **81,** 289–297.

Pierce, W. D. (1912). The insect enemies of cotton boll weevil. *USDA Bureau of Entomology Bulletin* No. 100.

Pollard, E. (1971). Hedges VI. Habitat diversity and crop pests: A study of *Brevicoryne brassicae* and its syrphid predators. *Journal of Applied Ecology* **8,** 751–780.

Potts, G. R. (1970). The effects of use of herbicides in cereals on the feeding ecology of the partridge. In *Proceedings of the 10th British Weed Control Conference*, pp. 299–302.

Potts, G. R. & Vickerman, G. P. (1974). Studies on the cereal ecosystem. *Advances in Ecological Research* **8,** 107–187.

Schneider, F. (1969). Bionomics and physiology of Aphidophagous Syrphidae. *Annual Review of Entomology* **14,** 103–124.

Smith, J. G. (1969). Some effects of crop background on populations of aphids and their natural enemies on brussels sprouts. *Annals of Applied Biology* **63,** 326–329.

Smith, J. G. (1976a). Influence of crop background on aphids and other phytophagous insects on brussels sprouts. *Annals of Applied Biology* **83,** 1–13.

Smith, J. G. (1976b). Influence of crop background on natural enemies of aphids on brussels sprouts. *Annals of Applied Biology* **83,** 15–29.

Southwood, T. R. E. & Cross, D. J. (1969). The ecology of the partridge III. Breeding success and the abundance of insects in natural habitats. *Journal of Animal Ecology* **38,** 497–509.

Speight, M. R. & Lawton, J. H. (1976). The influence of weed-cover on the mortality imposed on artificial prey by predatory ground beetles. *Oecologia* **23,** 211–223.

Speyer, E. R. & Parr, W. J. (1948). The tomato moth (*Diataraxia oleracea* L.). *Report of the Experimental Research Station, Cheshunt, for 1947*, pp. 41–64.

Stapley, J. H. (1973). Insect pests of coconuts in the Pacific Region. *Outlook on Agriculture* **7,** 211–217.

Stary, P. (1964). The foci of aphid parasites (Hymenoptera, Aphidiidae) in nature. *Ekologia polska* (A) **12,** 529–554.

Tahvanainen, J. O. & Root, R. B. (1972). The influence of vegetational diversity on the population ecology of a specialized herbivore, *Phyllotreta cruciferae* (Coleoptera: Chrysomelidae). *Oecologie* **10,** 321–346.

Taylor, L. R. (1977). Aphid forecasting and the Rothamsted Insect Survey. *Journal of the Royal Agricultural Society of England* **138,** 75–97.

Taylor, L. R. (1979). The Rothamsted Insect Survey; an approach to the theory and practice of pest forecasting in agriculture. In *Movements of Highly Mobile Insects;*

Concepts and Methodology in Research, pp. 148–185. G. G. Kennedy & R. L. Rabb. Raleigh: North Carolina State University.

Thurston, J. M. (1970). Some examples of weeds carrying pests and diseases of crops. In *Proceedings of the 10th British Weed Control Conference* **3,** 953–957.

Vickerman, G. P. (1974). Some effects of grass weed control on the arthropod fauna of cereals. In *Proceedings of the 12th British Weed Control Conference* **3,** 929–939.

Watson, M. A., Heathcote, G. D., Lauckner, F. B. & Sowray, P. A. (1975). The use of weather data and counts of aphids in the field to predict the incidence of yellowing viruses of sugar-beet crops in England in relation to the use of insecticides. *Annals of Applied Biology* **81,** 181–198.

Way, M. J. (1953). The relationship between certain ant species with particular reference to biological control of the coreid, *Theraptus* sp. *Bulletin of Entomological Research* **44,** 669–691.

Way, M. J. (1967). The nature and causes of annual fluctuations in numbers of *Aphis fabae* Scop. on field beans (*Vicia faba*). *Annals of Applied Biology* **59,** 175–188.

Way, M. J. (1977). Pest and disease status in mixed stands vs. monocultures; the relevance of ecosystem stability. In *Origins of Pest, Parasite, Disease and Weed Problems*, pp. 127–138. J. M. Cherrett & G. R. Sagar. Oxford: Blackwell.

Way, M. J. (1978). Integrated pest control with special reference to orchard problems. *Amos Memorial Lecture, Report of East Malling Research Station for 1977*, pp. 187–204.

Way, M. J. & Cammell, M. E. (1973). The problem of pest and diseases forecasting-possibilities and limitations as exemplified by work on the bean aphid, *Aphis fabae*. In *Proceedings of the 7th British Insecticide and Fungicide Conference* **3,** 933–954.

Way, M. J., Cammell, M. E., Alford, D. V., Gould, H. J., Graham, C. W., Lane, A., Light, W. I. St. G., Rayner, J. M., Heathcote, G. D., Fletcher, K. E. & Seal, K. (1977). Use of forecasting in chemical control of black bean aphid, *Aphis fabae* Scop., on spring-sown field beans, *Vicia faba* L. *Plant Pathology* **26,** 1–7.

Way, M. J., Cammell, M. E., Taylor, L. R. & Woiwod, I. P. (1981). The use of egg counts and suction trap samples to forecast the infestation of spring-sown field beans, *Vicia faba* L. by the black bean aphid, *Aphis fabae* Scop. *Annals of Applied Biology* **98,** 21–34.

Way, M. J., Glynne-Jones, D. G. & Johnson, C. G. (1956). Work in England on the effect of insecticides and other chemicals used in plant protection on beneficial insects and insect populations. In *Proceedings of the 5th Technical Meeting of the International Union for the Protection of Nature, Copenhagen 1954.*

Way, M. J., Murdie, G. & Galley, D. J. (1969). Experiments on integration of chemical and biological control of aphids on brussels sprouts. *Annals of Applied Biology* **63,** 459–475.

Webster, J. M. (1967). Some effects of 2,4-Dichlorophenooxyacetic acid herbicide on nematode-infested cereals. *Plant Pathology* **16,** 23–26.

Wood, B. J. (1971). Development of integrated control programmes for pests of tropical perennial crops in Malaysia. *Biological Control*, pp. 422–457. C. B. Huffaker. New York: Plenum.

Zandstra, B. H. & Motooka, P. S. (1978). Beneficial effects of weeds in pest management — a review. *PANS* **24,** 333–338.

Crop–weed–insect interactions and the development of pest-stable cropping systems

M A Altieri*
Department of Entomology, University of Georgia Coastal Plain Experiment Station, Tifton, Georgia 31794

Introduction

Traditional pest control disciplines have evolved separately and have divided the agroecosystem according to particular viewpoints. Some of these divisions are structural such as the separation of weeds, insects and crops. Others are functional aspects such as weed phenology, insect nutrition and crop productivity. As a result, approaches to pest management have tended to be simplistic and discipline-oriented.

Recently, however, some agricultural scientists have started to recognize that a crop system is a complex set of interacting subsystems with inputs, outputs and movement between them. For example, the presence of certain weeds within a crop can greatly influence the fauna and there are reports of decreased pest damage in weedy crops compared to that in weed-free monocultures (Pimentel, 1961; Guevara, 1962; Dempster, 1969; Tahvanainen and Root, 1972; Root, 1973; Smith, 1976; Saunders, unpublished). In many cases the reduced pest numbers resulted from an increase in populations of natural enemies, from chemical repellency or masking, from changes in the colonization background of the pest, or simply because the crop is less apparent to the pest (Way and Cammell, *1981*).

These findings indicate that the ecological assemblage of a specific field might be manipulated to encourage interactions that benefit the farmer. The potential of such interactions can be exploited to develop management strategies that improve the biological and economic stability of cropping systems. Despite supporting evidence the merit of using weeds to control pests is resisted due to the traditional concept of weeds as exclusively 'bad' and to the strong commercial pressures maintaining the current emphasis on control by pesticides.

This paper uses specific examples from Colombia and Florida to demonstrate that understanding the basic crop–weed–insect interactions in an area might suggest how agroecosystems can be adapted to minimize pest incidence in fields and over whole regions.

* Present address: Division of Biological Control, University of California, Berkeley, California 94720, USA

Regional manipulations

Weeds are so prevalent that they are an important component of agroecosystems and can be manipulated to manage pests and their natural enemies (Altieri and Whitcomb, 1979a). The uncultivated areas around many fields provide alternative food, breeding sites and shelter for many important beneficial insects (Dambach, 1948; van Emden, *1981*). In north Florida, predator density and diversity were greater in maize plots surrounded by annually burned pinelands and complex early successional weedy fields than in maize plots surrounded by sorghum and soybean fields (Table 1). Weed

Table 1 Predators of maize fields in north Florida

Maize	*Abundance (number of predators per 30 corn plants)*	*Diversity (number of species per 30 corn plants)*
Surrounded by complex weedy fields ($n = 36$)*	8.2a	4.3a
Surrounded by simple crop fields ($n = 36$)	4.3b	3.5b

In all tables the means followed by the same letter in each column are not significantly different according to Duncan's multiple range test ($P = 0.05$).
* n = number of samples.

species including goldenrod (*Solidago altissima*), camphorweed (*Heterotheca subaxillaris*), mexican tea (*Chenopodium ambrosioides*), sowthistle (*Sonchus* sp.), wild lettuce (*Lactuca* sp.), and pigweed (*Amaranthus* sp.) provide alternative prey for many predacious arthropods, thus serving as regional reservoirs of natural enemies. *See* Altieri and Whitcomb (1979b) for an example of a food web.

Manipulating cropping practices can help to maintain a natural balance of predator populations within particular agroecosystems. For example, ploughing date affects both the density and species composition of the subsequent vegetation. Experiments in Florida showed that native weeds varied with ploughing date and populations of herbivorous insects fluctuated according to the abundance of weed hosts (Altieri and Whitcomb, 1979b). Many chrysomelids and leafhoppers were collected in plots where preferred host plants were common, whereas few occurred in plots where preferred weed hosts were scarce. A leaf beetle (*Nodonota* sp.) was most numerous in plots ploughed in December, October and February, because these treatments enhanced the abundance of its preferred food supply, *Ambrosia artemisiifolia* (Fig. 1). The number of predacious arthropods feeding on the weed herbivores was proportional to the size of populations of their phytophagous prey. For example, October ploughing increased populations of

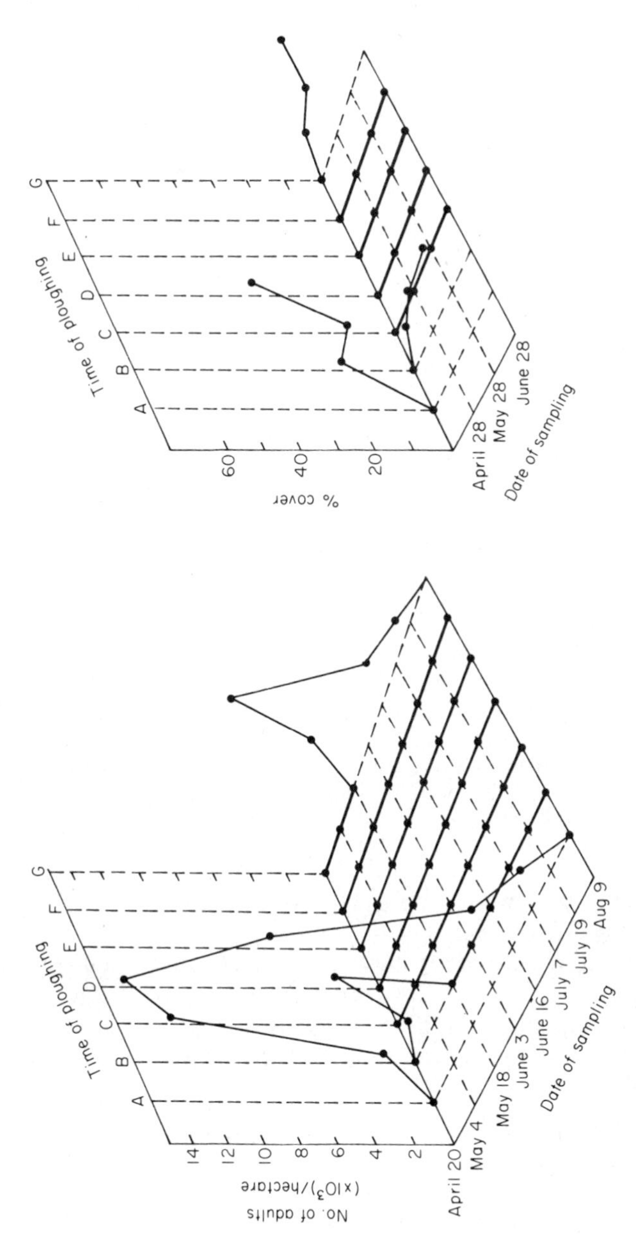

Figure 1 Effects of ploughing date on populations of the chrysomelid *Nodonota* sp. (left) and on the abundance of ragweed (*Ambrosia artemisiifolia*; right) in north Florida. Dates: A, December; B, February; C, Unploughed; D, April; E, June; F, August; G, October.

camphorweed and the predator–herbivore assemblage associated with this plant. Similarly, December ploughing encouraged the abundance of mexican tea and increased populations of 28 predatory insect and 9 spider species that fed on the herbivores of this weed. Figure 2 shows the mean relative

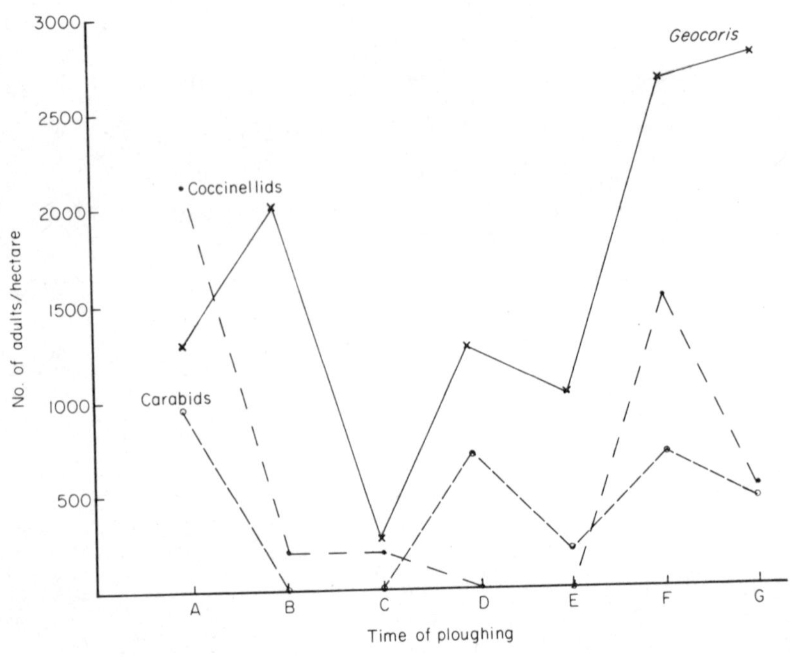

Figure 2 Relative annual abundance of three predator groups in north Florida fields ploughed at different times of the year. A, December; B, February; C, unploughed; D, April; E, June; F, August; G, October.

abundance of three predator families observed in various plots ploughed at different times in north Florida. Coccinellids and carabids were most numerous in plots ploughed in December because these treatments enhanced the abundance of goldenrod and mexican tea (Altieri and Whitcomb, 1979c). Both weeds support aphids, leaf beetles and leafhoppers which served as alternative food for predators. *Geocoris* sp. attained highest densities in the October plots.

Manipulations within fields

There are good possibilities of manipulating uncultivated areas bordering crops, but from the farmer's standpoint the effects of within-crop diversification on insect populations seem more tangible and may yield immediate results. Possible methods of manipulating weed composition and density within crop fields have been discussed previously (Altieri and Whitcomb, 1979a). When this is done to increase the biological control of certain pests, it is also essential to minimize competition with crops. Providing certain weeds as borders or alternate rows within fields is one possible approach.

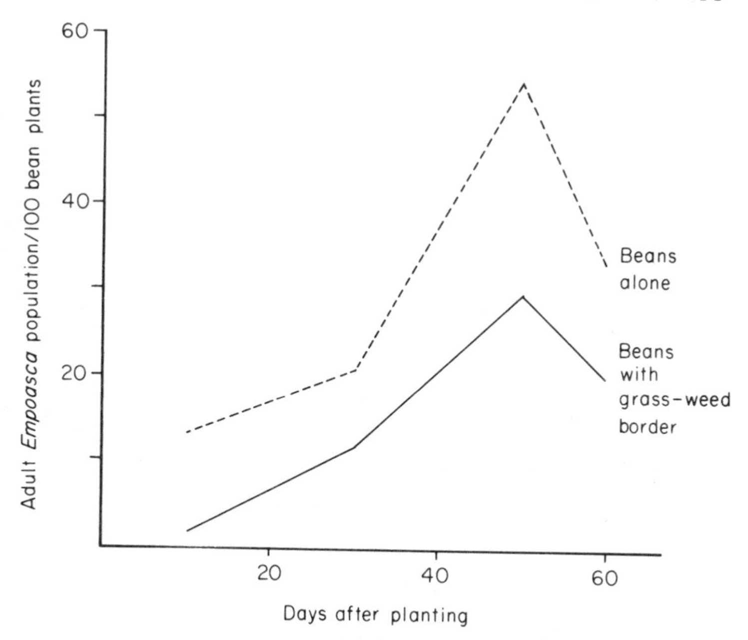

Figure 3 Effect of grass weed borders around bean plots (16 m^2) on populations of adult leafhoppers (*Empoasca kraemeri*).

In one Colombian experiment, borders of grass weeds (*Eleusine indica* and *Leptochloa filiformis*) 1 m wide were sown around bean plots 16 m^2 in area. Adult and nymphal populations of the leafhopper *Empoasca kraemeri*, the main bean pest of the Latin American tropics, were significantly reduced in plots with weed borders (Fig. 3). These seemed to exert a repellent or masking effect, decreasing the colonization and reproduction of *E. kraemeri* (Altieri *et al.*, 1977). The effects were still noticeable in larger plots (400 m^2) and similar effects were observed when beans were grown with pigweed (*Amaranthus dubius*) in alternate rows (Altieri, 1976). Population densities of a leafbeetle (*Diabrotica balteata*) were lower in the bean-pigweed plots than in the bean monocultures, apparently because pigweeds attracted leafbeetles and also because Carabidae (mainly *Lebia* spp.) were more numerous.

Experiments in north Florida during 1978 and 1979 tested the effects of weed diversity on the dynamics of maize insect pests and their predators (Altieri, 1979). Through various cultural manipulations (direct sowing, differential fertilization, early ploughing, etc.) characteristic weed communities were selectively established between the two central rows of plots (100 m²). The incidence of fall armyworm (*Spodoptera frugiperda*) was consistently higher in the weed-free habitats than in those with weeds. By contrast, corn earworm damage due to *Heliothis zea* was not affected by weed diversity, although significantly decreased by inter-planting soybean strips (Table 2).

Table 2 Damage by fall armyworm (*Spodoptera frugiperda*) and corn earworm (*Heliothis zea*) in simple and diversified maize systems in north Florida

	Fall armyworm (percentage plants damaged)		*Corn earworm (percentage ears damaged)*	
Maize system	*1978*	*1979*	*1978*	*1979*
Monoculture	31.5a	15.1a	32.7a	65.0a
Selected weed associations	16.5b	7.5b	32.0a	60.0a
Natural weed complex	15.8b	5.5c	32.2a	69.2a
Soybean strip	–	9.7b	–	53.3b

Table 3 General predators in weedy and weed-free maize fields in north Florida

	Number of individuals and species of predators per 30 corn plants			
	1978		*1979*	
Maize plantings	*Abundance*	*Diversity*	*Abundance*	*Diversity*
Monoculture	7.2a	3.9a	4.9a	3.4a
Selected weed associations	8.8b	4.3b	4.3ab	3.6a
Natural weed complex	8.6b	4.9b	3.7b	3.5a

This implies that in order to regulate populations of pests such as the corn earworm, it is important to select the right kind of plant diversity. In the 1978 experiments, predator abundance and diversity on maize plants were significantly higher in weed-diversified systems than in those kept weed free. Arthropod complexity seemed to parallel the trend in plant complexity. Conversely, in the 1979 experiment, total predator numbers on maize plants were slightly greater in monocultures than elsewhere, but there were no differences in the number of predator species (Table 3). These results were mainly due to the close proximity of the plots in 1979 which allowed insect predators to move easily between them. This problem seemed to be minimized in the 1978 experiment where plots were 50 m apart.

Conclusions

Cropping systems must be fully understood if they are to be changed to decrease the losses due to pests and diseases. An interdisciplinary approach is required to studies on the interrelationships between the various components of crop systems and their relationship with the surrounding matrix of vegetation.

This paper suggests that analysing crops, weeds and insects at a regional or field level can lead to improved pest management programmes. The examples given are somewhat experimental yet they indicate that weed scientists, entomologists, plant pathologists and agronomists should consider the implications of such findings and appreciate the possibilities of exploiting the beneficial effects of weeds. Clearly far more work is required on a much wider range of crops before comprehensive recommendations can be made to commercial growers in the United States and other developed countries. Change is also likely to be slow because of the risk of decreased yields and profits per unit area of land. However, energy shortages and galloping inflation may necessitate a reappraisal of current attitudes with increased emphasis on long-term environmental implications and energy conservation rather than on short-term financial aspects.

The situation is very different in the tropics where small farmers have exploited diversity in an empirical way for decades. For instance, Latin American peasant growers imitate the structure and species diversity of tropical forests by adopting mixed cropping systems. Combinations of tall and low annuals such as maize and beans, with a spreading ground cover of squash and weeds, are but one example of the ways in which subsistence farmers maintain diversity and minimize the risk of pest outbreaks whilst maximizing returns under low levels of technology. The advantages of such practices and their relevance for growers in developed countries are only now becoming apparent and merit far more attention than they have yet received.

The challenge for the future is to design stable and largely self-sustaining cropping systems that provide economic yields with minimal input of energy and resources. The manipulation of weeds on sound ecological principles is likely to be an integral part of such an approach.

Acknowledgements are due to many colleagues and institutions who have contributed personally and financially to the development of the studies. Particular recognition is given to the Centro Internacional de Agricultura Tropical, Tall Timbers Research Station, the Soil and Health Foundation, the University of Florida and John Ross.

References

Altieri, M. A. (1976). Regulation ecologica de plagas en agro-ecosistemas tropicales: un ejemplo mono y policultivos con malezas. *Tesis de Master*, Universidad Nacional de Colombia, Bogota.

Altieri, M. A. (1979). The design of pest stable corn agroecosystems based on the manipulation of insect populations through weed management. *PhD Dissertation*, University of Florida.

Altieri, M. A. & Whitcomb, W. H. (1979a). The potential use of weeds in the manipulation of beneficial insects. *HortScience* **14,** 12–18.

Altieri, M. A. & Whitcomb, W. H. (1979b). Manipulation of insect populations through seasonal disturbance of weed communities. *Protection Ecology* **1,** 185–202.

Altieri, M. A. & Whitcomb, W. H. (1979c). Predaceous arthropods associated with mexican tea (*Chenopodium ambrosioides* L.) in north Florida. *Florida Entomologist* **62,** 175–182.

Altieri, M. A., van Schoonhoven, Aart & Doll, J. D. (1977). The ecological role of weeds in insect pest management systems: A review illustrated with bean (*Phaseolus vulgaris* L.) cropping systems. *PANS* **23,** 185–206.

Dambach, C. A. (1948). *Ecology of Crop Field Borders*. Columbus: Ohio State University Press.

Dempster, J. P. (1969). Some effects of weed control on the numbers of the small cabbage white (*Pieris rapae* L.) on brussels sprouts. *Journal of Applied Ecology* **6,** 339–405.

van Emden, H. F. (*1981*). Wild plants in the ecology of insect pests. In *Pests, Pathogens and Vegetation*, pp. 251–261. J. M. Thresh. London: Pitman.

Guevara, J. C. (1962). Efecto de la practicas de siembra y de cultivos sobre plagas en maiz y frijol. *Fitotecnica Latinoamericana*, pp. 15–26.

Pimental, D. (1961). Species diversity and insect population outbreaks. *Annals of the Entomological Society of America* **54,** 76–86.

Root, R. B. (1973). Organization of a plant-arthropod association in simple and diverse habitats: The fauna of collards (*Brassica oleracea*). *Ecological Monographs* **43,** 95–124.

Smith, J. C. (1976). Influence of crop background on aphids and other phytophagous insects on brussels sprouts. *Annals of Applied Biology* **83,** 1–13.

Tahvanainen, J. C. & Root, R. B. (1972). The influence of vegetational diversity on the population ecology of a specialized herbivore, *Phyllotreta cruciferae* (Coleoptera: Chrysomelidae). *Oecologia* **10,** 321–346.

Way, M. J. & Cammell, M. E. (*1981*). Effects of weeds and weed control on invertebrate pest ecology. In *Pests, Pathogens and Vegetation*, pp. 443–458. J. M. Thresh. London: Pitman.

Weed hosts of aphid-borne viruses of vegetable crops in Florida

W C Adlerz
University of Florida IFAS, Agricultural Research Center Leesburg, Leesburg, Florida 32748

Introduction

Major problems have been encountered in the important vegetable growing areas of Florida due to the prevalence of aphid-borne viruses in many of the crops grown. Losses can be so great as to cause total crop failure and existing control measures have serious limitations.

This paper considers the crucial role played by weed hosts of the viruses and their vectors. Other recent reviews are already available dealing with relevant aspects of epidemiology or with the Florida situation (Carter, 1973; Duffus, 1971; Edwardson, 1974a, 1974b; Kring, 1972; Swenson, 1968; Thresh, *1981*).

Vegetable-growing in Florida

The Florida climate ranges from a temperate to sub-tropical transition zone in the north to the tropical Florida Keys. The chief determinants are: latitude, proximity to the Atlantic Ocean and Gulf of Mexico, and numerous inland lakes. The Gulf Stream warms the lower east coast (Bradley, 1972) and there is a low risk of killing frosts in the southern and coastal areas of the mainland (Greene *et al.*, 1969). These areas are used to grow frost-sensitive vegetables at times when this is impossible over much of North America. The main production is on the lower east coast from Palm Beach to Dade counties where the principal crops include *Phaseolus* bean, sweet corn, cucumber, squash, egg plant, pepper and potato for autumn, winter or spring harvest.

The annual rainfall averages 125–165 cm (50–65 in) but conditions tend to be dry during the vegetable-growing period and irrigation is often necessary (Bradley, 1972). Elaborate drainage schemes are also required to prevent flooding of the otherwise poorly drained soils. In south Florida the seepage irrigation and flood control systems necessitate extensive canal networks. These traverse Palm Beach county at 0.8 km (½ mile) intervals and laterals feed within-field irrigation ditches which are usually spaced at about 75–150 m.

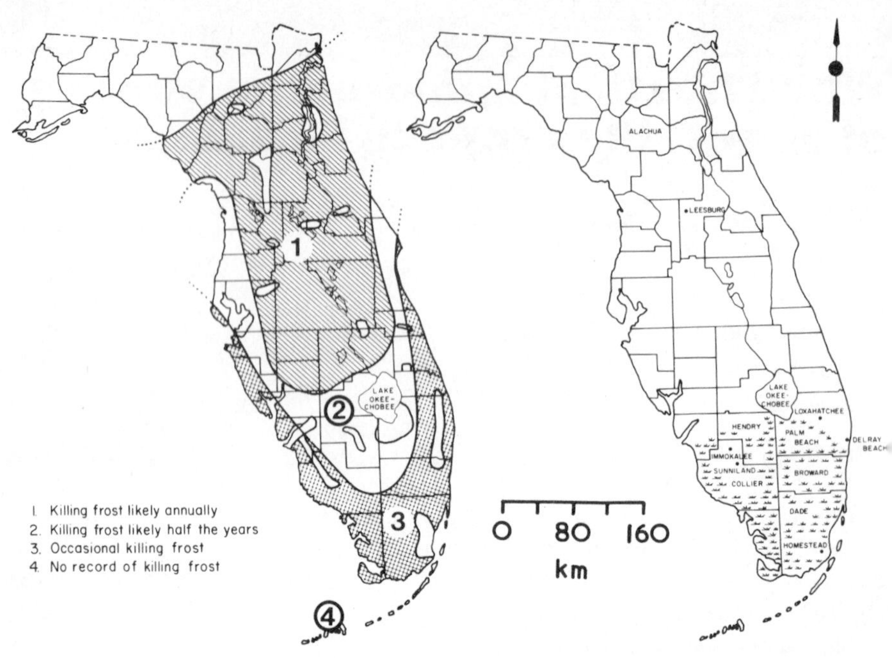

Figure 1 Frost hazard zones (left) and selected locations of crop production (right) in Florida.

Figure 2 A planting of spring cucumber in Palm Beach County, Florida alongside a permanent irrigation ditch lined with punk trees (*Melaleuca* spp.) used as a windbreak. Typically such ditches divide farms into small areas and become choked with weeds of many species including *Bidens pilosa* (as seen here with the white flowers) and *Momordica charantia*.

Vegetable viruses and their vectors

The major viruses affecting vegetable crops in Florida are members of the potyvirus group that are transmitted by aphids during brief feeding probes in plant tissue. The transmission process is so rapid that insecticidal control of migrant aphids does not effectively reduce virus spread. Many of these viruses have important weed hosts that thrive in the exceptionally favourable soil and climatic conditions.

Aphid flights most affecting vegetable production and virus spread are those peaking in December–January in Palm Beach county (Zitter, 1971; Zitter and Ozaki, 1973), in February–March at Homestead (Wolfenbarger, 1966) and usually in April in central and southwestern Florida (Adlerz, 1974a). Numbers of trapped aphids are lower in the southwest than on the lower east coat and in central Florida. *Myzus persicae* is a particularly important vector and it is common in all areas, comprising up to 98% of the aphid population on the lower east coast.

Weed hosts of viruses in Florida

The mild Florida climate permits certain annual plants to grow as perennials (Orsenigo and Zitter, 1971) and this greatly increases the number and species of plants that can act as primary virus sources. The more or less continuous growth of crop and weed hosts of viruses allows extensive spread between them and the development of large, stable virus reservoirs in weeds.

The major wild hosts of potato virus Y and tobacco etch virus are two annuals (*Solanum nigrum* and *S. gracile*) that grow as perennials in south Florida and are major sources for spread to sweet pepper (Simons, 1956; Zitter, 1971). Frost affects plant survival in central Florida where true perennial hosts such as *Physalis arenicola* and *S. aculeatissimum* become more important than annuals in cold winters (Anderson, 1959).

Two additional viruses of *Solanum* and *Physalis* weeds are pepper mottle virus and tomato yellows virus (Zitter, 1979). The latter affects tomato and potato in south-west Florida and is a virus of unknown affinities transmitted in the persistent (circulative) manner by aphids.

Potato virus Y was found in south Florida weeds of the Everglades in 1946 (Felix, 1946) and in central Florida weeds in 1957 (Anderson, 1959). It was found in pepper and tomato in south Florida only where potatoes had been grown, suggesting its introduction to the area in that crop (Simons *et al.*, 1956). The virus was found in the lower east and west coasts in 1957, again after potatoes were grown (Simons, 1959a).

Up to 1958 the main pepper viruses in the Everglades were potato virus Y and cucumber mosaic (Simons, 1958b). Tobacco etch was found in Palm Beach county for the first time in 1959 in tomatoes (Simons, 1959b). It became damaging to pepper in 1960 (Zitter, 1971) and, along with virus Y, was common in *S. nigrum* by 1970–71 (Zitter, 1971). Thus, although tobacco etch and potato virus Y were not found in east Palm Beach county until 1957 and 1959, they are now endemic in weeds and are major crop viruses. Pepper (Zitter, 1979) and tomato (Conover and Fulton, 1953) are most affected, whereas eggplant and potato are relatively unaffected (Simons, 1958b).

Cucumber mosaic virus has a wide host range of annual and perennial crop and wild plants (Wellman, 1935a, 1935b; Price, 1935). *Commelina diffusa*, the major weed host has been associated notably with infection in celery and pepper in central Florida and the Everglades (Wellman, 1937; Simons, 1958a, 1959b; Anderson, 1959; Zitter, 1970a). *C. diffusa* is an annual yet it is a source of virus throughout the winter, even in central Florida. It is renewed by seed and also vegetatively, so the viral reservoir is sustained even if plants are frost-damaged (Wellman, 1935a). Potato virus Y and tobacco etch have been invading *Solanum* and *Physalis* weeds in south Florida and they have become important in pepper and tomato, whereas the incidence of cucumber mosaic has been subsiding due partly to more effective use of herbicides along ditchbanks where *C. diffusa* is most common.

Momordica charantia is a very frost-sensitive annual cucurbit that grows as a perennial on the lower east coast. It is a major primary source of watermelon mosaic virus 1 (WMV-1) (Adlerz, 1972b) where squash and cucumber are grown during the winter. It is not of epidemiological importance in central Florida, where it dies with the first cold weather. It occurs on the lower west coast but does not flourish there near crops.

Melothria pendula is a perennial-rooted wild cucurbit and a natural host of WMV-1 as far north as Alachua County (Adlerz, 1969). It is the principal host for WMV-1 affecting spring-grown cucurbits only in south-west Florida, where the plant top usually survives the winter (Adlerz, 1969, 1972a). The top survives on the lower east coast also, but the plant is less abundant than the rampant *M. charantia* and is dwarfed by the latter's great size.

Annual weed hosts have been identified for most of the major vegetable viruses. In some cases the winter survival of weeds is not crucial because their main epidemiological importance is in carrying-over inoculum from spring to autumn crops. This is very significant in the epidemiology of watermelon mosaic virus 2 (WMV-2) in central Florida, where virus from spring watermelon and other cucurbits persists in weeds to infect autumn squash and cucumber (Adlerz, 1969). In other cases annual weeds have little epidemiological significance as with lettuce mosaic and celery mosaic. These have as their major sources, respectively, infected seed (Purcifull and Zitter, 1971) and celery (Zitter, 1970a), and not weeds. Several weeds identified elsewhere as important hosts of lettuce mosaic virus (Costa and Duffas, 1958) occur in Florida, but the virus has not been recovered from them.

Factors influencing the importance of weed hosts

Climate

Sometimes an abundance of virus-susceptible vegetation develops in frost-free areas in a way that is impossible in a colder climate. For example, *M. charantia* sprawls over other vegetation to a height of about 15 m and dominates the landscape in parts of the lower east coast but not farther north. In other cases, as with *Bidens* sp. and *Lepidium* sp., seedlings are produced throughout the year (Orsenigo and Zitter, 1971), ensuring a sequence of susceptible tissue for bidens mottle virus (Purcifull and Zitter, 1971).

Irrigation and flood control practices

The banks of lateral channels and drainage ditches provide abundant and especially favourable sites for the dense growth of weeds including species of epidemiological significance. Particular problems are encountered in south-west Florida due to *M. pendula*, which occurs naturally in the damp organic soils of the low woodland areas. Dense stands develop and thrive on the piles of rubble formed when woodland is first cleared, levelled and provided with drainage channels (Adlerz, 1969). There is a greatly increased incidence of WMV-1 in the weed stands which become a major threat to nearby crops. Such plantings would not otherwise be at great risk because less than 10% of Collier county farmland is in crops and fields in the area are usually isolated by as much as 16 km (10 miles) or more from similar holdings. Moreover, aphids are relatively few and the additional weed host *M. charantia* is not abundant.

Weeds as virus sources

Virus titre

The amount of virus in plant tissue that is available to aphid vectors is an important factor in virus spread. Virus titre has been determined more often in crop plants than in weeds. Simons (1958a) found that the transmission rate of potato virus Y from *Solanum* weeds to pepper by *Myzus persicae* was over 70%. This was comparable to transmission rates in crop plants by *M. persicae* of over 70% for Y (Simons, 1958a) and pepper mottle (Zitter, 1975), and by *Aphis gossypii* of over 70% for WMV-1 and WMV-2 (Adlerz, 1974a). High transmission efficiencies indicate a potential for rapid field spread (Simons, 1956) and all of these viruses spread rapidly at times.

Low aphid efficiency in transmitting virus from a major weed host is likely to retard the onset of infection. The transmission rate of the mottle virus of

Bidens to cultivated escarole (*Cichorium endiva*) with *M. persicae* averaged only 6% and between escarole plants 33% (Zitter, 1976). Infection of lettuce with bidens mottle from weed primary sources was (understandably) low in autumn plantings and was common only in spring plantings after several months (Zitter and Guzman, 1974).

Cucumber mosaic virus in *Commelina* showed no peak titre and transmission rates to pepper were low (0–5%) over a six-week period (Simons, 1958a). In pepper and celery, the virus reached peak titre two weeks after infection and then decreased rapidly. It was difficult to transmit from celery after 3–5 weeks (Zitter, 1970b; Simons, 1958a). In this case low transmission efficiencies reduced the spread of virus from weeds to crops and also within crops. Cucumber mosaic is no longer important in Florida (Zitter, 1977), and was not found during surveys of pepper in central or southern areas in 1970–72 (Zitter, 1971, 1972, 1973). The low titre in plants is undoubtedly an important contributing factor.

Proximity of weeds to crops

The main effect of having virus-infected weeds close to susceptible crops is that crops tend to be infected at an early stage allowing time for much secondary spread to occur whilst aphid populations are still increasing (Adlerz, 1974a). In south-west Florida WMV-1 can spread so readily from *M. pendula* to nearby watermelon that infection occurs in plantings a month or more before peak aphid flights (Adlerz, 1972a, 1974b). Similarly, on the lower east coast the spread of WMV-1 from nearby *M. charantia* to squash can be so rapid that plantings are often lost by late November (Adlerz, 1972b), about three weeks before peak winter aphid flights (Zitter, 1971; Zitter and Ozaki, 1973).

In Palm Beach County, in 1970–71 and 1971–72, tobacco etch, pepper mottle and potato virus Y spread from common *Solanum* weed sources (Zitter, 1971, 1973) to pepper two weeks before peak aphid flights (Zitter and Ozaki, 1973).

Weed abundance

Losses can occur extremely rapidly where inoculum sources are near crops and abundant. Squash fields bordered by *M. pendula* in south-west Florida have been almost totally infected with WMV-1 in as little as four weeks (Adlerz, 1972a). Similarly, pepper fields in Palm Beach county have been lost in six weeks (Zitter and Ozaki, 1973).

Weed and climatic effects on aphid flights

Weeds are important sources of *M. persicae*, but in south-eastern Florida

Genung and Orsenigo (1972) could not determine whether *M. persicae* populations built up on local weeds before infesting pepper and *S. nigrum*. Weeds and cultivated citrus are certainly important sources of *Aphis citricola* and weeds are the source of *Anuraphis middletonii* that spreads WMV-2 in central Florida (Adlerz, 1974a, 1978a).

The predominant wind direction in the spring over much of Florida is from the east (Bradley, 1972) and this influences virus spread. Winds consistently from one direction intensify secondary spread so that severe epidemics can develop even when early crop infection is low (Adlerz, 1978b).

Control measures

Chemical control of weeds

Since so many of the major viruses of vegetables have weed hosts, weed sanitation for disease control is commonly recommended (Zitter, 1970a, 1971; Purcifull and Zitter, 1971; Simons, 1956, 1957; Adlerz, 1969, 1972a). Wellman (1937) established that weed sanitation was effective against cucumber mosaic when he removed weeds from around celery fields. Herbicide use in vegetable production has since become a standard practice and the reduced incidence of cucumber mosaic may be one of the benefits derived.

Simons (1959a) reported that pepper yields were halved by virus on the lower east coast, where there was no effort to control nightshade. However, pepper losses were not decreased in 1970–71 despite pre-season weed control and attempts to establish fields away from weed hosts (Zitter, 1971). Efforts to reduce WMV-1 incidence on lower east coast squash by eliminating *M. charantia* have also failed to reduce crop losses. These continued although no other squash fields were nearby and growers removed physically or with herbicides all *M. charantia* plants from their own properties of up to 40.5 ha.

Significant losses may occur despite controlling weeds in crops because infected weeds along fences, roadsides and some ditchbanks are inaccessible to the sanitation effort. All ten weed sources of virus listed by Orsenigo and Zitter (1971) were common in non-crop as well as in crop areas and aphids have little difficulty in moving virus between wild plants and crops. Moreover, sanitation programmes that are effective in the autumn may fail in the spring when there is increased aphid activity in the warmer weather (Simons, 1957). In the spring WMV-1 has spread to a south-west Florida watermelon field across 9.6 km (6 miles) of land on which no virus source plants were found (Adlerz, unpublished).

As mentioned, attempts to control cucumber mosaic by eliminating weed host plants have succeeded, whereas similar attempts to control WMV-1 have failed. Ditchbank spraying to control *Commelina diffusa* is more widely and effectively carried out than programmes to reduce *M. charantia. M. charantia* is so pervasive on the lower east coast that control programmes by growers, however effective, do not significantly reduce area-wide inoculum. Transmission efficiency from viral weed hosts is low for cucumber mosaic and high for WMV-1, so increasing the distance from infected weeds to crop plants more effectively controls cucumber mosaic. Plant density is much lower for squash than for pepper or celery, so once the viruses are introduced, WMV-1 spreads to a high proportion of the plants sooner than cucumber mosaic. In addition, regardless of the size of individual holdings, squash plantings are made weekly so ensuring the continuous availability of small plants highly vulnerable to disease. Squash is inevitably the first crop to be lost to virus on the lower east coast each year.

Isolation of crops

Relocation of farms may be the first line of defence against losses from virus disease or it can be the last resort when other control methods have failed.

Watermelon growers found production too hazardous at Immokalee-Sunniland in Collier county, where *M. pendula* plants were abundant and mosaic losses were consistently severe (Adlerz, 1972a). Overall losses were reduced when growers moved to other parts of the county to achieve greater isolation and avoid *M. pendula* (Adlerz, 1972a). Plantings now are commonly separated by up to 16 km (10 miles) of mainly uncultivated land or pasture.

In the Delray Beach area, where *M. charantia* is abundant, losses of squash plantings due to WMV-1 begin in November. In Loxahatchee, about 37 km (23 miles) away, *M. charantia* is less common and WMV-1 is less damaging, causing a 'disease of attrition' on a large (162 ha) squash farm that operates from mid-September to April. Opportunity for this kind of control is limited, since Loxahatchee is on the edge of the frost-free area and few of the available sites combine freedom from frost and low incidence of weed hosts of virus.

The normal planting time for pepper on the lower east coast is September–early October, but plantings made in October can be virtually worthless or completely destroyed by mottle virus by January (Zitter, 1977, 1979). For spring crops some growers have moved to Collier and Hendry counties where fields are more isolated, aphid populations are lower, and virus is less damaging. Pepper plantings decreased in Palm Beach county from 2793 ha in 1966 to 850 in 1976, and increased in Collier and Hendry counties from 2247 ha to 3644 in the same period, mainly due to shifts in spring production.

Other control measures

Progress has been made in avoiding lettuce mosaic virus, using virus-free seed (Zitter and Guzman, 1974) and in controlling major pepper viruses with resistant varieties (Zitter, 1975).

Viruses that are transmitted in the non-persistent manner by aphids can be acquired from or inoculated to plants in single brief feeding probes and insecticidal control of virus spread is seldom successful. Nevertheless insecticides are applied routinely to many crop plants, especially Solanaceae, to prevent infestation by aphids. This does not prevent the spread of virus by aphids that fly through the field and visit plants briefly. Therefore alternative methods of control have been attempted.

Flying aphids can be deterred from landing on plants and transmitting virus by proper applications of aluminium, which acts as a repellent. Reflective aluminium foil applied to the soil beneath plants has been tested for virus control in Florida (Adlerz and Everett, 1968), but it has never been adopted commercially.

When correctly applied to plants, oil may form a coating through which virus cannot be transmitted by aphids. There is renewed interest in the use of oil sprays to decrease virus spread following successful tests on pepper and squash (Zitter and Ozaki, 1978). Growers report the harvest period of autumn-planted sweet pepper has been lengthened up to one month following weekly applications of 0.75% oil emulsion to inhibit virus transmission. Twice weekly applications are needed to significantly extend the harvest period for squash.

Concluding remarks

The lower east coast of Florida provides ideal climatic conditions for growing vegetables throughout the winter when most of the nation is out of production. Moreover, flooding of the otherwise poorly drained soil is prevented by drainage systems that are under national control and provide free water for seepage irrigation to maintain optimum soil moisture conditions.

Conditions optimal for farming are likewise optimal for the weed hosts of viruses and their vectors. Some weeds are extraordinarily abundant, and aphid flights are among the largest in the state, and mainly of *M. persicae* which is the best-known vector of plant viruses. A penalty of the intensive cropping systems adopted is that they facilitate virus spread to successive plantings of such crops as squash and pepper, and between crops as with pepper and tomato.

Control of those viruses having prevalent weed hosts as major sources has generally met with failure. Losses continue because growers cannot move away from the lower east coast to areas with fewer weeds, lower aphid populations and better isolation, without losing the main advantage of the favourable winter climate. Virus control techniques are still inadequate to realize the full production potential of Florida's unique lower east coast environment.

This paper is No. 2305 of the Florida Agricultural Experiment Station journal series.

References

Adlerz, W. C. (1969). Distribution of watermelon mosaic viruses 1 and 2 in Florida. *Florida State Horticultural Society Proceedings* **82,** 161–165.

Adlerz, W. C. (1972a). *Melothria pendula* plants infected with watermelon mosaic virus 1 as a source of inoculum for cucurbits in Collier County, Florida. *Journal of Economic Entomology* **65,** 1303–1306.

Adlerz, W. C. (1972b). *Momordica charantia* as a source of watermelon mosaic virus 1 for cucurbit crops in Palm Beach county, Florida. *Plant Disease Reporter* **56,** 563–564.

Adlerz, W. C. (1974a). Spring aphid flights and incidence of watermelon mosaic viruses 1 and 2 in Florida. *Phytopathology* **64,** 350–353.

Adlerz, W. C. (1974b). Wind effects on spread of watermelon mosaic virus 1 from local virus sources to watermelon. *Journal of Economic Entomology* **67,** 361–364.

Adlerz, W. C. (1978a). Secondary spread of watermelon mosaic virus 2 by *Anuraphis middletonii. Journal of Economic Entomology* **71,** 531–533.

Adlerz, W. C. (1978b). Watermelon mosaic virus 2 epidemics in Florida 1967–1977. *Journal of Economic Entomology* **71,** 596–597.

Adlerz, W. C. & Everett, P. H. (1968). Aluminum foil and white polyethylene mulches to repel aphids and control watermelon mosaic. *Journal of Economic Entomology* **61,** 1276–1279.

Anderson, C. W. (1959). A study of field sources and spread of five viruses of peppers in central Florida. *Phytopathology* **49,** 97–101.

Bradley, J. T. (1972). Climate of Florida. *United States Department of Commerce, Climatography of the United States* No. 60–68.

Carter, W. (1973). *Insects in Relation to Plant Disease*. New York, London, Sydney, Toronto: Wiley.

Conover, R. A. & Fulton, R. W. (1953). Occurrence of potato Y virus on tomatoes in Florida. *Plant Disease Reporter* **37,** 460–462.

Costa, A. S. & Duffas, J. E. (1958). Observations on lettuce mosaic in California. *Plant Disease Reporter* **42,** 583–586.

Duffus, J. S. (1971). Role of weeds in the incidence of virus diseases. *Annual Review of Phytopathology* **9,** 319–340.

Edwardson, J. R. (1974a). Some properties of the potato virus Y-group. *Florida Agricultural Experiment Stations Monograph Series* No. 4.

Edwardson, J. R. (1974b). Host ranges of viruses in the PVY-group. *Florida Agricultural Experiment Stations Monograph Series* No. 5.

Felix, E. L. (1946). Virus diseases of vegetables. *Florida Agricultural Experiment Stations Annual Report for 1946*, p. 191.

Genung, W. G. & Orsenigo, J. R. (1972). The wild host plants of green peach aphid *Myzus persicae* (Sulzer), the major pepper virus vector. *Belle Glade AREC Research Report* No. EV–1972–8.

Greene, R. E. L., Rose, G. N. & Brooke, D. L. (1969). Location of agricultural production in Florida. *Florida Agricultural Experiment Stations Bulletin* No. 733.

Kring, J. B. (1972). Flight behavior of aphids. *Annual Review of Entomology* **17,** 461–492.

Orsenigo, J. R. & Zitter, T. A. (1971). Vegetable virus problems in south Florida as related to weed science. *Florida State Horticultural Society Proceedings* **84,** 168–171.

Price, W. C. (1935). Classification of southern celery mosaic virus. *Phytopathology* **25,** 947–954.

Purcifull, D. E. & Zitter, T. A. (1971). Virus diseases affecting lettuce and endive in Florida. *Florida State Horticultural Society Proceedings* **84,** 165–168.

Simons, J. N. (1956). The pepper veinbanding mosaic virus in the everglades area of Florida. *Phytopathology* **46,** 53–57.

Simons, J. N. (1957). Effects of insecticides and physical barriers on field spread of pepper veinbanding mosaic virus. *Phytopathology* **47,** 139–145.

Simons, J. N. (1958a). Titers of three nonpersistent aphid-borne viruses affecting pepper in south Florida. *Phytopathology* **48,** 265–268.

Simons, J. N. (1958b). Virus diseases affecting vegetables in south Florida. *Florida State Horticultural Society Proceedings* **71,** 31–34.

Simons, J. N. (1959a). Potato virus Y appears in additional areas of pepper and tomato production in south Florida. *Plant Disease Reporter* **43,** 710–711.

Simons, J. N. (1959b). Viruses affecting vegetable crops in the everglades and adjacent areas of south Florida. *Florida Agricultural Experiment Stations Annual Report for 1959*, pp. 268–269.

Simons, J. N., Conover, R. A. & Walter, J. M. (1956). Correlation of occurrence of potato virus Y with areas of potato production in Florida. *Plant Disease Reporter* **40,** 531–533.

Swenson, K. G. (1968). Role of aphids in the ecology of plant viruses. *Annual Review of Phytopathology* **6,** 351–374.

Thresh, J. M. (*1981*). The role of weeds and wild plants in the epidemiology of plant virus diseases. In *Pests, Pathogens and Vegetation*, pp. 53–70. J. M. Thresh. London: Pitman.

Wellman, F. L. (1935a). Dissemination of southern celery mosaic virus on vegetable crops in Florida. *Phytopathology* **25,** 289–308.

Wellman, F. L. (1935b). The host range of southern celery mosaic virus. *Phytopathology* **25,** 377–404.

Wellman, F. L. (1937). Control of southern celery mosaic in Florida by removing weeds that serve as sources of mosaic infection. *United States Department of Agriculture Technical Bulletin* No. 548.

Wolfenbarger, D. O. (1966). Aphid trap collections over a three-year period from four southern Florida locations. *Journal of Economic Entomology* **59,** 953–954.

Zitter, T. A. (1970a). Cucumber mosaic and western celery mosaic-two aphid transmitted virus diseases of Florida celery. *Florida State Horticultural Society Proceedings* **83,** 188–191.

Zitter, T. A. (1970b). Titers of two virus diseases of celery affecting field spread. *Phytopathology* **60,** 1321.

Zitter, T. A. (1971). Virus diseases of pepper in South Florida. *Florida State Horticultural Society Proceedings* **84,** 177–183.

Zitter, T. A. (1972). Naturally occurring pepper virus strains in south Florida. *Plant Disease Reporter* **56,** 586–590.

Zitter, T. A. (1973). Further pepper virus identification and distribution studies in Florida. *Plant Disease Reporter* **57,** 991–994.

Zitter, T. A. (1975). Transmission of pepper mottle virus from susceptible and resistant pepper cultivars. *Phytopathology* **65,** 110–114.

Zitter, T. A. (1976). Properties and aphid transmission characteristics of two isolates of bidens mottle virus. *Belle Glade AREC Research Report* No. EV–1976–4.
Zitter, T. A. (1977). Epidemiology of aphid-borne viruses. In *Aphids as Virus Vectors*, pp. 385–412. K. F. Harris & K. Maramorosch. New York: Academic Press.
Zitter, T. A. (1979). Methods for controlling the most common vegetable viruses in south Florida. *Belle Glade AREC Research Report* No. EV–1979–8.
Zitter, T. A. & Guzman, V. L. (1974). Incidence of lettuce mosaic and bidens mottle viruses in lettuce and escarole fields in Florida. *Plant Disease Reporter* **58,** 1087–1091.
Zitter, T. A. & Ozaki, H. Y. (1973). Reaction of susceptible and tolerant pepper varieties to the pepper virus complex in south Florida. *Florida State Horticultural Society Proceedings* **86,** 146–152.
Zitter, T. A. & Ozaki, H. Y. (1978). Aphid-borne vegetable viruses controlled with oil spray. *Florida State Horticultural Society Proceedings* **91,** 287–289.

Interrelationships between wild host plant and aphid vector in the epidemiology of lettuce necrotic yellows

D K Martin and J W Randles
Hartley College of Advanced Education, Magill, South Australia and Plant Pathology Department, Waite Agricultural Research Institute, University of Adelaide, South Australia

Introduction

Lettuce necrotic yellows was first recognized in Victoria in 1954 and can cause severe losses in lettuce crops in both Australia (Stubbs and Grogan, 1963; Randles and Crowley, 1970) and New Zealand (Fry *et al.*, 1973), which are the only countries where the disease has been recognized. Epidemics occur in southern Australia during the autumn (March–July) and early summer (November and December) but in northern areas they appear less consistently.

Lettuce necrotic yellows virus (LNYV) is a member of the virus family Rhabdoviridae (Matthews, 1979; Francki and Randles, 1980). It is transmitted circulatively (Stubbs and Grogan, 1963) and transovarially (Boakye and Randles, 1974) in the aphid *Hyperomyzus lactucae*, and the only other recorded vector is *Hyperomyzus carduellinus* (Randles and Carver, 1971). The host range of LNYV is narrow, and in the field only *Sonchus oleraceus, S. hydrophilus, Lactuca serriola, Reichardia tingitana, Embergeria megalocarpa* and lettuce have been found to be naturally infected (Stubbs and Grogan, 1963; Randles and Carver, 1971). *S. oleraceus* (sowthistle) appears to be the major source of LNYV because eradication of the weed around and within lettuce crops is associated with reduced incidence of disease in crops (Stubbs *et al.*, 1963; I. S. Rogers and B. T. Baker, unpublished result), and the distribution of the other hosts is very restricted. Nevertheless, Randles and Carver (1971) have suggested that LNYV, which seems unrelated to other rhabdoviruses (Francki and Randles, 1980) may have originated in Australia in a native perennating host such as *E. megalocarpa* or *S. hydrophilus*, and may have become established in *S. oleraceus* after the introduction and naturalization of this new host and *Hyperomyzus lactucae.* The conclusion that *H. lactucae* is the major vector is supported by evidence that flights of this aphid are associated with high incidence of disease in lettuce 4–5 weeks later (Randles and Crowley, 1970).

We review here the biology of *S. oleraceus* and *H. lactucae*, and attempt to relate these to the epidemiology of LNYV in *S. oleraceus*, and the economic host species, lettuce.

Biology of *Sonchus oleraceus*

The importance of LNYV is directly related to the success of *S. oleraceus* as a widely distributed cosmopolitan weed. It grows on a variety of soils and quickly invades cultivated and bare soil or soil with sparse vegetation (Lewin, 1948). In South Australia it is most abundant in recently disturbed areas and along roadsides within the 375 mm isohyet (Martin, 1979). *S. oleraceus* occurs as erect, mature flowering plants or as immature, non-flowering rosettes (Fig. 1) and the relative proportions of the two forms vary throughout the year (Fig. 2(A)). In winter, the flowering plants support populations of *H. lactucae* (Fig. 2(B)), while the rosette forms, which are frost resistant (Lewin, 1948), later provide many new flowers as daylengths and temperatures increase during spring (Martin, 1979). Furthermore, germination and rapid growth of new plants in spring result in large numbers of flowering plants developing in early summer (Fig. 2(A)).

For Fig. 2 (opposite) approximately 110 thistles were counted on each sampling occasion. Aphid numbers and virus incidence were determined from a sub-sample of 30 plants. The data from each sub-area were standardized by conversion to numbers per 100 m^2. These standardized data were combined by multiplying each by a number which represented the proportion of the sub-area sampled in relation to the whole site. These numbers were then added to give the numbers of plants and aphids for each 100 m^2 of the whole site.

The duration of the rosette form is longest during the winter, and plants

Figure 1 Typical examples of three growth stages of *Sonchus oleraceus* (Left) miniature (non-flowering) form; (centre) and (right) mature flowering forms.

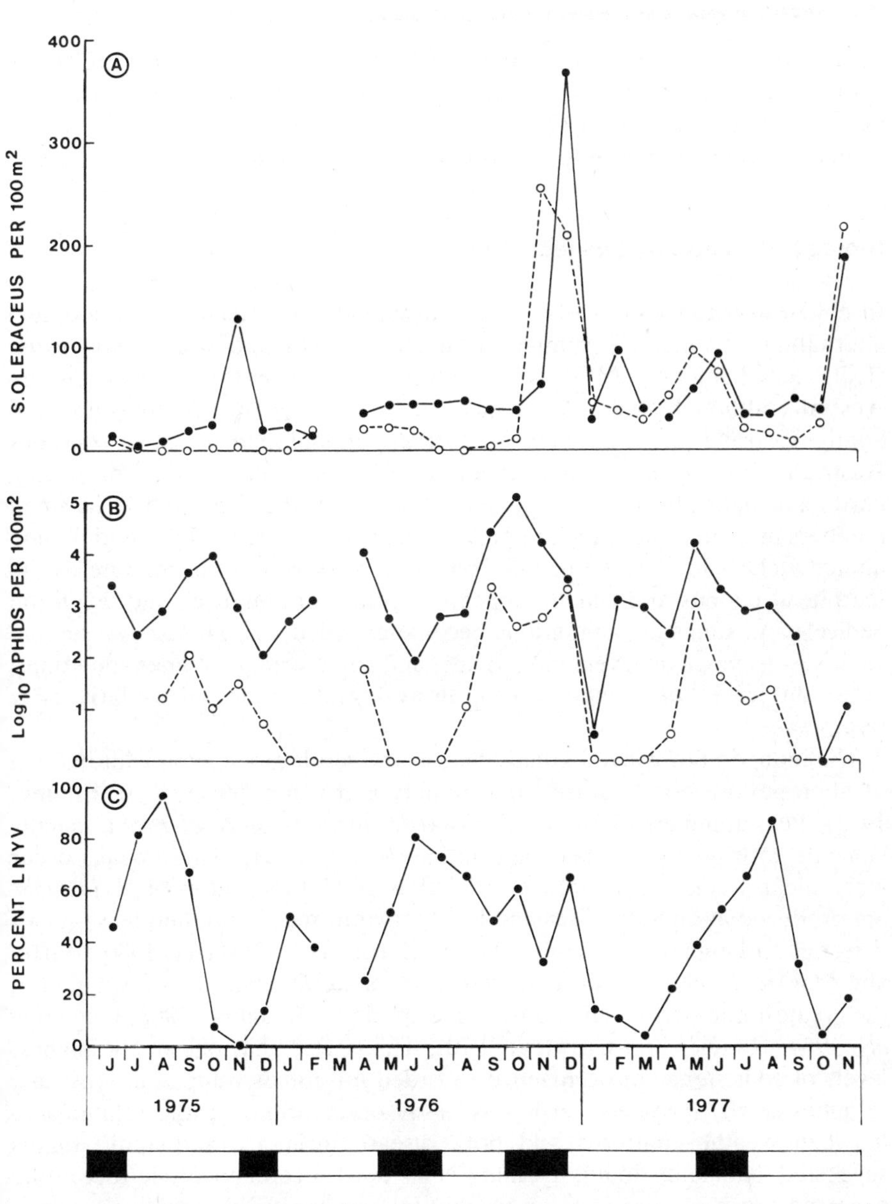

Figure 2 Field data collected over a 30 month period from four sub-areas around lettuce crops grown in the Adelaide foothills, South Australia.
(A) Numbers of *S. oleraceus* per 100 m^2:
●——● flowering forms favourable for aphid colonization; ○----○ rosette forms.
(B) Numbers of *H. lactucae* per 100 m^2: ●——● total; ○-----○ alate fourth instars.
(C) Percentage of thistles infected with LNYV.
The bar diagram (bottom) shows in black the main periods when infection with LNYV occurs.

germinating in autumn and winter generally remain as rosettes until early spring, when stem elongation and flowering occur. The duration of flowering becomes shorter as daylength and temperature increase. Some plants which commence flowering in autumn produce flowers throughout the winter.

Biology of *Hyperomyzus lactucae*

In cold European and North American climates *H. lactucae* is holocyclic, alternating between the summer host *S. oleraceus* and winter host *Ribes* (Hille Ris Lambers, 1949). Host alternation has not been recorded in Australia although male alatae have been observed in south-eastern and South Australia (Eastop, 1966; M. Carver, unpublished data). In southern Australia, *H. lactucae* is normally anholocyclic on *S. oleraceus*, reproducing parthenogenetically and viviparously throughout the year (albeit in small numbers in winter and mid-summer). *H. lactucae* are generally found on and immediately below the buds and flowers of *S. oleraceus* but migrate as the seed head pappus and achene appear. Colonies of aphids extend down the pedicels and stems as populations become crowded, but *H. lactucae* seldom colonizes leaves. However, individuals, and small young colonies sometimes occur on leaves of immature (non-flowering) *Sonchus*, particularly after rain.

The induction of winged aphid forms is the result of a complex interaction of photoperiod, temperature, food quality and aphid density (van Emden, 1972). Peak numbers of alate, 4th instar *H. lactucae* on *S. oleraceus* usually coincide with peaks in total aphid numbers (Fig. 2(B)). These peaks occur when mean weekly temperatures are below 20°C. Most alates of *H. lactucae* are produced at high densities and low temperatures (15°C), and least at low densities and high temperatures (26°C) (Martin, 1979; Maelzer, 1980; Martin and Maelzer, 1980). In southern Australia, alate *H. lactucae* are trapped in the greatest numbers during autumn and spring (O'Loughlin, 1963; Hughes *et al.*, 1964; Randles and Crowley, 1970). This species has one of the highest levels of consistency of occurrence recorded for aphids trapped in Australia (Hughes *et al.*, 1965). Nevertheless, early observations on the relationship between weather patterns and both disease incidence and aphid flights suggested that alatae do not fly, and LNYV does not spread into lettuce at the expected times of year if mean weekly temperatures fall outside the range 15–22°C, and weekly rainfall exceeds 5 mm (Randles and Crowley, 1970).

Epidemiology of LNYV in *Sonchus oleraceus*

The occurrence and importance of LNYV in lettuce can be considered as an incidental, or accidental, aspect of the ecology of LNYV in *S. oleraceus*. In

South Australia, thistles infected with LNYV and infested with *H. lactucae* are present throughout the year. Fluctuations occur in the number of *S. oleraceus* plants with flower heads (Fig. 2(A)), the number of alate *H. lactucae* present on these flowerheads (Fig. 2(B)) and the number of *S. oleraceus* plants with LNYV (Fig. 2(C)). Peak numbers of alate 4th instars are found on *S. oleraceus* flower heads in autumn (April–May) and spring (September–December) as shown in Fig. 2(B). These peaks coincide with maximum numbers of alate *H. lactucae* in traps (Randles and Crowley, 1970; Martin, 1979).

The incidence of infected *S. oleraceus* was high throughout a 30–month sampling period (Fig. 2(C)), except during the early summer of 1975 and the mid-summer of 1977. No obvious correlation between aphid numbers and disease incidence was apparent, possibly because of variation in the age structure of the plants, and varied weed control practices in the market garden area studied. The relationship between vector movement and disease incidence in *S. oleraceus* may perhaps only be determined by controlled sowings of the weed in the manner practised for lettuce (Randles and Crowley, 1970).

These observations support the view that LNYV is maintained in *S. oleraceus* as a widespread symptomless infection by the migration between, and colonization of, *S. oleraceus* plants by *H. lactucae*. The principal determinant of the populations of alate vectors appears to be the abundance of young flowerheads and the maximum rate of spread of LNYV occurs at seasons when flowerheads are most abundant.

Epidemiology of LNYV in lettuce

H. lactucae visits but does not colonize lettuce, and it is unable to acquire LNYV from infected lettuce (Stubbs and Grogan, 1963). The transmission characteristics are typical of a propagative virus and only aphids raised on infected *S. oleraceus* are thought to be involved in spread of LNYV (Boakye and Randles, 1974). Lettuce, therefore, is probably infected by viruliferous *H. lactucae* when seeking new hosts. Disease distribution apparently results only from primary spread of LNYV from *Sonchus*. The pattern of distribution in crops supports the view that secondary spread from infected lettuces does not occur and this would be expected because of the inability of aphids to breed on infected lettuce plants, which precludes completion of the temperature-dependent latent period of the virus (Boakye and Randles, 1974) before or during the migration phase.

The puzzling aspect of why lettuce becomes infected when it is not a host of *H. lactucae* has been investigated (Boakye, 1973; Boakye and Randles, 1974). The inoculation threshold for lettuce is 1–5 min, and starvation of aphids is necessary to induce them to feed on lettuce. Even then, they are reluctant to imbibe sap and finish probing earlier than on *S. oleraceus*. The requirement for such a period of starvation suggests that short-range spread of LNYV by alatae may be unimportant because a prolonged period of flight may be needed to overcome the settling or probing 'threshold'.

Epidemics occur regularly (Fig. 2) and would appear to be potentially predictable. However, no precise forecasting scheme has been developed, although growers now frequently avoid sowing crops in South Australia during September, and February–March to avoid the greatest risk of heavy losses (Randles and Crowley, 1970).

Several important epidemiological aspects of LNYV are incompletely understood. General observations suggest that the patterns of spread of LNYV in lettuce sometimes differ between regions. Stubbs and Grogan (1963) reported marginal spread in Victoria and this, together with the observation that eradication of *S. oleraceus* within approximately 150 m may reduce disease incidence from *c.* 75% to 6% (Stubbs *et al.*, 1963), implies that most spread of LNYV is over short distances. However, marginal patterns of spread have not been observed in South Australian lettuce crops. In a field trial designed to test the form of *H. lactucae* responsible for the spread of LNYV in lettuce, Boakye (1973) manipulated the morphological forms of *H. lactucae* on infected *S. oleraceus* plants placed within plots. A high 'background' incidence of about 40% disease in control plots was apparently related to the large number of alate *H. lactucae* trapped. Plots in which a source of additional alate viruliferous *H. lactucae* was placed showed no significantly higher incidence of LNYV. However, in plots where only apterous viruliferous aphids were maintained, disease incidence was around 80% and a gradient was shown from the aphid source plants, indicating that apterous aphids may supplement spread by alatae (Ribbands, 1963). Further investigations on disease distribution will need to take account of both long-range spread by alates and short-range spread by alates and/or apterae.

Strategies for control

The control of LNYV in lettuce could be directed towards controlling either the virus reservoir, the vector, or migration of the vector into lettuce. Early studies indicated that removal of *S. oleraceus* is associated with reduced incidence of disease in adjacent lettuce crops (*see* Introduction), but inadequate repetition of such trials leaves doubt as to the likely general effectiveness of this measure. In particular, aspects of the migration of viruliferous aphids are still too poorly understood for expensive weed control measures to be generally recommended.

Control directed towards the vector might be achieved either by the application of insecticides to *S. oleraceus* at the times of alate build-up, or by introducing biological control agents of *H. lactucae*. Parasites or predators, for example, might reduce aphid density to a level below that at which alates are induced. The present understanding of the population dynamics of *H. lactucae* (Martin, 1979) provides a good base for determining the effect of measures directed at modifying aphid populations. Such measures, and attempts to modify migratory behaviour, have yet to be studied.

Conclusions

From the epidemiological viewpoint, the LNYV system is ideal for studying the primary spread of a virus disease into crops, uncomplicated by the effects of secondary spread. The appearance and distribution of diseased plants can be considered to result solely from the activity of one aphid species, and because this flies so consistently in Australia (Hughes *et al.*, 1965), the disease is encountered regularly in the field.

The relative simplicity of the conceptual model of disease outbreaks of LNYV in lettuce makes this disease a useful model for studying host/virus/vector interrelationships. Until recently, the main effort in plant virus epidemiology has centred on the identification of vectors, availability of virus, mode of acquisition, and transmission by vectors. Recent studies on the growth and distribution of the virus source plant and on the population dynamics of the aphid vector, its relationship to, and interaction with, the source plant (Martin, 1979; Martin and Maelzer, 1980) make it possible to start quantifying the epidemiology of LNYV. It will now be possible for studies on disease control to test the impact of factors which influence the population dynamics of aphid vectors. There are still many facets of this 'model' which are not understood, particularly those related to vector behaviour.

References

Boakye, D. B. (1973). Transmission of lettuce necrotic yellows virus by *Hyperomyzus lactucae* (L.) (Homoptera: Aphididae): with special reference to aphid behaviour. *PhD Thesis*, University of Adelaide.

Boakye, D. B. & Randles, J. W. (1974). Epidemiology of lettuce necrotic yellows virus in South Australia. III. Virus transmission parameters, and vector feeding behaviour on host and non-host plants. *Australian Journal of Agricultural Research* **25,** 791–802.

Eastop, V. F. (1966). A taxonomic study of Australian Aphidoideae (Homoptera). *Australian Journal of Zoology* **14,** 399–592.

Emden, H. F. van (1972). *Aphid Technology*. London and New York: Academic Press.

Francki, R. I. B. & Randles, J. W. (1980). Rhabdoviruses infecting plants. In *Rhabdoviruses*, **III.**, pp. 135–165. D. H. L. Bishop. Florida: CRC Press.

Fry, P. R., Close, R. C., Procter, C. H. & Sunde, R. (1973). Lettuce necrotic yellows virus in New Zealand. *New Zealand Journal of Agricultural Research* **16,** 143–146.

Hille Ris Lambers, D. (1949). Contributions to a monograph of the Aphididae of Europe IV. *Temminckia* **8,** 182–323.

Hughes, R. D., Carver, M., Casimir, M., O'Loughlin, G. T. & Martyn, E. J. (1965). A comparison of the numbers and distribution of aphid species flying over eastern Australia in two successive years. *Australian Journal of Zoology* **13,** 823–839.

Hughes, R. D., Casimir, M., O'Loughlin, G. T. & Martyn, E. J. (1964). A survey of aphids flying over eastern Australia in 1961. *Australian Journal of Zoology* **12,** 174–200.

Lewin, R. A. (1948). Biological flora of British Isles. *Journal of Ecology* **36,** 203–223.

Maelzer, D. A. (1980). The capital ecology of aphid pest species in Australia. In *The ecology of Pests in Australia*, R. L. Kitching & R. E. Jones. Australia: CSIRO (in press).

Martin, D. K. (1979). The ecology of *Hyperomyzus lactucae* (L.) and the epidemiology of lettuce necrotic yellows virus. *PhD Thesis*, University of Adelaide.

Martin, D. K. & Maelzer, D. A. (1980). The potential size of colonies of the aphid *Hyperomyzus lactucae* (L.) determined by the influence of temperature and photoperiod on plant quality. *Oecologia* (in press).

Matthews, R. E. F. (1979). Classification and nomenclature of viruses. *Intervirology* **12,** 133–296.

O'Loughlin, G. T. (1963). Aphid trapping in Victoria. I. The seasonal occurrence of aphids in three localities and a comparison of two trapping methods. *Australian Journal of Agricultural Research* **14,** 61–69.

Randles, J. W. & Carver, M. (1971). Epidemiology of lettuce necrotic yellows virus in South Australia. II. Distribution of virus, host plants, and vectors. *Australian Journal of Agricultural Research* **22,** 231–237.

Randles, J. W. & Crowley, N. C. (1970). Epidemiology of lettuce necrotic yellows virus in South Australia. I. Relationship between disease incidence and activity of *Hyperomyzus lactucae* (L.). *Australian Journal of Agricultural Research* **21,** 447–453.

Ribbands, C. R. (1963). The spread of apterae of *Myzus persicae* (Sulz.) and of yellows viruses within a sugar-beet crop. *Bulletin of Entomological Research* **54,** 267–283.

Stubbs, L. L. & Grogan, R. G. (1963). Necrotic yellows: a newly recognised virus disease of lettuce. *Australian Journal of Agricultural Research* **14,** 439–459.

Stubbs, L. L., Guy, J. A. D. & Stubbs, K. J. (1963). Control of lettuce necrotic yellows virus disease by the destruction of common sowthistle (*Sonchus oleraceus*). *Australian Journal of Experimental Agriculture and Animal Husbandry* **3,** 215–218.

Section 7 **Commentary**

Commentary

F T Last
Institute of Terrestrial Ecology, Bush Estate, Penicuik, Midlothian, Scotland EH26 0QB

In considering my approach to the preparation of this final section, I decided to analyse in detail the limits set by the organizers when choosing the original title of the York symposium — *Weeds and wild plants in the ecology of crop pests and diseases*. In particular I am concerned with the definition of three words, *weed*, *crop* and *disease*.

What is a weed?

> A plant not valued for use or beauty,
> growing wild and rank, and regarded as
> cumbering the ground or hindering the
> growth of superior vegetation (Little *et al.*, 1964).

> A plant of no value and usually of rank
> growth especially one that tends to
> overgrow or choke out more desirable plants (Woolf, 1977).

To the layman these definitions may seem satisfactory, but to others concerned with plant diseases, whether attributable to attack by pests and/or pathogens, it is necessary to give a generous interpretation to 'hindering the growth of superior vegetation'. This interpretation must make allowance for the role of weeds as alternate or alternative hosts that are sometimes key elements in (a) the perpetuation of pests and pathogens during the, often inclement, intervals between one crop and the next, e.g. spindle, *Euonymus europaeus*, as an overwinter refuge for the bean aphid, *Aphis fabae*, and (b) the completion of life cycles of heteroecious fungi, e.g. barberry, *Berberis vulgaris*, as the host enabling the formation of the teleutospore stage of black stem rust of wheat, *Puccinia graminis*.

Weeds and weed control

Although the importance attached to weed control can be inferred from the fact that all areas of cereals, sugarbeet, top and soft fruits in Great Britain are sprayed with herbicides on average more than once per annum (Fryer, *1981*), I nevertheless suspect that crop managers, when considering the pros and cons, and costs of weed control, attach too little significance to impacts on pests and pathogens. With an ever-increasing awareness of the need for alternative sources of energy, would it be prudent to encourage weed growth (biomass) for fuel during the intervals between crops and would such practices be compatible with existing crop schedules? Could the simultaneous growth of traditional crops and crops of 'weeds' for fuel be countenanced, bearing in mind that the use of land may in future be judged in terms of energy efficiency instead of monetary effectiveness, as at present? What impact would the intentional, and probably selective, retention of 'fuel' weeds have upon the distribution of their own pests and pathogens and those of traditional crops?

The organizers of the symposium ambitiously asked the contributors to consider the role of weeds and wild plants in the ecology of *crop* pests and diseases, but perhaps we should also be addressing ourselves to the study of pests and pathogens of *natural* and mostly mixed assemblages of plants. Already such studies have facilitated significant advances in the sphere of biological control. Following the successful control of introduced prickly pear cacti, *Opuntia* spp., in Australia by the moth, *Cactoblastis cactorum* (Cussans, 1974), there has been a continued interest in the possible control of weeds by herbivores, e.g. flea beetles attacking alligatorweed (*Alternanthera philoxeroides*), snails, fishes including the grass carp (*Ctenopharyngodon idella*), etc. However, biological control, if it is to be achieved, seems to rest more dependably upon the use of plant pathogens. In Florida, isolates of endemic and exotic (introduced) fungi are being assessed for the control of water hyacinth (*Eichhornia crassipes*), with *Acremonium zonatum* and *Cercospora rodmanii* in the former category and *Uredo eichhorniae* and *Fusarium roseum* (*culmorum*) in the latter (Freeman *et al.*, 1978). However, the most notable exploitation of microbes has centred on the control of skeleton weed, *Chondrilla juncea*, a weed introduced into cultivated areas of SE Australia from the Mediterranean. In this instance, the use of both pests and pathogens has been explored with the introduction to Australia of the eriophyid mite, *Aceria chondrillae*, and the cecidomyid (gall midge), *Cystiphora schmidti*, whose destruction of skeleton weed was overshadowed by the efficacy of the rust fungus, *Puccinia chondrillina*. The strain of this fungus introduced from Vieste in southern Italy, is restricted to one, albeit the commonest narrow-leaved, of the three variants of skeleton weed recognized in Australia (Cullen *et al.*, 1973; Waterhouse, 1973). A second strain of *P. chondrillina*, this time attacking an intermediate form of skeleton weed, has recently been identified in western Turkey (Hasan, 1978). Together these pests and pathogens decrease the regeneration of overwintering tap-roots and minimize seed production (Hasan and Wapshere, 1973). While the use of the highly host-specific *P. chondrillina* for the control of skeleton weed seems to be socially acceptable the idea of using *Ceratocystis fagacearum*, the oak wilt pathogen, as suggested by French and Schroeder (1969) for selectively killing unwanted oaks, would be unthinkable in Europe.

Natural vegetation

Research thrives upon series of comparisons and contrasts. To me it is understandable, but somewhat disappointing, that the value to crop production of studying the pests and pathogens of natural vegetation has not been appreciated sooner. Perhaps this volume will serve to stimulate liaison between 'ecologists', on the one hand, and crop entomologists and plant pathologists, on the other. The former may, in future, accept more readily and overtly, that pests and pathogens can play a significant role in determining the recruitment and death of natural vegetation, so affecting Malthusian parameters. Similarly those concerned with protecting crops should take a leaf from the ecologist's book and recognize the merit of learning from natural assemblages with diverse arrays of genotypes of the same and/or different species. How much could be gained, when considering the ecophysiology of monocultures, from a knowledge of the many different micro-environments in mixed natural assemblages, not only in terms of direct effects on net photosynthesis and transpiration, but also indirectly upon the maintenance of micro-environments for pests and pathogens? Would a more careful analysis of the ecologist's plea for diversity have hastened the consideration of multiline plantings to minimize attacks by foliar pathogens of cereals? How far should the concept of diversity be taken? I will return to this subject later, but it might have been instructive to have had a contribution from one of the British tree pathologists struggling to explain, in the face of the devastation done since the late 1960s by the aggressive strains of *Ceratocystis ulmi* (the Dutch elm disease pathogen), the passive and historical acceptance of the widespread and often contiguous establishment of a restricted range of vegetatively-propagated genotypes of English elm (*Ulmus procera*).

Parasitic plants

Reverting to the original title of the symposium (now used as the sub-title of this volume), I would question the appropriateness of the word *disease*, which refers to a condition and not, as I suspect was intended, to a set of disease agents contrasting with pests, namely pathogens. This is not just a matter of semantics, as I, before seeing the detailed contents, thought that the range of disease-inducing agents to be considered, might include the extremely important group of parasitic and semi-parasitic plants of the Orobanchaceae, Scrophulariaceae and Loranthaceae, together with a discussion of allelopathy. Admittedly parasitic and semi-parasitic plants are unimportant in Britain, but, at a meeting focussed internationally, the role of weeds and wild plants in the population dynamics of *Orobanche* spp. and *Striga* spp. should not have been overlooked. In the past, studies of *Striga* spp., notably

S. hermonthica, have tended to be limited in extent and done in a piecemeal manner with insufficient regard to underlying principles (McGrath *et al.*, 1957). Should the control of this pathogen, by the manipulation of host susceptibility and the use of herbicides, be done without considering the effects of non-economic plants on the soil reserves of viable *Striga* seeds? The germination of *Striga* seeds is stimulated by root exudates which are produced by a wide variety of plants. Some of these resist subsequent invasion, whereas others are colonized, although amounts of subsequent damage differ greatly. On some compatible hosts *Striga* seedlings emerge above ground, grow prolifically at the expense of their hosts, and produce seeds copiously. On other hosts, on which developing *Striga* seedlings rarely emerge above ground, the flowering of the parasite is minimal and populations of viable seed are unlikely to be sustained.

With these different sets of inter-relations, which have their parallels in other host-parasite complexes, e.g. the clubroot pathogen, *Plasmodiophora brassicae*, and the stem nematode, *Ditylenchus dipsaci*, it is possible to countenance many ways of attempting the biological control, or more realistically the integrated control, of some parasites. Although attempts to breed crops of maize and sorghum that resist *Striga* spp. are meeting with some success, experience and ecological understanding indicate that the introduction of new varieties should form but one facet of a new approach to the problem in which other ways of minimizing soil-borne inocula are exploited concurrently, e.g. the selective retention of weeds that stimulate germination, but which are not parasitized, the use of 'susceptible' crops on which parasites do not flower, and the use of fully susceptible crops which, however, are repeatedly harvested at intervals to prevent the parasite forming seed (e.g. the cultivation of Sudan grass, *Sorghum sudanense*, for animal fodder). These are not new ideas and of course they are applicable to many different agents of disease substituting 'sporulate' and 'egg production' for 'flowering' when referring to fungi and nematodes, respectively. The different approaches were suggested many years ago at a time when treatments were either effective in their own right or were discarded with little regard being paid to an all-embracing 'holistic' approach.

As a forester, I am aware of the effects of members of the Loranthaceae (the mistletoe family), notably species of *Arceuthobium*, on practices adopted in NW America when harvesting native stands of pines and seeking new crops by natural regeneration. It is dangerous to leave scattered, infested specimens from which showers of *Arceuthobium* seeds may drop onto, and infect, regenerating pine seedlings, which are particularly sensitive to attack when young. However, more important in the British Isles, is the fact that afforestation has been affected by the classical example of allelopathy, namely the restricting effect of heather, *Calluna vulgaris*, on the growth of roots, and hence tops, of spruces, notably Sitka spruce (*Picea sitchensis*). The avoidance of Sitka 'check' (Handley, 1963) has advanced from the selective use of less 'desirable' timber species, through site-preparative silvicultural techniques (where turves of heather were inverted prior to planting), to the combined use of herbicides to check heather, and fertilizers to stimulate the growth of spruce. At this stage I might question what the editors intended by the word 'crop'.

Forestry

As is clear from the contents of this book, although not implicit in its title, the meeting was focussed on agriculture and horticulture; but weeds and wild plants also have an impact on the pests and pathogens of forests. Further, the predictably increasing areas of man-made forests will inevitably affect, directly and/or indirectly, the ecology of agriculture and horticulture. Bunting (*1981*) describes land use and land-use changes on a global scale, indicating that arable crops, permanent grass pastures and forest/woodland, respectively, occupy 12%, 28% and 28% of the areas of developed countries. He also states that the area of the world devoted to forestry and woodland remained unchanged between 1961/65 and 1977, whereas that given to arable usage increased by 5%. Of course, these figures integrate happenings in a diverse array of countries with a variety of environments, physical and socio-economic. Within England and Wales, the area of arable and grassland has decreased since 1935 from 10.1 to 9.6 million ha, a trend also reflected in a slight diminution in the area of rough grazing (Trask, *1981*). But what about forestry, a matter of cardinal importance with an increasing awareness of the pest status of deer (notably red deer, *Cervus elaphus*) and the pine beauty moth, *Panolis flammea*. The latter, earlier considered to be a relatively unimportant defoliator of the 'native' Scots pine (*Pinus sylvestris*), has recently done locally devastating damage to plantations of 'introduced' lodgepole pine (*P. contorta*). Moreover, the increasing extension of man-made forests into 'natural' upland rough grazings (misleadingly known as deer 'forests') has, with the concomitant provision of shelter, provided deer, as should have been expected, with a 'new' niche (Crooke, 1979).

What will happen to these pests, and possibly others, if the balance of land uses continues to change, as seems inevitable? The afforested area of Great Britain has more than doubled since 1919 from 3% to 7% (Table 1), but nonetheless Great Britain remains one of the least wooded of the developed regions. Its area of forests contrasts with the 19% and 29% of Belgium and the Federal Republic of Germany, respectively (Good, 1976). At the present, the financial 'Institutions' within Britain, particularly with the aim of ensuring the value of their long-term investments, are becoming interested in the acquisition of land for afforestation. Recently, the Centre for Agricultural Studies at the University of Reading argued the desirability of increasing the afforested area of Britain by 0.6-2.0 million ha during the next 50 years

Table 1 Areas in million ha of different land uses in Great Britain (England, Wales, Scotland) (after Callaghan and Jeffers, personal communication, 1980) (Percentages of total area of GB in italic type)

Total	Level 1	Level 2	Level 3	Level 4
TOTAL UK 24.4 (*100*)	Rural 22.6 (*92.6*)	Cultivated 13.5 (*55.3*)	Grassland 7.2 (*29.5*)	Leys 2.1 (*8.6*)
				Permanent pasture 5.1 (*20.9*)
			Arable 4.8 (*19.6*)	Cereals 3.7 (*15.1*)
				Root crops and vegetables 0.7 (*2.8*)
				Fallow 0.4 (*1.7*)
			Forest 1.4 (*5.9*)	Coniferous 1.4 (*5.8*)
				Coppice <0.1 (*0.1*)
			Orchards <0.1 (*0.3*)	
		Natural and semi-natural 9.1 (*37.3*)	Rough grazing 6.6 (*27.0*)	
			Woodland 0.6 (*2.5*)	Broadleaved 0.3 (*1.3*)
				Scrub etc. 0.3 (*1.2*)
			Inland water 0.3 (*1.2*)	
			Other semi-natural 1.6 (*6.6*)	
	Urban 1.8 (*7.4*)	Amenity 0.5 (*2.1*)		
		Other 1.3 (*5.3*)		

(Anon., 1980). In addition to these 'forces', the adoption of a rational approach to land-use within the European Community is likely to alter appreciably the patterns of production within Member States. Will milk and beef production be sustained in the UK? If not, will sheep replace cattle on the richer pastures and in the process release the less nutritious upland swards for other purposes, with forestry competing with amenity interests, nature conservation, water procurement, etc.? These indications and others suggest that forestry will become increasingly widespread with the greatest impact in Scotland. With increases of 0.6 or 2.0 million ha in Great Britain, it is estimated that the afforested area of the West Conservancy of Scotland would increase from 16%, at present, to 23% or 40%, respectively.

From past experiences, it is clear that the rational control of red deer and pine beauty moth is only likely to emerge from integrated studies of their natural and man-made habitats, e.g. 'native' Scots pinewoods and plantations of exotic lodgepole pine. The pest status of red deer varies to some extent with the species of forest tree — in Britain, lodgepole pine is more at risk than Sitka spruce — but additionally, it could depend upon forest design. When developing a mosaic of forest compartments planted at different dates and arranged to sustain yields within a reasonable area, is there likely to be an optimal planting area to minimize damage to the habitat, whether trees and/or ground vegetation, and maximize conservation and sporting interests, as well as meat production? Should areas planted on different occasions be randomly interspersed among each other? Bearing in mind that the answers to these questions will be strongly influenced by economic considerations, do we know sufficient of the behaviour of deer, including their occupancy of plantations of different ages, to suggest configurations that might ensure that damage could be minimized? What can be learnt from the natural food preferences of deer to ensure the planting of lures of attractive plants to facilitate the culling of excess animals which cannot be sustained without unacceptably damaging forest yields?

Considerations of this sort of course are also important in agriculture and horticulture. Much could, doubtless, be gained when considering the rational control of the rabbit in agriculture by reflecting upon the factors controlling the dispersion of this pest species from its 'natural' refuges in pockets of scrub woodland, etc. Matthews and Flegg (*1981*) discuss whether it is wise to plant orchards alongside existing woodland, so providing the flower bud-stripping bullfinch (*Pyrrhula pyrrhula*) with continuous cover. If it isn't, what separation is required to deter the movement of this bird from its natural shelter? This problem is not unlike that of deciding the width of the swathe to be burnt when managing moorland to maximize its 'use' by red grouse (*Lagopus lagopus*). This game bird is reluctant to fly far from the shelter of mature heather when feeding on succulent nutritious heather shoots regenerating after burning. In these instances the desire for nearby cover seems of overriding importance, as with the East African rodent, *Arvicanthis niloticus*, which seems more likely to ravage weedy, rather than weed-free crops of maize. But, of course, the control of vertebrate pests is often a complex affair involving emotive and/or political issues, whose solutions require commonsense, not scientific wizardry. There should be no surprise if damage is done by elephants, or for that matter other animals, invertebrate and vertebrate, when sections of their migration routes are taken into cultivation (Thresh, *1981*). In summary, it seems that far too much attention has been paid to the nature of the damage done to crops by vertebrate pests and far too little to augmenting our understanding of why these animals leave their natural habitats to become pests. This is surely the key to the problem.

To some extent this volume serves to provide an awareness and to reiterate facts that were already known, yet had not been assembled previously in quite the same way. Harlan's (*1981*) contribution reminds us that 15 of the world's most important annual crops originated from Savanna or Mediterranean regions. He also emphasizes the great adaptability of these plants, which became adapted to relatively short periods of growth delimited by periods of intense drought during their evolution. Remembering that trees in *natural* forests are unselected, and that those growing in man-made assemblages are unlikely to be the products of more than four cycles of man-imposed selection, one wonders what 'improvement' can be expected during the next century as we move away from the arboreal equivalents of primitive einkorn wheats. Foresters are naturally prudent and have tended to move trees between locations within the same climatic zone, e.g. Sitka spruce from the coastal belt of NW America (British Columbia and Alaska) to the western seaboard of Europe (both within the latitudinal range 50°–60°N), and *Pinus patula* from southern Mexico to southern India and NW Argentina (all *c.* 20° from the equator). Now, however, there are signs of greater enterprise with Monterey pine, *Pinus radiata* and the southern beeches, *Nothofagus* spp., from latitude 37°N and 35°S respectively, being grown in parts of Britain (50–58°N).

Is this a foretaste of events to come? In countenancing these introductions are we able to learn from factors controlling pests and parasites of indigenous near relatives and if so, bearing in mind the totally unexpected upsurge of pine beauty moth on lodgepole pine, how much value should be attached to such information? This is a matter of considerable debate although in the instance of *Pinus strobus*, the eastern white pine or Weymouth pine, there is no doubt that its exploitation in Britain was rejected because of the common occurrence of an alternate host, *Ribes nigrum*, of white pine blister rust *Cronartium ribicola* (Peace, 1962). In considering the possibly extended planting of *Nothofagus* spp. Welch (1981) concluded, tentatively from the scant evidence available from earlier scattered introductions, that few of the Coleoptera in the UK are potential pests, but that 26 species of Lepidoptera are to some extent leaf-feeders. Interestingly, all but two have been observed feeding on native *Quercus* spp., nine on native *Fagus* and one on *Castanea*, a naturalized alien. All three tree genera, like *Nothofagus*, belong to the Fagaceae.

In deciding to discuss the interpretation of *weed*, *crop* and *disease* I have, to some extent, extended the scope of this volume by referring to (i) the important and probably increasing role of forest crops, and (ii) wild plants as agents of allelopathic disease syndromes (Putnam and Duke, 1978). I might also have stressed the very important issues discussed by Atkinson and White (*1981*). Surprisingly, little is known about the effects of weeds on the growth of crop plants. For this reason, and because the study of disease depends upon an understanding of the growth of healthy plants, it was interesting to learn how weeds can deleteriously affect shoot extension and yields (quality and quantity) of apple trees. More importantly, however, Atkinson and White have tackled a problem which should be considered by many others concerned with root pests and pathogens. They have excavated root systems to show the effects of weeds on the distribution of apple roots. In so doing they have elucidated some of the mechanisms of plant competition, whilst not overlooking the simple fact that there are at least two sides to every story. As a result, it is necessary to balance the gain from diminished plant competition against the increased danger of damage from splash-dispersed fruit and rotting fungi, *Phytophthora* spp., in weed-free orchards (Harris, *1981*).

Hedgerows

In discussing forests and allelopathy I have already alluded to some of the major issues, but no discussion of land fragmentation and 'cover' would be complete without considering the pros and cons of hedgerows, as this is a matter of general concern. There is a considerable body of international literature indicating that hedges can significantly increase the degree of shelter and so increase the profits from crop plants and animals. The evidence in Britain is less persuasive and it seems that shelter is only justified in exceptional circumstances, as when seeking to improve early yields of strawberries and raspberries or to protect early flowering bulbs, notably narcissus in SW England. In this instance the value of protection outweighs the damage done by the root pathogen, *Rosellinia necatrix*, spreading from *Pittosporum* hedgerows (Wheeler, 1981). Without more evidence it is clear that the case for retaining hedgerows must be argued among those concerned with nature conservation, amenity, and land-use history and rationalization. Nonetheless, this has not deterred several contributors to this volume from re-stating some of the arguments. On the one hand hedgerows sustain populations of the parasites and predators that attack crop pests, e.g. the larval stage of the wasp, *Angitia fenestralis*, which infests the moth, *Plutella maculipennis*, a pest of brassicas (van Emden, *1981*). On the other, they maintain root-feeding virus-vector nematode species of *Longidorus*, *Tricho-*

dorus and *Xiphinema*. Hedgerows also harbour bird pests and pathogens, including the fireblight bacterium (*Erwinia amylovora*) of apple, pear, etc. Moreover, because of the aerodynamic effects of hedgerows and other barriers, airborne pests including virus vectors are deposited, and, therefore, tend to accumulate and spread most virus to leeward (Thresh, *1981*). In wishing to be tidy there is always a temptation to make broad generalizations, but even allowing for the volume of data already collected this would be unwise.

Concluding remarks

Natural vegetation affects the behaviour of crop pests in many different ways. Interestingly, Carlisle *et al.* (1965) found that the colour changes associated with the maturation of the desert locust, *Schistocerca gregaria*, and the initiation of yolk formation (vitellogenesis) were triggered by aromatic substances emitted at bud-break by myrrh (*Commiphora myrrhae*), but not by other desert scrub species of the Burseraceae, which include frankincense (*Boswellia carteri*) and balm of Gilead (*C. opobalsamum*). From subsequent laboratory tests the effect of myrrh was attributed to relatively common terpenoids: eugenol, α-pinene, β-pinene and limonene. In contrast, breeding of the destructive tropical bird, *Quelea quelea*, is triggered when the visual 'cue' of green vegetation is perceived (Marshall and Disney, 1957). The increasing and justified interest in sensory cues has been accompanied by a greater but still inadequate analysis of the stages of plant colonization by insects. The elimination of weeds to some extent may be counter-productive, as the benefit gained by eliminating plant competition is partially offset by increased colonization by aphids and/or other pests, which are not attracted by weedy crops or might have settled preferentially on weeds (Way and Cammell, *1981*). However, although some weeds may, by visual cues, attract early colonists, aphids may subsequently move to crop plants because weeds later become unattractive. The intricacies of the different mechanisms warrant more research and may prove to be as rewarding as the mycological analysis of the factor in grassland, namely the occurrence of the soil-inhabiting *Phialophora radicicola* var. *graminicola*, which retards the build-up of the take-all fungus, *Gaeumannomyces graminis*, when land is subsequently planted with cereals (Deacon, *1981*).

How is stability maintained? Why are natural stands seldom afflicted by epidemic disease? In posing these questions it is not presumed that the growth of 'natural' vegetation is not, or cannot be affected, for the growth of the common daisy, *Bellis perennis*, can be improved by applying fungicides (*see* Sagar, 1974). There is no need to reiterate all that has been written by Browning (*1981* and elsewhere). Instead, suffice it to stress that the reactions to attack by pathogens, and probably colonization by pests, shown by individual members of populations of undomesticated species, vary greatly (Harlan, *1981*). The reactions can be compared with those of present-day multiline crops with the added advantages that (i) undomesticated species usually occur as mixtures, and (ii) the configuration and disposition of leaves and roots of individuals of an undomesticated species are variable, probably affecting the persistence of moisture films and the effectiveness with which infectious propagules are dispersed and trapped (Dinoor, *1981*). As discussed by Wheeler (*1981*), events in natural assemblages can be partially mimicked by increasing the spaces between crop plants. Thus the incidence of *Sclerotium cepivorum* on onions and *Cercospora apii* on celery was decreased when spacings were widened. However, a great deal more remains to be learnt from studies of pathogen and pest incidence in natural vegetation. I commend the analytical approach adopted by Lapwood (1961a,b) when investigating the nature of horizontal resistance in potato cultivars to attack by the late blight fungus, *Phytophthora infestans*.

We tend to talk about the *evolution* of resistance or tolerance to attack by pests and pathogens, but is this what we really mean? I suspect we are sometimes guilty of sloppy thinking/expression. Having been concerned with the identification of plant genotypes able to tolerate large concentrations of copper and zinc in soil, I am aware that tolerant genotypes exist, albeit at a low frequency, in populations which have never been exposed to large concentrations of these substances (Bradshaw, 1976). Thus, in many instances, we should be referring to the build-up, and not the evolution, of tolerant genotypes. We tend to have other fixations as discussed by Browning (*1981*) in a plant pathological context. For this reason it is timely to be reminded, by more than one contributor, that many pathogens can attack a very wide range of hosts. Thus soil-borne pathogenic species of *Armillariella, Phytophthora, Pythium* and *Verticillium* are usually each able to colonize a great variety of domesticated and undomesticated plant species, albeit with different degrees of virulence (Sewell, *1981*). The same is equally true of 'specialized' parasites such as *Puccinia graminis avenae*, which can invade 71 host species in 36 genera of plants (Browning, *1981*).

If we were to adopt a less restricted (blinkered) attitude to our pest and pathogen problems, we would inevitably give greater consideration to the role of non-target species (how much do we know about the influence of buried weed seeds on the perpetuation of damping-off fungi?). We would take it for granted that weeds and wild plants are likely to be sources of pests and pathogens and, this being so, we would attempt to quantify their role in the epidemiology of pests and pathogens of crops. Much is known about the obligate alternate hosts of heteroecious fungi and their influence, by facilitating the formation of teleutospores, on the proliferation of specialized races (e.g. Dinoor (*1981*) discusses brown rust of barley, *Puccinia hordei*, and its liliaceous alternate hosts, *Ornithogalum* spp., and crown rust of oats, *Puccinia coronata* and its alternate buckthorn hosts *Rhamnus* spp.). On the other hand, few *quantitative* studies have been made of the part played by weeds and wild plants, sometimes symptomlessly infected, in the epidemiology of the vast numbers of viruses and other pathogens. This stricture applies less to the study of pests, possibly because there has long been a more enquiring attitude to their demography with the associated construction of life-tables. Without doubt, there is great scope, not the least because changing management practices are likely to alter the balance of weeds and wild plants in the future just as much as in the past. Around us we can see the effects of the herbicide-era with enhanced populations of graminaceous weeds, among which black grass, *Alopecurus myosuroides*, is of particular significance in perpetuating the ergot fungus *Claviceps purpurea* (Mantle, *1981*).

What does the future hold, bearing in mind that we should be experimenting with schemes of pest and pathogen control that are not energy intensive? Microbiologists should not overlook the role played by weeds and wild plants in sustaining populations of microbes that form (i) nitrogen-fixing nodules and galls, and (ii) the great variety of endo- and ecto- (sheathing)mycorrhizal associations. However, is it always desirable to sustain 'wild' populations of these microbes if they are less beneficial to crop plants than selected strains? While we should be considering methods of saving energy I can see that parallel and perhaps integrated studies of the effects of weeds and wild plants on pests and pathogens of crops and *vice versa* will have a more immediate impact on the ever-increasing efforts to establish agro-forestry, in an integrated attempt to maximize land use.

Weeds and wild plants will always exist: so will the need to study their influence on the ecology of crop pests and pathogens.

References

Anon. (1980). *Strategy for the U.K. forest industry*. Centre for Agricultural Strategy, Reading University.

Atkinson, D. & White, G. C. (*1981*). The effects of weeds and weed control on temperate fruit orchards and their environment. In *Pests, Pathogens and Vegetation*, pp. 415–428 J. M. Thresh. London: Pitman.

Bradshaw, A. D. (1976). Pollution and evolution. In *Effects of Air Pollutants on Plants*, pp. 135–159. T. A. Mansfield. Cambridge: Cambridge University Press.

Browning, J. A. (*1981*). The agroecosystem – natural ecosystem dichotomy and its impact on phytopathological concepts. In *Pests, Pathogens and Vegetation*, pp. 159–172. J. M. Thresh. London: Pitman.

Bunting, A. H. (*1981*). Changing patterns of land use: global trends. In *Pests, Pathogens and Vegetation*, pp. 23–37. J. M. Thresh. London: Pitman.

Carlisle, D. B., Ellis, P. E. & Betts, E. (1965). The influence of aromatic shrubs on sexual maturation in the desert locust *Schistocerca gregaria*. *Journal of Insect Physiology* **11,** 1541–1558.

Crooke, M. (1979). The development of populations of insects. In *The Ecology of Even-aged Forest Plantations*, pp. 209–217. E. D. Ford, D. C. Malcolm & J. Atterson. Cambridge: Institute of Terrestrial Ecology.

Cullen, J. M., Kable, P. F. & Catt, M. (1973). Epidemic spread of a rust imported for biological control. *Nature* **244**, 462–464.

Cussans, G. W. (1974). The biological contribution to weed control. In *Biology in Pest and Disease Control*, pp. 97–105. D. Price-Jones & M. E. Solomon. Oxford: Blackwells.

Deacon, J. W. (*1981*). The role of grasses in the ecology of take-all fungi. In *Pests, Pathogens and Vegetation*, pp. 191–198. J. M. Thresh. London: Pitman.

Dinoor, A. (*1981*). Epidemics caused by fungal pathogens in wild and crop plants. In *Pests, Pathogens and Vegetation*, pp. 143–158. J. M. Thresh. London: Pitman.

van Emden, H. F. (*1981*). Wild plants in the ecology of insect pests. In *Pests, Pathogens and Vegetation*, pp. 251–261. J. M. Thresh. London: Pitman.

Freeman, T. E., Charudattan, R. & Conway, K. E. (1978). Biological control of water weeds with plant pathogens. *Florida Water Resources Research Center OWRT Project* No. A-033-FLA.

French, D. W. & Schroeder, D. B. (1969). Oak wilt fungus, *Ceratocystis fagacearum*, as a selective silvicide. *Forest Science* **15**, 198–203.

Fryer, J. D. (*1981*). Weed control practices and changing weed problems. In *Pests, Pathogens and Vegetation*, pp. 403–414. J. M. Thresh. London: Pitman.

Good, J. E. G. (1976). Managing amenity tree resources in the United Kingdom. *Trees and Forests for Human Settlements, Proceedings of Pl.05-00 Symposium of IUFRO Vancouver, Canada, June and 16th IUFRO World Congress, Oslo, Norway, June*, pp. 158–176.

Handley, W. R. C. (1963). Mycorrhizal associations and *Calluna* heathland afforestation. *Forestry Commission Bulletin* 36.

Harlan, J. R. (*1981*). Ecological settings for the emergence of agriculture. In *Pests, Pathogens and Vegetation*, pp. 3–22. J. M. Thresh. London: Pitman.

Harris, D. C. (*1981*). Herbicide management in apple orchards and the fruit rot caused by *Phytophthora syringae*. In *Pests, Pathogens and Vegetation*, pp. 429–436. J. M. Thresh. London: Pitman.

Hasan, S. (1978). Further research on the host specialization of *Puccinia chondrillina* Bub. & Syd. *Abstract 3rd International Congress of Plant Pathology, Munich.*

Hasan, S. and Wapshere, A. J. (1973). The biology of *Puccinia chondrillina,* a potential biological control agent of skeleton weed. *Annals of Applied Biology* **74,** 325–332.

Hooper, M. D. (1981). Hedgerows as a resource. In *Forest and Woodland Ecology*, F. T. Last & A. S. Gardiner. Cambridge: Institute of Terrestrial Ecology (in press).

Lapwood, D. H. (1961a). Potato haulm resistance to *Phytophthora infestans*. I. Field assessment of resistance. *Annals of Applied Biology* **49,** 140–151.

Lapwood, D. H. (1961b). Potato haulm resistance to *Phytophthora infestans*. III. Lesion distribution and leaf destruction. *Annals of Applied Biology* **49,** 704–716.

Little, W., Fowler, H. W., Coulson, J. & Onions, C. T. (1964). *The Shorter Oxford English Dictionary*, 3rd edn., revised with addenda. Oxford: Clarendon Press.

McGrath, H., Shaw, W. C., Jansen, L. L., Lipscomb, B. R., Miller, P. R. & Ennis, W. B. (1957). Witchweed (*Striga asiatica*) — a new parasitic plant in the United

States. *Agricultural Research Service, U.S. Department of Agriculture, Special Publication* No. 10.

Mantle, P. G. (*1981*). Effects of weed grasses on the ecology and distribution of ergot (*Claviceps purpurea*). In *Pests, Pathogens and Vegetation*, pp. 437–442. J. M. Thresh. London: Pitman.

Marshall, A. J. & Disney, M. J. de S. (1957). Experimental induction of the breeding season in a xerophilous bird. *Nature* **180,** 647–649.

Matthews, N. J. & Flegg, J. J. M. (*1981*). Seeds, buds and bullfinches. In *Pests, Pathogens and Vegetation*, pp. 375–383. J. M. Thresh. London: Pitman.

Peace, T. R. (1962). *Pathology of Trees and Shrubs with Special Reference to Britain*. Oxford: Oxford University Press.

Putnam, A. R. & Duke, W. B. (1978). Allelopathy in agroecosystems. *Annual Review of Phytopathology* **16,** 431–451.

Sagar, G. R. (1974). On the ecology of weed control. In *Biology in Pest and Disease Control*, pp. 42–56. D. Price-Jones & M. E. Solomon. Oxford: Blackwells.

Sewell, G. W. F. (*1981*). Soil-borne fungal pathogens in natural vegetation and weeds of cultivation. In *Pests, Pathogens and Vegetation*, pp. 175–190. J. M. Thresh. London: Pitman.

Thresh, J. M. (*1981*). The role of weeds and wild plants in the epidemiology of plant virus diseases. In *Pests, Pathogens and Vegetation*, pp. 53–70. J. M. Thresh. London: Pitman.

Trask, A. B. (*1981*). Changing patterns of land use in England and Wales. In *Pests, Pathogens and Vegetation*, pp. 39–49. J. M. Thresh. London: Pitman.

Waterhouse, D. F. (1973). Insects and wheat in Australia. *Journal of the Australian Institute of Agricultural Science* **39,** 215–226.

Way, M. J. & Cammell, M. E. (*1981*). Effects of weeds and weed control on invertebrate pest ecology. In *Pests, Pathogens and Vegetation*, pp. 443–458. J. M. Thresh. London: Pitman.

Welch, R. C. (1981). Insects on exotic broadleaved trees of the Fagaceae, namely *Quercus borealis* and species of *Nothofagus*. In *Forest and Woodland Ecology*. F. T. Last & A. S. Gardiner. Cambridge: Institute of Terrestrial Ecology (in press).

Wheeler, B. E. J. (*1981*). The ecology of plant parasitic fungi. In *Pests, Pathogens and Vegetation*, pp. 131–142. J. M. Thresh. London: Pitman.

Woolf, H. B. (1977). (Chief Ed.) *Webster's New Collegiate Dictionary*. Springfield, Mass: G. & C. Merriam Co.

Subject indexes

Note: numbers in italics refer to complete papers.

Nematodes

Birds

Plants